QUANTUM PHYSICS

QUANTUM PHYSICS

Mircea S. Rogalski
Institute of Atomic Physics
Bucharest, Romania

and

Stuart B. Palmer
University of Warwick, UK

Gordon and Breach Science Publishers
Australia • Canada • China • France • Germany • India • Japan • Luxembourg • Malaysia
The Netherlands • Russia • Singapore • Switzerland

Copyright © 1999 OPA (Overseas Publishers Association) N.V. Published by license under the Gordon and Breach Science Publishers imprint.

All rights reserved.

No part of this book may be reproduced or utilized in any form or by any means, electronic or mechanical, including photocopying and recording, or by any information storage or retrieval system, without permission in writing from the publisher. Printed in Singapore.

Amsteldijk 166
1st Floor
1079 LH Amsterdam
The Netherlands

British Library Cataloguing in Publication Data

A catalogue record for this book is available from the British Library.

ISBN   90-5699-184-1 (Hardcover)

We wish to dedicate this book to the teachers who first introduced us
to the fascination of quantum physics,
Prof. Viorica Florescu at the University of Bucharest, and
Prof. Norman March at the University of Sheffield.

M. S. R.

S. B. P.

CONTENTS

Preface   xiii

PRELIMINARIES   1

P.1. Quantum Conditions   1
    Example P.1. The Planck quantum hypothesis   1
    Example P.2. The quantization rule for oscillatory motion   9
    Example P.3. The quantization rule for translational motion   10
    Example P.4. The quantization rule for elliptical motion   11
    Solved problems   13

P.2. The Particle Nature of Waves   15
    Example P.5. Inelastic collision of a photon with a free electron   18
    Example P.6. Elastic collision of a photon with a free electron   18
    Example P.7. Motion of waves and particles   20
    Solved problems   21

P.3. The Wave Nature of Particles   24
    Example P.8. The wave representation of an atomic electron   24
    Example P.9. Equation of the de Broglie wave propagation   25
    Example P.10. Phase velocity of the de Broglie wave   27
    Solved problems   29

Chapter 1   WAVE FUNCTIONS AND THE FIRST POSTULATE   33

1.1. Wave Packet Representation of a Particle   33
    Example 1.1. Wave packets   33
    Example 1.2. Statistical interpretation of position   40
    Solved problems   42

1.2. The Uncertainty Principle   44
    Example 1.3. The Heisenberg microscope   46
    Solved problems   47

	1.3. The Wave Function Space	49
	Example 1.4. The Dirac δ-function	51
	Example 1.5. Vector spaces	54
	Solved problems	58
	1.4. The First Postulate	59
	Example 1.6. Two-slit experiment with particles	61
	Example 1.7. The Dirac notation	62
	Solved problems	65
Chapter 2	OPERATORS AND THE SECOND POSTULATE	66
	2.1. The Second Postulate	66
	Example 2.1. Angular momentum in spherical polar coordinates	72
	Example 2.2. The Hamiltonian in a central force field	75
	Solved problems	77
	2.2. The Momentum Representation	83
	Example 2.3. Momentum representation of oscillatory motion	88
	Solved problems	90
	2.3. Linear Operators	93
	Example 2.4. Integral representation of operators	96
	Example 2.5. The rule of symmetrization in x_i and p_i	98
	Solved problems	100
	2.4. Hermitian Operators	104
	Example 2.6. Anti-Hermitian operators	107
	Example 2.7. Unitary operators	108
	Solved problems	108
Chapter 3	EIGENSTATES AND THE THIRD POSTULATE	111
	3.1. The Eigenvalue Problem	111
	Example 3.1. Degeneracy	114
	Example 3.2. The eigenvalue problem for momentum	119
	Example 3.3. The eigenvalue problem for position	120
	Solved problems	121
	3.2. The Third Postulate	124
	Example 3.4. The projection operator	126
	Solved problems	128
	3.3. The Matrix Eigenvalue Problem	129
	Example 3.5. Diagonalization of a matrix	133
	Solved problems	134

	3.4. Matrix Representations	137
	Solved problems	142
Chapter 4	COMMUTATION RELATIONS AND THE FOURTH POSTULATE	144
	4.1. Commutator Algebra	144
	Example 4.1. Functions of non-commuting operators	145
	Solved problems	148
	4.2. The Fourth Postulate	150
	Example 4.2. The Poisson brackets	150
	Example 4.3. Commutation relations for angular momentum	154
	Example 4.4. Unitary transformations of physical observables	156
	Solved problems	159
	4.3. Compatibility between Physical Observables	160
	Example 4.5. Removal of degeneracy	162
	Example 4.6. Minimum uncertainty states	167
	Solved problems	168
Chapter 5	TEMPORAL EVOLUTION AND THE FIFTH POSTULATE	171
	5.1. The Heisenberg Description	171
	Example 5.1. Temporal evolution of position and momentum	172
	Solved problems	175
	5.2. The Fifth Postulate	177
	Example 5.2. The continuity equation for probability	179
	Example 5.3. The classical limit of the time-dependent Schrödinger equation	181
	Solved problems	183
	5.3. The Equivalence between the Schrödinger and Heisenberg Descriptions	186
	Example 5.4. The energy-time uncertainty relation	188
	Solved problems	189
	5.4. The Schrödinger Equation for Stationary States	191
	Example 5.5. The classical limit of the time-independent Schrödinger equation	193
	Example 5.6. The energy representation of oscillatory motion	197
	Solved problems	199

x Contents

Chapter 6   ONE-DIMENSIONAL MOTION   202

6.1. Energy Eigenstates in One Dimension   202
Example 6.1. Motion in a constant potential energy   205
Example 6.2. The WKB approximation   206
Solved problems   210

6.2. The Energy Spectrum for Discontinuous Potentials   213
Example 6.3. Bound states in the square well potential energy   215
Example 6.4. The tunnel effect   219
Example 6.5. Energy bands in a one-dimensional array   223
Solved problems   225

6.3. The Linear Harmonic Motion   228
Example 6.6. The Hermite polynomials   232
Example 6.7. The harmonic oscillator in the Heisenberg description   235
Solved problems   238

Chapter 7   ELECTRON MOTION IN THE ATOM   242

7.1. Rotational Motion for a Single Particle   242
Example 7.1. Legendre polynomials   246
Solved problems   250

7.2. Radial Motion in Central Force Field   253
Example 7.2. Confluent hypergeometric functions   258
Example 7.3. Radial motion of a free particle   262
Solved problems   263

7.3. One-Electron Atoms   267
Solved problems   275

7.4. The Magnetic Moment of the Electron   276
Example 7.4. Classical motion in electric and magnetic fields   277
Example 7.5. One-electron atoms in a uniform magnetic field   280
Solved problems   282

Chapter 8   ANGULAR MOMENTUM   285

8.1. The Matrix Eigenvalue Problem for Angular Momentum   285
Example 8.1. Rotation properties of vector observables   286
Example 8.2. Spherical harmonic eigenstates of angular momentum   292
Example 8.3. Matrix representation of orbital angular momentum   296
Solved problems   298

	8.2. Electron Spin	301
	Solved problems	306
	8.3. Addition of Angular Momenta	308
	Example 8.4. Representation of angular momenta for $l = 1$ and $s = \frac{1}{2}$	311
	Example 8.5. The vector model of angular momentum	315
	Solved problems	317
	8.4. Spin Magnetic Moment	319
	Example 8.6. Spin-orbit interaction	320
	Solved problems	326
Chapter 9	**APPROXIMATE METHODS**	**329**
	9.1. Stationary State Perturbation Theory	329
	Example 9.1. The anharmonic oscillator	331
	Example 9.2. The Dalgarno perturbation method	333
	Example 9.3. Splitting of a twofold degenerate state	337
	Solved problems	338
	9.2. The Variational Method	341
	Example 9.4. Variational calculation of the ground state of hydrogen	343
	Example 9.5. Variational calculation of the first excited state of hydrogen	345
	Solved problems	346
	9.3. Time-Dependent Perturbation Theory	348
	Example 9.6. The Fermi golden rule	351
	Solved problems	353
Chapter 10	**MANY-PARTICLE SYSTEMS**	**357**
	10.1. The Pauli Exclusion Principle	357
	Example 10.1. The principle of indistinguishability	361
	Example 10.2. Wave functions for a two-particle system	365
	Solved problems	366
	10.2. The Helium Atom	371
	Example 10.3. Variational calculation of the ground state of helium	374
	Solved problems	378
	10.3. Multielectron Atoms	380
	Example 10.4. The Hartree-Fock approximation	384
	Example 10.5. Electron configuration of the low Z atoms	386
	Example 10.6. The jj coupling interaction	392
	Solved problems	394

xii   Contents

	10.4. The Shell Model of the Nucleus	397
	Example 10.7. Magnetic hyperfine interactions	403
	Solved problems	405
Chapter 11	**ATOMIC RADIATION**	408
	11.1. Radiative Transitions	408
	Example 11.1. Selection rules for electric dipole transitions	412
	Example 11.2. Polarization and intensity of atomic radiation	416
	Solved problems	418
	11.2. Spontaneous Emission	421
	Example 11.3. Amplification of radiation	425
	Solved problems	426
	11.3. Magnetic Resonance	428
	Example 11.4. Electron resonance transitions in hydrogen	432
	Solved problems	435
Chapter 12	**NUCLEAR RADIATION**	438
	12.1. Radioactive Decay Law	438
	Solved problems	440
	12.2. Alpha Decay	442
	Solved problems	445
	12.3. Beta Decay	447
	Solved problems	452
	12.4. Gamma Radiation	453
	Example 12.1. Nuclear gamma resonance	456
	Solved problems	463

Further Reading 467

Index 469

PREFACE

Quantum Physics is now over 70 years old and is covered by an array of fine textbooks, however there will always be a need for a new treatment, particularly for advanced students entering the field and for new researchers. Many students find difficulty in mastering the subject matter and often lack confidence in applying the well established principles of quantum physics to real problems. A particular limitation is the need on many occasions to search through many of the available books to find material of relevance to their needs.

The present book is written by two experimental condensed matter physicists and may therefore miss some of the subtle nuances that a theorist would value. However, students in the second half of a degree course may welcome our more pragmatic approach. We will assume that such students have already taken some elementary level courses in classical mechanics, electromagnetism and calculus, and have some familiarity with the ideas of modern physics. After a preliminary review of the origins of quantum physics, we present a gradual development of the quantum formalism, in a way that is different from most previous approaches. The first few chapters describe the basic postulates concerning wave functions, operators, measurement and temporal evolution. A concise and rather rigorous treatment is then applied to the fairly standard topics of one-dimensional and atomic motion, angular momentum and approximation theory. The postulates and special methods, which are always introduced in association with the appropriate mathematical tools, are used to derive the basic laws and their main consequences. The last three chapters, which are concerned with many-particle systems, atomic and nuclear radiation, are rather condensed, but treated in sufficient depth to make the book a useful reference for students studying atomic and nuclear physics courses. Aspects of present-day research are included throughout and designed to encourage the student of experimental physics, engineering and materials science to make this textbook a realistic introduction to more advanced or specialized texts.

Our main purposes are firstly to provide the necessary background of basic concepts and methods of quantum mechanics, and secondly to develop the ability to solve problems by providing a number of examples with either complete solutions or answers. The examples will supplement the core text and in addition illustrate how the formalism is applied to real physical situations. The problems are placed at appropriate points of the text to demonstrate how a particular system can be described quantum mechanically and in some cases will bring less obvious aspects to the reader's attention. We hope that the text provides a sound basis for serious learning, revising and exam preparation and an opportunity for the students to work through the material (with pencil and paper!) and to analyze the physical meaning of calculations.

The material is intended to provide full coverage of the subject matter for university lecture courses at the indicated level. The underpinning mathematics, which sometimes involves specialized techniques, is emphasized as special examples, making the book readable without reference to other textbooks. The formulae are in SI form throughout the book and should be used with SI units.

We wish to acknowledge the helpful suggestions from several of our colleagues, resulting in significant improvements of presentation. We are indebted to Mrs. Jennifer Rogalski for her unstinting support both with the figures and the preparation, in camera-ready form, of numerous drafts of the manuscript. The book would not have been possible without her. We are also grateful to the editorial staff of the Gordon and Breach Publishing Group from Reading UK for their continuing encouragement and support.

Mircea S. Rogalski

Stuart B. Palmer

PRELIMINARIES

P.1. Quantum Conditions

At the beginning of this century doubts were growing about the whole edifice of classical physics, including the laws of mechanics in their Lagrangian and Hamiltonian forms, of statistical thermodynamics, as formulated by Boltzmann and Gibbs, and of optics, namely Maxwell's and Fresnel's theory. This lead to the development of the concepts of quantum physics as the first step in accounting for a group of experiments, which did not fit into the established view of nature. The advocates of quantum physics postulated that physical quantities could take on only a discrete set of values, instead of the classical assumption of a continuum. The conceptual advance of a quantum hypothesis, where continuous physical quantities were replaced by discontinuous ones, was originally incorporated into theories which still preserved much of the classical physics. Moreover, the selection rules for the quantized physical states were defined such that the classical laws could be regarded as continuous approximations of the quantum laws for discontinuous physical processes.

EXAMPLE P.1. The Planck quantum hypothesis

In a cavity inside a body, where the walls are kept at constant temperature T, a state of thermodynamic equilibrium is experimentally obtained with respect to the exchange of electromagnetic radiation, commonly known as *thermal radiation*. The condition for equilibrium is that a balance should be reached between the rates of emission and absorption of radiation per unit area of the inner cavity surface. The energy density of the radiation in the cavity is dependent only on the wall temperature, $u = u(T)$, and can be therefore treated as a classical hydrostatic system, such that all the extensive thermodynamic variables of the radiation are proportional to the volume V of the cavity. For instance $U = Vu, S = Vs$ where the energy and entropy densities, u and s are functions of temperature only.

Since the thermal radiation has a continuous spectrum of frequencies extending over a wide range, it is convenient to characterize this mixture of monochromatic waves by a function called the *spectral energy density*, $u_\omega = u(\omega, T)$, which defines the radiation energy per unit volume and unit frequency interval as:

$$u(T) = \int u(\omega, T) d\omega \quad \text{or} \quad u(T) = \int u_\omega d\omega \tag{P.1}$$

2   Quantum Physics

It has been established by thermodynamic arguments that the spectral energy density always obeys the *Wien law*:

$$u(\omega,T) = \omega^3 F\left(\frac{\omega}{T}\right) \tag{P.2}$$

whatever the form of the function F. The energy of radiation in the cavity can be interpreted as the energy of the electromagnetic field, which can be represented by a wave function $\Psi(\vec{r},t)$, which satisfies the wave equation:

$$\nabla^2 \Psi = \frac{1}{c^2}\frac{\partial^2 \Psi}{\partial t^2} \tag{P.3}$$

provided the solutions $\Psi(\vec{r},t)$ obey the continuity conditions at the boundaries. The problem is analogous to that solved by classical mechanics, in one dimension, for the vibrating string, where only a set of *normal* frequencies ω_n are allowed, each corresponding to a *normal mode* of vibration. Finding the energy density distribution u_ω over the various frequencies implies the determination of the number of allowed independent electromagnetic oscillations $dN(\omega) = (dN/d\omega)\,d\omega$ in any range of frequency. If $\bar{E} = \langle E(\omega,T)\rangle$ denotes the average energy per electromagnetic oscillation at a frequency ω and a temperature T, the energy content of the cavity in a frequency interval $d\omega$ is:

$$V u(\omega,T)\,d\omega = \frac{dN(\omega)}{d\omega}\,d\omega\langle F(\omega,T)\rangle \quad \text{or} \quad u(\omega,T) = \frac{1}{V}\frac{dN(\omega)}{d\omega}\bar{E} = \frac{\omega^2}{c^3 \pi^2}\bar{E} \tag{P.4}$$

where $dN/d\omega = V\omega^2/c^3\pi^2$ is easily derived using statistical thermodynamic arguments.

Since an electromagnetic mode inside a cavity with conducting walls behaves like a harmonic oscillator, passing energy back and forth between electric and magnetic forms, Rayleigh and Jeans assumed that an ensemble of linear harmonic oscillators with the same frequency ω and energy given by:

$$E = \frac{p_x^2}{2m} + \frac{1}{2}m\omega^2 x^2 \tag{P.5}$$

can be used to derive the average energy $\bar{E}$ of a normal mode, according to the Boltzmann distribution formula:

$$\bar{E} = \langle E(\omega,T)\rangle = \frac{\int_{-\infty}^{\infty}\int_{-\infty}^{\infty} E e^{-E/k_B T}\,dx\,dp_x}{\int_{-\infty}^{\infty}\int_{-\infty}^{\infty} e^{-E/k_B T}\,dx\,dp_x} = k_B T \tag{P.6}$$

This result obeys the classical theorem of equipartition of energy, namely that the average energy of an ensemble of oscillators, at a given temperature T, is independent of the oscillation frequency ω. The average entropy $\overline{S}$ of a normal oscillator will be related to its average energy $\overline{E}$ by:

$$\left(\frac{\partial \overline{S}}{\partial \overline{E}}\right)_V = \frac{1}{T} = \frac{\text{const}}{\overline{E}} \tag{P.7}$$

so that:

$$\left(\frac{\partial^2 \overline{S}}{\partial \overline{E}^2}\right)_V = \frac{\text{const}}{\overline{E}^2} \tag{P.8}$$

Upon inserting the average energy (P.6) into Eq.(P.4), a spectral energy density expression, called the *Rayleigh-Jeans law*, follows as:

$$u_{RJ}(\omega, T) = \frac{\omega^2}{c^3 \pi^2} k_B T \tag{P.9}$$

The Rayleigh-Jeans law, plotted in Figure P.1, agrees with experimental data only for low frequencies. The characteristic experimental decrease in the spectral energy density at higher frequencies means, from the classical point of view, that the normal high frequency modes of electromagnetic oscillation contain less average energy $\overline{\varepsilon}$ than is expected from the theorem of equipartition of energy.

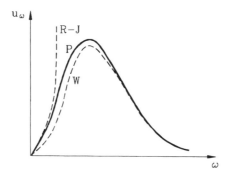

Figure P.1. Rayleigh-Jeans law, *R-J*, and Wien formula, *W*, referred to the experimental data, illustrated by the solid line *P*

As an approximation to the measured spectral energy distribution, represented in Figure P.1, Wien proposed the formula:

$$u_W(\omega, T) = c_1 \omega^3 e^{-c_2 \omega / T} \tag{P.10}$$

where the constants c_1 and c_2 are derived by fitting the exponential decrease of the spectral energy density found experimentally at high frequencies.

Comparison of the Wien formula (P.10) with the Rayleigh-Jeans law (P.9) allows us to assign, to a normal oscillator at high frequencies, the average energy:

4 Quantum Physics

$$\overline{E} = \text{const} \cdot \omega e^{-c_2 \omega / T} \tag{P.11}$$

which gives an underestimate of the real energy of normal electromagnetic modes. This arises because the Wien formula (P.10) does not account for the low frequency behaviour of u_ω. Equation (P.11) also reads:

$$T = \frac{\text{const} \cdot \omega}{\ln(\overline{E}/\omega)}$$

so that, at a given high frequency ω, we obtain:

$$\left(\frac{\partial \overline{S}}{\partial \overline{E}}\right)_V = \frac{1}{T} = \text{const} \cdot \ln \overline{E} \quad \text{or} \quad \left(\frac{\partial^2 \overline{S}}{\partial \overline{E}^2}\right)_V = \frac{\text{const}}{\overline{E}} \tag{P.12}$$

Planck removed the discrepancy between theory and experiment by using an interpolation between the low and high frequency properties of the normal electromagnetic oscillators in the cavity, as expressed by Eqs.(P.8) and (P.12), that is by assuming that:

$$\left(\frac{\partial^2 \overline{S}}{\partial \overline{E}^2}\right)_V = \frac{\alpha}{\overline{E}^2 + \beta \overline{E}} \tag{P.13}$$

where α, β are constant at a given frequency. Integration of this relation gives:

$$\left(\frac{\partial \overline{S}}{\partial \overline{E}}\right)_V = \ln\left(1 + \frac{\beta}{\overline{E}}\right)^{-\alpha/\beta}$$

and this result must be set equal to $1/T$, according to Eq.(P.7). It follows that:

$$\overline{E} = \frac{\beta}{e^{-\beta/\alpha T} - 1}$$

Therefore the spectral energy density (P.4) becomes:

$$u_\omega = \frac{\omega^2}{c^3 \pi^2} \frac{\beta}{e^{-\beta/\alpha T} - 1}$$

where the requirements of the Wien law (P.2) impose that $\beta \sim \omega$, so that finally we obtain the formula proposed by Planck in terms of two empirical constant parameters γ and δ:

$$u_\omega = \omega^2 \frac{\gamma \omega}{e^{\delta \omega / T} - 1} \tag{P.14}$$

which was found to be in agreement with experimental data for the whole range of frequencies. This empirical formula is readily derived, provided we accept the **Planck hypothesis** that:

The energy of a harmonic oscillator is an integer multiple of an energy quantum and is proportional to the oscillator frequency:

$$E_n = n\hbar\omega, \qquad n = 0, 1, 2, \ldots \qquad (P.15)$$

Hence, in a system of N oscillators, of average energy $\overline{E}$, the total energy $N\overline{E}$ is not distributed in a continuously divisible manner, as assumed in classical physics, but in discrete multiples of $\hbar\omega$. According to the basic assumption (P.15), the Boltzmann distribution formula (P.6) for the average energy becomes:

$$\overline{E} = \langle E(\omega, T)\rangle = \frac{\sum_{n=0}^{\infty} E_n e^{-E_n/k_BT}}{\sum_{n=0}^{\infty} e^{-E_n/k_BT}} \qquad (P.16)$$

Setting $y = 1/k_BT$ this relation reduces to:

$$\overline{E} = -\frac{\partial}{\partial y}\left(\ln \sum_{n=0}^{\infty} e^{-yE_n}\right)$$

Using equation (P.15) for E_n, the summation can be written as:

$$\sum_{n=0}^{\infty} e^{-ny\hbar\omega} = \sum_{n=0}^{\infty}(e^{-y\hbar\omega})^n = \sum_{n=0}^{\infty} z^n = \frac{1}{1-z} = \frac{1}{1-e^{-y\hbar\omega}}$$

and the average energy is given by:

$$\overline{E} = \langle E(\omega, T)\rangle = \frac{\hbar\omega}{e^{\hbar\omega/k_BT} - 1} \qquad (P.17)$$

which, in contrast to Eq.(P.6), is a function of both T and ω. The quantization condition (P.15) therefore invalidates the theorem of equipartition of energy. Inserting Eq.(P.17) into Eq.(P.4) we obtain the spectral energy density as:

$$u_P(\omega, T) = \frac{\omega^2}{c^3\pi^2} \frac{\hbar\omega}{e^{\hbar\omega/k_BT} - 1} \qquad (P.18)$$

which is the same as Eq.(P.14), but now a physical significance is assigned to the empirical constants γ and δ. Equation (P.18) is known as the *Planck radiation formula*. In the low frequency range, where $\hbar\omega/k_BT \ll 1$, we can use:

$$e^{\hbar\omega/k_BT} \cong 1 + \frac{\hbar\omega}{k_BT}$$

and the Planck radiation formula reduces to the Rayleigh-Jeans law, Eq.(P.9), whereas in the high frequency range, where the unity is negligibly small in the denominator of Eq.(P.18), so obtaining:

6 Quantum Physics

$$u_W(\omega, T) = \frac{\hbar}{c^3 \pi^2} \omega^3 e^{-\hbar\omega/k_B T} \tag{P.19}$$

which has the form of the Wien formula (P.10). ☙

The Planck idea that the energy of a harmonic oscillator, which is in equilibrium with electromagnetic radiation, may only take on the values $n\hbar\omega$, implies a discrete set of energy levels, separated by $\hbar\omega$. In other words, the emission and absorption processes should result in a discrete set of frequencies for the emitted or absorbed thermal radiation, according to:

$$\hbar\omega_{mn} = h\nu_{mn} = E_m - E_n \tag{P.20}$$

where ω_{mn} denotes an integral multiple of ω. Generalizing the Planck hypothesis and applying it to the mechanics of the atom, Bohr made use of quantum assumptions in order to explain the discrete radiation spectrum emitted by one electron atoms, which consists of intense lines, of definite frequencies, which can be clustered into several series that fit the empirical formula:

$$\nu_{mn} = R\left(\frac{1}{n^2} - \frac{1}{m^2}\right), \quad m > n \geq 1 \tag{P.21}$$

where m and n are integers and R is the Rydberg constant. The significant fact is that the frequency of the observed spectral lines of each series can be represented by the difference between two terms, one of which, T_n, is fixed and the other, T_m, is a variable. Similar series formulae have been found for the emission spectra of alkali metal atoms. Furthermore, in addition to the lines represented in the various series (P.21), there are other lines not included in Eq.(P.21). However, it was observed that the frequencies of all spectral lines also correspond to the differences between two quadratic terms:

$$\nu_{mn} = T_n - T_m \tag{P.22}$$

and this is called the *Ritz combination principle*. These results contradict the classical picture of an electron orbiting around a nucleus, radiating electromagnetic waves and leading to an unstable atom. Since the electron frequency of revolution will change smoothly, the emission spectrum should be continuous. A successful interpretation of the discrete spectrum of one electron atoms was proposed on the basis of the *Bohr postulates*:

> 1. **Postulate of stationary states:** *an electron moves only in certain permissible orbits which are stationary states, in the sense that no radiation is emitted. The condition for such states is that the orbital angular momentum of the electron equals an integer times h.*

Although this postulate contradicts the macroscopic requirement that an accelerated charge must radiate energy, the reality of stationary states has received direct experimental support from the well-known Franck-Hertz experiment. In the simple case of an electron in a circular orbit of radius r, the quantization condition reads:

$$m_e v r = n\hbar \qquad (P.23)$$

where m_e is the mass of the electron, v its velocity and n is a positive integer. The dynamical stability of the circular orbit requires that:

$$\frac{Ze^2}{4\pi\varepsilon_0 r^2} = \frac{m_e v^2}{r} \quad \text{or} \quad \frac{Ze_0^2}{r^2} = \frac{m_e v^2}{r} \qquad (P.24)$$

where Ze is the nuclear charge of the one electron atom and $e_0 = e/\sqrt{4\pi\varepsilon_0}$. Equations (P.23) and (P.24) may be solved for v and r:

$$v_n = \frac{Z}{n}\frac{e_0^2}{\hbar}, \quad r_n = \frac{n^2}{Z}\frac{\hbar^2}{m_e e_0^2} \qquad (P.25)$$

so that the orbital energy of state n will be written as:

$$E_n = \frac{m_e v^2}{2} - \frac{Ze_0^2}{r_n} = -\frac{Z^2}{n^2}\frac{m_e e_0^4}{2\hbar^2} = -13.6\frac{Z^2}{n^2}(eV) \qquad (P.26)$$

2. **Postulate of discrete transitions:** *emission and absorption of radiation occurs only when an electron makes a transition from one stationary state to another. The radiation has a definite frequency v_{mn} given by the condition:*

$$h v_{mn} = E_m - E_n \qquad (P.27)$$

The condition (P.27) states that radiation of frequency:

$$v_{mn} = Z^2 \frac{m_e e_0^4}{4\pi\hbar^3}\left(\frac{1}{n^2} - \frac{1}{m^2}\right)$$

is emitted when the electron drops from the mth to the nth state, where $m > n \geq 1$, so explaining the emission spectra of one electron atoms, Eq.(P.21), and therefore the empirical combination principle, Eq.(P.22). An analogy between the frequency condition (P.27) and the Planck assumption (P.20) concerning the discrete emission and absorption of radiation by oscillators is immediate. Both state that processes involving energy exchange at the microscopic scale must be described in discontinuous steps. A coherent theory should make such microscopic discontinuous processes consistent with the classical theory, which involves macroscopic quantities and processes which are all continuous. Bohr has formulated, in this respect, the *correspondence principle*, which states that quantum mechanics must give the same results as classical mechanics in the limit of large quantum numbers ($n \to \infty$) and low quantum jumps ($\hbar \to 0$). In other words, the results of classical theory are valid in such limiting cases, suggesting a way toward the quantum formulation of physical laws.

A simple illustration is given by a comparison between the classical and Bohr model predictions of the line spectra of one electron atoms. The classical electron, when confined to the periodic motion in one stationary orbit, should emit radiation with frequencies equal to those of its rotation, obtained from Eqs.(P.25) as:

$$v_{rot} = \frac{v_n}{2\pi r_n} = Z^2 \frac{m_e e_0^4}{2\pi \hbar^3} \frac{1}{n^3}$$

Assuming that a Bohr transition $n+1 \to n$ occurs, Eq.(P.28) gives the frequency of the emitted radiation as:

$$v_B = Z^2 \frac{m_e e_0^4}{4\pi \hbar^3} \left[\frac{1}{n^2} - \frac{1}{(n+1)^2} \right] = \left(Z^2 \frac{m_e e_0^4}{2\pi \hbar^3} \right) \frac{2n+1}{2n^2(n+1)^2}$$

and it is clear that $v_B \to v_{rot}$ if $n \to \infty$ if as expected. By applying the correspondence principle, some quantum mechanical results can be derived through inductive reasoning on the grounds of their classical mechanical counterparts. However, because the inductive arguments are not always rigorous, results derived in this way need to be confirmed by direct experimental evidence.

The definition of stationary states by quantum conditions in terms of quantum numbers is a basic procedure, followed by Sommerfeld in an extension of the Bohr theory, which takes into account the classical solution for the motion of one particle in a central force field. Since the orbital angular momentum $L = mvr$ is a constant of motion in a circular orbit, the quantization condition (P.23) may be written as:

$$2\pi L = nh \quad \text{or} \quad \int_0^{2\pi} L\, d\varphi = nh$$

For periodic mechanical systems, described by generalized coordinates x_i and momenta p_i, this expression was formulated as a *quantization rule* in the form:

$$S = \oint p_i dx_i = n_i h \tag{P.28}$$

which reads: *the action integral of any periodic motion equals an integer times h*. The integral is over the range of variation of the position coordinate x_i during a period of the motion. The rule of quantization (P.28) is general, in the sense that it is valid for any mechanical motion.

EXAMPLE P.2. The quantization rule for oscillatory motion

Using Eq.(P.5) for the total energy of an oscillatory motion in one dimension, we may express the Hamiltonian as:

$$H(x, p_x) = \frac{p_x^2}{2m} + \frac{m\omega^2 x^2}{2} = E \quad \text{or} \quad \frac{p_x^2}{2mE} + \frac{x^2}{2E/m\omega^2} = 1$$

It follows that a representative point (p_x, x) which defines a state, often called a phase, of an oscillatory motion in one dimension, describes an ellipse with semimajor axis $a = \sqrt{2mE}$ and semiminor axis $b = \sqrt{2E/m\omega^2}$, in a two-dimensional space, known as the phase space, as shown in Figure P.2.

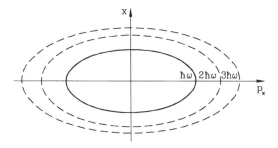

Figure P.2. Elliptical representation of an oscillatory motion in phase space

The area πab of the ellipse, namely $\pi ab = \pi\sqrt{2mE}\sqrt{2E/m\omega^2} = (2\pi/\omega)E$ where ω is the angular frequency, may alternatively be evaluated by the action integral S, which is a line integral along the closed trajectory. The rule of quantization (P.28), which shows that the action integral is a constant of motion, may be then expressed in terms of the energy and angular frequency of the oscillatory motion as:

$$S = \oint p_x dx = \pi ab = \frac{2\pi}{\omega} E = nh$$

which is the same condition as $E = n\hbar\omega$, given by Eq.(P.15). In other words, the Planck quantum condition is included as a special case in the quantization rule (P.28). ♦

Since each stationary state of a linear harmonic oscillator corresponds to an area of $\hbar\omega$ in phase space, the series of allowed elliptical representations of oscillatory motion, illustrated in Figure P.2, divides phase space into equal areas of $\hbar\omega$. Therefore the Planck quantum hypothesis requires the quantization of phase space, and hence, a basic change from classical statistical mechanics, where phase space is assumed to be continuous.

EXAMPLE P.3. The quantization rule for translational motion

Consider a free particle of mass m, moving in one dimension inside a potential well with infinite walls, defined by the potential energy function illustrated in Figure P.3:

$$U(x) = 0, \quad \text{for} \quad 0 < x < \alpha$$

$$= \infty, \quad \text{for} \quad x < 0 \quad \text{and} \quad x > \alpha$$

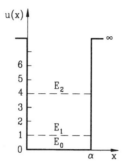

Figure P.3. The square well with infinite walls

The particle will move with constant momentum p_x between $x = 0$ and $x = \alpha$ and will reverse its direction of motion whenever it collides with the potential barrier. Thus, Eq.(P.28) reads:

$$\oint p_x dx = 2 p_x \int_0^\infty dx = nh$$

hence:

$$p_x = n \frac{h}{2\alpha} \tag{P.29}$$

It follows that the quantization rule (P.28) yields for the free particle the discrete energy levels:

$$E = \frac{p_x^2}{2m} = n^2 \frac{h^2}{8m\alpha^2}$$

which are indicated in Figure P.3. ☙

It is noteworthy that the quantization condition for momentum, Eq.(P.29), can be interpreted in terms of a stationary wave $\Psi(x,t) = A \sin kx \sin(\omega t + \varphi)$ restricted to a segment of length α, and has nodes at both ends: $\Psi(0,t) = \Psi(\alpha,t) = 0$. The allowed values of the wave number k, as derived from $\sin k_n \alpha = 0$, are $k_n = n\pi/\alpha$, are so yielding the allowed wavelength values:

$$\lambda = \frac{h}{p_x} \qquad (P.30)$$

In other words, for a free particle in a square potential well with infinite walls, we may associate a wavelength λ to each allowed value of momentum.

EXAMPLE P.4. The quantization rules for elliptical motion

Applying the quantization rule (P.28) to the momenta $p_r = m\dot{r}$, $p_\alpha = mr^2\dot{\alpha}$, which are conjugated, as defined in analytical mechanics, to the coordinates r and α of an elliptical electron motion around the nucleus (see Figure P.4 (a)), we have:

$$\oint p_\alpha d\alpha = \int_0^{2\pi} L\, d\alpha = 2\pi L = lh$$

$$\oint p_r dr = n_r h \qquad (P.31)$$

where both the azimuthal quantum number $n_\alpha = l$ and the radial quantum number n_r are integers, and $p_\alpha = L$ is the angular momentum, which is a constant of motion.

The radial momentum p_r can be derived from the constant energy equation which, in the case of the electron orbit under the central electrostatic force, reads:

$$E = \frac{m_e \dot{r}^2}{2} + \frac{L^2}{2m_e r^2} - \frac{Ze_0^2}{r} = \frac{1}{2m_e}\left(p_r^2 + \frac{L^2}{r^2}\right) - \frac{Ze_0^2}{r}$$

with $E < 0$, as required for bound elliptical orbits. Hence:

$$p_r = m_e \dot{r} = \pm\sqrt{2m_e\left(E + \frac{Ze_0^2}{r} - \frac{L^2}{2m_e r^2}\right)}$$

and we must consider the positive value of p_r when r increases from r_{min} to r_{max} and the negative value along the other half of the elliptical path. Hence, the action integral can be evaluated as:

$$\oint p_r dr = 2\int_{r_{min}}^{r_{max}} \sqrt{2m_e\left(E + \frac{Ze_0^2}{r} - \frac{L^2}{2m_e r^2}\right)}\, dr = -2\pi L + 2\pi Ze_0^2 \sqrt{\frac{m_e}{-2E}}$$

Inserting this result into Eq.(P.31) we obtain:

$$2\pi Ze_0^2 \sqrt{\frac{m_e}{-2E}} = (n_r + l)h = nh \qquad (P.32)$$

where the principal quantum number is written as $n = n_r + l$. Solving for E, we obtain the same energy levels as given by Eq.(P.26), without any restriction to circular orbits.

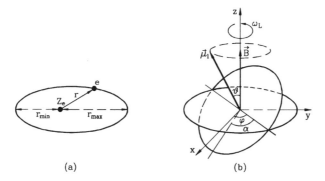

Figure P.4. Elliptical electron motion

Any elliptic orbit contains the nucleus in its plane, whose orientation is defined by a polar angle measured from the z-axis. The projection $L_z = L\cos\theta$ of the orbital angular momentum is also a constant, for a given orbit, and we may set $p_\varphi = L_z$ where φ is the azimuthal angle about the z-axis. The quantization rule (P.31) for p_φ reads:

$$\oint p_\varphi d\varphi = 2\pi p_\varphi = mh \tag{P.33}$$

where m is an integer. A discrete set of orientations θ is allowed by the equation:

$$\cos\theta = \frac{p_\varphi}{p_\alpha} = \frac{L_z}{L} = \frac{m}{k}, \quad -k < m < k$$

so that the Sommerfeld model accounts for a *spatial quantization* of the orbital angular momentum. Since the spatial orientation of orbits is apparent when a magnetic field is present, m is called the *magnetic quantum number*. This effect is classically assigned to an interaction of an applied field $\vec{B}$ with a magnetic moment $\vec{\mu}$, associated with the circular electron path, of area S:

$$\mu_l = IS = -ev\pi r^2 = -\frac{e\omega}{2}r^2 = -\frac{e}{2m_e}L \quad \text{or} \quad \vec{\mu}_l = -\frac{e}{2m_e}\vec{L} \tag{P.34}$$

where L stands for the angular momentum. The action of a magnetic field $\vec{B}$ gives a torque on $\vec{\mu}_l$ which is equal to the rate of change of the angular momentum:

$$\frac{d\vec{L}}{dt} = \vec{\mu}_l \times \vec{B} = -\frac{e}{2m_e}\vec{L} \times \vec{B}$$

Taking $|d\vec{L}| = L\sin\theta \, d\varphi$, as represented in Figure P.4 (b), we obtain:

$$L\sin\theta \frac{d\varphi}{dt} = \frac{e}{2m_e} LB\sin\theta \quad \text{or} \quad \omega_L = \frac{e}{2m_e} B$$

The angular velocity $\omega_L = d\varphi/dt$ is known as the *Larmor frequency*. It is a measure of the precession motion of the angular momentum vector about the z-axis, at constant polar orientation θ. An energy $\hbar\omega_L$ can be formally associated with the Larmor precession, on the basis of the Planck hypothesis (P.15). ♣

Problem P.1.1. The internal energy U of a crystal lattice is defined to be the total vibrational energy of a collection of N three dimensional ions at thermal equilibrium. Calculate the lattice heat capacity $C_V = (\partial U/\partial T)_V$ for the Einstein model which assumes that all the vibration modes have the same frequency ω_0.

(Solution): A system of N ions has $3N$ degrees of freedom and the lattice is therefore equivalent to $3N$ independent quantum oscillators, with average energy given by Eq.(P.17), so that:

$$U = 3N\langle E\rangle = 3N \frac{\hbar\omega_0}{e^{\hbar\omega_0/k_BT}-1}$$

By definition, the lattice heat capacity is:

$$C_V = 3N\hbar\omega_0 \frac{e^{\hbar\omega_0/k_BT}}{(e^{\hbar\omega_0/k_BT}-1)^2} \frac{\hbar\omega_0}{k_BT^2} = 3Nk_B\left(\frac{\hbar\omega_0}{k_BT}\right)^2 \frac{e^{\hbar\omega_0/k_BT}}{(e^{\hbar\omega_0/k_BT}-1)^2}$$

$$= 3Nk_B\left(\frac{\theta}{T}\right)^2 \frac{e^{\theta/T}}{(e^{\theta/T}-1)^2}$$

where $\theta = \hbar\omega_0/k_B$ is called the Einstein temperature. At high temperatures, where $T \gg \theta$, we can approximate:

$$e^{\theta/T} \cong 1 + \theta/T \quad \text{or} \quad e^{\theta/T} - 1 \cong \theta/T$$

so that C_V reduces to a temperature independent value $C_V \to 3Nk_B$ known as the Dulong-Petit experimental law. At low temperatures, where $T \ll \theta$, the unity may be neglected from the denominator, yielding:

$$C_V \to 3Nk_B\left(\frac{\theta}{T}\right)^2 e^{-\theta/T}$$

14 Quantum Physics

which is in poor agreement with the experimental T^3 dependence.

Problem P.1.2. Give the Planck radiation formula in terms of wavelength. Find the value λ_{max} for which the spectral density $u(\lambda,T)$ attains its maximum value.

(Solution): It follow from the fact that:

$$u(\lambda,T) = u(\omega,T)\left|\frac{d\omega}{d\lambda}\right| = u(\omega,T)\frac{2\pi c}{\lambda^2}$$

and from Eq.(P.18), that we may write:

$$u(\lambda,T) = \frac{\hbar}{2c^3\pi^3}\left(\frac{2\pi c}{\lambda}\right)^3 \frac{1}{e^{\hbar c/k_B T\lambda}-1}\left(\frac{2\pi c}{\lambda^2}\right) = \frac{8\pi\hbar c}{\lambda^5}\frac{1}{e^{\hbar c/k_B T\lambda}-1}$$

$$= \frac{8\pi(k_B T)^5}{\hbar^4 c^4}\frac{x^5}{e^x-1}$$

where $x = \hbar c/k_B T\lambda$. The maximum value of $u(\lambda,T)$ is determined by setting $df/dx = 0$, where $f(x) = x^5/(e^x-1)$. This is the same as setting $d\ln f/dx = 0$, which reads:

$$\frac{d}{dx}\left[5\ln x - \ln(e^x - 1)\right]_{x=x_0} = 0 \quad \text{or} \quad \frac{5}{x_0} = \frac{e^{x_0}}{e^{x_0}-1} \quad \text{or} \quad \frac{x_0}{5} = 1 - e^{-x_0}$$

The transcendental equation can only be solved by successive approximations for x_0, to obtain:

$$x_0 = 4.965 \quad \text{or} \quad \lambda_{max} = \left(\frac{\hbar c}{4.965 k_B}\right)\frac{1}{T}$$

This result is known as the Wien displacement law, where the Wien constant has the SI value of 2.898×10^{-3} m/K.

Problem P.1.3. A ball of mass m is elastically bouncing on a flat surface in a gravitational field. Use the Sommerfeld quantization rule to find the allowed energy levels of the ball.

(Solution): The total energy of the ball, which is a constant of motion, is given in terms of the height z from the surface as $E = p_z^2/2m + mgz$, where g is the acceleration of gravity. The linear momentum results as:

$$p_z = \sqrt{2m(E-mgz)}$$

and it is obvious that the admissible range of heights for the ball is $z \leq z_{max} = E/mg$, where z_{max} corresponds to the turning point where $p_z = 0$. The quantization rule, Eq.(P.28), written for a cycle of motion, reads:

$$2 \int_0^{E/mg} \sqrt{2m(E-mgz)}\, dz = nh$$

Using the integral formula, which is easily proved by differentiation:

$$\int (a+bz)^{\frac{1}{2}} dz = \frac{2}{3b}(a+bz)^{\frac{3}{2}}$$

it follows that:

$$\frac{3}{4}\sqrt{2m}\,\frac{E^{\frac{3}{2}}}{mg} = nh \quad \text{or} \quad E_n = \left(\frac{3n}{4}gh\sqrt{\frac{m}{2}}\right)^{\frac{2}{3}}$$

Problem P.1.4. Use the Sommerfeld quantization rule to determine the energy levels of a particle of mass m which freely rotates on a plane circular orbit of radius r (the rigid plane rotator).

(Answer): $E_n = n^2 \dfrac{h^2}{2mr^2}$

Problem P.1.5. Assuming that the Bohr quantization condition (P.23) might be applied to the angular momentum of the Earth, find the corresponding quantum number n. Derive the relative change in energy for an increase in n by unity.

(Answer): $n = 2.52 \times 10^{74}$, $\Delta E = 2.08 \times 10^{-41}\, J$

P.2. The Particle Nature of Waves

Einstein proposed that electromagnetic radiation exists in discrete quanta whilst propagating in space as well as during the emission or absorption processes. The thermal radiation inside a cavity might thus be regarded, from the point of view of entropy, as obeying the canonical distribution of n molecules of an ideal gas in a volume V, which gives:

16 Quantum Physics

$$S = ns_0 + nk_B \ln V + \frac{3}{2} nk_B \ln T \tag{P.35}$$

The radiation entropy S can be expanded in terms of the spectral entropy density s_ω, where s_ω is the entropy per unit volume in the frequency range between ω and $\omega + d\omega$:

$$S = V \int_0^\infty s_\omega \, d\omega \tag{P.36}$$

where $s_\omega = s(u_\omega, \omega)$ must depend on the spectral energy density u_ω, defined in Eq.(P.1), which can be conveniently rewritten as:

$$U = V \int_0^\infty u_\omega \, d\omega \tag{P.37}$$

Since the conditions for equilibrium, $\delta S = 0$, $\delta U = 0$, must hold simultaneously, the integrands δs_ω and δu_ω and must be proportional, that is:

$$\delta s_\omega = C \delta u_\omega \quad \text{or} \quad \frac{\partial s_\omega}{\partial u_\omega} \delta u_\omega = C \delta u_\omega \quad \text{i.e.} \quad \frac{\partial s_\omega}{\partial u_\omega} = C$$

Upon differentiation of Eqs.(P.36) and (P.37) we obtain the relation:

$$dS = V \int_0^\infty \frac{\partial s_\omega}{\partial u_\omega} du_\omega \, d\omega = CV \int_0^\infty du_\omega \, d\omega = C \, dU \tag{P.38}$$

which might be compared with the definition of the entropy change in an infinitesimal process. The interpretation of dU as the reversible added heat, allows us to identify C with the reciprocal of the temperature T, which gives:

$$C = \frac{1}{T} = \frac{\partial s_\omega}{\partial u_\omega} \tag{P.39}$$

The temperature can then be eliminated between Eq.(P.39) and the Wien formula (P.10), by writing:

$$\frac{1}{T} = -\frac{k_B}{\hbar \omega} \ln \frac{c^3 \pi^3 u_\omega}{\hbar \omega} \quad \text{or} \quad \frac{\partial s_\omega}{\partial u_\omega} = -\frac{k_B}{\hbar \omega} \ln \frac{c^3 \pi^3 u_\omega}{\hbar \omega}$$

so that, on integrating, we obtain the spectral entropy density as:

$$S_\omega = -\frac{k_B u_\omega}{\hbar \omega}\left(\ln\frac{c^3\pi^2 u_\omega}{\hbar \omega^3} - 1\right) \tag{P.40}$$

A similar relation holds between the spectral entropy $S_\omega = V s_\omega$ and the spectral energy $U_\omega = V u_\omega$ of the radiation:

$$S_\omega = -\frac{k_B U_\omega}{\hbar \omega}\left(\ln\frac{c^3\pi^2 U_\omega}{\hbar \omega^3 V} - 1\right) \tag{P.41}$$

so that the entropy change in the spectral range between ω and $\omega + d\omega$, associated with a reversible change of volume from V to V_0, can be written as:

$$\Delta S_\omega d\omega = \frac{k_B U_\omega d\omega}{\hbar \omega}\ln\frac{V}{V_0}$$

Einstein emphasized that this relation becomes identical with $\Delta S = nk_B \ln(V/V_0)$ and describes the n-particle gas behaviour, according to Eq.(P.34), provided that:

$$\frac{U_\omega d\omega}{\hbar \omega} = n \quad \text{or} \quad U_\omega d\omega = n\hbar \omega \tag{P.42}$$

The similarity of this relation to Eq.(P.15), assumed in the Planck theory, has been interpreted as an argument for the discrete nature of electromagnetic radiation. In other words, an energy:

$$E = \hbar\omega \tag{P.43}$$

is associated with each quantum of radiation, later called the *photon*. The amount of momentum associated with a photon was set according to the Maxwell theory, where the energy-to-momentum ratio of a plane electromagnetic wave is c, which gives:

$$p = \hbar\frac{\omega}{c} = \hbar k = \frac{h}{\lambda} \tag{P.44}$$

The Einstein statement concerning the corpuscular nature of radiation led to immediate explanations for the photoelectric and Compton effects, in terms of collisions of photons with atomic electrons. The photoelectric effect occurs through the inelastic collision of a photon with an electron bound in the atom, where radiative energy is transferred to bound electrons, liberating them from their host atoms. The kinetic energy of an electron produced by photoemission, measured by a retarding potential applied to just stop it, is simply given by the conservation of energy:

$$eV = \frac{mv^2}{2} = E - B$$

where E is the radiative energy, and B is the binding energy of the electron. The wave theory of light has provided no explanation either for the measured linear dependence of V on the frequency ω of the incident light, or for the threshold frequency ω_0 below which no electrons are emitted. If we set the photon energy as given by Eq.(P.43), the conservation of energy takes the experimentally confirmed form:

$$eV_0 = \hbar\omega - W = \hbar(\omega - \omega_0)$$

where the threshold frequency ω_0 is expressed as: $\omega_0 = B/\hbar$. The probability of the photoelectric effect is expected to be greater the more strongly the electron is bound, because a free electron cannot absorb a photon.

EXAMPLE P.5. Inelastic collision of a photon with a free electron

Let us apply the requirement of the conservation of energy and momentum to the hypothetical process of absorption of a photon by a free electron at rest. If the electron acquires a recoil velocity v on absorption, the conservation laws read:

$$\hbar\omega + m_e c^2 = \frac{m_e}{\sqrt{1-v^2/c^2}} c^2, \quad \hbar\frac{\omega}{c} = \frac{m_e}{\sqrt{1-v^2/c^2}} v$$

Eliminating $\hbar\omega/c$ between the two equations gives:

$$\frac{1-v/c}{\sqrt{1-v^2/c^2}} = 1$$

Since both possible solutions, $v/c = 0$ and $v/c = 1$, are unrealistic, it follows that a free electron cannot absorb a photon. ◊

Scattering by free electrons is a common interaction connected with the absorption of electromagnetic radiation. If the electromagnetic energy is large compared with the ionization energy, the atomic electron can be treated as free. The characteristic features of the scattered radiation have been explained by Compton, on the basis of an elastic collision of a photon with a free electron, where both energy and momentum are conserved.

EXAMPLE P.6. Elastic collision of a photon with a free electron

If a photon with energy E undergoes a collision with an electron of rest mass m_e, and is scattered through an angle θ, as represented in Figure P.5, the conservation of energy and momentum read:

$$E + m_e c^2 = E' + (p_e^2 c^2 + m_e^2 c^4)^{\frac{1}{2}}, \quad \vec{p} = \vec{p}' + \vec{p}_e$$

where $\vec{p}, \vec{p}'$ are photon momenta before and after the collision and the relativistic form of the recoil energy of the electron is appropriate for high energy collisions.

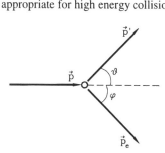

Figure P.5. Electron-photon interaction in the Compton effect

By eliminating the momentum $\vec{p}_e$ of the recoil electron between the two equations, we have:

$$EE'(1-\cos\theta) = m_e c^2 (E - E') \quad \text{or} \quad \frac{1}{E'} - \frac{1}{E} = \frac{1}{m_e c^2}(1-\cos\theta)$$

Provided E, E' are given by Eq.(P.43), we obtain the frequency of the scattered photon as:

$$\omega = \omega_0 \frac{m_e c^2}{m_e c^2 + \hbar\omega_0 (1-\cos\theta)}$$

It is seen that the frequency of the photon is not changed ($\omega = \omega_0$) either if it is scattered in the forward direction ($\theta = 0$), or if the collision is non-relativistic ($\hbar\omega_0 \ll m_e c^2$). In the relativistic case, where $\hbar\omega_0 \gg m_e c^2$, we can neglect $m_e c^2$ in the denominator to obtain the frequency ω of the scattered photon independent of ω_0:

$$\omega = \frac{m_e c^2}{\hbar(1-\cos\theta)}$$

It is common practice to express the Compton scattering in terms of the wavelength increase:

$$\Delta\lambda = \lambda' - \lambda = \frac{h}{m_e c}(1-\cos\theta)$$

which cannot be accounted for by the wave theory. ✥

Scattering experiments also confirm that the scattered photon is unidirectional, in agreement with the corpuscular theory, rather than a spherical wave, as predicted by the classical wave theory. This result finds support in the formal similarity between the law of wave propagation and the classical laws governing the motion of a particle.

EXAMPLE P.7. Motion of waves and particles

Wave motion obeys the *principle of least time*, in terms of which the Snell law of refraction defines the quickest route for a wave to travel from P_1 to P_2, Figure P.6 (a), across the boundary between two media, with velocities v_1 and v_2 respectively. If x and x_0 denote the position coordinates, with respect to B_1, of the points O and B_2, the time required from P_1 to P_2 is expressed in terms of the variable position x of the point O between B_1 and B_2 in the form:

$$t = \frac{P_1 O}{v_1} + \frac{P_2 O}{v_2} = \frac{n_1}{c}\sqrt{D_1^2 + x^2} + \frac{n_2}{c}\sqrt{D_2^2 + (x_0 - x)^2} = \frac{1}{c}(n_1 l_1 + n_2 l_2)$$

The condition that the time is a minimum reads:

$$\frac{dt}{dx} = n_1 \frac{x}{l_1} - n_2 \frac{x_0 - x}{l_2} = 0 \quad \text{or} \quad \frac{\sin\theta_1}{\sin\theta_2} = \frac{n_2}{n_1} = \frac{v_1}{v_2}$$

From the configuration illustrated in Figure P.6 (b) it is clear that, in a continuous medium, the wave propagation may be regarded as consisting of a series of refractions, which implies that the principle of least time should be formulated as a requirement for the integral:

$$t = \frac{1}{c}\int_{P_1}^{P_2} n\, dl \tag{P.45}$$

to be a minimum. If t is a minimum, any real or virtual variation δt should be equal to zero, which leads to the so-called variational formulation of the principle of least time as:

$$\delta \int_{P_1}^{P_2} n\, dl = 0 \quad \text{or} \quad \delta \int_{P_1}^{P_2} \frac{1}{v}\, dl = 0 \quad \text{or also} \quad \delta \int_{P_1}^{P_2} \frac{1}{\lambda}\, dl = 0 \tag{P.46}$$

The motion of a particle of energy E across the boundary between two media, where the potential energy is constant and equal to U_1 and U_2 respectively, can be described in terms of momenta derived from the law of conservation of energy as:

$$p_1 = \sqrt{2m(E - U_1)}, \qquad p_2 = \sqrt{2m(E - U_2)}$$

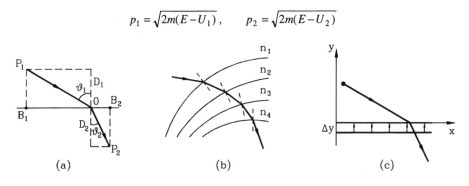

Figure P.6. Refraction of waves (a), (b) and particles (c)

As suggested in Figure P.6 (c), the abrupt change in the potential energy implies the existence of a normal force F_y at the boundary:

$$\frac{U_2 - U_1}{\Delta y} = -\frac{\Delta U}{\Delta y} \to F_y \quad \text{if} \quad \Delta y \to 0$$

which may influence the components of momentum along the y-axis only. In other words, the components of momentum along the boundary are conserved:

$$p_1 \sin\theta_1 = p_2 \sin\theta_2 \quad \text{or} \quad \frac{\sin\theta_1}{\sin\theta_2} = \frac{\sqrt{2m(E-U_2)}}{\sqrt{2m(E-U_1)}}$$

This law becomes identical to the Snell law for wave refraction, if the refractive index is replaced by the momentum:

$$n \leftrightarrow \sqrt{2m(E-U)} \quad \text{or} \quad \frac{1}{\lambda} \leftrightarrow p \quad (P.47)$$

The similarity between the propagation of waves and the motion of particles suggests that the trajectory of a particle in a force field may be determined by the requirement for the action integral:

$$S = \int_{P_1}^{P_2} p \, dl$$

which is obtained by substituting Eq.(P.46) into Eq.(P.45), to be a minimum. The variational formulation of this result:

$$\delta S = \delta \int_{P_1}^{P_2} p \, dl = \delta \int_{P_1}^{P_2} \sqrt{2m(E-U)} \, dl = 0 \quad (P.48)$$

is known as the *principle of least action*. ♦

Problem P.2.1. Use the particle nature of waves to derive the Doppler shift of a photon, emitted with frequency ω by a source of rest mass m_0 and velocity v.

(Solution): Let ω_0 be the frequency of a photon emitted by a fixed source, the rest mass of which is thus changed from m_0 to m_0':

$$m_0 c^2 = m_0' c^2 + \hbar \omega_0$$

If the same source moves with velocity $\vec{v}$, the photon is emitted with a frequency ω and the source suffers a recoil, such as its velocity changes to $\vec{v}'$. Conservation of energy and momentum now reads:

$$\frac{m_0 c^2}{\sqrt{1-v^2/c^2}} = \frac{m_0' c^2}{\sqrt{1-v'^2/c^2}} + \hbar\omega, \quad \vec{p} = \vec{p}' + \vec{p}_\gamma$$

where $\vec{p}, \vec{p}'$ are source momenta before and after radiation and $p_\gamma = E_\gamma/c = \hbar\omega/c$ is the momentum of the emitted photon. Assuming a configuration similar to that given in Figure P.5, where $\vec{p}_e$ is replaced by $\vec{p}_\gamma$, we obtain:

$$\frac{m_0 v}{\sqrt{1-v^2/c^2}} = \frac{m_0' v'}{\sqrt{1-v'^2/c^2}} \cos\theta + \frac{\hbar\omega}{c}\cos\varphi$$

It is a straightforward matter to eliminate m_0', v' and θ from the laws of conservation, yielding:

$$\frac{m_0 c^2 \omega}{\sqrt{1-v^2/c^2}}\left(1 - \frac{v}{c}\cos\varphi\right) - m_0 c^2 \omega_0 + \frac{\hbar\omega_0^2}{2} = 0$$

If the rest mass m_0 of the source is large enough, the last term may be neglected, so giving:

$$\omega = \frac{\sqrt{1-v^2/c^2}}{1-(v/c)\cos\varphi}\omega_0$$

For $\varphi = 0$ (forward emission) there is a Doppler shift towards higher frequencies: $\omega = \omega_0\sqrt{(1+v/c)(1-v/c)} > \omega_0$. For $\varphi = \pi$ (backward emission), the emitted frequency is shifted towards lower values: $\omega = \omega_0\sqrt{(1-v/c)(1+v/c)}$. A similar shift $\omega = (1-v^2/c^2)^{1/2}\omega_0 < \omega_0$ is predicted for $\varphi = \pi/2$. In the classical limit $v \ll c$, where v^2/c^2 can be neglected, we obtain:

$$\omega = \frac{\omega_0}{1-(v/c)\cos\varphi}$$

and no shift is predicted for $\varphi = \pi/2$.

Problem P.2.2. Scattering of light by solids can be described as a process of emission or absorption of a sound quantum, called a phonon, by a photon. Assuming the corpuscular nature of phonons, defined by the energy $E = \hbar\omega_s$ and momentum $p = \hbar\omega_s/v_s$, which are expressed in terms of the sound wave frequency

ω_s and velocity v_s in a way analogous to that postulated for photons, calculate the frequency ω_s of a phonon associated with scattering of a photon of frequency ω.

(Solution): Conservation of energy and momentum in a photon scattering process reads:

$$\hbar\omega = \hbar\omega' \pm \hbar\omega_s, \qquad \vec{p} = \vec{p}' \pm \vec{p}_s$$

where $\omega, \vec{p}$ and $\omega', \vec{p}'$ are the frequencies and momenta of the incident and scattered photon and $\vec{p}_s$ is the phonon momentum. The plus and minus signs indicate emission and absorption of a phonon respectively. It is convenient to use the configuration of Figure P.5, where $\vec{p}_e$ is replaced by $\vec{p}_s$, so that projecting the law of conservation of momentum onto the direction of $\vec{p}$ and onto the perpendicular direction gives:

$$\frac{\hbar\omega}{c/n} = \frac{\hbar\omega'}{c/n}\cos\theta \pm \frac{\hbar\omega_s}{v_s}\cos\varphi, \qquad 0 = \frac{\hbar\omega'}{c/n}\sin\theta + \frac{\hbar\omega_s}{v_s}\sin\varphi$$

where n is the refractive index of the solid. Eliminating ω' and φ from the laws of conservation gives:

$$\left(\frac{1}{v_s^2} - \frac{n^2}{c^2}\right)\omega_s^2 - \frac{4n^2\omega}{c^2}(\omega + \omega_s)\sin^2\frac{\theta}{2} = 0$$

We can neglect ω_s in the last term, since $\omega_s \ll \omega$, and also $(n/c)^2$ by comparison with $1/v_s^2$, since $v_s \ll c/n$, so obtaining:

$$\omega_s = \frac{2n\omega v_s}{c}\sin\frac{\theta}{2}$$

Problem P.2.3. Find the energy acquired by the recoil electron and the direction φ of the recoil motion in the Compton effect.

(Answer): $E = \dfrac{\hbar^2\omega^2}{mc^2 + \hbar\omega}, \qquad \tan\varphi = \dfrac{mc^2}{mc^2 + \hbar\omega}\cos\dfrac{\theta}{2} / \sin\dfrac{\theta}{2}$

Problem P.2.4. Given the Compton wavelength $\Lambda = h/m_e c$, find the wavelength change when light is scattered by protons of mass M.

(Answer): $\Delta\lambda = 2\dfrac{m_e}{M}\Lambda\sin^2\dfrac{\theta}{2}$

P.3. The Wave Nature of Particles

Although the similarity between the form of the trajectories and rays, as established by the principles of least time and least action, Eqs.(P.45) and (P.48), is of geometrical nature only, it suggests that the motion of a particle can be described, in a formal manner, by a wave. For simplicity, we may consider a monochromatic wave, which obeys the general equation of propagation, a particular case of which is Eq.(P.3):

$$\nabla^2 \Psi = \frac{1}{u^2} \frac{\partial^2 \Psi}{\partial t^2} \qquad (P.49)$$

where $u = \omega/k$ is the phase velocity. Substituting the usual representation of a monochromatic wave which is periodic in time, $\Psi(\vec{r},t) = \psi(\vec{r}) e^{-i\omega t}$, into Eq.(P.49) the wave equation for $\psi(\vec{r})$ reads:

$$\nabla^2 \psi(\vec{r}) + k^2 \psi(\vec{r}) = 0 \qquad (P.50)$$

Since the spherical wave solutions of Eq.(P.50) are unsuitable for the representation of the motion of a particle, we may solve this equation in Cartesian coordinates to find $\psi(\vec{r}) = A e^{i\vec{k}\cdot\vec{r}}$, so obtaining the plane wave representation:

$$\Psi(\vec{r},t) = A e^{i(\vec{k}\cdot\vec{r} - \omega t)} \qquad (P.51)$$

It is clear that the frequency ω of this wave should correspond to the energy E of the particle and the wave number $k = |\vec{k}| = 2\pi/\lambda$ will correspond to the momentum $p = mv$.

EXAMPLE P.8. The wave representation of an atomic electron

According to the postulate of discrete transitions, Eq.(P.27), the electromagnetic radiation emitted or absorbed when an electron makes a transition from one stationary state to another can be written as:

$$\Psi(\vec{r},t) = \psi(\vec{r}) e^{-i(E_m - E_n)t/\hbar} = \psi(\vec{r}) e^{iE_n t/\hbar} e^{-iE_m t/\hbar}$$

where the two factors are each assigned to a stationary state of definite energy. Therefore it is natural to assume that an electron in a stationary state can be represented by a monochromatic wave function, of the form (P.51), provided the frequency ω is expressed in terms of energy as required by the Einstein relation (P.43).

However, a matter wave, which is suitable to represent the periodic orbital motion of the electron, is required to be always in phase with the original wave, after having completed an orbit, because otherwise it will vanish due to interference. In terms of the wave theory, this means that the length of the orbit must be an integral multiple of the wavelength, namely:

$$\oint \frac{dr}{\lambda} = n$$

This condition becomes identical to the rule of quantization for the orbital motion, Eq.(P.31), which reads $\oint (p/h)dr = n$, if we assume $\lambda = h/p$, which means that the wave number k should be expressed in terms of momentum in a form similar to that derived for a photon, Eq.(P.44). ✣

The assumption first made by de Broglie, that the electrons are accompanied by matter waves, which are regarded as localized with the particle, implies that, in order to represent the motion of a particle with energy E and momentum p, the plane wave (P.51) should be written as:

$$\Psi(\vec{r},t) = Ae^{i(\vec{p}\cdot\vec{r}-Et)/\hbar} \tag{P.52}$$

The wavefront moves along the $\vec{p} = p\vec{e}_p$ direction, such that the distance between two points is λ if the corresponding phase change is 2π. In other words $\delta\vec{r} = \lambda\vec{e}_p$ if $\vec{p}\cdot\delta\vec{r}/\hbar = 2\pi$ which yields the matter wavelength:

$$\lambda = \frac{h}{p} \tag{P.53}$$

known as the *de Broglie hypothesis*. The matter or *de Broglie wave*, the rays of which represent the trajectories of a particle of energy E and momentum p, provides an explanation for the rules of quantization, which have been previously postulated without any physical basis.

EXAMPLE P.9. Equation of the de Broglie wave propagation

We expect the matter wave (P.52) to be a solution of a general equation, which should be independent of the E and p parameters of a particle. Assuming a particle which is moving in the field of a potential energy $U(\vec{r})$, the two parameters are related by:

$$E = \frac{p^2}{2m} + U(\vec{r}) = \frac{1}{2m}(p_x^2 + p_y^2 + p_z^2) + U(\vec{r}) \tag{P.54}$$

so that the required equation is readily obtained by substituting E and the components of momentum, as derived from Eq.(P.52) in terms of the wave function Ψ. Differentiating Eq.(P.52) gives:

$$\frac{\partial \Psi}{\partial t} = -\frac{i}{\hbar} E\Psi \quad \text{or} \quad E\Psi = i\hbar \frac{\partial \Psi}{\partial t}$$

$$\frac{\partial \Psi}{\partial x} = \frac{i}{\hbar} p_x \Psi, \quad \frac{\partial^2 \Psi}{\partial x^2} = \left(\frac{i}{\hbar} p_x\right)^2 \Psi \quad \text{or} \quad p_x^2 \Psi = -\hbar^2 \frac{\partial^2 \Psi}{\partial x^2}$$

and similar expressions for $p_y^2 \Psi$ and $p_z^2 \Psi$. If now Eq.(P.54) is multiplied by Ψ, and then use is made of the previous relations, we obtain:

$$i\hbar \frac{\partial \Psi}{\partial t} = -\frac{\hbar^2}{2m}\left(\frac{\partial^2 \Psi}{\partial x^2}+\frac{\partial^2 \Psi}{\partial y^2}+\frac{\partial^2 \Psi}{\partial z^2}\right)+U(\vec{r})\Psi = -\frac{\hbar^2}{2m}\nabla^2 \Psi + U(\vec{r})\Psi \qquad (P.55)$$

which is known as the *time-dependent Schrödinger equation* for the motion of a particle in a force field. If the total energy E is a constant of motion, and thus the magnitude of momentum is time-independent, Eq.(P.55) admits solutions $\Psi(\vec{r},t) = \psi(\vec{r})e^{-iEt/\hbar}$, which on substitution allow us to reduce it to an equation for $\psi(\vec{r})$ of the form:

$$E\psi(\vec{r}) = -\frac{\hbar^2}{2m}\nabla^2 \psi(\vec{r}) + U(\vec{r})\psi(\vec{r}) \qquad (P.56)$$

called the *time-independent Schrödinger equation* for one particle. It may be rewritten as:

$$\nabla^2 \psi(\vec{r}) + \frac{2m}{\hbar^2}[E - U(\vec{r})]\psi(\vec{r}) = \nabla^2 \psi(\vec{r}) + \left(\frac{p}{\hbar}\right)^2 \psi(\vec{r}) = 0$$

where use has been made of Eq.(P.54). In other words, Eq.(P.56) has the form of the classical wave equation (P.50), provided the wave number is expressed as required by the de Broglie hypothesis. ✥

The de Broglie wavelength (P.53) associated with electrons has been measured experimentally, providing a strong reason to retain the de Broglie concept of matter waves. For low energy electrons, λ corresponds to the atomic spacing of most crystalline solids. If Eq.(P.53) were correct, namely $\lambda = h/\sqrt{2m_e E} = h/\sqrt{2m_e eV} = \sqrt{15/V}$ (nm), such electrons should exhibit diffraction effects. A diffraction pattern for low energy electrons scattered from a nickel crystal was observed by Davisson and Germer. An angular plot of the intensity of the scattered electron current is given in Figure P.7, showing a distinct peak at $\theta = 50°$, for 54 eV electrons of wavelength $\lambda = 0.167$ nm. The classical condition of constructive interference, $r_1 - r_2 = m\lambda$, when applied to beams reflecting from neighbouring crystal planes, in the configuration shown in Figure P.7, becomes:

$$m\lambda = 2\frac{d}{\cos(\theta/2)} - 2d \tan(\theta/2)\sin(\theta/2) = 2d \cos(\theta/2) = a \sin\theta$$

where $d = a \sin(\theta/2)$ is the distance between the reflecting planes and $a = 0.215$ nm is the lattice constant of the *Ni* crystal. The calculated value of $\theta = 51°$ for $m = 1$ agrees very well with the results described above.

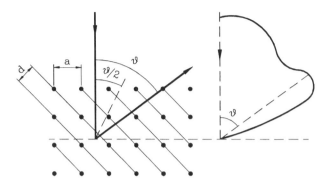

Figure P.7. Reflection of electron waves from a crystal and the corresponding intensity of the electron current

The concept of the matter wavelength, as defined by Eq.(P.53), holds for both relativistic and non-relativistic particles, provided that the appropriate form of momentum is used. In special relativity, the phase $\vec{k} \cdot \vec{r} - \omega t$ of the plane monochromatic wave (P.51) can be interpreted as the scalar product of the position four-vector $r_\mu = (\vec{r}, ict)$ and the wave four-vector $k_\mu = (\vec{k}, i\omega/c)$. By requiring that k_μ should be proportional to the four-momentum $p_\mu = (\vec{p}, iE/c)$, it follows that both relations $\vec{k} \sim \vec{p}$ and $\omega \sim E$ must hold for the same scalar coefficient. Since, on the grounds of the Planck hypothesis, it is natural to assume $E = \hbar\omega$, we may then infer that the space-like components of the two four-vectors must also be related by $\vec{p} = \hbar\vec{k}$, which is the same as the de Broglie hypothesis (P.53), now formulated in terms of the relativistic momentum.

However, in the calculation of the phase velocity of matter waves, $u = \omega/k$, a distinction appears between the relativistic and non-relativistic cases, which shows that the de Broglie frequency, in contrast to the de Broglie wavelength, is an inadequate concept, so pointing out the limits of the formal description of the motion of a particle by monochromatic matter waves.

EXAMPLE P.10. Phase velocity of the de Broglie wave

In the relativistic case, the relations between the corpuscular and wave parameters, Eqs.(P.43) and (P.53), take the form:

$$E = mc^2 = \frac{m_0 c^2}{\sqrt{1 - v^2/c^2}} = \hbar\omega, \quad \vec{p} = m\vec{v} = \frac{m_0 \vec{v}}{\sqrt{1 - v^2/c^2}} = \hbar\vec{k}$$

Substituting these results into the relativistic momentum-energy relation, which shows that the norm $\sum_\mu p_\mu^2$ of the four-momentum p_μ, that is defined similarly to the norm of the three-

28   Quantum Physics

dimensional vector $\vec{p}$, is invariant under Lorentz transformation, namely $p^2 - E^2/c^2 = -m_0^2 c^2$ one obtains the relativistic dispersion relation for the matter waves:

$$\frac{\omega^2}{c^2} = k^2 + \frac{m_0^2 c^2}{\hbar^2}$$

The phase velocity is readily derived as:

$$u = \frac{\omega}{k} = c\left(1 + \frac{m_0^2 c^2}{h^2}\lambda^2\right)^{\frac{1}{2}}$$

which obviously exceeds c for all material particles with $m_0 \neq 0$. However, there is no violation of the principle of special relativity, which states that the velocity of material particles may not exceed c, because the velocity of propagation u of the phase waves associated with a particle is not to be confused with the velocity v of this particle. The last result may be reformulated in terms of v as:

$$u = c\left(1 + \frac{m_0^2 c^2}{m^2 v^2}\right)^{\frac{1}{2}} = c\left[1 + (1 - v^2/c^2)\frac{c^2}{v^2}\right]^{\frac{1}{2}} = \frac{c^2}{v}$$

In the non-relativistic case, Eq.(P.54) holds, and the dispersion relation for the de Broglie waves reads:

$$\hbar\omega = \frac{\hbar^2 k^2}{2m} + U(\vec{r})$$

Thus, the phase velocity u of the wave associated with a non-relativistic particle may be expressed in terms of the velocity v of this particle by:

$$u = \frac{\omega}{k} = \frac{\hbar k}{2m} + \frac{U(\vec{r})}{\hbar k} = \frac{v}{2} + \frac{U(\vec{r})}{mv}$$

This means, for a free particle, for example, that the phase velocity is one-half of the particle velocity, and therefore it is not of any physical significance. ☙

Using the expressions for the relativistic change in energy and in the magnitude of the momentum:

$$dE = d\left(\frac{m_0 c^2}{\sqrt{1 - v^2/c^2}}\right) = \frac{m_0 v}{(1 - v^2/c^2)^{3/2}} dv$$

$$dp = d\left(\frac{m_0 v}{\sqrt{1 - v^2/c^2}}\right) = \frac{m_0}{(1 - v^2/c^2)^{3/2}} dv$$

the velocity v of a material particle can be obtained in terms of the wave parameters as:

$$v = \frac{dE/dv}{dp/dv} = \frac{dE}{dp} = \frac{d\omega}{dk} \qquad (P.57)$$

This form defines, in the classical wave theory, the *group velocity* of a continuous superposition of monochromatic waves, Eqs.(P.51) or (P.52), called a wave packet. Hence, Eq.(P.57) suggests that the mechanics of a material particle might be described in terms of an associated wave packet.

Problem P.3.1. Find the relation between the phase and group velocities of a wave packet obtained by superposing three waves:

$$\Psi = Ae^{i(kx-\omega t)}, \quad \Psi_+ = \frac{A}{2}e^{i[(k+dk)x-(\omega+d\omega)t]} \text{ and } \Psi_- = \frac{A}{2}e^{i[(k-dk)x-(\omega-d\omega)t]}$$

all travelling in the x-direction.

(Solution): Superposition of the three waves gives:

$$\Psi + \Psi_+ + \Psi_- = Ae^{i(kx-\omega t)}\left[1 + \frac{1}{2}e^{i(xdk-td\omega)} + \frac{1}{2}e^{-i(xdk-td\omega)}\right]$$

$$= Ae^{i(kx-\omega t)}[1+\cos(xdk-td\omega)] = 2A\cos^2\left(\frac{dk}{2}x - \frac{d\omega}{2}t\right)e^{i(kx-\omega t)}$$

The wave front of the harmonic wave, defined by $kx - \omega t = $ const, or $x = \omega t/k + $ const propagates with phase velocity $u = \omega/k$, while the wave packet envelope is modulated by the slowly varying square cosine function. Each point of the envelope, for instance its maximum, which is defined by $xdk - td\omega = 0$, propagates with the so-called group velocity v, which is:

$$v = \frac{x}{t} = \frac{d\omega}{dk}$$

Substituting $\omega = ku$ yields the relation between the two velocities as $v = u + k\,du/dk$.

Quantum Physics

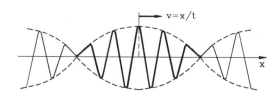

An alternative equation connecting the phase and group velocities can be derived in terms of wavelength $\lambda = 2\pi/k$, using:

$$\frac{du}{dk} = \frac{du}{d\lambda}\frac{d\lambda}{dk} = -\frac{\lambda^2}{2\pi}\frac{du}{d\lambda}$$

so that we obtain the so-called Rayleigh formula:

$$v = u - \lambda\frac{du}{d\lambda}$$

Problem P.3.2. Derive the phase velocity and the frequency of the de Broglie waves from the requirement that their group velocity, given by the Rayleigh formula, should be equal to the velocity v of a material particle.

(Solution): If the group velocity $v = u - \lambda du/d\lambda$ is assumed to be the velocity of a material particle, then the wavelength λ should be written as $\lambda = h/p = h/mv$. For a non-relativistic particle m is a constant, so that:

$$\frac{du}{d\lambda} = \frac{du}{dv}\frac{dv}{d\lambda} = -\frac{v}{\lambda}\frac{du}{dv} \quad \text{or} \quad \lambda\frac{du}{d\lambda} = -v\frac{du}{dv}$$

and we have:

$$v = u + v\frac{du}{dv} = \frac{d}{dv}(uv)$$

It follows that:

$$uv = \frac{v^2}{2} + \text{const} \quad \text{or} \quad u = \frac{v}{2} + \frac{\text{const}}{v}$$

and this is a result which has been previously derived in a different manner. We further obtain ω in the form:

$$\omega = uk = \frac{2\pi}{\lambda}u = 2\pi\frac{mv}{h}\left(\frac{v}{2} + \frac{\text{const}}{v}\right) = \frac{1}{\hbar}\left(\frac{mv^2}{2} + \text{const}\right)$$

or:

$$\hbar\omega = \frac{mv^2}{2} + \text{const}$$

which corresponds to Eq.(P.43) written for a non-relativistic material particle.

Problem P.3.3. Find the equation of the de Broglie wave propagation for a particle of mass m, energy E and momentum p related by the relativistic momentum-energy relation.

(Solution): If we associate the plane wave given by Eq.(P.52) to the motion of a particle of energy E and momentum p, we obtain:

$$\frac{\partial^2 \Psi}{\partial t^2} = -\frac{E^2}{\hbar^2}\Psi, \quad \nabla^2 \Psi = -\frac{p^2}{\hbar^2}\Psi$$

Substituting into the relativistic momentum-energy relation, which reads:

$$\left(\frac{E^2}{c^2} - p^2\right)\Psi = m^2 c^2 \Psi$$

we obtain:

$$-\frac{\hbar^2}{c^2}\frac{\partial^2 \Psi}{\partial t^2} + \hbar^2 \nabla^2 \Psi = m^2 c^2 \Psi$$

which is usually written in the form:

$$\left(\nabla^2 - \frac{1}{c^2}\frac{\partial^2}{\partial t^2}\right)\Psi = \frac{m^2 c^2}{\hbar^2}\Psi$$

and is known as the Klein-Gordon equation.

Problem P.3.4. Find the energy required for a beam of electrons, incident at a glancing angle $\alpha = 30°$ on the surface of a crystal (lattice constant $a = 2\text{Å}$), to undergo first order Bragg reflection from planes parallel to the surface.

(Answer): $E = \dfrac{h^2}{2ma^2 \sin^2 2\alpha} \approx 10\,\text{eV}$

Quantum Physics

Problem P.3.5. Derive the most probable de Broglie wavelength for the molecules of a perfect gas at a given temperature T, from their Maxwell distribution function written in terms of the de Broglie wavelength.

(Answer): $\lambda_p = \dfrac{h}{2\sqrt{mk_B T}}$

Problem P.3.6. Find the temperature required to allow a thermalized beam of hydrogen atoms, to produce a diffraction peak at $\theta = 90°$ from a crystal with lattice constant $a = 2$ Å.

(Answer): $T = \dfrac{h^2}{3mk_B a^2} = 160$ K

1. WAVE FUNCTIONS AND THE FIRST POSTULATE

1.1. Wave Packet Representation of a Particle

A consistent description of the motion of a particle by wave propagation implies not only a geometrical equivalence between a trajectory and a ray but also a kinematical equivalence between the time rates at which the particle or wave is traversing its path. This may be tentatively achieved by using a wave configuration, known as a wave packet, represented as a continuous superposition of monochromatic waves. If the wave packet can be regarded as a point, having zero amplitude everywhere except in a very small region, its motion along a trajectory might be interpreted as the motion of a single particle. The representation of a particle in space and time by means of a highly localized wave packet is a basic assumption of wave mechanics. The behaviour of the wave packet as a whole will account for experiments where particle-like properties are observed, whereas individual wavelength components of the packet will appear in experiments which depend upon interference and diffraction effects.

EXAMPLE 1.1. Wave packets

It is well-known that a harmonic wave can be expanded as a linear combination of sine and cosine functions, which reads:

$$\Psi(\xi) = A\cos(k\xi + \varphi) = a\cos k\xi + b\sin k\xi = a\cos\frac{2\pi}{\lambda}\xi + b\sin\frac{2\pi}{\lambda}\xi$$

where k is written in terms of a wavelength λ. By a procedure somewhat similar to curve fitting, where an nth degree polynomial is required to fit n points and to determine the expansion coefficients, any periodic function can be represented as the superposition of harmonic terms of frequencies $f_1, f_2 = 2f_1, f_3 = 3f_1, \ldots$ where $f_1 = 1/\lambda$. If we try to fit all points, by letting $n \to \infty$, the representation takes the form:

$$\Psi(\xi) = \frac{1}{2}a_0 + \sum_{n=1}^{\infty} a_n \cos\frac{2\pi n\xi}{\lambda} + \sum_{n=1}^{\infty} b_n \sin\frac{2\pi n\xi}{\lambda} \tag{1.1}$$

called the *Fourier series*. The form (1.1) converges to the original function at all points, provided $\Psi(\xi)$ and its first derivative are continuous within each period, conditions which are satisfied for all functions useful in physics. A more convenient alternative to the sine and cosine series is to use the complex notation which transforms Eq.(1.1) into:

$$\Psi(\xi) = \frac{1}{2}a_0 + \frac{1}{2}\sum_{n=1}^{\infty}(a_n + ib_n)e^{i2\pi n\xi/\lambda} + \frac{1}{2}\sum_{n=1}^{\infty}(a_n - ib_n)e^{-i2\pi n\xi/\lambda}$$

where the last term can be rewritten in the form:

$$\frac{1}{2}\sum_{n=-1}^{-\infty}(a_n - ib_n)e^{i2\pi n\xi/\lambda}$$

If the *complex Fourier coefficients* are now introduced as:

$$\Phi_n = \frac{1}{2}(a_n + ib_n), \quad n > 0$$

$$= \frac{1}{2}a_0, \quad n = 0$$

$$= \frac{1}{2}(a_n - ib_n), \quad n < 0$$

Eq.(1.1) takes the compact form:

$$\Psi(\xi) = \sum_{n=-\infty}^{\infty}\Phi_n e^{i2\pi n\xi/\lambda} \qquad (1.2)$$

Since the exponential functions form an orthogonal set:

$$\int_0^\lambda e^{i2\pi m\xi/\lambda} e^{-i2\pi n\xi/\lambda} d\xi = 0 \quad (m \neq n)$$

$$= \lambda \quad (m = n) \qquad (1.3)$$

we obtain from Eq.(1.2), by the method of term by term integration, the complex coefficients in the form:

$$\Phi_n = \frac{1}{\lambda}\int_0^\lambda \Psi(\xi) e^{-i2\pi n\xi/\lambda} d\xi \qquad (1.4)$$

The concept of *monochromatic waves*, of one unique frequency, is associated with periodic functions representing disturbances which indefinitely repeat themselves in space and time. In contrast, real waves have a finite extension in space and time and require a representation by non-periodic functions. If we let the wavelength λ of a periodic wave function $\Psi(\xi)$ approach infinity, so isolating a *single non-periodic pulse*, the wave number $k = 2\pi/\lambda$ becomes infinitesimal and the Fourier coefficients take values of the continuous function

1. Wave Functions and the First Postulate   35

$\Phi(2\pi/\lambda) = \Phi(k)$. Treating the limit of the sum in Eq.(1.2) as an integral, the wave function of a non-periodic pulse takes the form:

$$\Psi(\xi) = \lim_{\lambda \to \infty} \sum_{n=-\infty}^{\infty} \Phi_n e^{ink\xi} = \int_{-\infty}^{\infty} \Phi(k) e^{ik\xi} dk \qquad (1.5)$$

which is called the *Fourier integral formula*. Substituting the complex Fourier coefficients Φ_n, as given by Eq.(1.4), Eq.(1.5) can alternatively be written in the form:

$$\Psi(\xi) = \lim_{\lambda \to \infty} \frac{1}{\lambda} \sum_{n=-\infty}^{\infty} \int_0^\lambda \Psi(u) e^{ikn(\xi-u)} du = \frac{1}{2\pi} \int_{-\infty}^{\infty} dk \int_{-\infty}^{\infty} \Psi(u) e^{ik(\xi-u)} du$$

$$= \frac{1}{2\pi} \int_{-\infty}^{\infty} e^{ik\xi} dk \int_{-\infty}^{\infty} \Psi(u) e^{-iku} du$$

which compared to Eq.(1.5), defines the so-called Fourier spectrum as:

$$\Phi(k) = \frac{1}{2\pi} \int_{-\infty}^{\infty} \Psi(u) e^{-iku} du = \frac{1}{2\pi} \int_{-\infty}^{\infty} \Psi(\xi) e^{-ik\xi} d\xi \qquad (1.6)$$

The function $\Phi(k)$ is called the *Fourier transform* of $\Psi(\xi)$. Since the two functions $\Psi(\xi)$ and $\Phi(k)$, as given by Eqs.(1.5) and (1.6), are very nearly symmetrically related, they are said to form a *pair of Fourier transforms*. It is convenient to reformulate Eqs.(1.5) and (1.6) in the symmetrical form:

$$\Psi(\xi) = \frac{1}{\sqrt{2\pi}} \int_{-\infty}^{\infty} \Phi(k) e^{ik\xi} dk$$

$$\Phi(k) = \frac{1}{\sqrt{2\pi}} \int_{-\infty}^{\infty} \Psi(\xi) e^{-ik\xi} d\xi \qquad (1.7)$$

The Fourier integrals allow the spectrum analysis of *continuous waves* and therefore the description of waves of arbitrary profile, without any reference to the wave equation. A harmonic wave of finite extension in space and time may be written as:

$$\Psi(\xi) = A(\xi) e^{ik_0 \xi} \qquad (1.8)$$

where the envelope $A(\xi)$ is zero outside a specified range of values $\xi = x \pm vt$. The function (1.8) is said to describe a *wave packet* since the harmonic pulse can always be represented as a continuous superposition of harmonic waves, given by the Fourier integral (1.5). The *Gaussian wave packet* is given by the envelope:

$$A(\xi) = A e^{-a\xi^2}$$

represented in Figure 1.1 (a), which is symmetrically centred at $\xi = 0$. Its shape is characterized by the central height A and the width parameter a. The amplitude is diminished by a factor $1/e$ over an interval $\Delta\xi = 2/\sqrt{a}$ which is taken as the *effective width* of the pulse. Substituting Eq.(1.8) written for a Gaussian amplitude, the spectral density (1.6) becomes:

$$\Phi(k) = \frac{A}{2\pi} \int_{-\infty}^{\infty} e^{-a\xi^2 + i(k_0 - k)\xi} d\xi$$

$$= \frac{A}{2\pi} e^{-(k-k_0)^2/4a} \int_{-\infty}^{\infty} e^{-a[\xi + i(k-k_0)/2a]^2} d\xi = \frac{A}{2\pi} e^{-(k-k_0)^2/4a} \int_{-\infty}^{\infty} e^{-au^2} du$$

where it was convenient to set $u = \xi + i(k - k_0)/2a$. The last integral has the Gaussian form, so that we obtain:

$$\Phi(k) = \frac{A}{\sqrt{4\pi a}} e^{-(k-k_0)^2/4a} \qquad (1.9)$$

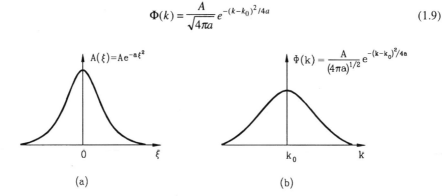

(a)   (b)

Figure 1.1. The Gaussian pulse and its Fourier transform

Hence, the Fourier transform (1.9) of the Gaussian pulse is another Gaussian function, represented in Figure 1.1 (b). The effective width of the spectrum:

$$\Delta k = 4\sqrt{a} = \frac{8}{\Delta\xi} \quad \text{or} \quad \Delta\xi \Delta k = 8 \qquad (1.10)$$

is inversely related to the effective width $\Delta\xi$ of the pulse, which means that a narrow pulse has a wide distribution of frequencies. The range in space, defined by the pulse width $\Delta\xi$ and the range in frequency, given by the bandwidth Δk are inversely related, as shown by Eq.(1.10). Although the constant might vary with the shape of the pulse, it is found that:

$$\Delta\xi \Delta k \geq 1 \qquad (1.11)$$

is always true. Equation (1.11) is known as the *bandwidth theorem*, which states that the product of the allowable range of two conjugated parameters has a lower limit, not equal to zero. If the range in space $\Delta\xi$ is infinite, the range in the wave number Δk is zero and one wave number defines the wave, which therefore is monochromatic.

1. Wave Functions and the First Postulate 37

The parameters $\omega = vk$ and $t = \xi/v$ (at a given position x) form another conjugate pair, obeying the bandwidth theorem (1.11) which reads:

$$\Delta t \Delta \omega \geq 1 \tag{1.12}$$

If the range in time Δt, available for the pulse passage through a given point, is small, the range in frequency $\Delta \omega$ is required to be large and vice-versa. ☙

Since the expression for de Broglie wave propagation is a linear differential equation, it will always admit solutions in the form of a superposition of monochromatic waves. For simplicity we limit our discussion to motions in one dimension, and consider the wave configuration:

$$\Psi(x,t) = \sum_n \Phi_n e^{in(kx-\omega t)} \tag{1.13}$$

that obviously satisfies Eq.(P.55), which in one dimension reads:

$$i\hbar \frac{\partial}{\partial t}\left(\sum_n \Phi_n e^{in(kx-\omega t)}\right) = -\frac{\hbar^2}{2m}\frac{\partial^2}{\partial x^2}\left(\sum_n \Phi_n e^{in(kx-\omega t)}\right) + U(x)\sum_n \Phi_n e^{in(kx-\omega t)}$$

For n ranging from $-\infty$ to ∞, Eq.(1.13) can be regarded as a Fourier series representation of a periodic function $\Psi(x,t)$, whereas a single pulse of finite extension in space and time can be isolated by making the period approach infinity, or $k \to 0$. The Fourier coefficients Φ_n may then be written as a continuous function $\Phi(k)$, and treating the limit of the sum in Eq.(1.13) as an integral, the wave function of a localized pulse takes the form:

$$\Psi(x,t) = \lim_{k\to 0}\sum_{n=-\infty}^{\infty} \Phi_n e^{in(kx-\omega t)} = \int_{-\infty}^{\infty}\Phi(k)e^{i(kx-\omega t)}dk \tag{1.14}$$

which is a Fourier integral, as defined by Eq.(1.5), representing a superposition of a continuum of plane waves.

The matter wave packet is expected to be highly localized in the neighbourhood of the particle position, since the component waves interfere destructively except in a limited region. Consider the Gaussian packet, the spectrum of which, Eq.(1.9), is:

$$\Phi(k) = e^{-\alpha(k-k_0)^2} = e^{-\alpha(\Delta k)^2} \tag{1.15}$$

where, for simplicity, we have set $\alpha = 1/4a$ and the amplitude has been omitted. In a dispersive medium, which transmits harmonic waves of differing frequency at different velocities, it is convenient to expand ω as a function of k in a Taylor series about the wave number k_0 of a wave packet which is defined by Eq.(1.14):

38   Quantum Physics

$$\omega = \omega_0 + v_g(k - k_0) + \zeta(k - k_0)^2 + \ldots \qquad (1.16)$$

where:

$$v_g = \left(\frac{d\omega}{dk}\right)_{k=k_0}, \quad \zeta = \frac{1}{2}\left(\frac{d^2\omega}{dk^2}\right)_{k=k_0} \qquad (1.17)$$

Taking the linear approximation of ω in Eq.(1.16), and substituting into Eq.(1.14) gives:

$$\Psi(x,t) = e^{i(k_0 x \pm \omega_0 t)} \int_{-\infty}^{\infty} \Phi(k) e^{i(k-k_0)(x \pm v_g t)} dk \qquad (1.18)$$

Since v_g is not dependent on k, Eq.(1.18) takes the equivalent form:

$$\Psi(x,t) = A(x \pm v_g t) e^{ik_0(x \pm v_0 t)}$$

which describes a harmonic wave of constant frequency ω_0 whose envelope $A(x \pm v_g t)$ is given in terms of the spectral density $\Phi(k)$ of the wave packet. In a frame moving with velocity $\pm v_g$ along the x-axis, the envelope appears unchanged. It follows that each point on the envelope, for instance its maximum, propagates with a velocity v_g, defined by Eq.(1.17), and called the *group velocity*.

The quadratic term in the Taylor series expansion of the dispersion relation does not contribute directly to the group velocity, causing instead a spatial spreading of the wave packet as it progresses in a dispersive medium. In the second order approximation of ω, Eq.(1.14) becomes:

$$\Psi(x,t) = e^{i(k_0 x - \omega_0 t)} \int_{-\infty}^{\infty} e^{i\Delta k(x-v_g t)} e^{-(\alpha+i\zeta t)(\Delta k)^2} dk$$

$$= e^{i(k_0 x - \omega_0 t)} \int_{-\infty}^{\infty} e^{-(\alpha+i\zeta t)[\Delta k - i(x-v_g t)/2(\alpha+i\zeta t)]^2} e^{-(x-v_g t)^2/4(\alpha+i\zeta t)} dk$$

$$= e^{i(k_0 x - \omega_0 t)} e^{-(x-v_g t)^2/4(\alpha+i\zeta t)} \int_{-\infty}^{\infty} e^{-(\alpha+i\zeta t)u^2} du$$

The last integral has the Gaussian form, so that we finally obtain:

$$\Psi(x,t) = \left(\frac{\pi}{\alpha+i\zeta t}\right)^{\frac{1}{2}} e^{-(x-v_g t)^2/4(\alpha+i\zeta t)} e^{i(k_0 x - \omega_0 t)} \qquad (1.19)$$

1. Wave Functions and the First Postulate

The function $\Psi(x,t)$ gives the coordinate representation of a travelling wave packet of phase velocity $u = \omega_0/k_0$ and amplitude:

$$|\Psi(x,t)|^2 = \left(\frac{\pi^2}{\alpha^2 + \zeta^2 t^2}\right)^{\frac{1}{2}} e^{-\alpha(x-v_0 t)^2/2(\alpha^2+\zeta^2 t^2)} \quad (1.20)$$

A spreading of the Gaussian wave packet, indicated by Eq.(1.20) and illustrated in Figure 1.2, occurs whenever the amplitude (or group) velocity v_g is different from the phase velocity u. This seems incompatible with our expectation that the matter waves associated with particles should remain highly localized.

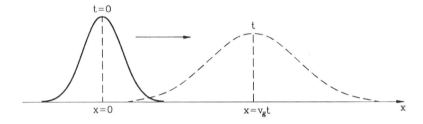

Figure 1.2. The spreading of the Gaussian matter wave packet

The de Broglie assumption was to assign the group velocity v_g to a classical particle of momentum p_x, namely:

$$v_g = \frac{d\omega}{dk} = \frac{p_x}{m} \quad (1.21)$$

If a particle with energy E obeys the Einstein relation (P.43) for electromagnetic radiation, that is:

$$E = \hbar\omega = \frac{p_x^2}{2m} + U(x) \quad \text{or} \quad \omega = \frac{p_x^2}{2m\hbar} + \frac{U(x)}{\hbar} \quad (1.22)$$

it follows that Eq.(1.21) only holds for the choice:

$$p_x = \hbar k \quad (1.23)$$

which is the same as the de *Broglie hypothesis*, Eq.(P.53).

The matter wave function (1.14) can be now expressed in terms of the corpuscular properties E and p_x, given by Eqs.(1.22) and (1.23), as:

$$\Psi(x,t) = \frac{1}{\sqrt{2\pi\hbar}} \int_{-\infty}^{\infty} \Phi(p_x) e^{i(p_x x - Et)/\hbar} dp_x = \frac{1}{\sqrt{2\pi\hbar}} \int_{-\infty}^{\infty} \Phi(p_x,t) e^{ip_x x/\hbar} dp_x \quad (1.24)$$

where the constant is obtained by normalizing $\Psi(x,t)$ in view of Eq.(1.3). As expected from Eq.(P.55), provided E is a constant of motion, the function $\Psi(x,t)$ is a solution to the equation:

$$i\hbar \frac{\partial \Psi(x,t)}{\partial t} = \frac{1}{\sqrt{2\pi\hbar}} \int_{-\infty}^{\infty} \Phi(p_x) E e^{i(p_x x - Et)/\hbar} dp_x$$

which is the same as:

$$i\hbar \frac{\partial \Psi(x,t)}{\partial t} = \frac{1}{\sqrt{2\pi\hbar}} \int_{-\infty}^{\infty} \Phi(p_x) \frac{p_x^2}{2m} e^{i(p_x x - Et)/\hbar} dp_x + \frac{1}{\sqrt{2\pi\hbar}} U(x) \int_{-\infty}^{\infty} \Phi(p_x) e^{i(p_x x - Et)/\hbar} dp_x$$

$$= -\frac{\hbar^2}{2m} \frac{\partial^2 \Psi(x,t)}{\partial x^2} + U(x)\Psi(x,t) \tag{1.25}$$

and will be associated with the motion of a particle, in coordinate representation.

Since, according to classical ideas, we expect that matter waves should travel without dispersion, the de Broglie concept of matter waves, formulated in terms of the wave packet representation of the motion of particles, implies that we must modify the classical concepts concerning the meaning of position and velocity of a particle. The wave function (1.24), if interpreted as a physical wave, cannot itself be identified with a material particle, which is expected to be partly reflected and partly transmitted at a boundary, because no particle may be split into two parts by reflection and transmission. $\Psi(x,t)$ is essentially a complex function which accounts for the spreading of the matter wave, illustrated in Figure 1.2. Its modulus is therefore large where the particle may be localized and small elsewhere. An indirect interpretation was suggested in terms of the distribution pattern rather than the wave which creates it. This is known as the *Born assumption* and states that:

$$P(x,t)dx = |\Psi(x,t)|^2 dx \tag{1.26}$$

where $P(x,t)dx$ is the *probability* of finding the particle described by $\Psi(x,t)$ at a point between x and $x + dx$ at a time t. In this description, the particle remains indivisible and without dimension, but its position is no longer exactly known. Therefore the spreading of the matter wave is associated with a decrease in the *probability density* $P(x,t)$.

EXAMPLE 1.2. Statistical interpretation of position

If a particle is represented by a wave packet, its position along the x-axis is a continuous variable, with values inside the segment between the intersections of the wave packet envelope with the x-axis. It is convenient to express the result of repeated measurements of the position in terms of a small parameter a, such that the x values occur N_0 times between 0 and a, N_1 times

1. Wave Functions and the First Postulate   41

between a and $2a, \ldots, N_x$ times between x and $x + a$. The probability that a measurement yields a value x between x and $x + a$ can be then defined by:

$$P_x = \frac{N_x}{N}$$

and may be represented by a rectangular area $P_x = aP(x)$, as shown in Figure 1.3 (a), where the area under the wave packet envelope should be normalized to unity:

$$\sum P_x = \sum \frac{N_x}{N} = 1 \qquad (1.27)$$

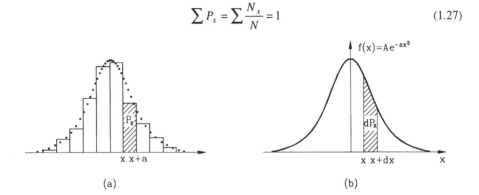

(a)   (b)

Figure 1.3. Geometrical representation of the probability density

In the limit where $a = dx \to 0$, Figure 1.3 (b), we may define the elementary probability dP_x in terms of the probability density $P(x)$, as:

$$dP_x = P(x)dx$$

such that the normalization condition reads:

$$\int_{-\infty}^{\infty} dP_x = \int_{-\infty}^{\infty} P(x)dx = 1$$

We can define an average value, or *expectation value*, of the position variable by considering either a large number of measurements on the same particle, or a single measurement on an assembly of identical particles. The expectation value of position is thus given by the weighted average:

$$\langle x \rangle = \int_{-\infty}^{\infty} x \, dP_x = \int_{-\infty}^{\infty} xP(x)dx \qquad (1.28)$$

and therefore an arbitrary function of the position variable has the expectation value:

$$\langle f(x) \rangle = \int_{-\infty}^{\infty} f(x)P(x)dx \qquad (1.29)$$

Quantum Physics

The difference between a given measurement of x and the expectation value of x is called the *deviation* of x:

$$\delta x = x - \langle x \rangle$$

Since the expectation value of the deviation is always zero, it is common practice to introduce the mean-square deviation:

$$\langle (\delta x)^2 \rangle = \int_{-\infty}^{\infty} (x - \langle x \rangle)^2 P(x) dx = \langle x^2 \rangle - \langle x \rangle^2$$

and to discuss the measurement of position in terms of the quantity:

$$\Delta x = \sqrt{\langle x^2 \rangle - \langle x \rangle^2}$$

called the *standard deviation*, or *dispersion*.

The interpretation of the motion of a particle in terms of a probability wave rather than a real wave reduces to the classical approach in the limiting case where the wave packet reduces to a point, and therefore the material particle has a definite position. However, in contrast to classical mechanics, where a definite position implies a definite velocity, which is the time derivative of the position, and hence implies definite momentum and energy of the particle, the statistical interpretation of the wave function requires that not only the position but also the velocity, momentum and energy are all considered from a probabilistic point of view.

Problem 1.1.1. Given the spectral density of a matter wave packet in the form of a symmetrical exponential:

$$\Phi(k) = A e^{-\Gamma |k|}$$

find the form of the wave function $\Psi(x)$.

(Solution): In view of Eqs.(1.7) we have:

$$\Psi(x) = \frac{A}{\sqrt{2\pi}} \int_{-\infty}^{\infty} e^{-\Gamma|k|} e^{ikx} dk = \frac{A}{\sqrt{2\pi}} \int_{0}^{\infty} e^{-\Gamma k} e^{ikx} dk - \frac{1}{\sqrt{2\pi}} \int_{0}^{\infty} e^{-\Gamma k} e^{-ikx} dk$$

$$= \frac{A}{\sqrt{2\pi}} \left(\frac{1}{\Gamma - ix} + \frac{1}{\Gamma + ix} \right) = \frac{A}{\sqrt{2\pi}} \frac{2\Gamma}{x^2 + \Gamma^2}$$

1. Wave Functions and the First Postulate   43

The function $\Psi(x)$ is said to be a Lorentzian and usually defines the spectral line profile. The quantity Γ represents one half of the width of the $\Psi(x)$ profile at one half its maximum amplitude, found at $x = 0$.

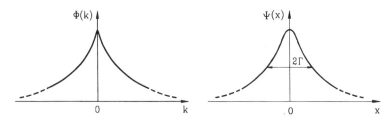

Problem 1.1.2. Show that, for a spherically symmetric wave function $\Psi(\vec{r}) = \Psi(r)$, the spectral density $\Phi(\vec{k})$ also has spherical symmetry.

(Solution): The pair of Fourier transforms, Eqs.(1.7), can be easily generalized for three dimensions in the form:

$$\Psi(\vec{r}) = \frac{1}{(2\pi)^{3/2}} \int_{-\infty}^{\infty}\int_{-\infty}^{\infty}\int_{-\infty}^{\infty} \Phi(\vec{k}) e^{i\vec{k}\cdot\vec{r}} d\vec{k}$$

$$\Phi(\vec{k}) = \frac{1}{(2\pi)^{3/2}} \int_{-\infty}^{\infty}\int_{-\infty}^{\infty}\int_{-\infty}^{\infty} \Psi(\vec{r}) e^{-i\vec{k}\cdot\vec{r}} d\vec{r}$$

where $d\vec{k}$ and $d\vec{r}$ denote the volume elements in $\vec{k}$-space and $\vec{r}$-space respectively. In our case, it is convenient to introduce spherical polar coordinates such that $d\vec{r} = r^2 \sin\theta\, dr\, d\theta\, d\varphi$. Choosing, for simplicity, the z-axis along the $\vec{k}$ direction, we have $\vec{k}\cdot\vec{r} = kr\cos\theta$, so that:

$$\Phi(\vec{k}) = \frac{1}{(2\pi)^{3/2}} \int_0^\infty \Psi(r) r^2 dr \int_0^{2\pi} d\varphi \int_0^\pi e^{-ikr\cos\theta} \sin\theta\, d\theta$$

The last integral can be written as:

$$\int_0^\pi e^{-ikr\cos\theta} d(\cos\theta) = \frac{1}{ikr}(e^{ikr} - e^{-ikr}) = 2\frac{\sin kr}{kr}$$

As a result, $\Phi(\vec{k})$ is obtained as a function of $k = |\vec{k}|$ only:

$$\Phi(\vec{k}) = \Phi(k) = \frac{2}{(2\pi)^{3/2} k} \int_0^\infty \Psi(r)(\sin kr) r\, dr$$

44   Quantum Physics

Problem 1.1.3. Calculate the spectral density $\Phi(k)$ of the wave packet represented by the square pulse function, defined by:

$$\Psi(x) = A \quad |x| < a$$
$$= 0 \quad |x| > a$$

(Answer): $\Phi(k) = \dfrac{2Aa}{\sqrt{2\pi}} \dfrac{\sin ka}{ka}$

Problem 1.1.4. Find the spectral density $\Phi(k)$ for the wave packet described by the function:

$$\Psi(x) = A e^{ik_0 x} \quad |x| < a$$
$$= 0 \quad |x| > a$$

(Answer): $\Phi(k) = \dfrac{2Aa}{\sqrt{2\pi}} \dfrac{\sin(k_0 - k)a}{(k_0 - k)a}$

Problem 1.1.5. Find a relation between the spectral density $\Phi(k)$ of the wave function $\Psi(x)$ and the spectral density $\Phi'(k)$ of $d\Psi(x)/dx$.

(Answer): $\Phi'(k) = ik\Phi(k)$

1.2. The Uncertainty Principle

The experimental support for the wave-particle relations $E = \hbar\omega$ and $p = \hbar k$, ascribes to the same entity attributes which are considered mutually exclusive in classical physics. This has required that we reconsider the classical concepts concerning the meaning of simultaneous position x and its conjugate momentum p_x. The basic difference between the classical and quantum interpretation of these concepts was demonstrated by Heisenberg in a statistical examination of de Broglie wave spreading. The Gaussian representation (1.20) associated with a particle becomes at a given instant of time conveniently chosen as $t = 0$, $|\Psi(x)|^2 = (\pi/\alpha) e^{-x^2/2\alpha}$ or, in normalized form:

1. Wave Functions and the First Postulate

$$|\Psi(x)|^2 = \frac{1}{\sqrt{2\pi\alpha}} e^{-x^2/2\alpha} \tag{1.30}$$

where the normalization constant was obtained from:

$$\int_{-\infty}^{\infty} C|\Psi(x)|^2 \, dx = C\frac{\pi}{\alpha} \int_{-\infty}^{\infty} e^{-x^2/2\alpha} dx = C\frac{\pi}{\alpha}\sqrt{2\pi\alpha} = 1$$

The degree of spreading of the wave about the point $x = 0$ is given by the standard deviation, which reads:

$$\Delta x = \sqrt{\langle x^2 \rangle - \langle x \rangle^2} = \sqrt{\langle x^2 \rangle} \tag{1.31}$$

where $\langle x \rangle = 0$. Hence the mean square value can be evaluated by Eq.(1.29) to be:

$$\langle x^2 \rangle = \frac{1}{\sqrt{2\pi\alpha}} \int_{-\infty}^{\infty} x^2 e^{-x^2/2\alpha} dx = \frac{1}{\sqrt{2\pi\alpha}} \frac{1}{2}\sqrt{\pi(2\alpha)^3} = \alpha$$

so that the spreading of the matter wave packet can be written as $\Delta x = \sqrt{\alpha}$.

The wave number representation of the same wave packet is given by the Gaussian distribution associated with the spectral density function (1.15):

$$|\Phi(k)|^2 = e^{-2\alpha(k-k_0)^2} \tag{1.32}$$

Since the spreading of k-values about the mean value k_0 is given by a standard deviation Δk, it is easily verified that, by a similar procedure, we obtain from Eq.(1.32):

$$\Delta k = \frac{1}{2\sqrt{\alpha}} \tag{1.33}$$

Thus, the product of standard deviations reads:

$$\Delta x \Delta k \geq \frac{1}{2} \tag{1.34}$$

and the equality sign holds only for Gaussian wave packets where the product takes a minimum value. In view of the de Broglie hypothesis (1.23), the relation (1.34) can also be written as:

$$\Delta x \Delta p_x \geq \frac{\hbar}{2} \tag{1.35}$$

EXAMPLE 1.3. The Heisenberg microscope

Consider an optical microscope used to determine the position of a particle, which is located by observing the photons scattered from it. In the configuration of Figure 1.3, the particle is illuminated along the x-axis and the scattered photon recoils through the objective lens, which subtends an angle of 2φ at the point where the particle is situated. If the wavelength of the light is λ, the precision with which the particle can be located is given by $\Delta x \sim \lambda / \sin \varphi$. Since the scattered photon may propagate in any direction α, within the angle 2φ, to enter the microscope, the components of its momentum along the x and y axes (see Figure 1.4) are undetermined in the range:

$$-\frac{h}{\lambda}\sin\varphi < \frac{h}{\lambda}\sin\alpha < \frac{h}{\lambda}\sin\varphi, \quad \frac{h}{\lambda}\cos\varphi < \frac{h}{\lambda}\cos\alpha < 1$$

By conservation of momentum, the motion of the particle is determined by the components of the recoil momentum:

$$p_x = \frac{h}{\lambda}(1-\sin\alpha), \quad p_y = -\frac{h}{\lambda}\cos\alpha$$

which are then uncertain by:

$$\Delta p_x = 2\frac{h}{\lambda}\sin\varphi, \quad \Delta p_y = \frac{h}{\lambda}(1-\cos\varphi)$$

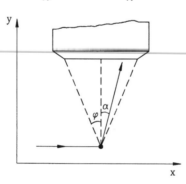

Figure 1.4. Configuration of the Heisenberg microscope

We thus obtain:

$$\Delta x \Delta p_x \sim 2h, \quad \Delta x \Delta p_y \sim \frac{1}{2}h\varphi$$

where use has been made of $\sin\varphi \sim \varphi$, $\cos\varphi \sim 1-\varphi^2/2$, which are valid for small values of φ. It follows that the exact determination of both the position and momentum of a particle is possible, if the two quantities are measured along different axes, x and p_y, and we are making $\varphi \to 0$. In contrast, this becomes impossible if position and its conjugate momentum along the same direction, x and p_x, are involved. The calculated product of uncertainties is greater than the theoretical minimum value obtained in Eq.(1.35). ☙

1. Wave Functions and the First Postulate

Constraints similar to those derived in Eq.(1.35) apply to any generalized coordinate and corresponding generalized momentum. It is for instance a straightforward matter to show that, for a particle which moves along the x-axis, with energy E given in terms of momentum p_x by the relativistic energy-momentum relation:

$$E^2 = c^2 p_x^2 + m_0^2 c^4 = m^2 c^4$$

we have the uncertainty in energy:

$$E \Delta E = c^2 p_x \Delta p_x \quad \text{or} \quad \Delta E = \frac{p_x}{m} \Delta p_x = v_x \Delta p_x$$

If Δt is a measure of the uncertainty in the localization in time of this particle, we may write $\Delta x = v_x \Delta t$ and substituting into Eq.(1.35) we obtain the energy-time uncertainty relation:

$$\Delta E \, \Delta t \geq \frac{\hbar}{2} \qquad (1.36)$$

The expressions (1.35) and (1.36) illustrate the **Heisenberg uncertainty principle**, which states that: *the product of uncertainties in the measurement of two canonically conjugate variables must be greater than or equal to $\hbar/2$.*

The uncertainty principle can also be interpreted in the sense that it is not possible even to define simultaneously the conjugate variables q and p_q in the classical sense. Hence, the state of a particle cannot be specified by both its coordinate and conjugate momenta at a given instant of time, and therefore the classical equations of motion, which can be solved for $q(t)$ and $p_q(t)$, so giving definite values of position and momentum of the particle, do not have an exact meaning. It follows that we must define either the coordinates of a state or its momenta.

Problem 1.2.1. Use the uncertainty principle to estimate the ground state energy of a harmonic oscillator in one dimension, the energy of which is given by Eq.(P.5).

(Solution): Since the total energy E of a harmonic oscillator is a constant of motion, we may write Eq.(P.5) in terms of the expectation values of position and linear momentum, in the form:

$$E = \frac{1}{2m}\langle p_x^2 \rangle + \frac{1}{2} m \omega^2 \langle x^2 \rangle$$

If we consider, for simplicity, oscillations about the point $x = 0$, where $p_x = 0$, we have $\langle x \rangle = \langle p_x \rangle = 0$, and the Heisenberg uncertainty relation reads:

$$\langle x^2 \rangle \langle p_x^2 \rangle \geq \frac{\hbar^2}{4}$$

The total energy thus becomes:

$$E \geq \frac{\hbar^2}{8m\langle x^2 \rangle} + \frac{1}{2}m\omega^2 \langle x^2 \rangle$$

It is easily verified that the right hand side has a minimum for $\langle x^2 \rangle = \hbar/2m\omega$, which implies that:

$$E \geq \frac{1}{2}\hbar\omega$$

In other words, the ground state energy of the harmonic oscillator in one dimension is $\hbar\omega/2$, a result that will be confirmed later on by rigorous calculation.

Problem 1.2.2. Estimate the size of the ground state electron orbit in the Bohr atom by using the Heisenberg uncertainty principle.

(Solution): For an electron moving in a circular orbit of radius r, about a nucleus of nuclear charge Ze, the total energy is given by:

$$E = \frac{p^2}{2m_e} - \frac{Ze_0^2}{r} = \frac{1}{2m_e}\langle p^2 \rangle - Ze_0^2 \left\langle \frac{1}{r} \right\rangle$$

and the uncertainty relation, Eq.(1.35), reads:

$$\langle p^2 \rangle \langle r^2 \rangle \geq \frac{\hbar^2}{4}$$

Using the approximations: $\langle r^2 \rangle \equiv \langle r \rangle^2$, $\langle 1/r \rangle = 1/\langle r \rangle$, which only are valid to an order of magnitude, the total energy becomes:

$$E \geq \frac{\hbar^2}{8m_e \langle r \rangle^2} - \frac{Ze_0^2}{\langle r \rangle}$$

The right-hand side is then minimized by taking:

$$\langle r \rangle = \frac{\hbar^2}{4Zm_e e_0^2}$$

This value is four times less the Bohr radius for the ground state of one-electron atoms, Eq.(P.25), so only estimating the order of magnitude of the ground state orbit size. As a result, an energy four times higher than that determined from the Bohr model, Eq.(P.26), is obtained.

Problem 1.2.3. Use the de Broglie hypothesis $p_x = \hbar k$, to estimate the magnitude of the uncertainty product $\Delta x \Delta p_x$ for the wave packet $\Psi(x)$ and its spectral density $\Phi(k)$, calculated in Problem 1.1.4.

(Answer): $\quad \Delta x \Delta p_x = \hbar$

Problem 1.2.4. The wave function $\Psi(x)$ defined in Problem 1.1.4 can be written in terms of the conjugated variables $\omega = uk$ and $t = x/u$ in the form:

$$\Psi(t) = e^{i\omega_0 t}, \quad |t| \le T$$

$$= 0, \quad |t| > T$$

Find the spectral density $\Phi(\omega)$ and use the Einstein relation $E = \hbar \omega$ to estimate the magnitude of the uncertainty product $\Delta E \Delta t$.

(Answer): $\quad \Delta E \Delta t = \hbar$

1.3. The Wave Function Space

Let us assume that the state of a particle, of energy E and momentum p_x, may be described by the de Broglie wave function $\Psi(x,t)$, as introduced in Eq.(1.24). A physical significance has been assigned to this wave function by Eq.(1.26), in the sense that $|\Psi(x,t)|^2 dx$ is the *probability* of finding the particle described by $\Psi(x,t)$ at a point between x and $x + dx$ at a time t. As the total probability of finding the particle somewhere on the x-axis must be unity, the wave function must be *normalized* to unity, that is:

$$\int_{-\infty}^{\infty} P(x,t)dx = \int_{-\infty}^{\infty} |\Psi(x,t)|^2 \, dx = 1 \tag{1.37}$$

50 Quantum Physics

Therefore, the state of the particle, at a given instant of time, is represented by a square-integrable function of position, in other words by a continuous single-valued complex function of the real variable x such that the integral $\int |\Psi(x)|^2 dx$ exists for integration over a given interval. Let us choose $t = 0$ in Eq.(1.24), so obtaining an expression for the wave function $\Psi(x)$ and its Fourier transform as follows:

$$\Psi(x) = \frac{1}{\sqrt{2\pi\hbar}} \int_{-\infty}^{\infty} \Phi(p_x) e^{ip_x x/\hbar} dp_x \qquad (1.38)$$

$$\Phi(p_x) = \frac{1}{\sqrt{2\pi\hbar}} \int_{-\infty}^{\infty} \Psi(x) e^{-ip_x x/\hbar} dx \qquad (1.39)$$

Consider an infinite set of monochromatic plane waves, each having a definite wave number $p_x/\hbar$ that belong to a continuous set of values:

$$u_{p_x}(x) = \frac{1}{\sqrt{2\pi\hbar}} e^{ip_x x/\hbar} \qquad (1.40)$$

Since none of the functions u_{p_x} can be expressed in terms of the others, they may be regarded as a set of basic functions, and therefore Eq.(1.38) may be interpreted as an expansion of the wave function in terms of a basis of monochromatic plane waves:

$$\Psi(x) = \int_{-\infty}^{\infty} \Phi(p_x) u_{p_x}(x) dp_x \qquad (1.41)$$

The Fourier coefficients (1.39), which are the components of the wave function relative to the infinite basis set, can then be formally written as a *scalar product*, defined as:

$$\Phi(p_x) = \int_{-\infty}^{\infty} u_{p_x}^*(x) \Psi(x) dx = \langle u_{p_x} | \Psi \rangle \qquad (1.42)$$

The definition introduced by Eq.(1.42) can be used consistently for the scalar product of the monochromatic plane waves, which are linearly independent:

$$\langle u_{p_x} | u_{p_x'} \rangle = \int_{-\infty}^{\infty} u_{p_x}^*(x) u_{p_x'}(x) dx = \frac{1}{2\pi} \int_{-\infty}^{\infty} e^{i(p_x' - p_x)x/\hbar} d\left(\frac{x}{\hbar}\right) = \delta(p_x' - p_x) \qquad (1.43)$$

Eq.(1.43) is known as the *orthonormality condition* for the monochromatic plane waves, and is expressed in terms of the δ-function which was deliberately introduced by Dirac. The δ-function is not an ordinary function, although it may be written in terms of continuous functions by making use of appropriate limiting procedures.

EXAMPLE 1.4. The Dirac δ-function

The δ-function is defined by the properties:

$$\delta(x) = 0 \quad \text{if} \quad x \neq 0$$

$$\int_{-\infty}^{\infty} f(x)\delta(x)dx = f(0)$$
(1.44)

where a correspondence is established between a continuous function differentiable at $x = 0$ and the number $f(0)$. From Eqs.(1.44) several useful properties follow:

$$\int_{-\infty}^{\infty} \delta(x)dx = 1$$
(1.45)

$$\int_{-\infty}^{\infty} f(x')\delta(x-x')dx' = f(x)$$

and also:

$$\delta(-x) = \delta(x)$$

$$x\delta(x) = 0$$

$$\delta(\alpha x) = \frac{1}{|\alpha|}\delta(x)$$
(1.46)

$$\delta(x^2 - \alpha^2) = \frac{1}{2|\alpha|}[\delta(x-\alpha) + \delta(x+\alpha)]$$

$$f(x)\delta(x-\alpha) = f(\alpha)\delta(x-\alpha)$$

the proof of which is obtained by multiplying each term by a continuous differentiable function $f(x)$ and then integrating over all space. The properties of the derivative $\delta'(x)$ of $\delta(x)$, namely:

$$\int_{-\infty}^{\infty} \delta'(x)dx = 0$$

$$\int_{-\infty}^{\infty} f(x)\delta'(x)dx = -\left[\frac{df(x)}{dx}\right]_{x=0}$$

$$\delta'(x) = -\delta'(-x)$$

$$x\delta'(x) = -\delta(x)$$
(1.47)

52 Quantum Physics

are easily justified by the same procedure.

In view of Eqs.(1.44) and (1.45), $\delta(x)$ may be interpreted as an even function which is zero for $x \neq 0$ and tends to infinity as x tends to zero, in such a way that the normalization condition, Eq.(1.45), holds. Consider, for instance, the function defined by:

$$\delta(x) = \frac{1}{2\pi} \int_{-\infty}^{\infty} e^{ip_x x/\hbar} d\left(\frac{p_x}{\hbar}\right) = \frac{1}{2\pi} \int_{-\infty}^{\infty} e^{ip_x x} dp_x = \lim_{\xi \to \infty} \frac{1}{2\pi} \int_{-\xi}^{\xi} e^{ip_x x} dp_x = \lim_{\xi \to \infty} \frac{\sin \xi x}{\pi x}$$

and let us examine its behaviour for both small and large x. At the limit as $x \to 0$ we have:

$$\lim_{x \to 0} \frac{\sin \xi x}{\pi x} = \frac{\xi}{\pi} \lim_{x \to 0} \frac{\sin \xi x}{\xi x} = \frac{\xi}{\pi}$$

which means that $\delta(0) = \lim_{\xi \to \infty} (\xi/\pi) \to \infty$, and the amplitude becomes infinite. For increasing x, the amplitude falls off as $1/|x|$, while $\sin(\xi x)/\pi x$ oscillates with a period $2\pi/\xi$ which becomes infinitesimally narrow as $\xi \to \infty$. It follows that $\delta(x)$ approaches zero everywhere except for a peak of infinite amplitude and infinitesimal width at the origin. The integral over all space is, however, unity, as predicted by Eq.(1.45):

$$\int_{-\infty}^{\infty} \lim_{\xi \to \infty} \frac{\sin \xi x}{\pi x} dx = \lim_{\xi \to \infty} \frac{2}{\pi} \int_0^{\infty} \frac{\sin \xi x}{x} dx = \frac{2}{\pi}\left(\frac{\pi}{2}\right) = 1$$

The pair of Fourier transforms, Eqs.(1.38) and (1.39), imply that:

$$\Psi(x) = \frac{1}{2\pi} \int_{-\infty}^{\infty} \int_{-\infty}^{\infty} \Psi(x') e^{ip_x(x-x')/\hbar} dx' d\left(\frac{p_x}{\hbar}\right) = \int_{-\infty}^{\infty} \Psi(x') \left[\frac{1}{2\pi} \int_{-\infty}^{\infty} e^{ip_x(x-x')/\hbar} d\left(\frac{p_x}{\hbar}\right)\right] dx'$$

$$= \int_{-\infty}^{\infty} \Psi(x') \delta(x-x') dx'$$

which therefore obeys the requirements (1.45) for the δ-function. ♪

The monochromatic plane waves (1.40), which form an orthonormal set, as defined by Eq.(1.43), also represent a *complete* basis of functions, in the sense that there exists no non-zero plane wave function which is orthogonal to every member of the set. The *closure relation*, which defines a complete set, is also expressible in terms of the δ-function in the form:

$$\int_{-\infty}^{\infty} u_{p_x}(x) u_{p_x}^*(x') dp_x = \frac{1}{2\pi} \int_{-\infty}^{\infty} e^{i(x-x')p_x/\hbar} d\left(\frac{p_x}{\hbar}\right) = \delta(x-x') \tag{1.48}$$

Both the orthonormality condition, Eq.(1.43), and the closure relation, Eq.(1.48), may be regarded as a generalization, to the case of a continuum of functions, of the familiar properties of a discrete set of functions that generate a vector space (see Example 1.5).

The orthonormality relation (1.43) allows us to introduce the *scalar product* of two wave functions expanded in terms of the same orthonormal basis, Eq.(1.40), namely $\Psi(x)$, defined by either Eq.(1.38) or Eq.(1.41), and:

$$\Theta(x) = \frac{1}{\sqrt{2\pi\hbar}} \int_{-\infty}^{\infty} \Gamma(p'_x) e^{ip'_x x/\hbar} dp'_x = \int_{-\infty}^{\infty} \Gamma(p'_x) u_{p'_x}(x) dp'_x$$

where the Fourier coefficients are given by the scalar product:

$$\Gamma(p'_x) = \int_{-\infty}^{\infty} u^*_{p'_x}(x) \Theta(x) dx$$

The scalar product can be successively written as:

$$\langle \Theta | \Psi \rangle = \int_{-\infty}^{\infty} \Theta^*(x) \Psi(x) dx = \int_{-\infty}^{\infty} dp_x \int_{-\infty}^{\infty} \Gamma^*(p'_x) \Phi(p_x) dp'_x \int_{-\infty}^{\infty} u^*_{p'_x}(x) u_{p_x}(x) dx$$

$$= \int_{-\infty}^{\infty} dp_x \int_{-\infty}^{\infty} \Gamma^*(p'_x) \Phi(p_x) \delta(p_x - p'_x) dp'_x = \int_{-\infty}^{\infty} \Gamma^*(p_x) \Phi(p_x) dp_x$$

so that, as a particular case, we have:

$$\langle \Psi | \Psi \rangle = \int_{-\infty}^{\infty} |\Phi(p_x)|^2 dp_x$$

The square root of this quantity is called the *norm* of the wave function $\Psi(x)$ and is denoted by $|\Psi(x)|$.

The representation (1.41) of the wave function $\Psi(x)$ in terms of the monochromatic plane waves u_{p_x}, where p_x is a continuous variable, constitutes the generalization of a linear combination of n linearly independent functions u_i which define an n-dimensional function space:

$$\Psi(x) = \sum_{i=1}^{n} \Phi_i u_i(x) \qquad (1.49)$$

This implies that the space of the square-integrable wave functions has properties similar to those of a vector space, the expression of which follows from the previous description by the correspondence rules:

54   Quantum Physics

$$i \leftrightarrow p_x, \quad \sum_i \leftrightarrow \int dx, \quad \delta_{ij} \leftrightarrow \delta(x-x')$$

and will be required in the general formulation of quantum theory.

EXAMPLE 1.5. Vector spaces

A *complex vector space* is a set of objects, called vectors, that possess the properties of addition and multiplication by scalar quantities defined by the familiar properties established for the ordinary two- or three-dimensional vectors, namely:

1. *Vector addition* states that for every pair of vectors $\vec{\xi}$ and $\vec{\eta}$ there exists a unique vector sum $\vec{\xi} + \vec{\eta}$, that has the following properties:

   (i) commutativity: $\vec{\xi} + \vec{\eta} = \vec{\eta} + \vec{\xi}$

   (ii) associativity: $(\vec{\xi} + \vec{\eta}) + \vec{\varsigma} = \vec{\xi} + (\vec{\eta} + \vec{\varsigma})$

   (iii) there exists a null vector $\vec{0}$ such that $\vec{\xi} + \vec{0} = \vec{\xi}$ for every vector $\vec{\xi}$

   (iv) there exists a vector $-\vec{\xi}$ for every vector $\vec{\xi}$ such that $\vec{\xi} + (-\vec{\xi}) = \vec{0}$

2. *Multiplication by scalars*, which in the general case may be regarded as complex numbers, states that for every vector $\vec{\xi}$ and every scalar a there exists a unique vector $a\vec{\xi}$ that has the properties:

   (i) associativity: $(ab)\vec{\xi} = a(b\vec{\xi})$

   (ii) multiplication by unity: $1\vec{\xi} = \vec{\xi}$

   (iii) distributivity with respect to the addition of scalars: $(a+b)\vec{\xi} = a\vec{\xi} + b\vec{\xi}$

   (iv) distributivity with respect to vector addition: $a(\vec{\xi} + \vec{\eta}) = a\vec{\xi} + a\vec{\eta}$

A vector space is called *unitary* if one postulates that with every pair of vectors $\vec{\xi}$ and $\vec{\eta}$ there is associated a complex number $\langle \vec{\xi} | \vec{\eta} \rangle$, called the *scalar product*, defined by the properties:

   (i) $\langle \vec{\xi} | \vec{\eta} \rangle = \langle \vec{\eta} | \vec{\xi} \rangle^*$, so that $\langle \vec{\xi} | \vec{\xi} \rangle$ is a real number

   (ii) $\langle \vec{\xi}, \vec{\xi} \rangle \geq 0$, where the equality holds only if $\vec{\xi} = \vec{0}$

   (iii) *linearity* with respect to the second factor:

$$\langle \vec{\xi} | a\vec{\eta} + b\vec{\varsigma} \rangle = a \langle \vec{\xi} | \vec{\eta} \rangle + b \langle \vec{\xi} | \vec{\varsigma} \rangle$$

A direct consequence of the previous properties is the *antilinearity* of the scalar product with respect to the first factor:

$$\langle a\vec{\eta} + b\vec{\varsigma} | \vec{\xi} \rangle = a^* \langle \vec{\eta} | \vec{\xi} \rangle + b^* \langle \vec{\varsigma} | \vec{\xi} \rangle$$

1. Wave Functions and the First Postulate

The norm of a vector $\vec{\xi}$ is defined by the positive square root:

$$|\vec{\xi}| = \sqrt{\langle \vec{\xi}|\vec{\xi}\rangle}$$

and allows us to *normalise* a vector $\vec{\xi}$, that is to form the vector $\hat{\xi} = \vec{\xi}/|\vec{\xi}|$ of unit norm, such that:

$$\langle \hat{\xi}|\hat{\xi}\rangle = 1$$

If into the obvious inequality $\langle \hat{\xi} - a\hat{\eta}|\hat{\xi} - a\hat{\eta}\rangle \geq 0$, or $1 + |a|^2 \geq a\langle \hat{\xi}|\hat{\eta}\rangle + a^*\langle \hat{\eta}|\hat{\xi}\rangle$, one substitutes $a = \langle \hat{\eta}|\hat{\xi}\rangle$, it follows that:

$$|\langle \hat{\xi}|\hat{\eta}\rangle|^2 \leq 1 \quad \text{or} \quad |\langle \vec{\xi}|\vec{\eta}\rangle|^2 \leq \langle \vec{\xi}|\vec{\xi}\rangle\langle \vec{\eta}|\vec{\eta}\rangle \tag{1.50}$$

a result which is known as the *Schwarz inequality*. It is noteworthy that Eq.(1.50) implies a simple geometrical interpretation, provided the scalar product is a real number. In this case we may define the quantity:

$$Q = \frac{\langle \vec{\xi}|\vec{\eta}\rangle}{|\vec{\xi}||\vec{\eta}|}$$

for which the Schwartz inequality gives $-1 \leq Q \leq 1$, and hence Q may be interpreted as cosine of an angle α. It follows that the scalar product takes the form:

$$\langle \vec{\xi}|\vec{\eta}\rangle = |\vec{\xi}||\vec{\eta}|\cos\alpha$$

which is similar to the definition of the dot product, in the three-dimensional vector space, in terms of the length of two vectors and the angle between them. Two non-zero vectors are said to be *orthogonal* provided that:

$$\langle \vec{\xi}|\vec{\eta}\rangle = 0$$

A set of normalized vectors $\vec{u}_1, \vec{u}_2, \ldots, \vec{u}_n$ which are mutually orthogonal obey the *orthonormality* condition:

$$\langle \vec{u}_i|\vec{u}_j\rangle = \delta_{ij} \tag{1.51}$$

where the Kronecker delta δ_{ij} is defined to be unity if $i = j$ and zero if $i \neq j$. Equation (1.51) implies that the vectors $\vec{u}_i$ form a linearly independent set, in the sense that the equation:

$$\sum_{i=1}^{n} a_i \vec{u}_i = 0$$

holds only if all scalars a_i are zero. The set of n orthonormal vectors $\vec{u}_i$ form a *basis* of an n-dimensional vector space provided any vector $\vec{\xi}$ of this space may be represented as a linear combination:

$$\vec{\xi} = \sum_{i=1}^{n} \xi_i \vec{u}_i \tag{1.52}$$

where the complex numbers ξ_i, called the components of $\vec{\xi}$ relative to the vector basis are uniquely determined. Assuming that there also exists the components ξ'_i of the same vector $\vec{\xi}$, we have:

$$\sum_{i=1}^{n} \xi'_i \vec{u}_i = \sum_{i=1}^{n} \xi_i \vec{u}_i \quad \text{or} \quad \sum_{i=1}^{n} (\xi'_i - \xi_i) \vec{u}_i = 0$$

which means that $\xi'_i = \xi_i$, because the vectors $\vec{u}_i$ are linearly independent. Hence, we obtain a *representation* of any vector $\vec{\xi}$ by an ordered set of complex components $\{\xi_1, \xi_2, ..., \xi_n\}$ relative to the basis $\vec{u}_i$. Each component is expressible from Eq.(1.52) in terms of the scalar product:

$$\langle \vec{u}_j | \vec{\xi} \rangle = \sum_{i=1}^{n} \xi_i \langle \vec{u}_j | \vec{u}_i \rangle = \sum_{i=1}^{n} \xi_i \delta_{ji} = \xi_j \tag{1.53}$$

The norm of a vector $\vec{\xi}$ has the form:

$$|\vec{\xi}|^2 = \langle \vec{\xi} | \vec{\xi} \rangle = \sum_{i=1}^{n} \xi_i^* \xi_j \langle \vec{u}_i | \vec{u}_j \rangle = \sum_{i=1}^{n} \xi_i^* \xi_j \delta_{ij} = \sum_{i=1}^{n} |\xi_i|^2 \tag{1.54}$$

It is obvious that the representation given by Eqs.(1.52), (1.53) and (1.54) in terms of a basis (1.51), in an n-dimensional vector space, is the generalization of the expansion in the three-dimensional space of an ordinary vector:

$$\vec{v} = \sum_{i=x,y,z} v_i \vec{e}_i, \quad |\vec{v}|^2 = \sum_{i=x,y,z} |v_i|^2$$

where $\vec{e}_x, \vec{e}_y$, and $\vec{e}_z$ form an orthonormal basis:

$$\vec{e}_i \cdot \vec{e}_j = \delta_{ij}$$

such that the components v_i of the vector $\vec{v}$ read:

$$v_i = \vec{e}_i \cdot \vec{v}$$

1. Wave Functions and the First Postulate

A further immediate generalization of the n-dimensional vector space is to consider an infinite dimensional space, where a set of infinitely many orthonormal vectors $\vec{u}_i$ is assumed to exist, such that any vector of the space may be expanded as:

$$\vec{\xi} = \sum_{i=1}^{\infty} \xi_i \vec{u}_i \tag{1.55}$$

The existence of the norm:

$$|\vec{\xi}|^2 = \sum_{i=1}^{n} |\xi_i|^2 \tag{1.56}$$

requires that the series converges, so ensuring that the representation (1.55) is uniquely determined. Assuming the existence of a vector:

$$\Delta\vec{\xi} = \vec{\xi} - \sum_{i=1}^{\infty} \xi_i \vec{u}_i$$

we obtain, in view of Eq.(1.56), that $\Delta\vec{\xi}$ is a null vector:

$$|\Delta\vec{\xi}|^2 = \langle \Delta\vec{\xi} | \Delta\vec{\xi} \rangle = |\vec{\xi}|^2 - \sum_{i=1}^{\infty} |\xi_i|^2 = 0 \tag{1.57}$$

The infinite orthonormal set of vectors $\vec{u}_i$ is then complete, so constituting a basis which generates a vector space called a *Hilbert space*.

A special case of a Hilbert space is the class of square-integrable functions, of importance in quantum mechanics, which can be generated by assuming the existence of a set of basic functions, introduced by the correspondence rule:

$$\vec{u}_i \leftrightarrow u_i(x)$$

and therefore obeying an orthonormality condition similar to that defined by Eq.(1.51):

$$\langle u_i | u_j \rangle = \int_{-\infty}^{\infty} u_i^*(x) u_j(x) dx = \delta_{ij} \tag{1.58}$$

Any function of this class will be expanded as a convergent series, which is analogous to that considered in Eq.(1.52):

$$\Psi(x) = \sum_{i=1}^{\infty} \Phi_i u_i(x) \tag{1.59}$$

where the coefficients are expressed by scalar products, as in Eq.(1.53):

58   Quantum Physics

$$\Phi_i = \langle u_i | \Psi \rangle = \int_{-\infty}^{\infty} u_i^*(x)\Psi(x)dx \qquad (1.60)$$

The last two equations imply that:

$$\Psi(x) = \sum_{i=1}^{\infty} \Phi_i u_i(x) = \sum_{i=1}^{\infty} \langle u_i, \Psi \rangle u_i(x) = \sum_{i=1}^{\infty} \left[\int_{-\infty}^{\infty} u_i^*(x')\Psi(x')dx' \right] u_i(x)$$

$$= \int_{-\infty}^{\infty} \Psi(x') \left[\sum_{i=1}^{\infty} u_i(x) u_i^*(x') \right] dx'$$

A comparison with the definition of the Dirac $\delta(x-x')$-function, Eq.(1.45), shows that the representation of the wave functions in terms of the orthonormal set of functions $u_i(x)$ requires that:

$$\sum_{i=1}^{\infty} u_i(x) u_i^*(x') = \delta(x-x') \qquad (1.61)$$

This is called the *closure relation* and ensures that the orthonormal set of functions $u_i(x)$ is *complete*, and therefore forms a *basis of representation* for the class of the square-integrable functions. ♛

Problem 1.3.1. Show that the Dirac δ-function can be represented by $\delta(x) = \lim\limits_{\alpha \to 0} \dfrac{1}{\alpha\sqrt{\pi}} e^{-x^2/\alpha^2}$.

(Solution):   We have to show that the function so defined obeys Eqs.(1.44). For an arbitrary function $f(x)$ we successively obtain

$$\int_{-\infty}^{\infty} f(x)\delta(x)dx = \lim_{\alpha \to 0} \int_{-\infty}^{\infty} f(x) \frac{1}{\alpha\sqrt{\pi}} e^{-x^2/\alpha^2} dx = \lim_{\alpha \to 0} \frac{1}{\sqrt{\pi}} \int_{-\infty}^{\infty} f(\alpha u) e^{-u^2} du$$

$$= \frac{1}{\sqrt{\pi}} \int_{-\infty}^{\infty} \lim_{\alpha \to 0} f(\alpha u) e^{-u^2} du = \frac{1}{\sqrt{\pi}} \int_{-\infty}^{\infty} f(0) e^{-u^2} du$$

$$= f(0) \frac{1}{\sqrt{\pi}} \int_{-\infty}^{\infty} e^{-u^2} du = f(0)$$

1. Wave Functions and the First Postulate

where the last integral has been calculated according to the Gaussian integral formula.

Problem 1.3.2. Find the Fourier transform $\Phi_\delta(k)$ of the Dirac $\delta(x)$ function.

(Solution): According to the definition of the Fourier transforms, Eq.(1.7), we have:

$$\Phi_\delta(k) = \frac{1}{\sqrt{2\pi}} \int_{-\infty}^{\infty} \delta(x) e^{-ikx} dx$$

Substituting the definition of the δ-function, Eq.(1.44), we obtain the spectral density $\Phi_\delta(k)$ as a constant function:

$$\Phi_\delta(k) = \frac{1}{\sqrt{2\pi}}$$

The reciprocal of this equation has the form (1.7) which reads:

$$\delta(x) = \frac{1}{\sqrt{2\pi}} \int_{-\infty}^{\infty} \Phi_\delta(k) e^{ikx} dk = \frac{1}{\sqrt{2\pi}} \int_{-\infty}^{\infty} e^{ikx} dx$$

and leads to the representation of the Dirac δ-function introduced before.

Problem 1.3.3. Show that the function:

$$\delta(x) = \lim_{\alpha \to 0} \frac{\alpha}{\pi} \frac{1}{x^2 + \alpha^2}$$

is a valid representation of the Dirac δ-function.

1.4. The First Postulate

The representation of a particle by a complex wave function can be extended to dynamical systems, such as atoms and molecules. Since the modulus squared of the wave function, which represents the state of a dynamical system, is regarded as a probability density, there are certain restrictions to be imposed on the square-integrable state function, usually formulated as the following *rule*:

A state function and its derivative which describe a dynamical system must be everywhere finite and continuous.

The probabilistic interpretation of the state function, which is not a direct observable, implies that a dynamical system might be potentially in a number of physical states. The detailed knowledge of physical states is not possible in quantum mechanics, in contrast to the classical concept of state. Furthermore, we may always postulate that there generally exists an infinite number of mutually exclusive states $\Psi_1, \Psi_2, \ldots$, for a dynamical system.

According to the classical picture, these states are selected as a result of repeated measurements on the same dynamical system, or of a single measurement on an assembly of identical systems, as discussed in Example 1.2. Since it is supposed possible to determine the state of each member of the assembly, the general state of the system might be represented by:

$$\Psi = f_1 \Psi_1 + f_2 \Psi_2 + \ldots \tag{1.62}$$

where the coefficients f_i are determined by the fraction of members of the assembly that occupy the various states. The quantum states of a dynamical system are, however, described by wave functions which obey the linear equation (1.25), and therefore, if two state functions $\Psi_1(x,t)$ and $\Psi_2(x,t)$ are solutions, so is the function:

$$\Psi(x,t) = a_1 \Psi_1(x,t) + a_2 \Psi_2(x,t)$$

where the coefficients are complex numbers. This result is generalized as the **principle of superposition of states:**

If n particular states of a dynamical system are described by the wave functions $\Psi_1(x,t)$, $\Psi_2(x,t), \ldots$, their linear superposition:

$$\Psi(x,t) = \sum_i a_i \Psi_i(x,t) \tag{1.63}$$

where the coefficients a_i are complex numbers, describes a new state of the system.

It follows that, although the norm $|\Psi(x,t)|$ is the physically significant quantity, the phase factor of the solution cannot be ignored, since the modulus of $\Psi(x,t)$ depends on the relative phases of $\Psi_i(x,t)$. Such a behaviour has been confirmed by measurements of electron diffraction in the two-slit configuration, the optical analogue of which are the experiments based on superposition of two coherent beams obtained by the division of a wavefront. Both results are determined by the phase relation between the two parts of the wave function, each associated with one of the slits, in the two-slit experiment.

1. Wave Functions and the First Postulate

EXAMPLE 1.6. Two-slit experiment with particles

Experiments have been performed with beams of electrons, incident on two slits of dimensions and spacing of the same order of magnitude as the de Broglie wavelength.

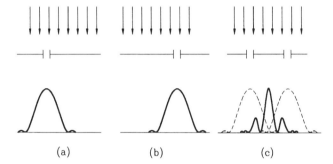

Figure 1.5. Two-slit experiment with electrons

The resulting interference pattern was recorded using a fluorescent screen. Figure 1.5 shows as solid lines the observed patterns if either slit is blocked off, (a) and (b), and if both slits are open, (c). In the latter case, where the single-slit pattern is given in dotted line for comparison, there are obvious interference effects, which are also produced after a sufficient time if only one electron at a time is fired at the slits. If the dynamical states of the electrons are described by the wave functions Ψ_1 and Ψ_2 in Figure 1.5 (a) and (b) respectively, the interference effects can be explained on the grounds of the principle of superposition of states, by taking:

$$\Psi_{12} = \Psi_1 + \Psi_2$$

when both slits are open. As a consequence, the intensity of the two-slit pattern can be interpreted in terms of the probability density of finding the particle on the screen, namely:

$$|\Psi_{12}|^2 = |\Psi_1|^2 + |\Psi_2|^2 + 2 Re(\Psi_1^* \Psi_2)$$

where the last term on the right hand side accounts for the interference pattern, in agreement with the experimental result. ✣

If the mutually exclusive states of a dynamical system form an orthonormal set:

$$\langle \Psi_i | \Psi_j \rangle = \int_{-\infty}^{\infty} \Psi_i^*(x,t) \Psi_j(x,t) dx = \delta_{ij} \qquad (1.64)$$

which may be also assumed to be complete, the principle of superposition of states (1.63) can be interpreted as an *expansion theorem* which states that any state function Ψ can be expanded in terms of a complete set of orthonormal functions Ψ_i:

$$\Psi = \sum_{i=1}^{\infty} a_i \Psi_i \qquad (1.65)$$

where the coefficients are given by:

$$a_i = \langle \Psi_i | \Psi \rangle \qquad (1.66)$$

The state of a dynamical system may then be represented as a vector in a Hilbert space, such that Ψ and $c\Psi$ define the same state for any complex number c. It is common practice to define the state by a vector of unit norm:

$$|\Psi|^2 = \langle \Psi | \Psi \rangle = \sum_{i=1}^{\infty} |a_i|^2 = 1 \qquad (1.67)$$

This result is similar to the normalization condition (1.27) for the probability of measurement. Hence, in the case of an assembly of identical non-interacting particles, the fraction of the assembly in state Ψ_i is $|a_i|^2$, which represents the quantum counterpart of the coefficients f_i in Eq.(1.62), interpreted in terms of the classical concept of state. Given a second state vector Φ in the same wave function space, expressed as:

$$\Phi = \sum_i b_i \Psi_i$$

the *scalar product* of the two state vectors can be defined as follows:

$$\langle \Psi | \Phi \rangle = \int_{-\infty}^{\infty} \Psi^*(x,t) \Phi(x,t) dx = \sum_i a_i^* b_i \qquad (1.68)$$

The first postulate of quantum mechanics refers to the existence of a wave function that depends upon position coordinates and time and describes as completely as possible the physical properties of a given state in the coordinate representation.

> **Postulate 1:** *Every state of a dynamical system is described by a normalized wave function with state vector properties.*

It is common practice to express the properties of the state vectors in the so-called Dirac notation, which has the advantage of providing a general formulation that applies to any dynamical system.

EXAMPLE 1.7. The Dirac notation

A state vector is completely specified by its components in a given orthonormal basis and all vector operations (addition, scalar multiplication and scalar product) are carried out in terms of these components rather then involving the basis vectors. As a result, a convenient representation for a n-dimensional state vector consists of a column vector ($n \times 1$ matrix) obtained by collecting these components:

1. Wave Functions and the First Postulate

$$\Psi = \sum_{i=1}^{n} a_i \varphi_i \quad \text{or} \quad \Psi = \begin{pmatrix} a_1 \\ a_2 \\ \ldots \\ a_n \end{pmatrix} \quad (1.69)$$

$$\Phi = \sum_{i=1}^{n} b_i \varphi_i \quad \text{or} \quad \Phi = \begin{pmatrix} b_1 \\ b_2 \\ \ldots \\ b_n \end{pmatrix} \quad (1.70)$$

When we form the scalar product (1.68), we have to multiply the elements of a row vector ($1 \times n$ matrix) onto those of the column vector:

$$\langle \Psi | \Phi \rangle = (a_1^* a_2^* \ldots a_n^*) \begin{pmatrix} b_1 \\ b_2 \\ \ldots \\ b_n \end{pmatrix} = a_1^* b_1 + a_2^* b_2 + \ldots + a_n^* b_n \quad (1.71)$$

This shows that the conjugated function should be represented by a row vector, obtained by transposition of the column vector (1.69) and complex-conjugation of all scalar coefficients. The two matrices:

$$\Psi = \begin{pmatrix} a_1 \\ a_2 \\ \ldots \\ a_n \end{pmatrix} \quad \text{and} \quad \Psi^+ = (a_1^* a_2^* \ldots a_n^*)$$

are said to be *adjoints* to each other, and belong to different wave function spaces, said to be dual to each other, whose elements are in one-to-one correspondence. Following Dirac, a convenient notation for a state vector Ψ is the *ket* vector $|\Psi\rangle$, when it makes its entry as the second factor in the scalar product (1.71), and the *bra* vector $\langle \psi |$, whenever it appears as the first factor:

$$|\Psi\rangle = \begin{pmatrix} a_1 \\ a_2 \\ \ldots \\ a_n \end{pmatrix}, \quad \langle \Psi | = (a_1^* a_2^* \ldots a_n^*) \quad (1.72)$$

These expressions can be expanded in terms of the basis vectors, which also are of two kinds:

$$|\varphi_i\rangle \equiv |i\rangle = \begin{pmatrix} 0 \\ \ldots \\ 1 \\ \ldots \\ 0 \end{pmatrix} \quad (1.73)$$

64   Quantum Physics

and:

$$\langle \varphi_i | \equiv \langle i | = (0 \ldots 1 \ldots 0) \tag{1.74}$$

Hence Eqs.(1.72) can be written as:

$$|\Psi\rangle = \sum_{i=1}^{n} a_i |i\rangle \tag{1.75}$$

$$\langle \Psi | = \sum_{i=1}^{n} a_i^* \langle i | \tag{1.76}$$

where the basis vectors are orthonormal according to Eq.(1.58):

$$\langle i | j \rangle = \delta_{ij} \tag{1.77}$$

Taking the scalar product of both sides of Eq.(1.75) with $\langle j |$, we obtain the expansion coefficients:

$$\langle j | \Psi \rangle = \left\langle j \Big| \sum_{i=1}^{n} a_n |i\rangle \right\rangle = \sum_{i=1}^{n} a_i \langle j | i \rangle = \sum_{i=1}^{n} a_i \delta_{ji} = a_j$$

so that Eq.(1.75) becomes:

$$|\Psi\rangle = \sum_{i=1}^{n} |i\rangle \langle i | \Psi \rangle \tag{1.78}$$

Similarly, from Eq.(1.76) we have:

$$\langle \Psi | j \rangle = a_j^*$$

so that:

$$\langle \Psi | = \sum_{i=1}^{n} \langle \Psi | i \rangle \langle i | \tag{1.79}$$

In view of the probability density (1.26), the expectation value of an arbitrary function of position, in the state $\Psi(x,t)$ of a given system, Eq.(1.29), can be written as:

$$\langle f(x) \rangle = \int_{-\infty}^{\infty} f(x) P(x,t) dx = \int_{-\infty}^{\infty} \Psi^*(x,t) f(x) \Psi(x,t) dx = \langle \Psi | f \Psi \rangle = \langle \Psi | f | \Psi \rangle \tag{1.80}$$

The advantage of the Dirac notation for the integrals of Eq.(1.77) is that we do not have to write out all the integration variables. ♨

1. Wave Functions and the First Postulate

Problem 1.4.1. Prove the triangle inequality:

$$|(\Psi + \Phi)| \leq |\Psi| + |\Phi|$$

and find the conditions where the equality holds.

(Solution): We may write:

$$|(\Psi + \Phi)|^2 = \langle \Psi + \Phi | \Psi + \Phi \rangle = \langle \Psi | \Psi \rangle + \langle \Psi | \Phi \rangle + \langle \Phi | \Psi \rangle + \langle \Phi | \Phi \rangle$$

$$= \langle \Psi | \Psi \rangle + 2\,\text{Re}\langle \Psi | \Phi \rangle + \langle \Phi | \Phi \rangle \leq \langle \Psi | \Psi \rangle + 2|\langle \Psi | \Phi \rangle| + \langle \Phi | \Phi \rangle$$

where use has been made of the property $Re\langle \Psi | \Phi \rangle \leq |\langle \Psi | \Phi \rangle|$ which always holds for a complex number. From the Schwartz inequality, Eq.(1.50), it follows that:

$$|(\Psi + \Phi)|^2 \leq \langle \Psi | \Psi \rangle + 2\sqrt{\langle \Psi | \Psi \rangle}\sqrt{\langle \Phi | \Phi \rangle} + \langle \Phi | \Phi \rangle = |\Psi|^2 + 2|\Psi||\Phi| + |\Phi|^2$$

$$= (|\Psi| + |\Phi|)^2$$

and this shows the validity of the inequality. The equality only holds for either $|\Psi\rangle = 0, |\Phi\rangle = 0$ or $|\Psi\rangle = a|\Phi\rangle$, where a is a complex number.

Problem 1.4.2. Show that the parallelogram law:

$$|(\Psi + \Phi)|^2 + |(\Psi - \Phi)|^2 = 2|\Psi|^2 + 2|\Phi|^2$$

is valid in the wave function space.

Problem 1.4.3. Show that the scalar product of two wave functions can be expressed in terms of the norm in the form:

$$4\langle \Psi | \Phi \rangle = |(\Psi + \Phi)|^2 - |(\Psi - \Phi)|^2 - i|(\Psi + i\Phi)|^2 + i|(\Psi - i\Phi)|^2$$

2. OPERATORS AND THE SECOND POSTULATE

2.1. The Second Postulate

Physical observables which are measured in a particular state of a dynamical system can be conveniently represented in classical mechanics by mathematical variables, on the grounds that all have definite values in any classical state. In contrast, the statistical interpretation of wave functions describing the quantum states implies that there are uncertainties in the measurement of any two observables which can be represented by canonically conjugate variables, such as position q and momentum p_q. This requires any measured dynamical observable A in a quantum state to be associated with an expectation value $\langle A \rangle$.

In the coordinate representation of a wave function we may simply use Eq.(1.80) to calculate of the expectation values of the physical observables, provided that we find a suitable representation of these variables in terms of the coordinates. An obvious choice is to represent the position observable x by the coordinate variable x, so that the expectation value of the position x in a state Ψ results from Eq.(1.80) as:

$$\langle x \rangle = \int_{-\infty}^{\infty} x|\Psi(x,t)|^2 \, dx = \int_{-\infty}^{\infty} \Psi^*(x,t) x \Psi(x,t) dx = \langle \Psi | \hat{x} \Psi \rangle = \langle \Psi | \hat{x} | \Psi \rangle \quad (2.1)$$

This is the same as the expectation value of the position Eq.(1.28), defined in terms of the probability density for the presence of a particle at a given point with abscissa x. It is common practice to discuss Eq.(2.1) in terms of an operator $\hat{x}$, representing a multiplicative law by which we associate every state $\Psi(x,t)$ of a dynamical system with another state $\hat{x}\Psi(x,t)$ that belongs to the same wave function space. It is therefore considered that the position observable x is associated with an operator $\hat{x}$ which consists of multiplication of the wave function by the position variable x:

$$\hat{x}\Psi(x,t) = x\Psi(x,t) \quad (2.2)$$

On the same grounds, the physical observables which are classically expressed in terms of powers of the position variable, x^k, have the expectation value:

2. Operators and the Second Postulate 67

$$\langle x^k \rangle = \int_{-\infty}^{\infty} x^k |\Psi(x,t)|^2 \, dx = \int_{-\infty}^{\infty} \Psi^*(x,t) x^k \Psi(x,t) dx = \langle \Psi | \hat{x}^k \Psi \rangle = \langle \Psi | \hat{x}^k | \Psi \rangle$$

It follows that any physical observable, such as the potential energy, that is classically represented by a function of the position variable $f(x)$, for which the Taylor expansion exists:

$$f(x) = \sum_{k=0}^{\infty} \frac{1}{k!} \left[\frac{d^k f(x)}{dx^k} \right]_{x=0} x^k \qquad (2.3)$$

will have a quantum representation in the form of a multiplicative operator $\hat{f}(x)$:

$$\langle f(x) \rangle = \int_{-\infty}^{\infty} \hat{f}(x) |\Psi(x,t)|^2 \, dx = \int_{-\infty}^{\infty} \Psi^*(x,t) \hat{f}(x) \Psi(x,t) dx = \langle \Psi | \hat{f} \Psi \rangle = \langle \Psi | \hat{f} | \Psi \rangle \quad (2.4)$$

If we now consider the x-component of the linear momentum, p_x, which classically is expressed in terms of the position coordinate by:

$$p_x = mv_x = m \frac{dx}{dt}$$

its expectation value, as given by Eq.(1.80), becomes:

$$\langle p_x \rangle = m \frac{d}{dt} \langle x \rangle = m \frac{d}{dt} \int_{-\infty}^{\infty} \Psi^*(x,t) x \Psi(x,t) dx = m \int_{-\infty}^{\infty} \left(\frac{\partial \Psi^*}{\partial t} x \Psi + \Psi^* x \frac{\partial \Psi}{\partial t} \right) dx \quad (2.5)$$

The partial derivative $\partial x/\partial t$ was omitted from the integral sign, since both the position coordinate and time are independent variables and the state vector $\Psi(x,t)$ itself only varies with time. Inserting both the equation of motion of a particle, Eq.(1.25), and its complex conjugate into Eq.(2.5) we then obtain:

$$\langle p_x \rangle = -\frac{1}{2} i\hbar \int_{-\infty}^{\infty} \left(\frac{\partial^2 \Psi^*}{\partial x^2} x \Psi - \Psi^* x \frac{\partial^2 \Psi}{\partial x^2} \right) dx$$

It is a straightforward matter to show that:

$$\frac{\partial^2 \Psi^*}{\partial x^2} x\Psi - \Psi^* x \frac{\partial^2 \Psi}{\partial x^2} = \frac{\partial}{\partial x}\left(\frac{\partial \Psi^*}{\partial x} x\Psi \right) - \frac{\partial}{\partial x}(\Psi^* \Psi) + \Psi^* \frac{\partial \Psi}{\partial x} - \frac{\partial}{\partial x}\left(\Psi^* x \frac{\partial \Psi}{\partial x} \right) + \Psi^* \frac{\partial \Psi}{\partial x}$$

and hence, the result of the integration is:

68   Quantum Physics

$$\langle p_x \rangle = -\frac{1}{2}i\hbar\left[\left(\frac{\partial \Psi^*}{\partial x}x\Psi - \Psi^*\Psi - \Psi^*x\frac{\partial \Psi}{\partial x}\right)_{-\infty}^{\infty} + \int_{-\infty}^{\infty} 2\Psi^*\frac{\partial \Psi}{\partial x}dx\right]$$

Since the wave functions are zero at infinity, where the probability density is considered to be vanishingly small, the expectation value of the x component of the linear momentum reduces to:

$$\langle p_x \rangle = \int_{-\infty}^{\infty} \Psi^*(x,t)\left(-i\hbar\frac{\partial}{\partial x}\right)\Psi(x,t)dx$$

$$= \int_{-\infty}^{\infty} \Psi^*(x,t)\hat{p}_x\Psi(x,t)dx = \langle \Psi|\hat{p}_x\Psi\rangle = \langle \Psi|\hat{p}_x|\Psi\rangle \qquad (2.6)$$

The quantity denoted by:

$$\hat{p}_x = -i\hbar\frac{\partial}{\partial x} \qquad (2.7)$$

represents a differential law by which we associate every state vector $\Psi(x,t)$ of a dynamical system with another state vector $\hat{p}_x\Psi(x,t)$. It is then said that the x-component of the linear momentum is associated with the differential operator $\hat{p}_x$, Eq.(2.6), which represents this observable in terms of the position coordinate x.

We may further consider classical observables expressed in terms of powers of the linear momentum components, such as kinetic energy in one dimension:

$$\frac{p_x^2}{2m} = E - U(x)$$

the expectation value of which reads:

$$\frac{1}{2m}\langle p_x^2\rangle = \int_{-\infty}^{\infty} \Psi^*(x,t)E\Psi(x,t)dx - \int_{-\infty}^{\infty} \Psi^*(x,t)U(x)\Psi(x,t)dx$$

Provided E is a constant of motion, we may use Eq.(1.25) to obtain:

$$\int_{-\infty}^{\infty} \Psi^*(x,t)E\Psi(x,t)dx = i\hbar\int_{-\infty}^{\infty} \Psi^*(x,t)\frac{\partial \Psi(x,t)}{\partial t}dx$$

$$= -\frac{\hbar^2}{2m}\int_{-\infty}^{\infty} \Psi^*(x,t)\frac{\partial^2\Psi(x,t)}{\partial x^2}dx + \int_{-\infty}^{\infty} \Psi^*(x,t)U(x)\Psi(x,t)dx$$

so that the expectation value reduces to:

$$\langle p_x^2 \rangle = \int_{-\infty}^{\infty} \Psi^*(x,t)\left(-i\hbar\frac{\partial}{\partial x}\right)^2 \Psi(x,t)dx = \int_{-\infty}^{\infty} \Psi^*(x,t)\hat{p}_x^2 \Psi(x,t)\, dx$$

The results obtained for $\langle p_x \rangle$ and $\langle p_x^2 \rangle$ suggest that it might be possible to infer that:

$$\langle p_x^k \rangle = \int_{-\infty}^{\infty} \Psi^*(x,t)\left(-i\hbar\frac{\partial}{\partial x}\right)^k \Psi(x,t)dx = \langle \Psi| \hat{p}_x^k \Psi \rangle = \langle \Psi| \hat{p}_x^k |\Psi \rangle \qquad (2.8)$$

is valid for any $k \geq 1$. As a result, classical observables, which are given by an arbitrary function $g(p_x)$, which has a Taylor expansion:

$$g(p_x) = \sum_{k=0}^{\infty} \frac{1}{k!}\left(\frac{d^k g}{dp_x^k}\right)_{p_x=0} p_x^k \qquad (2.9)$$

will be similarly represented by a differential operator $\hat{g}(-i\hbar\partial/\partial x)$, such that:

$$\langle g(p_x) \rangle = \int_{-\infty}^{\infty} \Psi^*(x,t)\hat{g}\left(-i\hbar\frac{\partial}{\partial x}\right)\Psi(x,t)dx = \langle \Psi| \hat{g}(p_x)\Psi \rangle = \langle \Psi| \hat{g}(p_x)|\Psi \rangle \quad (2.10)$$

The multiplicative and differential operators, which are mostly encountered in quantum mechanics, are all linear operators, which transform a linear combination of wave functions into a linear combination of different wave functions. It is clear that operators like $\hat{A}\Psi = \Psi + a$, where a is a non-zero complex number, or $\hat{A}\Psi = \Psi^2$, are not linear over the wave function space. We limit ourselves to assume that any physical observable A will be represented in quantum mechanics by a linear operator $\hat{A}$, which defines a correspondence between two state functions of the same dynamical system. The expectation value of a quantum operator $\hat{A}$ in a state $\Psi(x,t)$ of a given system will be defined by:

$$\langle \hat{A} \rangle = \int_{-\infty}^{\infty} \Psi^*(x,t)\hat{A}\Psi(x,t)dx = \langle \Psi| \hat{A}\Psi \rangle = \langle \Psi| \hat{A}|\Psi \rangle \qquad (2.11)$$

Equation (2.11) represents an obvious generalization of Eqs.(2.1), (2.4), (2.6) and (2.10), provided we can find a way of expressing the operator $\hat{A}$ in terms of coordinates and we consider that the expectation value of the operator, $\langle \hat{A} \rangle$, represents the average value of the observable A in the state $\Psi(x,t)$, that is the average result of a quantum measurement on a system, in the given state. In other words, we will assume that:

70   Quantum Physics

$$\langle A \rangle = \langle \hat{A} \rangle \qquad (2.12)$$

It is clear that the expectation value $\langle x \rangle$ given by Eq.(2.1) is real, because integration of a real function always gives a real result, and this is also valid for the potential energy $U(x)$, Eq.(2.4), provided it is expressed by a real function of coordinates. It can be shown that $\langle p_x \rangle$ expressed by Eq.(2.6) is also real, because:

$$\langle p_x \rangle = \int_{-\infty}^{\infty} \Psi^* \left(-i\hbar \frac{\partial \Psi}{\partial x} \right) dx = -i\hbar \int_{-\infty}^{\infty} \left(\Psi^* \frac{\partial \Psi}{\partial x} + \frac{\partial \Psi^*}{\partial x} \Psi \right) dx + \int_{-\infty}^{\infty} \Psi \left(i\hbar \frac{\partial}{\partial x} \right) \Psi^* dx$$

$$= -i\hbar \int_{-\infty}^{\infty} \frac{\partial}{\partial x} (\Psi^* \Psi) dx + \langle p_x \rangle^* = -i\hbar (\Psi^* \Psi)_{-\infty}^{\infty} + \langle p_x \rangle^* = \langle p_x \rangle^*$$

where use has been made of the property that state functions vanish at infinity. The same is true for kinetic energy, the expectation value of which reduces to a real integral:

$$\frac{1}{2m} \langle p_x^2 \rangle = -\frac{\hbar^2}{2m} \int_{-\infty}^{\infty} \Psi^* \frac{\partial^2 \Psi}{\partial x^2} dx = -\frac{\hbar^2}{2m} \left(\Psi^* \frac{\partial \Psi}{\partial x} \right)_{-\infty}^{\infty} + \frac{\hbar^2}{2m} \int_{-\infty}^{\infty} \left(\frac{\partial \Psi^*}{\partial x} \right) \left(\frac{\partial \Psi}{\partial x} \right) dx$$

$$= \frac{\hbar^2}{2m} \int_{-\infty}^{\infty} \left| \frac{\partial \Psi}{\partial x} \right|^2 dx$$

Since the expectation value of any physical observable $\langle A \rangle$ should be real in any state of a dynamical system, it will be required that $\langle \hat{A} \rangle$ is also real, which implies, in view of Eq.(2.11) that:

$$\int_{-\infty}^{\infty} \Psi^*(x,t) \hat{A} \Psi(x,t) dx = \left(\int_{-\infty}^{\infty} \Psi^*(x,t) \hat{A} \Psi(x,t) dx \right)^* = \int_{-\infty}^{\infty} \left[\hat{A} \Psi(x,t) \right]^* \Psi(x,t) dx$$

or, in the Dirac notation:

$$\langle \Psi | \hat{A} \Psi \rangle = (\langle \Psi | \hat{A} \Psi \rangle)^* = \langle \hat{A} \Psi | \Psi \rangle \qquad (2.13)$$

Operators which obey Eq.(2.13) are called Hermitians, and we will further assume that only such linear operators should be associated with physical observables.

As a result of these considerations, we are now in a position to formulate the second postulate, which is a general statement about how physical observables are represented mathematically in quantum mechanics.

Postulate 2. *Every physical observable may be represented by a Hermitian linear operator. The operators representing the*

2. Operators and the Second Postulate

position and momentum of a particle in one dimension are x_i and $-i\hbar \partial/\partial x_i$ respectively, where x_i stands for Cartesian coordinates. Operators representing other physical observables bear the same functional relation to these as do the corresponding classical quantities to the classical position and momentum variables.

The states of three-dimensional dynamical systems are defined in the coordinate space by wave functions $\Psi(\vec{r},t)$, which in Cartesian coordinates will be normalized according to:

$$\int_{-\infty}^{\infty}\int_{-\infty}^{\infty}\int_{-\infty}^{\infty} \Psi^*(x,y,z,t)\Psi(x,y,z,t)\,dx\,dy\,dz = 1 \qquad (2.14)$$

which is a simple extension of Eq.(1.37). As a consequence, the usual operators representing three-dimensional physical observables, given by classical functions $f(x_i)$ and $g(p_i)$, take the following forms:

1. *position* operator $\hat{\vec{r}}$:

$$\vec{r} = x\vec{e}_x + y\vec{e}_y + z\vec{e}_z \rightarrow \hat{\vec{r}} = \vec{r}$$

2. *linear momentum* operator $\hat{\vec{p}}$:

$$\vec{p} = p_x\vec{e}_x + p_y\vec{e}_y + p_z\vec{e}_z \rightarrow \hat{\vec{p}} = -i\hbar\left(\vec{e}_x\frac{\partial}{\partial x} + \vec{e}_y\frac{\partial}{\partial y} + \vec{e}_z\frac{\partial}{\partial z}\right) = -i\hbar\nabla \qquad (2.15)$$

3. *kinetic energy* operator $\hat{T}$:

$$T = \frac{p^2}{2m} = \frac{1}{2m}\left(p_x^2 + p_y^2 + p_z^2\right) \rightarrow \hat{T} = -\frac{\hbar^2}{2m}\left(\frac{\partial^2}{\partial x^2} + \frac{\partial^2}{\partial y^2} + \frac{\partial^2}{\partial z^2}\right) = -\frac{\hbar^2}{2m}\nabla^2 \qquad (2.16)$$

4. *Hamiltonian* operator $\hat{H}$:

$$H = \frac{p^2}{2m} + U(x,y,z) \rightarrow \hat{H} = -\frac{\hbar^2}{2m}\nabla^2 + U(x,y,z) \qquad (2.17)$$

5. *angular momentum* operator:

72 Quantum Physics

$$\vec{L} = \vec{r} \times \vec{p} = \begin{vmatrix} \vec{e}_x & \vec{e}_y & \vec{e}_z \\ x & y & z \\ p_x & p_y & p_z \end{vmatrix}$$

so that the operators representing its components are:

$$L_x = yp_z - zp_y \to \hat{L}_x = -i\hbar\left(y\frac{\partial}{\partial z} - z\frac{\partial}{\partial y}\right)$$

$$L_y = zp_x - xp_z \to \hat{L}_y = -i\hbar\left(z\frac{\partial}{\partial x} - x\frac{\partial}{\partial z}\right) \qquad (2.18)$$

$$L_z = xp_y - yp_x \to \hat{L}_z = -i\hbar\left(x\frac{\partial}{\partial y} - y\frac{\partial}{\partial x}\right)$$

and the magnitude of angular momentum is represented by the operator:

$$\hat{L}^2 = \hat{L}_x^2 + \hat{L}_y^2 + \hat{L}_z^2 \qquad (2.19)$$

EXAMPLE 2.1. Angular momentum in spherical polar coordinates

For describing dynamical systems with spherical symmetry, it is convenient to introduce the unit vectors $\vec{e}_r, \vec{e}_\theta$ and $\vec{e}_\varphi$ in the directions of increasing r, θ and φ, the spherical polar coordinate. Expressions of the operators $\hat{L}_x, \hat{L}_y, \hat{L}_z$ and $\hat{L}^2$, which are appropriate for physical interpretation, should be written, in this case, in terms of r, θ and φ. From Figure 2.1 it is apparent that:

$$x = r\sin\theta\cos\varphi, \quad y = r\sin\theta\sin\varphi, \quad z = r\cos\theta \qquad (2.20)$$

which gives:

$$r^2 = x^2 + y^2 + z^2, \quad \theta = \arctan\frac{\sqrt{x^2+y^2}}{z}, \quad \varphi = \arctan\frac{y}{x} \qquad (2.21)$$

and the volume element has the form:

$$dV = r^2\sin\theta\,dr\,d\theta\,d\varphi = \rho(r,\theta)\,dr\,d\theta\,d\varphi \qquad (2.22)$$

As a result, the normalization condition for wave functions $\Psi(\vec{r},t) = \Psi(r,\theta,\varphi,t)$ becomes:

$$\int_0^\infty \int_0^\pi \int_0^{2\pi} \Psi^*(r,\theta,\varphi)\Psi(r,\theta,\varphi)r^2\sin\theta\,dr\,d\theta\,d\varphi = 1 \qquad (2.23)$$

2. Operators and the Second Postulate

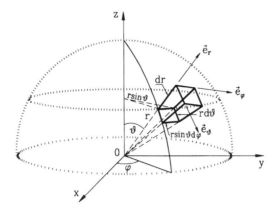

Figure 2.1. Configuration of spherical polar coordinates

From Eqs.(2.20) it follows that:

$$\frac{\partial r}{\partial x} = \sin\theta \cos\varphi \qquad \frac{\partial \theta}{\partial x} = \frac{1}{r}\cos\theta \cos\varphi \qquad \frac{\partial \varphi}{\partial x} = -\frac{1}{r}\frac{\sin\varphi}{\sin\theta}$$

$$\frac{\partial r}{\partial y} = \sin\theta \sin\varphi \qquad \frac{\partial \theta}{\partial y} = \frac{1}{r}\cos\theta \sin\varphi \qquad \frac{\partial \varphi}{\partial y} = \frac{1}{r}\frac{\cos\varphi}{\sin\theta} \qquad (2.24)$$

$$\frac{\partial r}{\partial z} = \cos\theta \qquad \frac{\partial \theta}{\partial z} = -\frac{1}{r}\sin\theta \qquad \frac{\partial \varphi}{\partial z} = 0$$

which yields:

$$\frac{\partial}{\partial x} = \frac{x}{r}\frac{\partial}{\partial r} + \frac{1}{\sin\theta}\frac{zx}{r^3}\frac{\partial}{\partial \theta} - \frac{y/x^2}{(1+y^2/x^2)}\frac{\partial}{\partial \varphi}$$

or, in view of Eqs.(2.20):

$$\frac{\partial}{\partial x} = \sin\theta \cos\varphi \frac{\partial}{\partial r} + \frac{\cos\theta \cos\varphi}{r}\frac{\partial}{\partial \theta} - \frac{\sin\varphi}{r\sin\theta}\frac{\partial}{\partial \varphi}$$

In a similar manner we obtain:

$$\frac{\partial}{\partial y} = \sin\theta \sin\varphi \frac{\partial}{\partial r} + \frac{\cos\theta \sin\varphi}{r}\frac{\partial}{\partial \theta} + \frac{\cos\varphi}{r\sin\theta}\frac{\partial}{\partial \varphi}$$

and also:

$$\frac{\partial}{\partial z} = \cos\theta \frac{\partial}{\partial r} - \frac{\sin\theta}{r}\frac{\partial}{\partial \theta}$$

Quantum Physics

Substituting the expressions for the partial derivatives into Eq.(2.18) yields:

$$\hat{L}_z = -i\hbar r \sin\theta \cos\varphi \left(\sin\theta \sin\varphi \frac{\partial}{\partial r} + \frac{\cos\theta \sin\varphi}{r} \frac{\partial}{\partial \theta} + \frac{\cos\varphi}{r\sin\theta} \frac{\partial}{\partial \varphi} \right)$$

$$+ i\hbar r \sin\theta \sin\varphi \left(\sin\theta \cos\varphi \frac{\partial}{\partial r} + \frac{\cos\theta \cos\varphi}{r} \frac{\partial}{\partial \theta} - \frac{\sin\varphi}{r\sin\theta} \frac{\partial}{\partial \varphi} \right)$$

$$= -i\hbar(\cos^2\varphi + \sin^2\varphi)\frac{\partial}{\partial \varphi} = -i\hbar \frac{\partial}{\partial \varphi} \qquad (2.25)$$

and similarly:

$$\hat{L}_x = i\hbar \left[\sin\varphi \frac{\partial}{\partial \theta} + \frac{\cos\varphi}{\tan\theta} \frac{\partial}{\partial \varphi} \right] \qquad (2.26)$$

$$\hat{L}_y = i\hbar \left[-\cos\varphi \frac{\partial}{\partial \theta} + \frac{\sin\varphi}{\tan\theta} \frac{\partial}{\partial \varphi} \right] \qquad (2.27)$$

The product of the last two operators depends on the order in which the operators are applied, namely:

$$\hat{L}_x \hat{L}_y = -\hbar^2 \cos\varphi \sin\varphi \frac{\partial^2}{\partial \theta^2} + \hbar^2 \sin^2\varphi \left(\frac{1}{\tan\theta} \frac{\partial^2}{\partial \varphi \partial \theta} - \frac{1+\tan^2\theta}{\tan^2\theta} \frac{\partial}{\partial \varphi} \right)$$

$$- \hbar^2 \frac{\cos\varphi}{\tan\theta} \left(-\sin\varphi \frac{\partial}{\partial \theta} + \cos\varphi \frac{\partial^2}{\partial \varphi \partial \theta} - \frac{\cos\varphi}{\tan\theta} \frac{\partial}{\partial \varphi} - \frac{\sin\varphi}{\tan\theta} \frac{\partial^2}{\partial \varphi^2} \right)$$

is different from:

$$\hat{L}_y \hat{L}_x = -\hbar^2 \cos\varphi \sin\varphi \frac{\partial^2}{\partial \theta^2} - \hbar^2 \cos^2\varphi \left(\frac{1}{\tan\theta} \frac{\partial^2}{\partial \varphi \partial \theta} - \frac{1+\tan^2\theta}{\tan^2\theta} \frac{\partial}{\partial \varphi} \right)$$

$$- \hbar^2 \frac{\sin\varphi}{\tan\theta} \left(\cos\theta \frac{\partial}{\partial \theta} + \sin\varphi \frac{\partial^2}{\partial \varphi \partial \theta} - \frac{\sin\varphi}{\tan\theta} \frac{\partial}{\partial \varphi} + \frac{\cos\varphi}{\tan\theta} \frac{\partial^2}{\partial \varphi^2} \right)$$

by:

$$\hat{L}_x \hat{L}_y = \hat{L}_y \hat{L}_x - \hbar^2 \frac{\partial}{\partial \varphi} = \hat{L}_y \hat{L}_x - i\hbar \hat{L}_z \qquad (2.28)$$

As a consequence, it will be more convenient to calculate $\hat{L}^2$ by reformulating Eq.(2.19) as:

$$\hat{L}^2 = \hat{L}_x^2 + \hat{L}_y^2 + \hat{L}_z^2 - i\left(\hat{L}_x \hat{L}_y - \hat{L}_y \hat{L}_x\right) + \hbar \hat{L}_z$$

$$= \hat{L}_z^2 + \hbar \hat{L}_z + (\hat{L}_x + i\hat{L}_y)(\hat{L}_x - i\hat{L}_y) = \hat{L}_z^2 + \hbar \hat{L}_z + \hat{L}_+ \hat{L}_- \quad (2.29)$$

where the new operators $\hat{L}_+, \hat{L}_-$ which has been introduced assume simpler forms in spherical polar coordinates:

$$\hat{L}_+ = \hat{L}_x + i\hat{L}_y = i\hbar e^{i\varphi}\left(-i\frac{\partial}{\partial \theta} + \frac{\cos\theta}{\sin\theta}\frac{\partial}{\partial \varphi}\right) \quad (2.30)$$

$$\hat{L}_- = \hat{L}_x - i\hat{L}_y = i\hbar e^{-i\varphi}\left(i\frac{\partial}{\partial \theta} + \frac{\cos\theta}{\sin\theta}\frac{\partial}{\partial \varphi}\right) \quad (2.31)$$

and their product reduces to:

$$\hat{L}_+ \hat{L}_- = -\hbar^2\left(\frac{\partial^2}{\partial \theta^2} + \frac{1}{\tan\theta}\frac{\partial}{\partial \theta} + \frac{1}{\tan^2\theta}\frac{\partial^2}{\partial \varphi^2} - i\frac{\partial}{\partial \varphi}\right) \quad (2.32)$$

Substituting Eq.(2.32) into Eq.(2.29) we finally obtain:

$$\hat{L}^2 = -\hbar^2\left(\frac{\partial^2}{\partial \theta^2} + \frac{1}{\tan\theta}\frac{\partial}{\partial \theta} + \frac{1}{\sin^2\theta}\frac{\partial^2}{\partial \varphi^2}\right) = -\hbar^2\left[\frac{1}{\sin\theta}\frac{\partial}{\partial \theta}\left(\sin\theta\frac{\partial}{\partial \theta}\right) + \frac{1}{\sin^2\theta}\frac{\partial^2}{\partial \varphi^2}\right] \quad (2.33)$$

Equations (2.25), (2.30), (2.31) and (2.33) are extensively used in the quantum description of atomic and nuclear systems. ♦

EXAMPLE 2.2. The Hamiltonian in a central force field

In a central force field, where the potential energy has spherical symmetry, $U(\vec{r}) = U(r)$, the appropriate Hamiltonian operator should be expressed in terms of spherical polar coordinates, starting from Eq.(2.17), which reads:

$$H = \frac{1}{2m}\vec{p}\cdot\vec{p} + U(r)$$

It is convenient firstly to find an appropriate form for the quantum operator associated with the linear momentum according to Eq.(2.15), and the problem is then reduced to that of writing the gradient operator in terms of r, θ and φ. It is clear from Figure 2.1 that the coordinate increments, for small displacements along the directions $\vec{e}_r, \vec{e}_\theta$ and $\vec{e}_\varphi$, are given by:

$$\frac{\partial \vec{r}}{\partial r} = \vec{e}_r, \quad \frac{\partial \vec{r}}{\partial \theta} = r\vec{e}_\theta, \quad \frac{\partial \vec{r}}{\partial \varphi} = r\sin\theta\vec{e}_\varphi$$

respectively, so that the components of the gradient operator are obtained by considering in turn the increment along each direction, while the other two coordinates are held constant:

$$\vec{e}_r \cdot \nabla = \frac{d}{|d\vec{r}/\partial r|dr} = \frac{d}{dr}\bigg|_{\theta,\varphi=ct} = \frac{\partial}{\partial r}$$

$$\vec{e}_\theta \cdot \nabla = \frac{d}{|\partial \vec{r}/\partial \theta|d\theta} = \frac{1}{r}\frac{d}{d\theta}\bigg|_{r,\varphi=ct} = \frac{1}{r}\frac{\partial}{\partial \theta}$$

$$\vec{e}_\varphi \cdot \nabla = \frac{d}{|\partial \vec{r}/\partial \varphi|d\varphi} = \frac{1}{r\sin\theta}\frac{d}{d\varphi}\bigg|_{r,\theta=ct} = \frac{1}{r\sin\theta}\frac{\partial}{\partial \varphi}$$

Substituting these results into Eq.(2.15) we obtain:

$$\hat{p} = -i\hbar\nabla = -i\hbar\sum_{q=r,\theta,\varphi}\vec{e}_q(\vec{e}_q\cdot\nabla) = -i\hbar\left(\vec{e}_r\frac{\partial}{\partial r}+\vec{e}_\theta\frac{1}{r}\frac{\partial}{\partial \theta}+\vec{e}_\varphi\frac{1}{r\sin\theta}\frac{\partial}{\partial \varphi}\right) \quad (2.34)$$

It follows that:

$$\hat{H} = -\frac{\hbar^2}{2m}\left(\vec{e}_r\frac{\partial}{\partial r}+\vec{e}_\theta\frac{1}{r}\frac{\partial}{\partial \theta}+\vec{e}_\varphi\frac{1}{r\sin\theta}\frac{\partial}{\partial \varphi}\right)\left(\vec{e}_r\frac{\partial}{\partial r}+\vec{e}_\theta\frac{1}{r}\frac{\partial}{\partial \theta}+\vec{e}_\varphi\frac{1}{r\sin\theta}\frac{\partial}{\partial \varphi}\right)+U(r) \quad (2.35)$$

and we should use the partial derivatives of the unit vectors, which obviously are all functions of position, and can be derived, using Figure 2.1, in the form:

$$\frac{\partial \vec{e}_r}{\partial r}=0, \quad \frac{\partial \vec{e}_r}{\partial \theta}=\vec{e}_\theta, \quad \frac{\partial \vec{e}_r}{\partial \varphi}=\vec{e}_\varphi\sin\theta$$

$$\frac{\partial \vec{e}_\theta}{\partial r}=0, \quad \frac{\partial \vec{e}_\theta}{\partial \theta}=-\vec{e}_r, \quad \frac{\partial \vec{e}_\theta}{\partial \varphi}=\vec{e}_\varphi\cos\theta$$

$$\frac{\partial \vec{e}_\varphi}{\partial r}=0, \quad \frac{\partial \vec{e}_\varphi}{\partial \theta}=0, \quad \frac{\partial \vec{e}_\varphi}{\partial \varphi}=-\vec{e}_r\sin\theta-\vec{e}_\theta\cos\theta$$

Substituting these equations into Eq.(2.35) yields:

$$\hat{H}=-\frac{\hbar^2}{2m}\left(\frac{\partial^2}{\partial r^2}+\frac{2}{r}\frac{\partial}{\partial r}+\frac{1}{r^2}\frac{\partial^2}{\partial \theta^2}+\frac{1}{r^2}\frac{\cos\theta}{\sin\theta}\frac{\partial}{\partial \theta}+\frac{1}{r^2\sin^2\theta}\frac{\partial^2}{\partial \varphi^2}\right)+U(r)$$

$$=-\frac{\hbar^2}{2m}\left[\frac{1}{r^2}\frac{\partial}{\partial r}\left(r^2\frac{\partial}{\partial r}\right)+\frac{1}{r^2\sin\theta}\frac{\partial}{\partial \theta}\left(\sin\theta\frac{\partial}{\partial \theta}\right)+\frac{1}{r^2\sin^2\theta}\frac{\partial^2}{\partial \varphi^2}\right]+U(r) \quad (2.36)$$

Comparison of Eqs.(2.36) and (2.33) shows that:

$$\hat{H}=-\frac{\hbar^2}{2m}\frac{1}{r^2}\frac{\partial}{\partial r}\left(r^2\frac{\partial}{\partial r}\right)+\frac{\hat{L}^2}{2mr^2}+U(r) \quad (2.37)$$

Note that, from Eq.(2.20), the components of velocity can be written as:

$$\dot{x} = \frac{dx}{dt} = \dot{r}\sin\theta\cos\varphi + r\cos\theta\cos\varphi\dot{\theta} - r\sin\theta\sin\varphi\dot{\varphi}$$

$$\dot{y} = \frac{dy}{dt} = \dot{r}\sin\theta\sin\varphi + r\cos\theta\sin\varphi\dot{\theta} + r\sin\theta\cos\varphi\dot{\varphi}$$

$$\dot{z} = \frac{dz}{dt} = \dot{r}\cos\theta - r\sin\theta\dot{\theta}$$

and, hence, kinetic energy reads:

$$T = \frac{m}{2}(\dot{x}^2 + \dot{y}^2 + \dot{z}^2) = \frac{m}{2}(\dot{r}^2 + \dot{r}^2\dot{\theta}^2 + r^2\sin^2\theta\dot{\varphi}^2)$$

The linear momentum components conjugated to r, θ and φ are respectively:

$$p_r = m\dot{r}, \quad p_\theta = mr^2\dot{\theta}, \quad p_\varphi = mr^2\sin^2\theta\,\dot{\varphi}$$

and lead to the classical Hamiltonian:

$$H = T(p_r, p_\theta, p_\varphi) + U(r) = \frac{1}{2m}\left(p_r^2 + \frac{1}{r^2}p_\theta^2 + \frac{1}{r^2\sin^2\theta}p_\varphi^2\right) + U(r)$$

Replacing the linear momenta by differential operators written as $\hat{p}_r = -i\hbar\,\partial/\partial r$, $\hat{p}_\theta = -i\hbar\,\partial/\partial\theta$ and $\hat{p}_\varphi = -i\hbar\,\partial/\partial\varphi$, the Hamiltonian can be written as:

$$\hat{H} = -\frac{\hbar^2}{2m}\left(\frac{\partial^2}{\partial r^2} + \frac{1}{r^2}\frac{\partial^2}{\partial\theta^2} + \frac{1}{r^2\sin^2\theta}\frac{\partial^2}{\partial\varphi^2}\right) + U(r)$$

which is clearly different from the correct result, Eq.(2.36). The linear momentum components written in the form $-i\hbar\,\partial/\partial x_i$ for coordinates other than Cartesian, as restricted by Postulate 1, should therefore be avoided. ∎

Problem 2.1.1. Show that, for a one-dimensional dynamical system whose total energy is a constant of motion, the expectation value of the product of operators $\hat{x}\hat{p}_x$ is time-independent. Use this property to derive the quantum equation corresponding to the classical virial theorem:

$$\langle 2T \rangle = n\langle U \rangle$$

78 Quantum Physics

where $T = p_x^2/2m$, and $U(x)$ is assumed to be an homogenous function of position of the order n.

(Solution): The time evolution of the expectation value can be written, as shown in Eq.(2.5), in the form:

$$\frac{d}{dt}\langle x\hat{p}_x \rangle = \int_{-\infty}^{\infty} \frac{\partial \Psi^*(x,t)}{\partial t} x\hat{p}_x \Psi(x,t)dx + \int_{-\infty}^{\infty} \Psi^*(x,t)x\hat{p}_x \frac{\partial \Psi(x,t)}{\partial t} dx$$

since only the wave function has an explicit time dependence. If the total energy E is a constant of motion, Eqs.(1.24) gives:

$$i\hbar \frac{\partial \Psi(x,t)}{\partial t} = E\Psi(x,t), \quad i\hbar \frac{\partial \Psi^*(x,t)}{\partial t} = -E\Psi^*(x,t)$$

which yields $d\langle x\hat{p}_x \rangle/dt = 0$. Substituting both the wave equation (1.25), written as:

$$i\hbar \frac{\partial \Psi(x,t)}{\partial t} = \frac{\hat{p}_x^2}{2m} \Psi(x,t) + U(x)\Psi(x,t)$$

and its complex conjugate, we obtain:

$$i\hbar \frac{d}{dt}\langle x\hat{p}_x \rangle = -\int_{-\infty}^{\infty} \Psi^*(x,t) \left[\frac{\hat{p}_x^2}{2m} + U(x)\right] x\hat{p}_x \Psi(x,t)$$

$$+ \int_{-\infty}^{\infty} \Psi^*(x,t) x\hat{p}_x \left[\frac{\hat{p}_x^2}{2m} + U(x)\right] \Psi(x,t)dx = 0$$

It follows that:

$$\int_{-\infty}^{\infty} \Psi^*(x,t)\left[\frac{\hat{p}_x^2}{2m} + U(x)\right] x\hat{p}_x \Psi(x,t)dx = \int_{-\infty}^{\infty} \Psi^*(x,t)x\hat{p}_x \left[\frac{\hat{p}_x^2}{2m} + U(x)\right] \Psi(x,t)dx$$

or, in the Dirac notation:

$$\langle \Psi | \left[\frac{\hat{p}_x^2}{2m} + U(x)\right] x\hat{p}_x | \Psi \rangle = \langle \Psi | x\hat{p}_x \left[\frac{\hat{p}_x^2}{2m} + U(x)\right] | \Psi \rangle$$

Using the simple property:

$$(\hat{p}_x x)\Psi(x,t) = -i\hbar \frac{\partial}{\partial x}[x\Psi(x,t)] = -i\hbar x \frac{\partial \Psi(x,t)}{\partial x} - i\hbar \Psi(x,t) = (x\hat{p}_x - i\hbar)\Psi(x,t)$$

2. Operators and the Second Postulate

we successively obtain :

$$\frac{\hat{p}_x^2}{2m} x\hat{p}_x|\Psi\rangle = \frac{\hat{p}_x}{2m}(x\hat{p}_x - i\hbar)\hat{p}_x|\Psi\rangle$$

$$= \left(\frac{\hat{p}_x x}{2m}\hat{p}_x^2 - \frac{i\hbar}{2m}\hat{p}_x^2\right)|\Psi\rangle = \left(x\hat{p}_x \frac{\hat{p}_x^2}{2m} - \frac{i\hbar}{m}\hat{p}_x^2\right)|\Psi\rangle$$

We also have:

$$x\hat{p}_x[U(x)\Psi(x,t)] = xU(x)\hat{p}_x\Psi(x,t) - i\hbar x\frac{\partial U(x)}{\partial x}\Psi(x,t)$$

and substituting these results we finally obtain:

$$\langle\Psi|\frac{\hat{p}_x^2}{m}|\Psi\rangle = \langle\Psi|x\frac{\partial U(x)}{\partial x}|\Psi\rangle$$

This can be written in the form:

$$\langle 2\hat{T}\rangle = \left\langle x\frac{\partial U}{\partial x}\right\rangle$$

If U is an homogenous function of x of the order n, it follows that $\langle 2\hat{T}\rangle = n\langle U\rangle$.

Problem 2.1.2. Find the coordinate representation of the operator $\hat{p}_r$, associated with the radial component p_r of linear momentum, by comparing the Hamiltonian operator, Eq.(2.37), with its classical counterpart for a central force field.

(Solution): The radial component of linear momentum is expressed in classical mechanics by the dot product of $\vec{e}_r$ and $\vec{p}$:

$$p_r = \vec{e}_r \cdot \vec{p} = \frac{1}{r}\vec{r}\cdot\vec{p} = \frac{1}{r}(xp_x + yp_y + zp_z)$$

The classical Hamiltonian, Eq.(2.17), may be written in terms of the variables p_r and L, by using the identity:

$$p^2 = p_x^2 + p_y^2 + p_z^2$$

$$= \frac{1}{r^2}\left[(xp_x + yp_y + zp_z)^2 + (yp_z - zp_y)^2 + (zp_x - xp_z)^2 + (xp_y - yp_x)^2\right]$$

$$= p_r^2 + \frac{1}{r^2}(L_x^2 + L_y^2 + L_z^2) = p_r^2 + \frac{1}{r^2}L^2$$

Substituting into Eq.(2.17) gives:

$$H = \frac{1}{2m}\hat{p}_r^2 + \frac{1}{2mr^2}\hat{L}^2 + U(r)$$

Comparison with Eq.(2.37) indicates the representation:

$$\hat{p}_r^2 = -\hbar^2 \frac{1}{r^2}\frac{\partial}{\partial r}\left(r^2 \frac{\partial}{\partial r}\right)$$

Since, for any radial function $\Psi(r)$, we may write:

$$\hat{p}_r^2 \Psi(r) = -\hbar^2\left[\frac{\partial^2 \Psi(r)}{\partial r^2} + \frac{2}{r}\frac{\partial \Psi(r)}{\partial r}\right] = -\hbar^2\left(\frac{\partial}{\partial r} + \frac{1}{r}\right)\left[\frac{\partial \Psi(r)}{\partial r} + \frac{\Psi(r)}{r}\right]$$

$$= -\hbar^2\left(\frac{\partial}{\partial r} + \frac{1}{r}\right)\left(\frac{\partial}{\partial r} + \frac{1}{r}\right)\Psi(r)$$

it follows that we have:

$$\hat{p}_r = -i\hbar\left(\frac{\partial}{\partial r} + \frac{1}{r}\right)$$

This shows that the rule of associating an operator of the form $-i\hbar \partial/\partial x_i$ to the linear momentum component conjugated to x_i is not applicable for the pair of conjugated variables r, p_r.

Problem 2.1.3. Use the definitions of the linear and angular momentum operators, Eqs.(2.15) and (2.19), to derive the identity:

$$\hat{p}^2 = \hat{p}_r^2 + \frac{1}{r^2}\hat{L}^2$$

which is similar to the classical relation between the observables.

(Solution):   It is a straightforward matter to show that Eq.(2.19), which in Cartesian coordinates reads:

$$\hat{L}^2 = -\hbar^2\left(y\frac{\partial}{\partial z} - z\frac{\partial}{\partial y}\right)\left(y\frac{\partial}{\partial z} - z\frac{\partial}{\partial y}\right) - \hbar^2\left(z\frac{\partial}{\partial x} - x\frac{\partial}{\partial z}\right)\left(z\frac{\partial}{\partial x} - x\frac{\partial}{\partial z}\right)$$

$$-\hbar^2\left(x\frac{\partial}{\partial y} - y\frac{\partial}{\partial x}\right)\left(x\frac{\partial}{\partial y} - y\frac{\partial}{\partial x}\right)$$

may also be written as:

$$\hat{L}^2 = -\hbar^2 \left[x^2 \left(\frac{\partial^2}{\partial y^2} + \frac{\partial^2}{\partial z^2} \right) + y^2 \left(\frac{\partial^2}{\partial z^2} + \frac{\partial^2}{\partial x^2} \right) + z^2 \left(\frac{\partial^2}{\partial x^2} + \frac{\partial^2}{\partial y^2} \right) \right]$$

$$+ \hbar^2 \left[2xy \frac{\partial^2}{\partial x \partial y} + 2yz \frac{\partial^2}{\partial y \partial z} + 2zx \frac{\partial^2}{\partial z \partial x} + 2x \frac{\partial}{\partial x} + 2y \frac{\partial}{\partial y} + 2z \frac{\partial}{\partial z} \right]$$

If we use the expansion:

$$(\hat{\vec{r}} \cdot \hat{\vec{p}})^2 = -\hbar^2 \left(x \frac{\partial}{\partial x} + y \frac{\partial}{\partial y} + z \frac{\partial}{\partial z} \right) \left(x \frac{\partial}{\partial x} + y \frac{\partial}{\partial y} + z \frac{\partial}{\partial z} \right)$$

$$= -\hbar^2 \left[x^2 \frac{\partial^2}{\partial x^2} + y^2 \frac{\partial^2}{\partial y^2} + z^2 \frac{\partial^2}{\partial z^2} + 2xy \frac{\partial^2}{\partial x \partial y} + 2yz \frac{\partial^2}{\partial y \partial z} + 2zx \frac{\partial^2}{\partial z \partial x} + x \frac{\partial}{\partial x} + y \frac{\partial}{\partial y} + z \frac{\partial}{\partial z} \right]$$

where use has been made of Eq.(2.15), the addition of the last two relations yields:

$$\hat{L}^2 + (\hat{\vec{r}} \cdot \hat{\vec{p}}) = -\hbar^2 (x^2 + y^2 + z^2) \left(\frac{\partial^2}{\partial x^2} + \frac{\partial^2}{\partial y^2} + \frac{\partial^2}{\partial z^2} \right) + \hbar^2 \left(x \frac{\partial}{\partial x} + y \frac{\partial}{\partial y} + z \frac{\partial}{\partial z} \right)$$

$$= r^2 \hat{p}^2 + i\hbar (\hat{\vec{r}} \cdot \hat{\vec{p}})$$

It follows that the linear momentum can be expressed as:

$$\hat{p}^2 = \frac{1}{r^2} \left[\hat{L}^2 + (\hat{\vec{r}} \cdot \hat{\vec{p}})^2 - i\hbar (\hat{\vec{r}} \cdot \hat{\vec{p}}) \right]$$

The radial part of the previous relation can be rewritten by using Eq.(2.34) for $\hat{\vec{p}}$, which gives:

$$\hat{\vec{r}} \cdot \hat{\vec{p}} = -i\hbar r \frac{\partial}{\partial r} = r\hat{p}_r + i\hbar$$

so that:

$$\hat{p}^2 = \frac{1}{r^2} \left[\hat{L}^2 + (\hat{\vec{r}} \cdot \hat{\vec{p}}) r \hat{p}_r \right] = \frac{1}{r^2} \left[\hat{L}^2 + (r\hat{p}_r)(r\hat{p}_r) + i\hbar r \hat{p}_r \right]$$

It is easily verified that:

$$(\hat{p}_r r) \Psi = \hat{p}_r (r\Psi) = -i\hbar \left(\frac{\partial}{\partial r} + \frac{1}{r} \right) (r\Psi)$$

82   Quantum Physics

$$= -i\hbar r \frac{\partial \Psi}{\partial r} - 2i\hbar \Psi = -i\hbar r\left(\frac{\partial}{\partial r} + \frac{1}{r}\right)\Psi - i\hbar \Psi = (r\hat{p}_r - i\hbar)\Psi$$

so that we finally obtain:

$$\hat{p}^2 = \frac{1}{r^2}\left[\hat{L}^2 + (r^2 \hat{p}_r - i\hbar r)\hat{p}_r + i\hbar r\hat{p}_r\right] = \frac{1}{r^2}\hat{L}^2 + \hat{p}_r$$

Problem 2.1.4. When there is symmetry about an axis of a dynamical system, the wave functions are conveniently expressed in terms of cylindrical polar coordinates r, θ, z. Using:

$$x = r\cos\theta, \qquad y = r\sin\theta, \qquad z = z$$

and the coordinate increments:

$$\frac{\partial \vec{r}}{\partial r} = \vec{e}_r, \qquad \frac{\partial \vec{r}}{\partial \theta} = r\vec{e}_\theta, \qquad \frac{\partial \vec{r}}{\partial z} = \vec{e}_z$$

which define the volume element:

$$dV = r\,d\theta\,dr\,dz = \rho(r)\,d\theta\,dr\,dz$$

find the Hamiltonian operator representation in cylindrical polar coordinates.

(Answer): $\quad \hat{H} = -\dfrac{\hbar^2}{2m}\left[\dfrac{1}{r}\dfrac{\partial}{\partial r}\left(r\dfrac{\partial}{\partial r}\right) + \dfrac{1}{r^2}\dfrac{\partial^2}{\partial \theta^2} + \dfrac{\partial^2}{\partial z^2}\right] + U(r, \theta, z)$

Problem 2.1.5. A state of a harmonic oscillator in one-dimension is described by the wave function $\Psi(x) = Ae^{-\alpha x^2}$ where $\alpha = m\omega/2\hbar$. Find the normalization constant and calculate the expectation value of the energy in this state.

(Answer): $\quad A = \left(\dfrac{m\omega}{\pi\hbar}\right)^{\frac{1}{4}}, \quad \langle E \rangle = \dfrac{1}{2}\hbar\omega$

2.2. The Momentum Representation

The form of the operators, given by Postulate 2, derives from the basic definitions of the position and momentum operators in the coordinate representation, where the state function $\Psi(x,t)$ of a dynamical system is so defined that $|\Psi(x,t)|^2 dx$ gives the probability of finding the system at a point between x and $x + dx$ at a time t. We have seen, however, that the wave function $\Psi(x,t)$ can be expressed in terms of E and p_x by Eq.(1.24), so that the normalization condition can be reformulated as:

$$\int_{-\infty}^{\infty} \Psi^*(x,t)\Psi(x,t)dx = \int_{-\infty}^{\infty} \Psi^*(x,t)\left[\frac{1}{\sqrt{2\pi\hbar}}\int_{-\infty}^{\infty}\Phi(p_x,t)e^{ip_x x/\hbar}dp_x\right]dx$$

$$= \int_{-\infty}^{\infty} \Phi^*(p'_x,t)\left[\frac{1}{2\pi}\int_{-\infty}^{\infty} e^{i(p_x-p'_x)x/\hbar}d\left(\frac{x}{\hbar}\right)\right]\Phi(p_x,t)dp'_x dp_x$$

$$= \int_{-\infty}^{\infty}\int_{-\infty}^{\infty} \Phi^*(p'_x,t)\Phi(p_x,t)\delta(p_x - p'_x)dp'_x dp_x$$

$$= \int_{-\infty}^{\infty} \Phi^*(p_x,t)\Phi(p_x,t)dp_x = 1 \qquad (2.38)$$

where use has been made of Eq.(1.43). It is obvious that $\Phi(p_x,t)$ is the Fourier transform of $\Psi(x,t)$, defined as:

$$\Phi(p_x,t) = \frac{1}{\sqrt{2\pi\hbar}}\int_{-\infty}^{\infty}\Psi(x)e^{-i(p_x x - Et)/\hbar}dx = \frac{1}{\sqrt{2\pi\hbar}}\int_{-\infty}^{\infty}\Psi(x,t)e^{-ip_x x/\hbar}dx \qquad (2.39)$$

which describes the momentum distribution in a quantum state. Equation (2.38) suggests the possibility of introducing a probability density expressed in terms of the normalized *momentum wave functions*, Eq.(2.39), which plays, in momentum space, a role analogous to that of $\Psi(x,t)$ in coordinate space. In other words, if we know $\Phi(p_x,t)$ we may calculate the expectation value of an arbitrary physical observable, provided it is represented by an operator written as a function of the variables p_i of the momentum space and their partial derivatives. It is straightforward to reformulate Eq.(2.6) as:

$$\langle p_x \rangle = \int_{-\infty}^{\infty} \Psi^*(x,t)\left(-i\hbar\frac{\partial}{\partial x}\right)\Psi(x,t)dx$$

$$= \int_{-\infty}^{\infty} \Psi^*(x,t) \left[\frac{1}{\sqrt{2\pi\hbar}} \int_{-\infty}^{\infty} \Phi(p_x,t) \left(-i\hbar \frac{\partial}{\partial x} \right) e^{ip_x x/\hbar} dp_x \right] dx$$

$$= \int_{-\infty}^{\infty} \Psi^*(x,t) \left[\frac{1}{\sqrt{2\pi\hbar}} \int_{-\infty}^{\infty} p_x \Phi(p_x,t) e^{ip_x x/\hbar} dp_x \right] dx$$

$$= \int_{-\infty}^{\infty} \int_{-\infty}^{\infty} \Phi^*(p_x',t) p_x \Phi(p_x,t) \left[\frac{1}{2\pi} \int_{-\infty}^{\infty} e^{i(p_x - p_x')x/\hbar} d\left(\frac{x}{\hbar} \right) \right] dp_x' dp_x$$

$$= \int_{-\infty}^{\infty} \int_{-\infty}^{\infty} \Phi^*(p_x',t) p_x \Phi(p_x,t) \delta(p_x - p_x') dp_x' dp_x$$

$$= \int_{-\infty}^{\infty} \Phi^*(p_x,t) p_x \Phi(p_x,t) dp_x \tag{2.40}$$

From Eq.(2.40) it is clear that, in the momentum representation, a natural choice is to associate the observable p_x with the multiplicative operator:

$$\hat{p}_x = p_x \tag{2.41}$$

This rule can be extended to the momentum representation of physical observables, such as kinetic energy T, expressible in terms of powers of p_x, for which we obtain, according to Eq.(2.8):

$$\langle p_x^k \rangle = \int_{-\infty}^{\infty} \Psi^*(x,t) \left(-i\hbar \frac{\partial}{\partial x} \right)^k \Psi(x,t) dx$$

$$= \int_{-\infty}^{\infty} \Psi^*(x,t) \left(-i\hbar \frac{\partial}{\partial x} \right)^k \left[\frac{1}{\sqrt{2\pi\hbar}} \int_{-\infty}^{\infty} \Phi(p_x,t) e^{ip_x x/\hbar} dp_x \right] dx$$

$$= \int_{-\infty}^{\infty} \Psi^*(x,t) \left[\frac{1}{\sqrt{2\pi\hbar}} \int_{-\infty}^{\infty} p_x^k \Phi(p_x,t) e^{ip_x x/\hbar} dp_x \right] dx$$

$$= \int_{-\infty}^{\infty} \int_{-\infty}^{\infty} \Phi^*(p_x',t) p_x^k \Phi(p_x,t) \left[\frac{1}{2\pi} \int_{-\infty}^{\infty} e^{i(p_x - p_x')x/\hbar} d\left(\frac{x}{\hbar} \right) \right] dp_x' dp_x$$

$$= \int_{-\infty}^{\infty} \int_{-\infty}^{\infty} \Phi^*(p'_x,t) p_x^k \Phi(p_x,t) \delta(p_x - p'_x) \, dp'_x dp_x$$

$$= \int_{-\infty}^{\infty} \int_{-\infty}^{\infty} \Phi^*(p_x,t) p_x^k \Phi(p_x,t) \, dp_x \tag{2.42}$$

The rule of representation for momentum, Eq.(2.41), can be generalized, in view of Eq.(2.42), to physical observables classically expressed by arbitrary functions of p_x, since any function of the variable p_x in momentum space, conjugated with the position variable x in coordinate space, may be expanded in a Taylor series in the form given by Eq.(2.9). It follows immediately that we may write:

$$\langle g(p_x) \rangle = \int_{-\infty}^{\infty} \Psi^*(x,t) \hat{g}\left(-i\hbar \frac{\partial}{\partial x}\right) \Psi(x,t) dx = \int_{-\infty}^{\infty} \Phi^*(p_x,t) \hat{g}(p_x) \Phi(p_x,t) \, dp_x \tag{2.43}$$

If we now consider the position observable x, we obtain by combining Eqs.(2.1) and (1.24):

$$\langle x \rangle = \int_{-\infty}^{\infty} \Psi^*(x,t) x \Psi(x,t) dx = \int_{-\infty}^{\infty} \Psi^*(x,t) \left[\frac{1}{\sqrt{2\pi\hbar}} \int_{-\infty}^{\infty} \Phi(p_x,t) x e^{ip_x x/\hbar} dp_x \right] dx$$

$$= \int_{-\infty}^{\infty} \Psi^*(x,t) \left[\frac{1}{\sqrt{2\pi\hbar}} \int_{-\infty}^{\infty} e^{ip_x x/\hbar} \left(i\hbar \frac{\partial}{\partial p_x} \right) \Phi(p_x,t) dp_x \right] dx$$

$$= \int_{-\infty}^{\infty} \int_{-\infty}^{\infty} \Phi^*(p'_x,t) \left(i\hbar \frac{\partial}{\partial p_x} \right) \Phi(p_x,t) \left[\frac{1}{2\pi} \int_{-\infty}^{\infty} e^{i(p_x - p'_x)x/\hbar} d\left(\frac{x}{\hbar}\right) \right] dp'_x dp_x$$

$$= \int_{-\infty}^{\infty} \int_{-\infty}^{\infty} \Phi^*(p'_x,t) \left(i\hbar \frac{\partial}{\partial p_x} \right) \Phi(p_x,t) \delta(p_x - p'_x) \, dp'_x dp_x$$

$$= \int_{-\infty}^{\infty} \Phi^*(p_x,t) \left(i\hbar \frac{\partial}{\partial p_x} \right) \Phi(p_x,t) \, dp_x \tag{2.44}$$

Use has been made of the property:

$$\int_{-\infty}^{\infty} \Phi(p_x,t) x e^{ip_x x/\hbar} dp_x = \int_{-\infty}^{\infty} \Phi(p_x,t) \left(-i\hbar \frac{\partial}{\partial p_x} \right) e^{ip_x x/\hbar} dp_x$$

$$= -i\hbar \left[\Phi(p_x,t) e^{ip_x x/\hbar} \right]_{-\infty}^{\infty} + \int_{-\infty}^{\infty} e^{ip_x x/\hbar} \left(i\hbar \frac{\partial}{\partial p_x} \right) \Phi(p_x,t) dp_x$$

$$= \int_{-\infty}^{\infty} e^{ip_x x/\hbar} \left(i\hbar \frac{\partial}{\partial p_x} \right) \Phi(p_x,t) dp_x \qquad (2.45)$$

where the momentum wave functions are vanishingly small at infinity. From Eq.(2.44) it follows that, in the momentum representation, the position observable x should be associated with the differential operator:

$$\hat{x} = i\hbar \frac{\partial}{\partial p_x} \qquad (2.46)$$

Furthermore, the momentum representation of physical observables which can be classically expressed in terms of powers of the position coordinates, x^k, will be given by operators of the form $(i\hbar \partial/\partial p_x)^k$, since we may successively write, in view of Eq.(2.45):

$$x^k \Psi(x,t) = x^{k-1} \frac{1}{\sqrt{2\pi\hbar}} \int_{-\infty}^{\infty} \Phi(p_x,t) x e^{ip_x x/\hbar} dp_x = x^{k-2} \frac{1}{\sqrt{2\pi\hbar}} \int_{-\infty}^{\infty} i\hbar \frac{\partial \Phi(p_x,t)}{\partial p_x} x e^{ip_x x/\hbar} dp_x$$

$$= x^{k-3} \frac{1}{\sqrt{2\pi\hbar}} \int_{-\infty}^{\infty} (i\hbar)^2 \frac{\partial^2 \Phi(p_x,t)}{\partial p_x^2} x e^{ip_x x/\hbar} dp_x$$

$$= \ldots = \frac{1}{\sqrt{2\pi\hbar}} \int_{-\infty}^{\infty} e^{ip_x x/\hbar} \left(i\hbar \frac{\partial}{\partial p_x} \right)^k \Phi(p_x,t) dp_x$$

Hence we obtain:

$$\langle x^k \rangle = \int_{-\infty}^{\infty} \Psi^*(x,t) x^k \Psi(x,t) dx$$

$$= \int_{-\infty}^{\infty} \int_{-\infty}^{\infty} \Phi^*(p_x',t) \left(i\hbar \frac{\partial}{\partial p_x} \right)^k \Phi(p_x,t) \left[\frac{1}{2\pi} \int_{-\infty}^{\infty} e^{i(p_x - p_x')x/\hbar} d\left(\frac{x}{\hbar}\right) \right] dp_x' dp_x$$

$$= \int_{-\infty}^{\infty} \Phi^*(p_x,t) \left(i\hbar \frac{\partial}{\partial p_x} \right)^k \Phi(p_x,t) dp_x \qquad (2.47)$$

Any function of position $f(x)$, with the Taylor expansion given by Eq.(2.3), will obey, in view of Eq.(2.47):

2. Operators and the Second Postulate

$$\langle f(x) \rangle = \int_{-\infty}^{\infty} \Psi^*(x,t)\hat{f}(x)\Psi(x,t)dx = \int_{-\infty}^{\infty} \Phi(p_x,t)\hat{f}\left(i\hbar\frac{\partial}{\partial p_x}\right)\Phi(p_x,t)dp_x \quad (2.48)$$

Therefore, a physical observable, like the potential energy, expressible as a function of position in the coordinate representation, will be associated in the momentum representation with an operator $\hat{f}(i\hbar\partial/\partial p_x)$. The pair of equations (2.43) and (2.48) proves the equivalence of the coordinate and the momentum representations. The rules of representation in momentum space, given in terms of coordinates x_i and their conjugated momenta p_i as:

$$\hat{x}_i = i\hbar\frac{\partial}{\partial p_i}, \quad \hat{p}_i = p_i \quad (2.49)$$

correspond to the rules established for the coordinate representation by Postulate 2, including the restriction that x_i should be Cartesian coordinates.

The equation of motion of a particle, which has the form (1.25) in the coordinate representation, may be reformulated for momentum space by using first Eq.(1.24) which directly gives:

$$i\hbar\frac{\partial\Psi(x,t)}{\partial t} = \frac{1}{\sqrt{2\pi\hbar}}\int_{-\infty}^{\infty} i\hbar\frac{\partial\Phi(p_x,t)}{\partial t}e^{ip_x x/\hbar}dp_x$$

$$-\frac{\hbar^2}{2m}\frac{\partial^2\Psi(x,t)}{\partial x^2} = \frac{1}{\sqrt{2\pi\hbar}}\int_{-\infty}^{\infty} \Phi(p_x,t)\frac{p_x^2}{2m}e^{ip_x x/\hbar}dp_x \quad (2.50)$$

and then the property:

$$U(x)\Psi(x,t) = \frac{1}{\sqrt{2\pi\hbar}}\int_{-\infty}^{\infty} e^{ip_x x/\hbar}\hat{U}\left(i\hbar\frac{\partial}{\partial p_x}\right)\Phi(p_x,t)dp_x \quad (2.51)$$

which is implied by Eq.(2.45). Substituting Eqs.(2.50) and (2.51) into Eq.(1.25) we obtain the momentum representation of the wave equation, provided E is a constant of motion, in the differential form:

$$i\hbar\frac{\partial\Phi(p_x,t)}{\partial t} = \frac{p_x^2}{2m}\Phi(p_x,t) + \hat{U}\left(i\hbar\frac{\partial}{\partial p_x}\right)\Phi(p_x,t) \quad (2.52)$$

EXAMPLE 2.3. Momentum representation of oscillatory motion

For linear harmonic motion parallel to the x-axis, the total energy E, which is a constant, has the form (P.5), such that the wave equation in the coordinate representation, Eq.(1.25), reads:

$$i\hbar \frac{\partial \Psi(x,t)}{\partial t} = -\frac{\hbar^2}{2m}\frac{d^2\Psi(x,t)}{dx^2} + \frac{m\omega^2 x^2}{2}\Psi(x,t)$$

A solution to this equation is the normalized wave function:

$$\Psi(x,t) = \Psi(x)e^{-iEt/\hbar} = \left(\frac{m\omega}{\pi\hbar}\right)^{\frac{1}{4}} e^{-m\omega x^2/2\hbar} e^{-iEt/\hbar}$$

which describes a state of the harmonic oscillator, with a Gaussian distribution in the x-axis direction, centred about the origin:

$$\langle x \rangle = 0, \quad \langle x^2 \rangle = \left(\frac{m\omega}{\pi\hbar}\right)^{\frac{1}{2}} \int_{-\infty}^{\infty} e^{-m\omega x^2/\hbar} x^2 dx = \frac{1}{2}\frac{\hbar}{m\omega}, \quad |\Psi(x,t)|^2 = |\Psi(x)|^2 = e^{-(m\omega/\hbar)x^2}$$

The function describing the same state in momentum space is the Fourier transform of $\Psi(x)$, which, using Eq.(1.39), can be written as:

$$\Phi(p_x) = \left(\frac{m\omega}{4\pi^3\hbar^3}\right)^{\frac{1}{4}} \int_{-\infty}^{\infty} e^{-m\omega x^2/2\hbar} e^{-ip_x x/\hbar} dx = \left(\frac{m\omega}{4\pi^3\hbar^3}\right)^{\frac{1}{4}} e^{-p_x^2/2m\hbar\omega} \int_{-\infty}^{\infty} e^{-m\omega(x-ip_x/m\omega)^2/2\hbar} dx$$

The integral reduces to a known Gaussian integral by setting $u = x - ip_x/m\omega$, so giving:

$$\Phi(p_x) = \frac{1}{(\pi\hbar m\omega)^{\frac{1}{4}}} e^{-p_x^2/2m\hbar\omega}$$

and, hence, the probability density for the measurement of p_x is:

$$|\Phi(p_x)|^2 = \frac{1}{(\pi\hbar m\omega)^{\frac{1}{2}}} e^{-p_x^2/m\hbar\omega}$$

Hence, for the Gaussian distribution of p_x we obtain:

$$\langle p_x \rangle = 0, \quad \langle p_x^2 \rangle = \frac{1}{(\pi\hbar m\omega)^{\frac{1}{2}}} \int_{-\infty}^{\infty} e^{-p_x^2/m\hbar\omega} p_x^2 dp_x = \frac{1}{2} m\hbar\omega$$

It is a straightforward matter to show, by direct substitution, that the normalized momentum wave function:

$$\Phi(p_x,t) = \frac{1}{(\pi\hbar m\omega)^{\frac{1}{4}}} e^{-p_x^2/2m\hbar\omega} e^{-iEt/\hbar}$$

is a solution of the one-dimensional wave equation in momentum space, Eq.(2.52), which reads:

$$i\hbar \frac{\partial \Phi(p_x,t)}{\partial t} = \frac{p_x^2}{2m}\Phi(p_x,t) - \frac{1}{2}m\omega^2\hbar^2 \frac{\partial^2 \Phi(p_x,t)}{\partial p_x^2}$$

and defines the same energy state, $E = \hbar\omega/2$. ♦

The wave equation in coordinate space, Eq.(1.25), which reads:

$$\left(-\frac{\hbar^2}{2m}\frac{\partial^2}{\partial x^2} - i\hbar\frac{\partial}{\partial t}\right)\Psi(x,t) = -U(x)\Psi(x,t)$$

can also be transformed into an integral equation in momentum space, by taking the Fourier transform of both sides, which gives:

$$\frac{1}{\sqrt{2\pi\hbar}} \int_{-\infty}^{\infty} \left[-\frac{\hbar^2}{2m}\frac{\partial^2 \Psi(x,t)}{\partial x^2}\right] e^{-ip_x x/\hbar} dx + \frac{1}{\sqrt{2\pi\hbar}} \int_{-\infty}^{\infty} \left[-i\hbar \frac{\partial \Psi(x,t)}{\partial t}\right] e^{-ip_x x/\hbar} dx$$

$$= -\frac{1}{\sqrt{2\pi\hbar}} \int_{-\infty}^{\infty} U(x)\Psi(x,t) e^{-ip_x x/\hbar} dx$$

The first integral on the left hand side can be integrated by parts, as follows:

$$\int_{-\infty}^{\infty} \left[-\hbar^2 \frac{\partial^2 \Psi(x,t)}{\partial x^2}\right] e^{-ip_x x/\hbar} dx$$

$$= -\hbar^2 \left[\frac{\partial \Psi(x,t)}{\partial x} e^{-ip_x x/\hbar}\right]_{-\infty}^{\infty} + \int_{-\infty}^{\infty} \left[-i\hbar \frac{\partial \Psi(x,t)}{\partial x}\right]\left(i\hbar\frac{\partial}{\partial x}\right) e^{-ip_x x/\hbar} dx$$

$$= \int_{-\infty}^{\infty} p_x \left[-i\hbar \frac{\partial \Psi(x,t)}{\partial x}\right] e^{-ip_x x/\hbar} dx$$

$$= -i\hbar p_x \left[\Psi(x,t) e^{-ip_x x/\hbar}\right]_{-\infty}^{\infty} + p_x \int_{-\infty}^{\infty} \Psi(x,t)\left(i\hbar\frac{\partial}{\partial x}\right) e^{-ip_x x/\hbar} dx$$

$$= p_x^2 \int_{-\infty}^{\infty} \Psi(x,t) e^{-ip_x x/\hbar} dx = \sqrt{2\pi\hbar}\, p_x^2 \Phi(p_x,t)$$

so that we obtain:

$$\left(\frac{p_x^2}{2m} - i\hbar\frac{\partial}{\partial t}\right)\Phi(p_x,t) + \frac{1}{\sqrt{2\pi\hbar}}\int_{-\infty}^{\infty} U(x)\Psi(x,t)e^{-ip_x x/\hbar}dx = 0 \qquad (2.53)$$

Using Eq.(1.24), the integral can be written in the form:

$$\frac{1}{\sqrt{2\pi\hbar}}\int_{-\infty}^{\infty} U(x)\Psi(x,t)e^{-ip_x x/\hbar}dx = \frac{1}{2\pi\hbar}\int_{-\infty}^{\infty}\int_{-\infty}^{\infty} U(x)\Phi(p'_x,t)e^{i(p'_x-p_x)x/\hbar}dp'_x dx$$

$$= \int_{-\infty}^{\infty} W(p_x - p'_x)\Phi(p'_x,t)dp'_x \qquad (2.54)$$

$W(p_x)$ appears as the Fourier transform of the potential energy, which is defined as:

$$W(p_x) = \frac{1}{2\pi\hbar}\int_{-\infty}^{\infty} U(x)e^{-ip_x x/\hbar}dx \qquad (2.55)$$

such that:

$$U(x) = \int_{-\infty}^{\infty} W(p_x)e^{ip_x x/\hbar}dp_x \qquad (2.56)$$

Substituting Eq.(2.54) into Eq.(2.53) we finally obtain the integral equation in momentum space:

$$\left(\frac{p_x^2}{2m} - i\hbar\frac{\partial}{\partial t}\right)\Phi(p_x,t) + \int_{-\infty}^{\infty} W(p_x - p'_x)\Phi(p'_x,t)dp'_x = 0 \qquad (2.57)$$

which will determine the unknown function $\Phi(p_x,t)$, in the case where the potential energy is a complicated function of coordinates, and therefore it is not convenient to use the differential wave equation in momentum space, Eq.(2.52).

Problem 2.2.1. Using the momentum representation, show that the time dependence of the expectation value of position $\langle x \rangle$ for a free particle can be written as:

$$\langle x \rangle = \langle x \rangle_{t=0} + vt$$

where $v = \langle p_x \rangle_{t=0}/m$.

(Solution): Since the total energy E is a constant of motion, we may write the wave function $\Phi(p_x,t)$, Eq.(2.39), in terms of $\Phi(p_x,0) = \Phi(p_x)$, given by Eq.(1.39), in the form:

$$\Phi(p_x,t) = \Phi(p_x)e^{-iEt/\hbar}$$

so obtaining the expectation value $\langle x \rangle$, Eq.(2.44), as:

$$\langle x \rangle = i\hbar \int_{-\infty}^{\infty} \Phi^*(p_x)e^{iEt/\hbar} \frac{\partial}{\partial p_x}\left[\Phi(p_x)e^{-iEt/\hbar}\right]dp_x$$

$$= i\hbar \int_{-\infty}^{\infty} \Phi^*(p_x)\frac{\partial \Phi(p_x)}{\partial p_x} dp_x + i\hbar \int_{-\infty}^{\infty} \Phi^*(p_x)\left(\frac{t}{i\hbar}\frac{\partial E}{\partial p_x}\right)\Phi(p_x)dp_x$$

$$= i\hbar \int_{-\infty}^{\infty} \Phi^*(p_x,0)\hat{x}\Phi(p_x,0)dp_x + \frac{t}{m}\int_{-\infty}^{\infty} \Phi^*(p_x,0) p_x \Phi(p_x,0)dp_x$$

where the operator $\hat{x}$ has been introduced as defined by Eq.(2.46), and for a free particle in one dimension we have:

$$\frac{\partial E}{\partial p_x} = \frac{p_x}{m}$$

It follows that:

$$\langle x \rangle = \langle x \rangle_{t=0} + \frac{1}{m}\langle p_x \rangle_{t=0}$$

Problem 2.2.2. Calculate the wave function and the energy E, assumed constant, for the motion of a particle of mass m in a potential energy:

$$U(x) = -\eta\delta(x)$$

where η is real positive constant, by solving the integral equation in momentum space, Eq.(2.57).

(Solution): The Fourier transform of the potential energy $U(x)$ is obtained from Eq.(2.55) in the form:

$$W(p_x - p'_x) = \frac{1}{2\pi\hbar} \int_{-\infty}^{\infty} e^{-i(p_x-p'_x)x/\hbar}(-\eta)\delta(x)dx = -\frac{\eta}{2\pi\hbar}$$

so that the integral term of Eq.(2.57) reduces to:

$$\int_{-\infty}^{\infty} W(p_x - p_x')\Phi(p_x',t)dp_x' = -\frac{\eta}{2\pi\hbar}\int_{-\infty}^{\infty}\Phi(p_x,t)dp_x = -\frac{\eta}{\sqrt{2\pi\hbar}}\Psi(0,t)$$

The integral on the right-hand side have been expressed in terms of the wave function $\Psi(0,t)$, written for the position $x=0$ in the coordinate representation, Eq.(1.24):

$$\Psi(0,t) = \frac{1}{\sqrt{2\pi\hbar}}\int_{-\infty}^{\infty}\Phi(p_x,t)dp_x$$

Substituting into Eq.(2.57), yields:

$$\left(\frac{p_x^2}{2m} - E\right)\Phi(p_x,t) - \frac{\eta}{\sqrt{2\pi\hbar}}\Psi(0,t) = 0$$

or:

$$\Phi(p_x,t) = \frac{1}{\sqrt{2\pi\hbar}}\frac{2m\eta}{p_x^2 - 2mE}\Psi(0,t)$$

It follows from Eq.(1.24) that:

$$\Psi(x,t) = \frac{m\eta}{\pi\hbar}\Psi(0,t)\int_{-\infty}^{\infty}\frac{e^{ip_x x/\hbar}}{p_x^2 - 2mE}dp_x$$

The result of integration is given by the formula:

$$\int_{-\infty}^{\infty}\frac{e^{-ip x}}{p^2 + k^2} = \frac{\pi}{k}e^{-k|x|}$$

so that we have:

$$\Psi(x,t) = \frac{m\eta}{\hbar\sqrt{-2mE}}\Psi(0,t)e^{-\sqrt{-2mE}\,|x|/\hbar}$$

It is clear that the energy E is restricted to be:

$$\hbar\sqrt{-2mE} = m\eta \quad \text{or} \quad E = -\frac{m\eta^2}{2\hbar^2}$$

so obtaining:

$$\Psi(x,t) = \Psi(0,t)e^{-m\eta|x|/\hbar^2}$$

Problem 2.2.3. The position of a particle on the x-axis is given by the wave function:

$$\Psi(x) = \frac{1}{\sqrt{\alpha}} e^{-|x|/\alpha}$$

Find the corresponding momentum wave function and calculate $\langle p_x \rangle$, $\langle p_x^2 \rangle$.

(Answer): $\Phi(p_x,t) = \left(\frac{2\alpha\hbar^3}{\pi}\right)^{\frac{1}{2}} \frac{e^{-iEt/\hbar}}{\alpha^2 p_x^2 + \hbar^2}$, $\langle p_x \rangle = 0$, $\langle p_x^2 \rangle = \frac{\hbar^2}{\alpha^2}$

Problem 2.2.4. Find the wave function $\Phi(p_x, t)$ which describes, in momentum space, the state of a harmonic oscillator with the coordinate representation:

$$\Psi(x,t) = 2\left(\frac{\alpha^3}{\pi}\right)^{\frac{1}{4}} x e^{-\alpha x^2/2} e^{-iEt/\hbar}$$

and calculate the energy of the state.

(Answer): $E = \frac{3}{2}\hbar\omega$

2.3. Linear Operators

We have seen that a quantum operator $\hat{A}$, which is associated with a physical observable A, defines a procedure to transform a given state function Ψ into a different state function Λ of the same dynamical system, and this can be formally written as:

$$\Lambda = \hat{A}\Psi \tag{2.58}$$

The operator $\hat{A}$ is said to be linear over the wave function space, if:

$$\hat{A}(a\Psi + a'\Psi') = a\hat{A}\Psi + a'\hat{A}\Psi' \tag{2.59}$$

for every wave functions Ψ, Ψ' and for all real and complex numbers a, a'. In other words, a linear operator defines a correspondence between two linear combinations of wave functions:

$$a\Psi + a'\Psi' \to a\Lambda + a'\Lambda'$$

where use has been made of Eq.(2.58) written for either Ψ, Λ or Ψ', Λ'. Note that operators do not generally have the property expressed by Eq.(2.58). For example, if $\hat{A}$ is defined by $\hat{A}\Psi = e^\Psi$ we have:

$$\hat{A}(a\Psi + a'\Psi') = e^{a\Psi + a'\Psi'}$$

whereas:

$$a\hat{A}\Psi + a'\hat{A}\Psi' = ae^\Psi + a'e^{\Psi'} \neq \hat{A}(a\Psi + a'\Psi')$$

The most simple linear operators consist of complex numbers a, for which Eq.(2.58) reduces to $\Lambda = a\Psi$ and we will often encounter two particular cases, namely the *identity operator*, $a = 1$, which describes multiplication of a wave function by unity and is defined by:

$$\Psi = \hat{I}\Psi \qquad (2.60)$$

and the *zero operator*, $a = 0$, which is defined by:

$$0 = \hat{0}\Psi \qquad (2.61)$$

for all Ψ in the wave function space.

If two linear operators $\hat{A}$ and $\hat{B}$ are such that, for every function Ψ on which both can operate, we always have:

$$\hat{A}\Psi = \hat{B}\Psi \qquad (2.62)$$

then $\hat{A}$ and $\hat{B}$ are said to be equal to each other, and it is common practice to adopt the notation of algebra:

$$\hat{A} = \hat{B} \qquad (2.63)$$

Such a relation is called an operator equation.

If the respective results of operating on Ψ with two linear operators $\hat{A}$ and $\hat{B}$ are added together, then the final result $\hat{A}\Psi + \hat{B}\Psi$ is a function Λ linearly dependent on Ψ, and hence a linear operator $\hat{C}$ always exists, such that $\Lambda = \hat{C}\Psi$. The sum of the two operators is defined by:

$$\hat{C} = \hat{A} + \hat{B} \qquad (2.64)$$

so that we have:

$$\hat{C}\Psi = (\hat{A} + \hat{B})\Psi = \hat{A}\Psi + \hat{B}\Psi \tag{2.65}$$

This notation can be extended to sum the results of operating on the same function with more than two operators, and it is a straightforward matter to prove the properties of commutativity:

$$\hat{A} + \hat{B} = \hat{B} + \hat{A} \tag{2.66}$$

and associativity:

$$\hat{A} + \hat{B} + \hat{C} = (\hat{A} + \hat{B}) + \hat{C} = \hat{A} + (\hat{B} + \hat{C}) \tag{2.67}$$

It follows that we may add equal operators to both sides of an operator equation without destroying the equality.

If we first operate on Ψ with a linear operator $\hat{B}$, getting the result $\hat{B}\Psi$, which is a linearly dependent function of Ψ, and next operate on this result with the operator $\hat{A}$, then the final result $\hat{A}(\hat{B}\Psi)$ of the two successive operations is again a linearly dependent function of Ψ, say Λ. Therefore, a linear operator $\hat{C}$ should exist, such that $\Lambda = \hat{C}\Psi$, and we will define the product:

$$\hat{C} = \hat{A}\hat{B} \tag{2.68}$$

provided that:

$$\hat{C}\Psi = \hat{A}\hat{B}\Psi = \hat{A}(\hat{B}\Psi) \tag{2.69}$$

The properties of associativity:

$$\hat{A}(\hat{B}\hat{C}) = (\hat{A}\hat{B})\hat{C} = \hat{A}\hat{B}\hat{C} \tag{2.70}$$

and distributivity with respect to addition:

$$(\hat{A} + \hat{B})\hat{C} = \hat{A}\hat{C} + \hat{B}\hat{C} \quad \text{or} \quad \hat{A}(\hat{B} + \hat{C}) = \hat{A}\hat{B} + \hat{A}\hat{C} \tag{2.71}$$

are always valid. The square of an operator $\hat{A}$ is $\hat{A}^2 = \hat{A}\hat{A}$ and similarly, for $n \geq 2$:

$$\hat{A}^n\Psi = \hat{A}(\hat{A}^{n-1}\Psi) = \hat{A}^{n-1}(\hat{A}\Psi) \quad \text{or} \quad \hat{A}^n = \hat{A}\hat{A}^{n-1} = \hat{A}^{n-1}\hat{A} \tag{2.72}$$

As complex numbers are particular cases of linear operators, we may consider the products $a\hat{A}$ and $\hat{A}a$, which are related by Eq.(2.59) in the form:

$$\hat{A}(a\Psi) = a(\hat{A}\Psi)$$

It follows that we always have:

$$(\hat{A}a)\Psi = (a\hat{A})\Psi \quad \text{or} \quad \hat{A}a = a\hat{A} \tag{2.73}$$

which means that the two products define equal operators. We say, according to Eq.(2.73), that complex numbers and linear operators always commute.

It is easily verified that the multiplicative and differential operators associated with the physical observables of principal interest, which are position, linear momentum, angular momentum and energy, all satisfy the previously defined properties of linear operators.

EXAMPLE 2.4. Integral representation of operators

Because integration over the wave function space also has the properties required for linear operators, many quantum operators are often described in an integral representation.

Consider the familiar case of the Fourier transform procedure for deriving momentum wave functions $\Phi(p_x)$ from coordinate wave functions $\Psi(x)$. We may reformulate Eq.(1.39) in terms of an operator $\hat{F}$, in the form:

$$\Phi = \hat{F}\Psi \tag{2.74}$$

where the operator $\hat{F}$ is defined by:

$$\Phi(p_x) = \int_{-\infty}^{\infty} \hat{F}(x, p_x)\Psi(x)dx \tag{2.75}$$

and comparison between Eqs.(2.75) and (1.39) gives the integral representation of $\hat{F}$ as:

$$\hat{F}(x, p_x) = \frac{1}{\sqrt{2\pi\hbar}} e^{-ip_x x/\hbar} \tag{2.76}$$

Similarly, the representation of the Dirac δ-function, Eq.(1.45), which reads:

$$\Psi(x) = \int_{-\infty}^{\infty} \delta(x-x')\Psi(x')dx' \tag{2.77}$$

can be reformulated in terms of the identity operator, Eq.(2.60), if the integral representation of the identity operator $\hat{I}$ is written in the form:

2. Operators and the Second Postulate

$$\Psi(x) = \int_{-\infty}^{\infty} \hat{I}(x, x')\Psi(x')dx' \qquad (2.78)$$

By comparing Eqs.(2.78) and (2.77), it is clear that:

$$\hat{I}(x, x') = \delta(x - x') \qquad (2.79)$$

and it is apparent that the integral representation of the identity operator is given by the Dirac δ-function. ♌

Although, from the properties formulated so far, there appears to be a complete similarity between the operator and the number algebra, there are two properties which make a distinction.

(i) *Non-commuting operators.* We say that two operators $\hat{A}$ and $\hat{B}$ commute when they satisfy:

$$\hat{A}\hat{B} = \hat{B}\hat{A} \qquad (2.80)$$

but often we will find that:

$$\hat{A}\hat{B} \neq \hat{B}\hat{A} \qquad (2.81)$$

A simple example is provided by the operators associated, by Postulate 2, with position and linear momentum in one dimension. For any wave function $\Psi(x_i)$ we have:

$$\frac{d}{dx_i}(x_i \Psi) = \Psi + x_i \frac{d\Psi}{dx_i}$$

so that:

$$\hat{p}_i(\hat{x}_i \Psi) = -i\hbar \Psi + \hat{x}_i(\hat{p}_i \Psi) \qquad \text{or} \qquad \hat{p}_i \hat{x}_i = \hat{x}_i \hat{p}_i - i\hbar$$

Similarly we have:

$$\frac{d}{dx_j}(x_i \Psi) = x_i \frac{d\Psi}{dx_j} \qquad \text{or} \qquad \hat{p}_j \hat{x}_i = \hat{x}_i \hat{p}_j$$

The two results can be written in the more compact form:

$$\hat{p}_j \hat{x}_i = \hat{x}_i \hat{p}_j - i\hbar \delta_{ij} \qquad (2.82)$$

EXAMPLE 2.5. The rule of symmetrization in x_i and p_i

It follows from Eq.(2.82) that there is an ambiguity in transcribing a classical function of a physical observable $A(x_i, p_i, t)$, into operator form, as established by Postulate 2. A simple rule of thumb to avoid this difficulty is to ensure that the terms of $A(x_i, p_i, t)$, containing products of functions of x_i and functions of p_i, be symmetrized in x_i and p_i. For example:

$$x_i p_i \to \frac{1}{2}(\hat{x}_i \hat{p}_i + \hat{p}_i \hat{x}_i)$$

$$x_i f(p_i) \to \frac{1}{2}\left[\hat{x}_i f(\hat{p}_i) + f(\hat{p}_i)\hat{x}_i\right] \tag{2.83}$$

$$p_i f(x_i) \to \frac{1}{2}\left[\hat{p}_i f(\hat{x}_i) + f(\hat{x}_i)\hat{p}_i\right]$$

Symmetrization implies that the quantum operators formed in this way are all Hermitian.

If the restriction imposed by this rule is applied to the classical functions expressed in terms of the curvilinear coordinates q_i and their conjugated momenta $A(q_i, p_i, t)$, it provides us with a method of deriving the form of the operator associated with the radial component of linear momentum. This can be obtained if we symmetrize the classical expression:

$$p_r = \vec{e}_r \cdot \vec{p} = \frac{\vec{r}}{r} \cdot \vec{p} \quad \to \quad \hat{p}_r = -\frac{1}{2}i\hbar\left(\nabla \cdot \frac{\vec{r}}{r} + \frac{\vec{r}}{r} \cdot \nabla\right)$$

where Eq.(2.15) has been used for $\vec{p}$. The standard form, derived before for $\hat{p}_r$, results from applying this operator to a wave function $\Psi(\vec{r})$:

$$\hat{p}_r \Psi(\vec{r}) = -\frac{i\hbar}{2}\nabla \cdot \left[\frac{\vec{r}}{r}\Psi(\vec{r})\right] - \frac{i\hbar}{2}\frac{\vec{r}}{r} \cdot \nabla\Psi(\vec{r}) = -i\hbar\frac{\vec{r}}{r} \cdot \nabla\Psi(\vec{r}) - \frac{i\hbar}{2}\frac{\Psi(\vec{r})}{r}\nabla \cdot \vec{r} - \frac{i\hbar}{2}\Psi(\vec{r})\vec{r} \cdot \nabla\left(\frac{1}{r}\right)$$

where it is easily verified that:

$$\frac{\vec{r}}{r} \cdot \nabla = \frac{\partial}{\partial r}, \quad \nabla \cdot \vec{r} = 3, \quad \nabla\left(\frac{1}{r}\right) = -\frac{\vec{r}}{r^3}$$

Substituting these results, we obtain:

$$\hat{p}_r \Psi(\vec{r}) = -i\hbar\left(\frac{\partial}{\partial r} + \frac{1}{r}\right)\Psi(\vec{r}) \quad \text{or} \quad \hat{p}_r = -i\hbar\left(\frac{\partial}{\partial r} + \frac{1}{r}\right)$$

This expression can be rewritten in terms of the function:

$$\rho(r, \theta) = \frac{dV}{dr\, d\theta\, d\varphi} = r^2 \sin\theta$$

2. Operators and the Second Postulate

where V is the volume element in spherical polar coordinates, by replacing $r = 2\rho/(\partial\rho/\partial r)$ so that:

$$\hat{p}_r = -i\hbar\left(\frac{\partial}{\partial r} + \frac{1}{2\rho}\frac{\partial\rho}{\partial r}\right)$$

In the same way, any operator representing a linear momentum component conjugated with a curvilinear coordinate q_i is obtained in the form:

$$\hat{p}_i = -i\hbar\left(\frac{\partial}{\partial q_i} + \frac{1}{2\rho}\frac{\partial\rho}{\partial q_i}\right) \quad (2.84)$$

For example, the linear momentum components conjugated with θ and φ are:

$$\hat{p}_\theta = -i\hbar\left(\frac{\partial}{\partial \theta} + \frac{1}{2}\cot\theta\right), \quad \hat{p}_\varphi = -i\hbar\frac{\partial}{\partial \varphi}$$

respectively. ⚜

From Eqs.(2.60) it follows that, if $\hat{A}$ and $\hat{B}$ are equal, then:

$$\hat{C}\hat{A} = \hat{C}\hat{B} \quad \text{and} \quad \hat{A}\hat{C} = \hat{B}\hat{C}$$

but $\hat{A}\hat{C}$ does not necessarily be equal to $\hat{C}\hat{B}$, nor does $\hat{B}\hat{C}$ necessarily be equal to $\hat{C}\hat{A}$. This illustrates the importance of observing the order in which the individual operations should be carried out, when the conventional notation is used for successive operation. It can be shown that the reciprocal of the property described by Eq.(2.73) for the product of numbers and linear operators is always valid: if a linear operator commutes with any other linear operator, it only can be a real or a complex number.

(ii) *Singular operators.* If two operators $\hat{A}$ and $\hat{B}$ are related by the equation:

$$\hat{A}\hat{B} = \hat{B}\hat{A} = \hat{I} \quad (2.85)$$

we say that they are reciprocal to each other and we write:

$$\hat{A}^{-1} = \hat{B}, \quad \hat{B}^{-1} = \hat{A} \quad (2.86)$$

An operator for which a reciprocal exists is called a *non-singular* operator and it can be inverted, such that:

$$\hat{A}\hat{A}^{-1} = \hat{I} \quad \text{or} \quad \hat{A}^{-1}\hat{A} = \hat{I} \quad (2.87)$$

100 Quantum Physics

The inversion is possible for every function Ψ, for if Eq.(2.58) holds, and $\hat{A}$ has a reciprocal, then Ψ can be reconstructed by means of $\hat{A}^{-1}$:

$$\Psi = \hat{A}^{-1}\Lambda$$

If there is some non-zero Ψ for which $\hat{A}\Psi = 0$, it is obvious that $\hat{A}$ has no reciprocal and then it is called a *singular* operator. However, while there is a reciprocal for all numbers, except zero, we often find non-zero singular operators. If $\hat{A}$ or $\hat{B}$ are two such operators, the equation:

$$\hat{A}\hat{B} = 0 \tag{2.88}$$

does not necessarily imply that either $\hat{A} = 0$ or $\hat{B} = 0$. Consider, for example, that $\hat{A}$ is the differential operator d/dx and $\hat{B}$ is an integral operator defined by:

$$\Lambda = \int_{-\infty}^{\infty} \Psi(x)dx = \hat{B}\Psi$$

where Λ is linearly dependent on Ψ and it is obviously a constant function. The product of the two operators gives the result expressed by Eq.(2.88), namely:

$$\hat{A}\hat{B}\Psi = \frac{d}{dx}(\hat{B}\Psi) = \frac{d\Lambda}{dx} = 0$$

In other words the product $\hat{A}\hat{B}$ is zero, and thus it is a singular operator, although each factor is a non-zero operator. It is easily verified that, if the product of two operators is non-singular, then each operator has a reciprocal.

Problem 2.3.1. Find an integral representation for the differential operator $\hat{D} = \partial/\partial x$ in terms of the Dirac δ-function.

(Solution): Consider a wave function $\Psi(x)$, which is written in terms of the δ-function as:

$$\Psi(x) = \int_{-\infty}^{\infty} \delta(x-x')\Psi(x')dx'$$

We have:

2. Operators and the Second Postulate   101

$$\hat{D}\Psi(x) = \frac{\partial \Psi(x)}{\partial x} = \int_{-\infty}^{\infty} \delta(x-x') \frac{\partial \Psi(x')}{\partial x'} dx'$$

and integrating by parts yields:

$$\hat{D}\Psi(x) = [\delta(x-x')\Psi(x')]_{-\infty}^{\infty} - \int_{-\infty}^{\infty} \left[\frac{\partial}{\partial x'}\delta(x-x')\right]\Psi(x')dx'$$

$$= \int_{-\infty}^{\infty} \left[\frac{\partial}{\partial x}\delta(x-x')\right]\Psi(x')dx'$$

where $\Psi(x')$ was assumed vanishingly small for $x \to \infty$. This equation should be compared with the integral representation, defined by:

$$\hat{D}\Psi(x) = \int_{-\infty}^{\infty} D(x,x')\Psi(x')dx'$$

to find that the integral representation of the operator $\hat{D}$ is the derivative of the Dirac δ-function:

$$D(x,x') = \frac{\partial}{\partial x}\delta(x-x')$$

Problem 2.3.2. Consider the motion of a particle of mass m, charge e and velocity $\vec{v}$ in an electromagnetic field, described by the scalar potential V and the vector potential $\vec{A}$. Find the Hamiltonian operator associated with the classical expression of the total energy:

$$H = \frac{1}{2m}(\vec{p} + e\vec{A})^2 - eV$$

in the coordinate representation.

(Solution): According to the requirement of symmetrization in q_i and p_i, if the classical Hamiltonian contains products of the form $p_i f(x_i)$ or $x_i f(p_i)$, these must be replaced by symmetrical expressions, which classically are equivalent, so that:

$$H[x_i, p_i, p_i f(x_i)] \to \hat{H}\left\{\hat{x}_i, \hat{p}_i, \frac{1}{2}[\hat{p}_i f(\hat{x}_i) + f(\hat{x}_i)\hat{p}_i]\right\}$$

In our case, we find the product:

$$(\vec{p} + e\vec{A})^2 = (p^2 + 2e\vec{p}\cdot\vec{A} + e^2 A^2)$$

where the product $\bar{p}\cdot\bar{A}$ has to be symmetrized, such that:

$$(\bar{p}+e\bar{A})^2 = p^2 + e(\bar{p}\cdot\bar{A}+\bar{A}\cdot\bar{p}) + e^2 A^2$$

As a result, the Hamiltonian operator reads:

$$\hat{H} = \frac{1}{2m}\hat{p}^2 + \frac{e}{2m}(\hat{\bar{p}}\cdot\bar{A}+\bar{A}\cdot\hat{\bar{p}}) + \frac{e^2}{2m}A^2 - eV$$

In the coordinate representation, with $\hat{\bar{p}} = -i\hbar\nabla$, we obtain:

$$\hat{H} = -\frac{\hbar^2}{2m}\nabla^2 - \frac{ie\hbar}{m}\bar{A}\cdot\nabla - \frac{ie\hbar}{2m}\nabla\cdot\bar{A} + \frac{e^2}{2m}A^2 - eV$$

where use has been made of the property:

$$\nabla\cdot(\bar{A}\Psi) = \bar{A}\cdot\nabla\Psi + (\nabla\cdot\bar{A})\Psi$$

Problem 2.3.3. Given two observables A and B that are represented by the operators $\hat{A}$ and $\hat{B}$, where $\hat{A}$ is assumed to be non-singular, show that the operator function $(\hat{A}-\gamma\hat{B})^{-1}$ can be expanded in powers of γ in a similar manner to the algebraic function $(A-\gamma B)^{-1}$, namely:

$$(A-\gamma B)^{-1} = A^{-1}\left(1-\gamma\frac{B}{A}\right)^{-1} = A^{-1}\sum_{k=0}^{\infty}\left(\frac{B}{A}\right)^k \gamma^k$$

(Solution): Consider the power expansion:

$$(\hat{A}-\gamma\hat{B})^{-1} = \sum_{k=0}^{\infty}\hat{C}_k\gamma^k$$

where $\hat{C}_k$ are operators to be determined. Multiplication of both sides by $\hat{A}-\gamma\hat{B}$ gives:

$$1 = (\hat{A}-\gamma\hat{B})\sum_{k=0}^{\infty}\hat{C}_k\gamma^k = (\hat{A}-\gamma\hat{B})\hat{C}_0 + \sum_{k=1}^{\infty}\hat{A}\hat{C}_k\gamma^k - \sum_{k=1}^{\infty}\hat{B}\hat{C}_k\gamma^{k+1}$$

$$= \hat{A}\hat{C}_0 + \sum_{k=0}^{\infty}\hat{A}\hat{C}_{k+1}\gamma^{k+1} - \sum_{k=0}^{\infty}\hat{B}\hat{C}_k\gamma^{k+1} = \hat{A}\hat{C}_0 + \sum_{k=0}^{\infty}(\hat{A}\hat{C}_{k+1}-\hat{B}\hat{C}_k)\gamma^{k+1}$$

2. Operators and the Second Postulate   103

This identity holds if and only if the coefficients of all powers of γ are zero, which yields:

$$\hat{C}_{k+1} = \hat{A}^{-1}\hat{B}\hat{C}_k$$

and also $\hat{C}_0 = \hat{A}^{-1}$. Substituting the successive expressions:

$$\hat{C}_1 = \hat{A}^{-1}\hat{B}\hat{A}^{-1}, \quad \hat{C}_2 = \hat{A}^{-1}\hat{B}\hat{A}^{-1}\hat{B}\hat{A}^{-1} = \hat{A}^{-1}(\hat{B}\hat{A}^{-1})^2, \ldots$$

into the power expansion, gives:

$$(\hat{A} - \gamma\hat{B})^{-1} = \hat{A}^{-1}\sum_{k=0}^{\infty}(\hat{B}\hat{A}^{-1})^k \gamma^k$$

which is analogous to the algebraic power expansion.

Problem 2.3.4. Show that the operator $\hat{A} = (x + d/dx)^2$ is a linear operator.

Problem 2.3.5. Show that the equation:

$$\hat{S}(\lambda)\Psi(x) = \Psi(x+\lambda)$$

defines a linear operator $\hat{S}(\lambda)$. Prove that:

$$\hat{S}(\lambda)\hat{S}(x) = \hat{S}(\lambda + \chi)$$

and that a reciprocal operator $\hat{S}^{-1}(\lambda)$ always exists, such that:

$$\hat{S}^{-1}(\lambda) = \hat{S}(-\lambda)$$

Problem 2.3.6. Calculate the linear momentum operators $\hat{p}_r, \hat{p}_\theta, \hat{p}_z$ corresponding to the cylindrical polar coordinates r, θ, z.

(Answer): $\quad \hat{p}_r = -i\hbar\left(\dfrac{\partial}{\partial r} + \dfrac{1}{2r}\right), \quad \hat{p}_\theta = -i\hbar\dfrac{\partial}{\partial \theta}, \quad \hat{p}_z = -i\hbar\dfrac{\partial}{\partial z}$

2.4. Hermitian Operators

For any Ψ, Λ and Φ which belong to the same wave function space, we may consider that the scalar product, which is linearly dependent on Φ^* and Ψ, can be written as:

$$\langle\Phi|\Lambda\rangle = \langle\Phi|\hat{A}\Psi\rangle = \langle\Phi|\hat{A}|\Psi\rangle = \int_{-\infty}^{\infty} \Phi^*(x,t)\hat{A}\Psi(x,t)dx \qquad (2.89)$$

where Λ is given by Eq.(2.58). If we assume that the representation of the scalar product in terms of Φ^* and Ψ, as given by Eq.(2.89), is unique, it follows that the complex conjugate $\langle\Lambda|\Phi\rangle$ should be linearly dependent of Ψ^* and Φ, and hence, a linear operator $\hat{B}$ should exist such that:

$$\langle\Lambda|\Phi\rangle = \left(\langle\Phi|\hat{A}\Psi\rangle\right)^* = \int_{-\infty}^{\infty} \Psi^*(x,t)\hat{B}\Phi(x,t)dx \qquad (2.90)$$

The correspondence between the two operators makes it possible to define $\hat{B}$ as the *adjoint* of $\hat{A}$, denoted for convenience by $\hat{B} = \hat{A}^+$. By combining Eqs.(2.90) and (2.89) we may call $\hat{A}^+$ the adjoint of $\hat{A}$, if and only if:

$$\int_{-\infty}^{\infty} \Psi^*(x,t)\left[\hat{A}^+\Phi(x,t)\right]dx = \int_{-\infty}^{\infty} \left[\hat{A}\Psi(x,t)\right]^*\Phi(x,t)dx \qquad (2.91)$$

which, in the Dirac notation, reads:

$$\langle\Psi|A^+\Phi\rangle = \langle\hat{A}\Psi|\Phi\rangle \qquad (2.92)$$

Taking the complex conjugate of both constituents of Eq.(2.92) gives:

$$\int_{-\infty}^{\infty}\left[\hat{A}^+\Phi(x,t)\right]^*\Psi(x,t)dx = \int_{-\infty}^{\infty}\Phi^*(x,t)\left[(\hat{A}^+)^+\Psi(x,t)\right]dx = \int_{-\infty}^{\infty}\Phi^*(x,t)\hat{A}\Psi(x,t)dx \quad (2.93)$$

Since the last two integrals are identical for any two wave functions $\Phi(x,t)$ and $\Psi(x,t)$, we conclude that:

$$(\hat{A}^+)^+ = \hat{A} \qquad (2.94)$$

In the particular case of a complex number a, which may replace the operator in Eq.(2.93), we obtain:

2. Operators and the Second Postulate 105

$$a^+ = a^* \qquad (2.95)$$

It is easily verified from Eq.(2.91) that:

$$(\hat{A} + \hat{B})^+ = \hat{A}^+ + \hat{B}^+ \qquad (2.96)$$

We may also use Eq.(2.91) for a product of two operators, which successively gives:

$$\int_{-\infty}^{\infty} \Psi^*(x,t)\left[(\hat{A}\hat{B})^+ \Phi(x,t)\right]dx = \int_{-\infty}^{\infty} \left[(\hat{A}\hat{B})\Psi(x,t)\right]^* \Phi(x,t)dx$$

$$= \int_{-\infty}^{\infty} \left[\hat{A}\left(\hat{B}\Psi(x,t)\right)\right]^* \Phi(x,t)dx = \int_{-\infty}^{\infty} \left[\hat{B}\Psi(x,t)\right]^* \hat{A}^+ \Phi(x,t)dx$$

$$= \int_{-\infty}^{\infty} \Psi^*(x,t)\left[\hat{B}^+\left(\hat{A}^+\Phi(x,t)\right)\right]dx = \int_{-\infty}^{\infty} \Psi^*(x,t)\left[(\hat{B}^+\hat{A}^+)\Phi(x,t)\right]dx$$

which means that:

$$(\hat{A}\hat{B})^+ = \hat{B}^+\hat{A}^+ \qquad (2.97)$$

We have seen that Postulate 2 imposes the condition of being Hermitian on every linear operator $\hat{A}$ which may be associated with a physical observable A, Eq.(2.13). The Hermitian operators can be also defined as self-adjoint operators, which are equal to their adjoints:

$$\hat{A}^+ = \hat{A} \qquad (2.98)$$

It follows from Eqs.(2.91) and (2.92) that a linear operator is Hermitian over the wave function space if and only if:

$$\int_{-\infty}^{\infty} \Psi^* \hat{A}\Phi dx = \int_{-\infty}^{\infty} (\hat{A}\Psi)^* \Phi dx \qquad (2.99)$$

or:

$$\langle \Psi | \hat{A}\Phi \rangle = \langle \hat{A}\Psi | \Phi \rangle \qquad (2.100)$$

We have shown that the linear operators associated with position, linear momentum, angular momentum and energy all satisfy Eq.(2.13), and it is clear that Eq.(2.100) is reduced to Eq.(2.13) when $\Phi = \Psi$. This implies that these quantum operators will also satisfy Eq.(2.100), since we may write Eq.(2.13) for the linear combination $\Psi + a\Phi$, which reads:

$$\langle \Psi + a\Phi | \hat{A}\Psi + a\hat{A}\Phi \rangle = \langle \hat{A}\Psi + a\hat{A}\Phi | \Psi + a\Phi \rangle \qquad (2.101)$$

In view of Eq.(2.13), Eq.(2.101) reduces to the form:

$$a\left(\langle \Psi | \hat{A}\Phi \rangle - \langle \hat{A}\Psi | \Phi \rangle\right) = a^*\left(\langle \Psi | \hat{A}\Phi \rangle - \langle \hat{A}\Psi | \Phi \rangle\right)$$

which shows that the product of the complex numbers a and $\langle \Psi | \hat{A}\Phi \rangle - \langle \hat{A}\Psi | \Phi \rangle$ is real. This can be valid for any values of a, as assumed by Eq.(2.101), if and only if $\langle \Psi | \hat{A}\Phi \rangle - \langle \hat{A}\Psi | \Phi \rangle$ is zero, which is the same as Eq.(2.100). It follows that a linear operator $\hat{A}$ cannot represent a physical observable unless it obeys the restrictive condition of Eq.(2.100) for any states Ψ and Φ which belong to the wave function space.

The properties of a Hermitian operator derive from Eq.(2.98) which defines it to be self-adjoint. By combining Eqs.(2.95) and (2.98) we obtain:

$$(a\hat{A})^+ = a^* \hat{A} \qquad (2.102)$$

which shows that the product of a Hermitian operator and a number is Hermitian, provided the number is real. Equation (2.96) implies, in view of Eq.(2.98), that the sum of two Hermitian operators is also a Hermitian operator:

$$(\hat{A} + \hat{B})^+ = \hat{A} + \hat{B} \qquad (2.103)$$

Similarly, Eq.(2.97) reduces to:

$$(\hat{A}\hat{B})^+ = \hat{B}\hat{A} \qquad (2.104)$$

which means that the product of two Hermitian operators is a Hermitian operator if and only if the two operators commute.

It is worth noting that, in the Dirac notation, we have:

$$|\hat{A}\Psi\rangle = \hat{A}|\Psi\rangle, \quad \langle \hat{A}\Psi | = \langle \Psi | \hat{A}^+ \qquad (2.105)$$

so that the definition of a Hermitian operator, Eq.(2.100), reads:

$$\langle \Psi | \hat{A} | \Phi \rangle = \langle \Psi | \hat{A}^+ | \Phi \rangle \qquad (2.106)$$

which is the same as Eq.(2.98).

EXAMPLE 2.6. Anti-Hermitian operators

A linear operator $\hat{A}$ is said to be anti-Hermitian over the wave function space if and only if:

$$\int_{-\infty}^{\infty} \Psi^*(x,t)\hat{A}\Phi(x,t)dx = -\int_{-\infty}^{\infty} \left[\hat{A}\Psi(x,t)\right]^* \Phi(x,t)dx$$

or:

$$\langle\Psi|\hat{A}\Phi\rangle = -\langle\hat{A}\Psi|\Phi\rangle \qquad (2.107)$$

For $\Psi = \Phi$, Eq.(2.107) reduces to:

$$\langle\hat{A}\rangle = \langle\Psi|\hat{A}\Psi\rangle = -\langle\hat{A}\Psi|\Psi\rangle = -\langle\hat{A}\rangle^*$$

showing that the expectation value of an anti-Hermitian operator is always a pure imaginary number. This implies that we simply obtain an anti-Hermitian multiplying by i a Hermitian operator, with a real expectation value.

It follows immediately that any linear operator $\hat{A}$ can be separated into Hermitian and anti-Hermitian parts:

$$\hat{A} = \hat{B} + i\hat{C} \qquad (2.108)$$

where both $\hat{B}$ and $\hat{C}$ are assumed to be Hermitian operators. We then have for every Ψ and Φ:

$$\langle\Psi|(\hat{B}+i\hat{C})\Phi\rangle = \langle\Psi|(\hat{B}+i\hat{C})|\Phi\rangle = \langle(\hat{B}+i\hat{C})^+\Psi|\Phi\rangle = \langle(\hat{B}-i\hat{C})\Psi|\Phi\rangle$$

where use has been made of Eq.(2.105). In view of Eqs.(2.92), it follows that the adjoint of $\hat{A}$ over the wave function space is:

$$\hat{A}^+ = \hat{B} - i\hat{C} \qquad (2.109)$$

From Eqs.(2.108) and (2.109) we have:

$$\hat{B} = \frac{1}{2}(\hat{A}+\hat{A}^+), \qquad \hat{C} = \frac{1}{2i}(\hat{A}-\hat{A}^+)$$

so that the Hermitian and anti-Hermitian parts of any linear operator can be always separated in the form:

$$\hat{A} = \frac{1}{2}(\hat{A}+\hat{A}^+) + \frac{1}{2}(\hat{A}-\hat{A}^+) \qquad (2.110)$$

This shows that Hermitian and anti-Hermitian operators play, with respect to linear operators, a role analogous to that of real and imaginary numbers with respect to complex numbers. ☙

EXAMPLE 2.7. Unitary operators

A linear operator $\hat{U}$ is called *unitary* if:

$$\hat{U}\hat{U}^+ = \hat{I} \tag{2.111}$$

Comparison with Eq.(2.85) shows that $\hat{U}$ and $\hat{U}^+$ are reciprocal to each other, and it follows that:

$$\hat{U}^+\hat{U} = \hat{I} \tag{2.112}$$

The product of two unitary operators is also unitary, because:

$$(\hat{U}\hat{V})^+ = \hat{V}^+\hat{U}^+ = \hat{V}^{-1}\hat{U}^{-1} = (\hat{U}\hat{V})^{-1}$$

where Eq.(2.97) has been used. It follows that the product $\hat{U}\hat{V}$ obeys the definition (2.111).

Unitary operators preserve the scalar product between the state vectors from the wave function space, since assuming:

$$\hat{U}\Psi = \Lambda$$

$$\hat{U}\Phi = \Theta$$

we successively have:

$$\langle \Psi | \Phi \rangle = \langle \Psi | \hat{I} | \Phi \rangle = \langle \Psi | \hat{U}^+\hat{U} | \Phi \rangle = \langle \hat{U}\Psi | \hat{U}\Phi \rangle = \langle \Lambda | \Theta \rangle \tag{2.113}$$

By taking $\Psi = \Phi$, it is obvious that the norm of the state functions is not changed under an unitary operation.

This suggest an analogy between unitary operations and the familiar rotations in the three dimensional space, which also preserve the length of vectors and their scalar products. ✤

Problem 2.4.1. Show that the parity operator $\hat{\Pi}$, defined by:

$$\hat{\Pi}\Psi(x,t) = \Psi(-x,t)$$

is Hermitian. Prove that $\hat{\Pi}$ is a unitary operator, that is $\hat{\Pi}\hat{\Pi}^+ = \hat{\Pi}^+\hat{\Pi} = \hat{I}$.

(Solution): We may use, for simplicity, the definition (2.13) of a Hermitian operator, which reads:

2. Operators and the Second Postulate   109

$$\langle\Psi|\hat{A}|\Psi\rangle = \langle\Psi|\hat{A}^+|\Psi\rangle$$

In the case of a parity operator, we have:

$$\langle\Psi|\hat{\Pi}|\Psi\rangle = \int_{-\infty}^{\infty}\Psi^*(x,t)\hat{\Pi}\Psi(x,t)dx$$

$$= \int_{-\infty}^{\infty}\Psi^*(x,t)\Psi(-x,t)dx = \left[\int_{-\infty}^{\infty}\Psi^*(-x,t)\hat{\Pi}\Psi(-x,t)dx\right]^*$$

$$= \int_{-\infty}^{\infty}\left[\hat{\Pi}\Psi(x,t)\right]^*\Psi(x,t)dx = \langle\hat{\Pi}\Psi|\Psi\rangle = \langle\Psi|\hat{\Pi}^+|\Psi\rangle$$

which implies that $\hat{\Pi}$ is Hermitian. We have immediately that:

$$\hat{\Pi}^2\Psi(x,t) = \hat{\Pi}\left[\hat{\Pi}\Psi(x,t)\right] = \hat{\Pi}\Psi(-x,t) = \Psi(x,t)$$

which shows that $\hat{\Pi}^2 = \hat{I}$, or $\hat{\Pi} = \hat{\Pi}^{-1}$. It follows that:

$$\hat{\Pi}^{-1} = \hat{\Pi}^+$$

which means that $\hat{\Pi}$ is a unitary operator.

Problem 2.4.2.   Show that the operators $\hat{L}_z = -i\hbar\partial/\partial\varphi$ and $e^{\hat{L}_z} = e^{-i\hbar\partial/\partial\varphi}$ are Hermitian.

(Solution):   We have immediately, integrating by parts, and using the periodicity of the spherical wave functions, that:

$$\langle\Psi|\hat{L}_z\Phi\rangle = \int_0^{2\pi}\Psi^*(r,\theta,\varphi)\left[-i\hbar\frac{\partial\Phi(r,\theta,\varphi)}{\partial\varphi}\right]d\varphi$$

$$= \left[-i\hbar\Psi^*(r,\theta,\varphi)\Phi(r,\theta,\varphi)\right]_0^{2\pi} + i\hbar\int_0^{2\pi}\Phi(r,\theta,\varphi)\frac{\partial\Psi^*(r,\theta,\varphi)}{\partial\varphi}d\varphi$$

$$= \int_0^{2\pi}\left[-i\hbar\frac{\partial\Psi(r,\theta,\varphi)}{\partial\varphi}\right]^*\Phi(r,\theta,\varphi)d\varphi = \langle\hat{L}_z\Psi|\Phi\rangle$$

and therefore $\hat{L}_z$ is Hermitian, according to Eq.(2.100). It follows, in view of Eq.(2.97), that any power of $\hat{L}_z$ is also Hermitian. By definition:

$$e^{-i\hbar \partial/\partial \varphi} = \sum_{k=0}^{\infty} \frac{1}{k!}\left(-i\hbar \frac{\partial}{\partial \varphi}\right)^k$$

and the hermiticity of the operator is obvious.

Problem 2.4.3. If $\hat{A}$ and $\hat{B}$ are Hermitian operators, show that:

$$\hat{S} = \hat{A}\hat{B} + \hat{B}\hat{A} \quad \text{and} \quad \hat{D} = i(\hat{A}\hat{B} - \hat{B}\hat{A})$$

are also Hermitian.

---

3. EIGENSTATES AND THE THIRD POSTULATE

3.1. The Eigenvalue Problem

The representation of physical observables by Hermitian linear operators was postulated assuming that an expectation value, defined by Eqs.(2.11) and (2.12), is associated with any observable A measured in a quantum state Ψ of a dynamical system. In view of measurement uncertainties, defined by the dispersion:

$$(\Delta A)^2 = \langle A^2 \rangle - \langle A \rangle^2 = \langle (A - \langle A \rangle)^2 \rangle = \langle (\delta A)^2 \rangle$$

the expectation value $\langle A \rangle$ is taken in a state Ψ if and only if $(\Delta A)^2$ vanishes. It is convenient to introduce the deviation operator:

$$\delta \hat{A} = \hat{A} - \langle A \rangle \hat{I} \qquad (3.1)$$

such that the condition of a vanishing dispersion for the observable A in a given state Ψ reads:

$$\langle \Psi | (\delta \hat{A})^2 \Psi \rangle = 0$$

Because $\delta \hat{A}$ is a Hermitian operator, we can also write:

$$\langle \delta \hat{A} \Psi | \delta \hat{A} \Psi \rangle = |\delta \hat{A} \Psi|^2 = 0$$

which means that the norm of $\delta \hat{A} \Psi$ is zero and this implies that we have a null function $\delta \hat{A} \Psi = 0$. In other words:

$$\hat{A} \Psi = \langle A \rangle \Psi \qquad (3.2)$$

It follows that the state function Ψ where a definite expectation value $\langle A \rangle$ of the observable A is observed, should be a solution of Eq.(3.2) rather than an arbitrary function. The action of the operator $\hat{A}$ on Ψ has to be simply that of rescaling. All the state functions which are solutions of Eq.(3.2) are called eigenfunctions of $\hat{A}$, and the constant factor $\langle A \rangle$ is said to be the corresponding *eigenvalue*. Hence we have shown that if $\langle (\delta\hat{A})^2 \rangle = 0$, the state function should be an eigenfunction of $\hat{A}$, which means that a physical observable only takes definite values in states represented by eigenfunctions of the associated quantum operator. Conversely, given a quantum operator $\hat{A}$ one can systematically determine all its eigenvalues and eigenfunctions, which are solutions of the so-called *eigenvalue equation*:

$$\hat{A}\Psi = \lambda\Psi \tag{3.3}$$

It is visible that the eigenvalue equation fixes the eigenfunctions only up to an overall scale factor, which may always be chosen such that the eigenfunctions be considered normalized. Then, in view of Eq.(2.11), Eq.(3.3) gives:

$$\langle \hat{A} \rangle = \int_{-\infty}^{\infty} \Psi^* \hat{A}\Psi dx = \lambda \int_{-\infty}^{\infty} \Psi^*\Psi dx = \lambda$$

If we further apply the operator $\hat{A}$ to Eq.(3.3), we obtain $\hat{A}^2\Psi = \lambda^2\Psi$ and this leads to:

$$\langle \hat{A}^2 \rangle = \int_{-\infty}^{\infty} \Psi^* \hat{A}^2 \Psi dx = \lambda^2 \int_{-\infty}^{\infty} \Psi^*\Psi dx = \lambda^2$$

It follows that:

$$(\Delta A)^2 = \langle \hat{A}^2 \rangle - \langle \hat{A} \rangle^2 = 0$$

and this shows that the dispersion vanishes if a physical observable is measured in states represented by eigenfunctions of the associated operator.

The requirement of the second postulate that physical observables should be represented by Hermitian operators results from their important property of having *real eigenvalues*. The proof of this statement is immediate if both sides of Eq.(3.3), where the eigenvalue λ is a constant, are multiplied by Ψ^* and then integrated to give:

$$\int_{-\infty}^{\infty} \Psi^* \hat{A}\Psi dx = \int_{-\infty}^{\infty} \Psi^* \lambda \Psi dx = \lambda \int_{-\infty}^{\infty} |\Psi|^2 dx$$

This result may also be expressed, in view of Eq.(2.99) written for $\Phi = \Psi$, in the form:

$$\int_{-\infty}^{\infty} \Psi^* \hat{A}\Psi dx = \int_{-\infty}^{\infty} (\hat{A}\Psi)^* \Psi dx = \lambda^* \int_{-\infty}^{\infty} |\Psi|^2 dx$$

3. Eigenstates and the Third Postulate

so that $\lambda = \lambda^*$ and λ has to be real.

The eigenvalue equation (3.3) always has solutions, and this can be shown by using the expansion (1.69) of the state vector Ψ in terms of a n-dimensional orthonormal basis:

$$\Psi = \sum_{i=1}^{n} a_i \varphi_i = \sum_{i=1}^{n} \langle \varphi_i | \Psi \rangle \varphi_i \qquad (3.4)$$

In a similar way, the vector $\hat{A}\Psi$ can be expanded as $\hat{A}\Psi = \sum_{j=1}^{n} a_j \hat{A}\varphi_j$ and this yields:

$$\hat{A}\Psi = \sum_{i=1}^{n} \langle \varphi_i | \hat{A}\Psi \rangle \varphi_i = \sum_{i=1}^{n} \langle \varphi_i | \sum_{j=1}^{n} a_j \hat{A}\varphi_j \rangle \varphi_i = \sum_{i=1}^{n}\sum_{j=1}^{n} \langle \varphi_i | \hat{A}\varphi_j \rangle a_j \varphi_i = \sum_{i=1}^{n}\sum_{j=1}^{n} A_{ij} a_j \varphi_i \quad (3.5)$$

where the coefficients A_{ij} are defined as:

$$A_{ij} = \langle \varphi_i | \hat{A}\varphi_j \rangle = \langle \varphi_i | \hat{A} | \varphi_j \rangle \qquad (3.6)$$

Substituting Eqs.(3.4) and (3.5) into Eq.(3.3) one obtains:

$$\sum_{i=1}^{n} \left(\sum_{j=1}^{n} A_{ij} a_j - \lambda a_i \right) \varphi_i = 0$$

which represents the following n linear homogenous equations for the a_i:

$$(A_{11} - \lambda)a_1 + A_{12}a_2 + \ldots + A_{1n}a_n = 0$$

$$A_{21}a_1 + (A_{22} - \lambda)a_2 + \ldots + A_{2n}a_n = 0$$

$$\ldots \qquad \ldots \qquad \ldots \qquad\qquad (3.7)$$

$$A_{n1}a_1 + A_{n2}a_2 + \ldots + (A_{nn} - \lambda)a_n = 0$$

This system of simultaneous equations has a nontrivial solution for the a_i provided that the eigenvalue λ satisfies the determinantal equation:

$$\det(A_{ij} - \lambda \delta_{ij}) = \begin{vmatrix} A_{11} - \lambda & A_{12} & \cdots & A_{1n} \\ A_{21} & A_{22} - \lambda & \cdots & A_{2n} \\ \cdots & \cdots & \cdots & \cdots \\ A_{n1} & A_{n2} & \cdots & A_{nn} - \lambda \end{vmatrix} = 0 \qquad (3.8)$$

This is an equation of nth-order in λ, called the *characteristic equation* of $\hat{A}$:

$$P_n(\lambda) = C_n\lambda^n + C_{n-1}\lambda^{n-1} + \ldots + C_1\lambda + C_0 = 0 \qquad (3.9)$$

Since the *characteristic polynomial* $P_n(\lambda)$ is determined by Eq.(3.3), which is valid for any basis of the state vector representation, its roots λ will be basis independent. For each λ there exists a non-zero state vector ψ, obtained by solving Eqs.(3.7) for the a_i, such that $\hat{A}\psi = \lambda\psi$, which means that λ is an eigenvalue. It is shown that the nth-order polynomial has n roots, hence every quantum operator will have n real eigenvalues, not necessarily distinct.

Theorem. *If a Hermitian operator possesses no repeated eigenvalues, its eigenfunctions form a mutually orthonormal set.*

Proof. Let λ_i, λ_j be distinct eigenvalues of a Hermitian operator $\hat{A}$, and ψ_i, ψ_j the corresponding eigenfunctions. Since the eigenvalues are real, we have:

$$\hat{A}\psi_i = \lambda_i\psi_i \qquad \text{and} \qquad (\hat{A}\psi_j)^* = \lambda_j\psi_j^*$$

It follows that:

$$\int_{-\infty}^{\infty} \psi_j^* \hat{A}\psi_i\, dx = \lambda_i \int_{-\infty}^{\infty} \psi_j^* \psi_i\, dx$$

$$\int_{-\infty}^{\infty} \psi_j^* \hat{A}\psi_i\, dx = \int_{-\infty}^{\infty} (\hat{A}\psi_j)^* \psi_i\, dx = \lambda_j \int_{-\infty}^{\infty} \psi_j^* \psi_i\, dx$$

Provided $\lambda_i \neq \lambda_j$, the last two equations lead to:

$$\int_{-\infty}^{\infty} \psi_j^* \psi_i\, dx = 0 \qquad \text{or} \qquad \langle \psi_j | \psi_i \rangle = 0$$

Hence, if the roots of the characteristic equation (3.9) are all distincts, the set of n corresponding eigenfunctions ψ_i represents an orthonormal basis in a n-dimensional wave function space.

EXAMPLE 3.1. Degeneracy

It may happen that there are repeated roots for the characteristic polynomial and, in this case, a single eigenvalue represents more than one eigenfunction. The phenomenon of k linearly independent eigenfunctions $\psi_i^{(1)}, \psi_i^{(2)}, \ldots, \psi_i^{(k)}$ corresponding to a particular eigenvalue λ_i is called k-fold *degeneracy* associated with λ_i:

3. Eigenstates and the Third Postulate 115

$$\hat{A}\psi_i^{(s)} = \lambda_i \psi_i^{(s)} \qquad (s = 1,2,\ldots,k) \qquad (3.10)$$

The earlier theorem was restricted to no repeated eigenvalues only because, in case of degeneracy, the fact that $\langle \psi_j | \psi_i^{(s)} \rangle = 0$ if $\lambda_i \neq \lambda_j$ for any value of s does not imply that $\psi_i^{(s)}$ are themselves mutually orthogonal. However, a set of k orthonormal eigenfunctions $\psi_i'^{(s)}$ can always be found by forming linear combinations $\sum_s c_s \psi_i^{(s)}$ which are also eigenfunctions corresponding to λ_i:

$$\hat{A}\left(\sum_s c_s \psi_i^{(s)}\right) = \lambda_i \left(\sum_s c_s \psi_i^{(s)}\right)$$

Consider first a pair of eigenfunctions $\psi_i^{(1)}, \psi_i^{(2)}$ such that:

$$\int_{-\infty}^{\infty} \left(\psi_i^{(1)}\right)^* \psi_i^{(2)} dx = \langle \psi_i^{(1)} | \psi_i^{(2)} \rangle \neq 0$$

If we choose $\varphi_i^{(1)} = \psi_i^{(1)}$ and $\varphi_i^{(2)} = \psi_i^{(2)} + c_{21}\varphi_i^{(1)}$ then we have:

$$\int_{-\infty}^{\infty} \left(\varphi_i^{(1)}\right)^* \varphi_i^{(2)} dx = \int_{-\infty}^{\infty} \left(\varphi_i^{(1)}\right)^* \psi_i^{(2)} dx + c_{21}\int_{-\infty}^{\infty} \left(\varphi_i^{(1)}\right)^* \varphi_i^{(1)} dx = 0$$

and the orthogonality of $\varphi_i^{(1)}$ and $\varphi_i^{(2)}$ holds provided that:

$$c_{21} = -\frac{\langle \varphi_i^{(1)} | \psi_i^{(2)} \rangle}{\left|\varphi^{(1)}\right|^2} \quad \text{or} \quad \varphi_i^{(2)} = \psi_i^{(2)} - \frac{\langle \varphi_i^{(1)} | \psi_i^{(2)} \rangle}{\left|\varphi^{(1)}\right|^2} \varphi_i^{(1)}$$

In a similar way, if we next replace $\psi_i^{(3)}$ by:

$$\varphi_i^{(3)} = \psi_i^{(3)} + c_{31}\varphi_i^{(1)} + c_{32}\varphi_i^{(2)}$$

then, using the orthogonality of $\varphi_i^{(1)}$ and $\varphi_i^{(2)}$, and $\varphi_i^{(1)}$ and $\varphi_i^{(3)}$ respectively, one obtains c_{31} and c_{32} such that the eigenfunction:

$$\varphi_i^{(3)} = \psi_i^{(3)} - \frac{\langle \varphi_i^{(1)} | \psi_i^{(3)} \rangle}{\left|\varphi_i^{(1)}\right|^2} \varphi_i^{(1)} - \frac{\langle \varphi_i^{(2)} | \psi_i^{(3)} \rangle}{\left|\varphi_i^{(2)}\right|^2} \varphi_i^{(2)}$$

is orthogonal to both $\varphi_i^{(1)}$ and $\varphi_i^{(2)}$. Following the same steps, we finally obtain the eigenfunction $\varphi_i^{(k)}$ which is orthogonal to all previous ones, in the form:

$$\varphi_i^{(k)} = \psi_i^{(k)} - \sum_{s=1}^{k-1} \frac{\langle \varphi_i^{(s)} | \psi_i^{(k)} \rangle}{|\varphi_i^{(s)}|^2} \varphi_i^{(s)}$$

This process, known as the *Gram-Schmidt orthogonalization process*, leads to k orthogonal eigenfunctions corresponding to a k-fold degenerated eigenvalue. An orthonormal set is then provided by the eigenfunctions $\psi_i'^{(s)} = \varphi_i^{(s)}/|\varphi_i^{(s)}|$, $s = 1,2,\ldots,k$, which are also orthonormal to any eigenfunction ψ_j for which $\lambda_j \neq \lambda_i$. In other words, the earlier theorem can be restated in the form:

> **Theorem.** *For every Hermitian operator there exists a basis formed by its orthonormal eigenfunctions.*

which applies to both the case of distinct eigenvalues and the degenerate case. The eigenfunctions corresponding to a degenerate eigenvalue, Eq.(3.10), will tacitly be assumed as orthonormal, and one may in general associate to λ_i the eigenstate:

$$\psi_i = \sum_{s=1}^{k} c_s \psi_i^{(s)} \qquad (3.11)$$

which obviously obeys Eq.(3.10). ❧

The discussion of the eigenvalue problem in a n-dimensional function space can be regarded as a model of finite dimensionality for the mathematical formalism of quantum mechanics. However, in general, a quantum operator may have an infinite number of real eigenvalues and therefore it is needed to generalize the preceding concepts to infinite dimensions. If the eigenvalues are plotted on the real axis, we then will expect to obtain an infinite discrete set of representative points λ_n, but the eigenvalues also may vary continuously along the real axis, such that the index n might be replaced by a continuous parameter λ. It may also happen that the eigenvalues are represented by both an infinite discrete set λ_n and a continuum of representative points over a certain range.

Consider first the case in which the eigenvalue problem can be solved for an infinite set of discrete eigenvalues:

$$\hat{A}\psi_n = \lambda_n \psi_n \qquad (3.12)$$

We will assume that the result proved for finite dimensions holds for the Hilbert space of the square-integrable functions, namely:

> *The eigenfunctions* $\psi_1, \psi_2, \ldots, \psi_n, \ldots$, *of a Hermitian operator, corresponding to eigenvalues* $\lambda_1, \lambda_2, \ldots, \lambda_n, \ldots$, *not necessarily distinct, form a complete basis.*

Hence, it is natural to suppose that a state vector $\Psi(x)$ can be expanded in the form (1.65), which reads:

3. Eigenstates and the Third Postulate   117

$$\Psi = \sum_n a_n \psi_n \qquad (3.13)$$

with coefficients given by the scalar products:

$$a_n = \int \psi_n^*(x)\Psi(x)dx \qquad (3.14)$$

Integration is assumed over a given domain (x_1, x_2) which eventually may be extended from $-\infty$ to ∞. Substituting the state vector expansion into Eq.(3.14) it follows that:

$$a_n = \int \psi_n^*(x) \left[\sum_m a_m \psi_m(x) \right] dx = \sum_m a_m \int \psi_n^*(x)\psi_m(x)dx$$

which only holds if the eigenfunctions form an orthonormal set:

$$\int \psi_n^*(x)\psi_m(x)dx = \delta_{nm} \qquad (3.15)$$

Conversely, substituting Eq.(3.14) into the state vector expansion (3.13) we have:

$$\Psi(x) = \sum_n \psi_n(x) \int \psi_n^*(x')\Psi(x')dx' = \int \Psi(x') \left[\sum_n \psi_n^*(x')\psi_n(x) \right] dx'$$

and hence, in view of Eq.(1.45), one obtains:

$$\int \Psi(x') \left[\delta(x-x') - \sum_n \psi_n^*(x')\psi_n(x) \right] dx' = 0$$

This leads to the closure relation for the orthonormal set of eigenfunctions:

$$\sum_n \psi_n^*(x')\psi_n(x) = \delta(x-x') \qquad (3.16)$$

which, if satisfied, implies that the eigenfunctions form a complete set, and therefore makes possible the expansion (3.13) for arbitrary Ψ.

If the eigenvalue spectrum is continuous, the eigenvalue equation reads:

$$\hat{A}\psi(x,\lambda) = \lambda\psi(x,\lambda) \qquad (3.17)$$

because the eigenfunctions depend on the continuous parameter λ. In this case, the expansion of an arbitrary state vector $\Psi(x)$ in terms of $\psi(x,\lambda)$, corresponding to Eq.(3.13), should be written as an integral over the range of λ, which eventually is extended from $-\infty$ to ∞:

$$\Psi(x) = \int a(\lambda)\psi(x,\lambda)d\lambda \qquad (3.18)$$

where the coefficients are similarly given by:

$$a(\lambda) = \int \psi^*(x,\lambda)\Psi(x)dx \qquad (3.19)$$

Substituting Eq.(3.18) into Eq.(3.19) we now have:

$$a(\lambda) = \int \psi^*(x,\lambda)\left[\int a(\mu)\psi(x,\mu)d\mu\right]dx = \int a(\mu)\left[\int \psi^*(x,\lambda)\psi(x,\mu)dx\right]d\mu$$

or, using for $a(\lambda)$ the Dirac δ-function representation, Eq.(1.45):

$$\int \psi^*(x,\lambda)\psi(x,\mu)dx = \delta(\lambda - \mu) \qquad (3.20)$$

This equation corresponds to Eq.(3.15) which defines an orthonormal set in the Hilbert space of square-integrable functions. According to the interpretation of Eq.(1.45), the right-hand side vanishes for $\lambda \neq \mu$, which is a way of showing the orthogonality of the eigenfunctions $\psi(x,\lambda)$. However, for $\lambda = \mu$, we obtain:

$$\int |\psi(x,\lambda)|^2 dx = \infty$$

and therefore these eigenfunctions are not square-integrable functions. In difference from the eigenfunctions $\psi_n(x)$, normalizable to unity, which are proper vectors of the Hilbert space defined by mathematicians, the eigenfunctions $\psi(x,\lambda)$ corresponding to a continuum of eigenvalues are normalizable to the Dirac δ-function and represent improper vectors. However, they should be included in a *generalized wave function space*, which is defined as a space of functions that can be either normalized to unity or to the Dirac δ-function.

If Eq.(3.19) is now substituted into Eq.(3.18), we obtain:

$$\Psi(x) = \int \psi(x,\lambda)\left[\int \psi^*(x',\lambda)\Psi(x')dx'\right]d\lambda = \int \Psi(x')\left[\int \psi^*(x',\lambda)\psi(x,\lambda)d\lambda\right]dx'$$

and so:

$$\int \psi^*(x',\lambda)\psi(x,\lambda)d\lambda = \delta(x - x') \qquad (3.21)$$

This is the closure relation for the eigenfunctions $\psi(x,\lambda)$, which is a generalization of Eq.(3.16).

EXAMPLE 3.2. The eigenvalue problem for momentum

Using the definition (2.7) of the operator $\hat{p}_x$ associated with the momentum observable of a particle moving along the x-axis, the eigenvalue problem reads:

$$-i\hbar \frac{d\psi(x)}{dx} = p_x \psi(x) \tag{3.22}$$

The solution of this equation is $\psi(x) = N e^{ip_x x/\hbar}$, where N is an arbitrary complex or real constant. Although a solution $\psi(x)$ exists irrespective of the complex or real values which p_x may assume, we are restricted by the Hermitian nature of the operator $\hat{p}_x$ to real eigenvalues p_x and to their corresponding eigenfunctions. Considering unrestricted x values from $-\infty$ to ∞, any real number $-\infty < p_x < \infty$ is an eigenvalue. In other words, the eigenvalue spectrum is continuous along the real axis, and the eigenfunctions, which are not square-integrable, should be denoted by $\psi(x, p_x)$. The constant N is obtained from Eq.(3.20) which reads:

$$\int_{-\infty}^{\infty} \psi^*(x, p_x) \psi(x, p_x') dx = |N|^2 \int_{-\infty}^{\infty} e^{i(p_x' - p_x)x/\hbar} dx = \delta(p_x' - p_x)$$

and reduces to $2\pi\hbar|N|^2 \delta(p_x' - p_x) = \delta(p_x' - p_x)$ where use has been made of Eq.(1.43). In view of the properties (1.45) of the Dirac δ-function, integration of the last equation over p_x' gives $2\pi\hbar|N|^2 = 1$, hence:

$$\psi(x, p_x) = \frac{1}{\sqrt{2\pi\hbar}} e^{ip_x x/\hbar} \tag{3.23}$$

These functions are the same as the monochromatic plane waves (1.40), and therefore represent an infinite set of eigenfunctions from the generalized wave function space, normalized according to Eq.(1.43), which also constitutes a complete basis, in view of Eq.(1.48). The momentum p_x is not quantified, so it can be precisely defined, and hence, there should be a complete uncertainty with respect to position. This corresponds to the fact that the eigenfunctions are not normalized to unity but to the Dirac δ-function, $|\psi(x, p_x)|^2$ is therefore not a probability density and the particle is not localized. The expansion of an arbitrary state vector $\Psi(x)$ in terms of $\psi(x, p_x)$, as defined by Eq.(3.18), takes the form of the Fourier expansion (1.38), where the coefficients $a(p_x)$, denoted by $\Phi(p_x)$, are given by Eq.(1.39), which is a particular case of Eq.(3.19).

If the motion of the particle is restricted to be in a finite interval $-\alpha \le x \le \alpha$, such that for $|x| > \alpha$ any function of position should vanish, it is natural to impose to the eigenfunctions $\psi(x)$ the condition of continuity $\psi(0) = \psi(\alpha) = 0$. This leads to a discrete eigenvalue spectrum:

$$p_x = n \frac{\pi\hbar}{\alpha}, \quad n = 0, \pm 1, \pm 2, \ldots$$

which represents an infinite set. In this case we should apply to the infinite set $\psi_n(x)$ the normalization condition (3.15), which reads:

$$\int_{-\alpha}^{\alpha} \psi_n^*(x)\psi_m(x)dx = N_n^* N_m \int_{-\alpha}^{\alpha} e^{i\pi x(n-m)/\alpha} dx = \delta_{nm}$$

and gives $2\alpha |N_n|^2 = 1$. It follows that:

$$\psi_n(x) = \frac{1}{\sqrt{2\alpha}} e^{in\pi x/\alpha}$$

The closure relation, Eq.(3.16), takes for these eigenfunctions the form:

$$\frac{1}{2\alpha} \sum_{n=-\infty}^{\infty} e^{in\pi(x-x')/\alpha} = \delta(x-x')$$

We see that the solution to the eigenvalue problem (3.22) is not uniquely determined by the quantum operator, but also depends on the specific conditions imposed on eigenfunctions. ♦

EXAMPLE 3.3. The eigenvalue problem for position

For a particle moving along the x-axis, the eigenvalue problem:

$$\hat{x}\psi(x) = \lambda\psi(x) \tag{3.24}$$

may be conveniently rewritten, in view of Eq.(2.2), in the form $(x-\lambda)\psi(x) = 0$, where x is the position variable. It follows that all the eigenvalues λ should be real, otherwise the first factor is not vanishing and therefore $\psi(x)$ is a null function for arbitrary x. If λ assumes a real value $\lambda = x'$, the corresponding eigenfunction vanishes for $x \neq x'$ and exhibits a singularity at the point $x = x'$. Since solutions of this kind should exist for any real value $-\infty < x' < \infty$, the eigenvalue spectrum is continuous, and the eigenfunction $\psi(x,x')$ should obey the normalization condition (3.20), which reads:

$$\int_{-\infty}^{\infty} \psi^*(x,x')\psi(x,x'')dx = \delta(x'-x'') \tag{3.25}$$

It follows that we may choose:

$$\psi(x,x') = \delta(x-x') \tag{3.26}$$

which obeys the condition (3.25) and satisfies the eigenvalue equation (3.24) for $\lambda = x'$. The closure relation (3.21) always holds in the form:

3. Eigenstates and the Third Postulate

$$\int_{-\infty}^{\infty} \psi^*(x'',x')\psi(x,x')dx' = \int_{-\infty}^{\infty} \delta(x''-x')\delta(x-x')dx' = \delta(x-x'')$$

In view of Eq.(1.45), the normalized wave function can be written as:

$$\Psi(x) = \int_{-\infty}^{\infty} \Psi(x')\delta(x-x')dx' = \int_{-\infty}^{\infty} \Psi(x')\psi(x,x')dx'$$

which corresponds to the expansion (3.18), with coefficients $a(x') = \Psi(x')$. ♦

Problem 3.1.1. Show that the characteristic equation $P_n(\lambda) = 0$ of the observable A is satisfied by the operator $\hat{A}$, namely $P_n(\hat{A}) = 0$.

(Solution): Assuming ψ_i an eigenvector of $\hat{A}$, corresponding to the eigenvalue λ_i, by successively applying the operator $\hat{A}$ on the left-hand side of Eq.(3.3) we obtain:

$$\hat{A}^k \psi_i = \lambda_i^k \psi_i$$

and it follows that for the characteristic polynomial $P_n(\hat{A})\psi_i = P_n(\lambda)\psi_i = 0$. Using the expansion (3.13) of an arbitrary state vector Ψ in terms of the eigenbasis, we may write:

$$P_n(\hat{A})\Psi = \sum_i a_i P_n(\hat{A})\psi_i = 0$$

and therefore $P_n(\hat{A}) = 0$ because it transforms any state vector into the null vector.

Problem 3.1.2. Find the eigenvalues and the eigenfunctions of the parity operator, defined by:

$$\hat{\Pi}\psi(x) = \psi(-x)$$

(Solution): We can write, using the definition of $\hat{\Pi}$:

$$\hat{\Pi}^2 \psi(x) = \hat{\Pi}\psi(-x) = \psi(x)$$

and, using the eigenvalue equation $\hat{\Pi}\psi(x) = \pi\psi(x)$, we have:

$$\hat{\Pi}^2\psi(x) = \hat{\Pi}\pi\psi(x) = \pi^2\psi(x)$$

It follows that $\pi^2 = 1$ which leads to the eigenvalues $\pi = \pm 1$. For $\pi = 1$, the eigenfunction should be an even function:

$$\hat{\Pi}\psi_+(x) = \psi_+(-x) = \psi_+(x)$$

and the eigenfunction corresponding to $\pi = -1$ is an odd function:

$$\hat{\Pi}\psi_-(x) = \psi_-(-x) = -\psi_-(-x)$$

In other words, the eigenfunctions of $\hat{\Pi}$ should have a definite parity. It may happen that the eigenstates $\psi(x)$ are degenerate, and then the relation $\psi(-x) = \pi\psi(x)$ is no longer necessary valid. In this case, we always may built on the symmetrical and antisymmetrical linear combinations:

$$\psi^s(k) = \frac{1}{\sqrt{2}}[\psi(x) + \psi(-x)]$$

$$\psi^a(x) = \frac{1}{\sqrt{2}}[\psi(x) - \psi(-x)]$$

which have a definite parity.

Problem 3.1.3. Solve the energy eigenvalue problem for a particle in a one-dimensional potential well $0 \le x \le a$, for which $U(x) = 0$ inside the well and $U(x) = \infty$ outside.

(Solution): The energy eigenvalue equation reads:

$$\hat{H}\psi = E\psi$$

where $\hat{H}$ is the Hamiltonian operator defined by Eq.(2.17). Inside the well we obtain:

$$-\frac{\hbar^2}{2m}\frac{d^2\psi}{dx^2} = E\psi \quad \text{or} \quad \frac{d^2\psi}{dx^2} + \frac{2mE}{\hbar^2}\psi = 0$$

Denoting $k = \sqrt{2mE}/\hbar$ we find the solution:

$$\psi = A\sin(kx + \varphi)$$

3. Eigenstates and the Third Postulate

which should be continuous at $x = 0$ and $x = \alpha$. Since, outside the well we have:

$$-\frac{\hbar^2}{2m}\frac{d^2\psi}{dx^2} + U(x)\psi = E\psi \quad \text{or} \quad \frac{1}{\psi}\frac{d^2\psi}{dx^2} = \frac{2m}{\hbar^2}[U(x) - E]$$

and hence, it follows that ψ must vanish when $U(x) = \infty$, we must take:

$$\psi(0) = A\sin\varphi = 0, \quad \psi(\alpha) = A\sin(k\alpha + \varphi) = 0$$

Therefore $\varphi = 0$ and k is restricted to the discrete values $k_n = n\pi/\alpha$ ($n = 1, 2, \ldots$). We find the same set of energy eigenvalues as in Example P.3:

$$E_n = \frac{\hbar^2 k^2}{2m} = n^2 \frac{\pi^2 \hbar^2}{2m\alpha^2}$$

The normalization of the corresponding eigenfunctions gives:

$$\int_{-\infty}^{\infty} \psi_n^*(x)\psi_n(x)dx = |A|^2 \int_0^{\alpha} \sin^2\frac{n\pi x}{\alpha} dx = 1, \quad \text{or} \quad \psi_n(x) = \sqrt{\frac{2}{\alpha}}\sin\frac{n\pi x}{\alpha}$$

Problem 3.1.4. Given the eigenvalues λ_n of the operator $\hat{A}$, find the eigenvalues of $\hat{A}^{-1}$.

(Answer): $1/\lambda_n$

Problem 3.1.5. Find the eigenvalues and eigenfunctions of the angular momentum operator $\hat{L}_z = -i\hbar\, d/d\varphi$.

(Answer): $\lambda_m = \pm m\hbar, \quad \psi_m(\varphi) = Ne^{im\varphi} \quad (m = 1, 2, \ldots)$

Problem 3.1.6. Find the eigenvalues and eigenfunctions of the differential operator:

$$\hat{A} = \frac{1}{r^2}\frac{d}{dr}\left(r^2 \frac{d}{dr}\right)$$

(Answer): $\lambda = -\alpha^2, \quad \psi(r,\alpha) = N(\sin\alpha r)/r$ for any real α.

3.2. The Third Postulate

The third postulate is concerned with the significance of eigenvalues and eigenfunctions of a quantum operator in predicting the result of measurements on the associated physical observable. The eigenvalue equation (3.2) indicates that the expected result $\langle A \rangle$ of a measurement on the observable A is necessarily an eigenvalue of the representative operator $\hat{A}$. If we assume that the eigenvalue spectrum includes both an infinite discrete set λ_n and a continuum of eigenvalues λ over a finite range, an arbitrary state of a dynamical system, represented by a normalized state function $\Psi(x)$ must be expanded in terms of all eigenfunctions of $\hat{A}$, in the form:

$$\Psi(x) = \sum_n a_n \psi_n(x) + \int a(\lambda)\psi(x,\lambda)d\lambda \qquad (3.27)$$

where $\psi_n(x)$ and $\psi(x,\lambda)$ are solution of Eqs.(3.12) and (3.17) respectively. Substituting Eq.(3.27) into Eq.(3.14), which defines the coefficients a_n, we have:

$$a_n = \int \psi_n^*(x) \left[\sum_m a_m \psi_m(x) + \int a(\lambda)\psi(x,\lambda)d\lambda \right] dx$$

$$= \sum_m a_m \int \psi_n^*(x)\psi_m(x)dx + \int a(\lambda)\left[\int \psi_n^*(x)\psi(x,\lambda)dx\right]d\lambda$$

$$= a_m \delta_{nm} + \int a(\lambda)\left[\int \psi_n^*(x)\psi(x,\lambda)dx\right]d\lambda$$

and this only is valid if the eigenfunctions corresponding to the discrete and continuous eigenvalue spectra are mutually orthogonal:

$$\int \psi_n^*(x)\psi(x,\lambda)dx = 0 \qquad (3.28)$$

The same result is obtained if $\Psi(x)$, given by Eq.(3.27), is substituted into Eq.(3.19), which expresses the $a(\lambda)$. Equation (3.28) indicates that the eigenfunctions $\psi(x,\lambda)$, corresponding to the continuos eigenvalue spectrum, should not be excluded from the complete vector basis required for the expansion (3.27), although they are not square-integrable functions.

It is convenient to introduce the scalar product of $\Psi(x)$ with another normalized state function $\Phi(x)$, expanded in terms of the same vector basis:

$$\Phi(x) = \sum_n b_n \psi_n(x) + \int b(\lambda)\psi(x,\lambda)d\lambda \qquad (3.29)$$

which reads:

3. Eigenstates and the Third Postulate

$$\int \Phi^*(x)\Psi(x)dx = \sum_n b_n^* \int \psi_n^*(x)\Psi(x)dx + \int b^*(\lambda)\left[\int \psi^*(x,\lambda)\Psi(x)dx\right]d\lambda$$

or:

$$\langle \Phi|\Psi\rangle = \sum_n b_n^* a_n + \int b^*(\lambda)a(\lambda)dx \qquad (3.30)$$

where use has been made of Eqs.(3.14) and (3.19). For $\Phi(x) \equiv \Psi(x)$ we obtain the normalization condition:

$$\sum_n |a_n|^2 + \int |a(\lambda)|^2 \, d\lambda = 1 \qquad (3.31)$$

If we apply the operator $\hat{A}$ to both sides of Eq.(3.27), the expansion becomes:

$$\hat{A}\Psi(x) = \sum_n a_n \lambda_n \psi_n(x) + \int a(\lambda)\lambda\psi(x,\lambda)d\lambda$$

and therefore, from Eq.(3.30):

$$\langle \Psi|\hat{A}\Psi\rangle = \sum_n |a_n|^2 \lambda_n + \int |a(\lambda)|^2 \lambda\, d\lambda \qquad (3.32)$$

We have seen before, Eq.(2.11), that $\langle \Psi|\hat{A}\Psi\rangle$ represents the expectation value of $\hat{A}$ in an arbitrary state Ψ, which is now expressed in terms of the eigenvalues and eigenfunctions of $\hat{A}$. The interpretation of both the expansion (3.32) and the normalization condition (3.31) is introduced as a postulate of quantum mechanics concerning the result of a measurement on a physical observable.

> **Postulate 3.** *The eigenvalues of a quantum operator represent the possible results of carrying out a measurement of the associated physical observable. If the system is in a state described by an eigenfunction Ψ, the probability of obtaining as a result of the measurement a discrete eigenvalue λ_n of the quantum operators is:*
>
> $$P(\lambda_n) = |a_n|^2 = |\langle \psi_n|\Psi\rangle|^2 \qquad (3.33)$$
>
> *where ψ_n is the normalized eigenfunction corresponding to λ_n. The probability to find a result in the interval $(\lambda, \lambda + d\lambda)$ from the continuous eigenvalue spectrum is:*

$$dP = P(\lambda)d\lambda = |a(\lambda)|^2\, d\lambda = |\langle \psi(x,\lambda)|\Psi\rangle|^2 d\lambda \qquad (3.34)$$

where $\psi(x,\lambda)$ is the generalized eigenfunction which corresponds to λ.

Hence, the postulate states the probability $P(\lambda_n) = |\langle \psi_n|\Psi\rangle|^2$ that the state function Ψ of the system is identical with the eigenfunction ψ_n corresponding to the discrete eigenvalue spectrum immediately after a physical measurement, and the similar probability density $P(\lambda) = |\langle \psi(x,\lambda)|\Psi\rangle|^2$ for the continuous eigenvalue spectrum. This is an explicit formulation of the statistical predictions of quantum mechanics on the measurement of physical observables, since Eq.(3.31) is interpreted as a normalization condition for probabilities:

$$\sum_n P(\lambda_n) + \int P(\lambda)d\lambda = 1 \qquad (3.35)$$

while the expectation value (3.32) reduces to the familiar weighted average:

$$\langle \Psi|\hat{A}\Psi\rangle = \sum_n \lambda_n P(\lambda_n) + \int \lambda\, P(\lambda)d\lambda = \langle A\rangle \qquad (3.36)$$

EXAMPLE 3.4. The projection operator

A simple geometrical interpretation of the third postulate can be formulated in terms of the familiar concepts of vector algebra. The expansion (3.27) of an arbitrary state function is written in the Dirac notation as:

$$|\Psi\rangle = \sum_n a_n|\psi_n\rangle + \int a(\lambda)|\psi(\lambda)\rangle d\lambda$$

such that in view of Eqs.(3.14) and (3.19), $a_n = \langle \psi_n|\Psi\rangle$ and $a(\lambda) = \langle \psi(\lambda)|\Psi\rangle$, we obtain:

$$|\Psi\rangle = \left(\sum_n |\psi_n\rangle\langle\psi_n| + \int |\psi(\lambda)\rangle\langle\psi(\lambda)|d\lambda\right)|\Psi\rangle$$

Because this result is true for all $|\Psi\rangle$, the brackets include the identity operator, which thus takes the form:

$$\hat{I} = \sum_n |\psi_n\rangle\langle\psi_n| + \int |\psi(\lambda)\rangle\langle\psi(\lambda)|d\lambda = \sum_n P_{\lambda_n} + \int P_\lambda d\lambda \qquad (3.37)$$

It is convenient to introduce the *projection operators* for the kets $|\psi_n\rangle$ and $|\psi(\lambda)\rangle$ in the form:

3. Eigenstates and the Third Postulate

$$\hat{P}_{\lambda_n}|\Psi\rangle = |\psi_n\rangle\langle\psi_n|\Psi\rangle = a_n|\psi_n\rangle \rightarrow \hat{P}_{\lambda_n} = |\psi_n\rangle\langle\psi_n|$$

$$\hat{P}_{\lambda}|\Psi\rangle = |\psi(\lambda)\rangle\langle\psi(\lambda)|\Psi\rangle = a(\lambda)|\psi(\lambda)\rangle \rightarrow \hat{P}_{\lambda} = |\psi(\lambda)\rangle\langle\psi(\lambda)|$$

(3.38)

In other words, P_{λ_n} and P_λ project out the component of any ket $|\Psi\rangle$ along the direction of the eigenkets $|\psi_n\rangle, |\psi(\lambda)\rangle$ respectively, and we can write:

$$|\Psi\rangle = \sum_n P_{\lambda_n}|\Psi\rangle + \int P_{\lambda}|\Psi\rangle d\lambda \qquad (3.39)$$

It is noteworthy that projection operators act similarly on the bras:

$$\langle\Psi|P_{\lambda_n} = \langle\Psi|\psi_n\rangle\langle\psi_n| = a_n^*\langle\psi_n|$$

$$\langle\Psi|P_{\lambda} = \langle\Psi|\psi(\lambda)\rangle\langle\psi(\lambda)| = a^*(x)\langle\psi(\lambda)|$$

It is obvious, in view of Eqs.(3.15) and (3.21) that:

$$P_{\lambda_n} P_{\lambda_m} = |\psi_n\rangle\langle\psi_n|\psi_m\rangle\langle\psi_m| = \delta_{nm} P_m$$

$$P_\lambda P_\mu = |\psi(\lambda)\rangle\langle\psi(\lambda)|\psi(\mu)\rangle\langle\psi(\mu)| = \delta(\lambda-\mu) P_\mu$$

$$P_{\lambda_n} P_\lambda = |\psi_n\rangle\langle\psi_n|\psi(\lambda)\rangle\langle\psi(\lambda)| = 0$$

Equation (3.37) shows that the sum of the projections of a state vector along all the directions (eigenkets) equals the vector itself, and constitutes a *completeness relation* for the eigenkets of a quantum operator. Then the third postulate can be interpreted in the sense that the probability of obtaining, as a result of the measurement, an eigenvalue λ_n or λ, equals the magnitude of the projection a_n or $a(\lambda)$ of the state vector $|\Psi\rangle$ along the direction of corresponding eigenkets $|\psi_n\rangle$ or $|\psi(\lambda)\rangle$ respectively. ☙

The third Postulate implies that the effect of the measurement of a physical observable A on a state $|\Psi\rangle$ of a dynamical system, yielding a definite result λ_n (or λ) is to change the state vector $|\Psi\rangle$ into an eigenstate $|\psi_n\rangle = \hat{P}_{\lambda_n}|\Psi\rangle$ or $|\psi(\lambda)\rangle = P_\lambda|\Psi\rangle$ corresponding to the eigenvalue obtained in the measurement. It the state after the measurement is normalized, we then may write:

$$|\Psi\rangle \rightarrow |\psi_n\rangle = \frac{\hat{P}_{\lambda_n}|\Psi\rangle}{\langle \hat{P}_{\lambda_n}\Psi|\hat{P}_{\lambda_n}\Psi\rangle^{\frac{1}{2}}}$$

$$|\Psi\rangle \rightarrow |\psi(\lambda)\rangle = \frac{\hat{P}_\lambda|\Psi\rangle}{\langle \hat{P}_\lambda\Psi|\hat{P}_\lambda\Psi\rangle^{\frac{1}{2}}}$$

(3.40)

where the projection operators have been introduced by Eq.(3.38). If the initial state of the dynamical system were an eigenstate of $\hat{A}$, it is left invariant by the measurement, and this is the only case which corresponds to the classical concept of measurement. In classical mechanics any state of a system is invariant under a measurement of a dynamical variable A, but only the eigenstates of $\hat{A}$ are not changed by quantum measurement of A.

Problem 3.2.1. Show that the completeness relation for an eigenbasis $|\psi_1\rangle, |\psi_2\rangle, \ldots |\psi_n\rangle, \ldots$ can be written, in terms of arbitrary state vectors Ψ and Φ, as:

$$\langle\Phi|\Psi\rangle = \sum_n \langle\Phi|\psi_n\rangle\langle\psi_n|\Psi\rangle$$

(Solution): If the relation holds for any Ψ and Φ, the eigenbasis forms a complete set, for there exists no independent state vector Λ orthogonal to it. Assuming the contrary, we may apply the equation in the particular case $\Psi = \Phi = \Lambda$, which gives:

$$\langle\Lambda|\Lambda\rangle = \sum_n \langle\Lambda|\psi_n\rangle\langle\psi_n|\Lambda\rangle = 0$$

which means that $\Lambda = 0$. Conversely, if the eigenbasis is complete, the given relation follows from the expansions:

$$|\Psi\rangle = \sum_n |\psi_n\rangle\langle\psi_n|\Psi\rangle \quad \text{and} \quad \langle\Phi| = \sum_n \langle\Phi|\psi_n\rangle\langle\psi_n|$$

which give:

$$\langle\Phi|\Psi\rangle = \sum_{m,n} \langle\Phi|\psi_m\rangle\langle\psi_m|\psi_n\rangle\langle\psi_n|\Psi\rangle = \sum_n \langle\Phi|\psi_n\rangle\langle\psi_n|\Psi\rangle$$

Problem 3.2.2. Consider a particle moving in one dimension inside the potential well described in Problem 3.1.3. Find the expectation values of position $\langle x \rangle$ and $\langle x^2 \rangle$ in the energy eigenstates $\psi_n(x) = \sqrt{2/\alpha}\,\sin(n\pi x/\alpha)$.

(Solution): Since the eigenvalue spectrum of position is continuous, the normalized energy eigenstates can be expanded, as shown in Example 3.3, as:

$$\psi_n(x) = \int_{-\infty}^{\infty} \psi_n(x')\delta(x-x')dx'$$

which has the form (3.18) with $a(x') = \psi_n(x')$. From Eq.(3.32) it follows that:

$$\langle x \rangle_n = \langle \psi_n | x | \psi_n \rangle = \int_{-\infty}^{\infty} |\psi_n(x)|^2 x\, dx = \frac{2}{\alpha} \int_0^{\alpha} x \sin^2 \frac{n\pi x}{\alpha} dx$$

$$= \frac{2}{\alpha} \int_0^{\alpha} \left(\frac{x}{2} - \frac{x}{2} \cos \frac{2n\pi x}{\alpha} \right) dx = \frac{1}{\alpha} \int_0^{\alpha} x\, dx - \frac{1}{\alpha} \int_0^{\alpha} x \cos \frac{2n\pi x}{\alpha} dx = \frac{\alpha}{2}$$

which means that the average position is the same in all the energy eigenstates. Similarly we have:

$$\langle x^2 \rangle_n = \langle \psi_n | x^2 | \psi_n \rangle = \frac{2}{\alpha} \int_0^{\alpha} x^2 \sin^2 \frac{n\pi x}{\alpha} dx = \frac{1}{\alpha} \int_0^{\alpha} x^2 dx - \frac{1}{\alpha} \int_0^{\alpha} x^2 \cos \frac{2n\pi x}{\alpha} dx$$

The second term, if integrated by parts, two times in turn, reduces to $\alpha^3/2n^2\pi^2$, so that:

$$\langle x^2 \rangle_n = \alpha^3 \left(\frac{1}{3} - \frac{1}{2n^2\pi^2} \right)$$

3.3. The Matrix Eigenvalue Problem

The eigenvalue problem can be reformulated using the matrix representation of any quantum operator in an appropriate function space defined by a complete basis of wave functions. Consider the linear relation between the components a_i, b_i of the n-dimensional state vectors (1.69) and (1.70) with respect to the basis vectors $|i\rangle$:

$$b_i = \sum_{k=1}^{n} A_{ik} a_k \qquad (3.41)$$

that contains n^2 coefficients A_{ik}. If these coefficients are regarded as the elements of a $n \times n$ matrix, Eq.(3.41) also reads:

$$\begin{pmatrix} b_1 \\ b_2 \\ \vdots \\ b_n \end{pmatrix} = \begin{pmatrix} A_{11} & A_{12} & \cdots & A_{1n} \\ A_{21} & A_{22} & \cdots & A_{2n} \\ \vdots & \vdots & \cdots & \vdots \\ A_{n1} & A_{n2} & \cdots & A_{nn} \end{pmatrix} \begin{pmatrix} a_1 \\ a_2 \\ \vdots \\ a_n \end{pmatrix} \qquad (3.42)$$

We can represent the square matrix as a linear operator $\hat{A}$ which transforms one wave function into another, by writing Eq.(3.42) in the form:

$$\Phi = \hat{A}\Psi \quad \text{or} \quad |\Phi\rangle = \hat{A}|\Psi\rangle \qquad (3.43)$$

The unit matrix, which is a diagonal matrix with 1's along the diagonal, will be represented as the identity operator $\hat{I}$, describing the multiplication of a wave function by unity, Eq.(2.60). In view of the expansion (3.4), which holds for an arbitrary state function in the form:

$$|\Psi\rangle = \sum_{i=1}^{n} |i\rangle\langle i|\Psi\rangle = \sum_{i=1}^{n} P_i|\Psi\rangle \qquad (3.44)$$

we may express the identity operator as a sum over projection operators, Eq.(3.37).

Conversely, for every linear operator $\hat{A}$ we can form a square matrix, in terms of a complete set of functions $|i\rangle$, by the scalar product of Eq.(3.43) with $\langle i|$, which gives:

$$\langle i|\Phi\rangle = \langle i|\hat{A}|\Psi\rangle = \langle i|\hat{A}\hat{I}|\Psi\rangle = \sum_{k=1}^{n} \langle i|\hat{A}|k\rangle\langle k|\Psi\rangle = \sum_{i=1}^{n} \langle i|\hat{A}|k\rangle a_k$$

Since $\langle i|\Phi\rangle = b_i$ this result is identical with that given by Eq.(3.41), provided we set:

$$A_{ik} = \langle i|\hat{A}|k\rangle \qquad (3.45)$$

which are called the *matrix elements* of the linear operator $\hat{A}$ in the particular complete set of functions $|i\rangle$. It is a straightforward matter to show that the rules of matrix algebra for addition, multiplication, associativity and distributivity apply to the matrix elements (3.45), following from their definition. The matrix representing the product of operators is the product of the matrices representing the operators:

$$(\hat{A}\hat{B})_{ik} = \langle i|\hat{A}\hat{B}|k\rangle = \langle i|\hat{A}\hat{I}\hat{B}|k\rangle = \sum_j \langle i|\hat{A}|j\rangle\langle j|\hat{B}|k\rangle = \sum_j A_{ij} B_{jk} \qquad (3.46)$$

In view of the definition (2.92) of the adjoint operator $\hat{A}^+$, we have:

$$(\hat{A}^+)_{ik} = \langle i|\hat{A}^+|k\rangle = \langle \hat{A}\varphi_i|\varphi_k\rangle = (\langle\varphi_k|\hat{A}\varphi_i\rangle)^* = (\langle\varphi_k|\hat{A}|\varphi_i\rangle)^* = (\hat{A})^*_{ki} \qquad (3.47)$$

which means that the adjoint operation consists of taking the transpose conjugate of the matrix representing $\hat{A}$. It follows that for Hermitian operators, Eq.(2.106):

$$A_{ik} = \langle i|\hat{A}|k\rangle = (\langle k|\hat{A}|i\rangle)^* = A^*_{ki} \qquad (3.48)$$

3. Eigenstates and the Third Postulate 131

In other words, if $\hat{A}$ is a *Hermitian operator, then its matrix representation in a complete basis is Hermitian.* Hence, the elements in the principal diagonal of a Hermitian matrix must be real. When all the elements are real, the matrix must be symmetric. We now have an equivalent way of writing the n linear homogenous equations (3.7) in the matrix form:

$$\begin{pmatrix} A_{11} & A_{12} & \cdots & A_{1n} \\ A_{21} & A_{22} & \cdots & A_{2n} \\ \cdots & \cdots & \cdots & \cdots \\ A_{n1} & A_{n2} & \cdots & A_{nn} \end{pmatrix} \begin{pmatrix} a_1 \\ a_2 \\ \cdots \\ a_n \end{pmatrix} = \lambda \begin{pmatrix} a_1 \\ a_2 \\ \cdots \\ a_n \end{pmatrix} \qquad (3.49)$$

Equation (3.49) is called the *matrix eigenvalue equation*, and the following theorem holds:

Theorem. *A Hermitian matrix is diagonal in a basis consisting of its orthonormal eigenfunctions and has its eigenvalues as the diagonal elements.*

Proof. Let λ_1 be a root of the characteristic polynomial $P_n(\lambda)$ to which it corresponds the eigenfunction ψ_1, from an n-dimensional wave function space, that may be assumed to be normalized to unity. If ψ is an arbitrary vector orthogonal to ψ_1, $\langle \psi | \psi_1 \rangle = 0$ then all vectors $\hat{A}\psi$ are also orthogonal to ψ_1, for:

$$\langle \hat{A}\psi | \psi_1 \rangle = \langle \psi | \hat{A} \psi_1 \rangle = \lambda_1 \langle \psi | \psi_1 \rangle = 0$$

and therefore the operator is restricted to the subspace of all vectors with null projection along the direction of ψ_1. This is an $n-1$ dimensional subspace, because any n-dimensional state function can be written in the form:

$$\Psi = a_1 \psi_1 + \psi = a_1 \psi_1 + \sum_{i=1}^{n-1} \alpha_i \psi'_i$$

where the basis vectors $\psi'_1, \psi'_2, \ldots, \psi'_{n-1}$ may all be chosen orthonormal to ψ_1. In other words the ψ'_i are chosen such as each one has a null projection along the direction of ψ_1. In this basis, Eq.(3.49) reads:

$$\begin{pmatrix} \lambda_1 & 0 & \cdots & 0 \\ 0 & A_{22} & \cdots & A_{2n} \\ \cdots & \cdots & \cdots & \cdots \\ 0 & A_{n2} & \cdots & A_{nn} \end{pmatrix} \begin{pmatrix} a_1 \\ \alpha_1 \\ \cdots \\ \alpha_{n-1} \end{pmatrix} = \lambda \begin{pmatrix} a_1 \\ \alpha_1 \\ \cdots \\ \alpha_{n-1} \end{pmatrix}$$

and the characteristic equation reduces to:

$$P_n(\lambda) = (\lambda_1 - \lambda) \begin{vmatrix} A_{22} - \lambda & \cdots & A_{2n} \\ \cdots & \cdots & \cdots \\ A_{n2} & \cdots & A_{nn} - \lambda \end{vmatrix}$$

$$= (\lambda_1 - \lambda)(c_{n-1}\lambda^{n-1} + \ldots + c_1\lambda + c_0) = (\lambda_1 - \lambda)P_{n-1}(\lambda) = 0$$

If $P_{n-1}(\lambda)$ has a root λ_2, not necessarily distinct from λ_1, and ψ_2 is the corresponding eigenfunction, we may introduce the $n-2$ dimensional subspace of wave vectors orthonormal to both ψ_1 and ψ_2 and the problem reduces in a similar way to the characteristic equation:

$$(\lambda_1 - \lambda)(\lambda_2 - \lambda)P_{n-2}(\lambda) = 0$$

with both λ_1 and λ_2 diagonal elements of the Hermitian matrix. It is obvious that finally we obtain by this procedure an n-dimensional basis such that Eq.(3.49) simplifies to:

$$\begin{pmatrix} \lambda_1 & 0 & \cdots & 0 \\ 0 & \lambda_2 & \cdots & 0 \\ \cdots & \cdots & \cdots & \cdots \\ 0 & 0 & \cdots & \lambda_n \end{pmatrix} \begin{pmatrix} a_1 \\ a_2 \\ \cdots \\ a_n \end{pmatrix} = \lambda \begin{pmatrix} a_1 \\ a_2 \\ \cdots \\ a_n \end{pmatrix} \quad (3.50)$$

Hence, the matrix representation of the Hermitian operator $\hat{A}$ in some orthonormal basis $|1\rangle, |2\rangle, \ldots, |n\rangle$, used in Eq.(3.49), becomes diagonal if $\hat{A}$ is represented in its own eigenbasis $|\psi_1\rangle, |\psi_2\rangle, \ldots, |\psi_n\rangle$, because the matrix elements (3.45) reduce to:

$$\langle \psi_i | \hat{A} | \psi_k \rangle = \lambda_i \langle \psi_i | \psi_k \rangle = \lambda_i \delta_{ik} \quad (3.51)$$

If $\hat{U}$ is the operator inducing this change of basis:

$$|\psi_k\rangle = \hat{U}|k\rangle$$

its matrix elements:

$$U_{ik} = \langle i|\hat{U}|k\rangle = \langle i|\psi_k\rangle \quad (3.52)$$

are the coefficients of the expansion of each eigenbasis vector:

$$|\psi_k\rangle = \sum_{i=1}^{n} U_{ik} |i\rangle, \qquad |k\rangle = \sum_{i=1}^{n} V_{ik} |\psi_i\rangle \quad (3.53)$$

3. Eigenstates and the Third Postulate

where the reciprocal change of basis results from:

$$V_{ik} = \langle \psi_i | k \rangle = \langle k | \psi_i \rangle^* = U_{ki}^* = (\hat{U}^+)_{ik}$$

This means that for any basis vector we have:

$$|k\rangle = \hat{U}^+ |\psi_k\rangle = \hat{U}^+ \hat{U} |k\rangle$$

In other words $\hat{U}^+ \hat{U}$ is the identity operator, see Eq.(2.112), and therefore an orthonormal basis in transformed into another by an unitary operation. The matrix elements (3.51) of the Hermitian operator $\hat{A}$ in its own eigenbasis:

$$\lambda_i \delta_{ik} = \langle \psi_i | \hat{A} | \psi_k \rangle = \langle i | \hat{U}^+ \hat{A} \hat{U} | k \rangle$$

are equals to the matrix elements of the operator $\hat{U}^+ \hat{A} \hat{U} = \hat{U}^{-1} \hat{A} \hat{U}$ in the orthonormal basis $|1\rangle, |2\rangle, ..., |n\rangle$. Conversely, in view of Eq.(3.53), the matrix elements A_{ik} in this basis, Eq.(3.45):

$$A_{ik} = \langle i | \hat{A} | k \rangle = \langle \psi_i | \hat{U} \hat{A} \hat{U}^+ | \psi_k \rangle \tag{3.54}$$

are the same as the matrix elements of $\hat{U} \hat{A} \hat{U}^+ = \hat{U} \hat{A} \hat{U}^{-1}$ in the eigenbasis of $\hat{A}$.

EXAMPLE 3.5. Diagonalization of a matrix

Consider a matrix of rank 3, representing a Hermitian operator $\hat{A}$ in the orthonormal basis $|1\rangle, |2\rangle, |3\rangle$:

$$(\hat{A}) = \begin{pmatrix} A_{11} & A_{12} & A_{13} \\ A_{21} & A_{22} & A_{23} \\ A_{31} & A_{32} & A_{33} \end{pmatrix}$$

which is to be transformed to a diagonal matrix ($D_{ik} = D_{kk} \delta_{ik}$):

$$(\hat{D}) = \begin{pmatrix} D_{11} & 0 & 0 \\ 0 & D_{22} & 0 \\ 0 & 0 & D_{33} \end{pmatrix}$$

Let $(\hat{U})$ be a unitary matrix so that:

$$(\hat{U}^{-1})(\hat{A})(\hat{U}) = (\hat{D}) \quad \text{or} \quad (\hat{A})(\hat{U}) = (\hat{U})(\hat{D})$$

134 Quantum Physics

or, in terms of the matrix elements:

$$\sum_{j=1}^{3} A_{ij} U_{jk} = \sum_{j=1}^{3} U_{ij} D_{jk} = U_{ik} D_{kk}$$

or explicitly:

$$A_{i1} U_{1k} + A_{i2} U_{2k} + A_{i3} U_{3k} = U_{ik} D_{kk} \quad (i,k = 1,2,3)$$

Setting $k = 1$, we obtain three linear homogenous equations for $i = 1, 2$ and 3:

$$A_{11} U_{11} + A_{12} U_{21} + A_{13} U_{31} = U_{11} D_{11}$$

$$A_{21} U_{11} + A_{22} U_{21} + A_{23} U_{31} = U_{21} D_{11}$$

$$A_{31} U_{11} + A_{32} U_{21} + A_{33} U_{31} = U_{31} D_{11}$$

and, by eliminating U_{11} and U_{21}, we have:

$$\begin{vmatrix} A_{11} - D_{11} & A_{12} & A_{13} \\ A_{21} & A_{22} - D_{11} & A_{23} \\ A_{31} & A_{32} & A_{33} - D_{11} \end{vmatrix} = 0$$

If we set $k = 2$ and $k = 3$ respectively, the same procedure yields:

$$\begin{vmatrix} A_{11} - D_{22} & A_{12} & A_{13} \\ A_{21} & A_{22} - D_{22} & A_{23} \\ A_{31} & A_{32} & A_{33} - D_{22} \end{vmatrix} = 0, \quad \begin{vmatrix} A_{11} - D_{33} & A_{12} & A_{13} \\ A_{21} & A_{22} - D_{33} & A_{23} \\ A_{31} & A_{32} & A_{33} - D_{33} \end{vmatrix} = 0$$

It follows that the diagonal elements D_{11}, D_{22} and D_{33} are the roots of the three-dimensional characteristic equation (3.8). For each eigenvalue, the linear homogenous equations can be solved for the U_{ik} giving the corresponding eigenfunction of $\hat{A}$, which is a column vector. Therefore, the unitary matrix $(\hat{U})$ is built out of the eigenvectors of $\hat{A}$, in the basis $|1\rangle, |2\rangle, |3\rangle$, as columns. ♌

Problem 3.3.1. Solve the eigenvalue problem for the matrix $(\hat{A}) = \begin{pmatrix} 0 & 1 & 0 \\ 1 & 0 & 0 \\ 0 & 0 & n \end{pmatrix}$.

(Solution):   We have the characteristic equation:

$$\begin{vmatrix} -\lambda & 1 & 0 \\ 1 & -\lambda & 0 \\ 0 & 0 & n-\lambda \end{vmatrix} = (n-\lambda)(\lambda^2 - 1) = 0$$

which gives $\lambda_1 = n, \lambda_2 = 1, \lambda_3 = -1$. Substituting in turn the λ_i into the matrix eigenvalue equation we obtain for $\lambda_1 = n$:

$$\begin{pmatrix} 0 & 1 & 0 \\ 1 & 0 & 0 \\ 0 & 0 & n \end{pmatrix} \begin{pmatrix} a_1 \\ a_2 \\ a_3 \end{pmatrix} = n \begin{pmatrix} a_1 \\ a_2 \\ a_3 \end{pmatrix}$$

or $a_1 = a_2 = 0$, that is the eigenvector should have the form:

$$\begin{pmatrix} 0 \\ 0 \\ a_3 \end{pmatrix}$$

If we use the normalization condition, and proceed in the same way for λ_2 and λ_3, we obtain:

$$|1\rangle = \begin{pmatrix} 0 \\ 0 \\ 1 \end{pmatrix}, \quad |2\rangle = \begin{pmatrix} \frac{1}{\sqrt{2}} \\ \frac{1}{\sqrt{2}} \\ 0 \end{pmatrix}, \quad |3\rangle = \begin{pmatrix} \frac{1}{\sqrt{2}} \\ -\frac{1}{\sqrt{2}} \\ 1 \end{pmatrix}$$

It is easy to see that these eigenvectors form the three columns of the $(\hat{U})$ matrix which determine the diagonalization of $(\hat{A})$. The diagonal representation of the operator $\hat{A}$ is:

$$(\hat{A}) = \begin{pmatrix} n & 0 & 0 \\ 0 & 1 & 0 \\ 0 & 0 & -1 \end{pmatrix}$$

in the basis of its normalized eigenvectors:

$$|\psi_1\rangle = \begin{pmatrix} 1 \\ 0 \\ 0 \end{pmatrix}, \quad |\psi_2\rangle = \begin{pmatrix} 0 \\ 1 \\ 0 \end{pmatrix}, \quad |\psi_3\rangle = \begin{pmatrix} 0 \\ 0 \\ 1 \end{pmatrix}$$

Problem 3.3.2. Solve the eigenvalue problem for the matrix $(\hat{A}) = \begin{pmatrix} 0 & 1 & 1 \\ 1 & 0 & 1 \\ 1 & 1 & 0 \end{pmatrix}$.

(Solution): The characteristic equation:

$$\begin{vmatrix} -\lambda & 1 & 1 \\ 1 & -\lambda & 1 \\ 1 & 1 & -\lambda \end{vmatrix} = (\lambda+1)^2(\lambda-2) = 0$$

shows that we have degeneracy. For $\lambda_1 = 2$, it is immediate that we may take:

$$|1\rangle = \frac{1}{\sqrt{3}} \begin{pmatrix} 1 \\ 1 \\ 1 \end{pmatrix}$$

For $\lambda_2 = \lambda_3 = -1$, we have the condition for coefficients:

$$a_1 + a_2 + a_3 = 0, \quad a_1' + a_2' + a_3' = 0$$

with the possible, but not unique, choice:

$$\begin{pmatrix} 1 \\ -1 \\ 0 \end{pmatrix}, \begin{pmatrix} 1 \\ 0 \\ -1 \end{pmatrix}$$

Since this vectors are not orthogonal, we may take linear combinations to form the orthogonal pair:

$$\begin{pmatrix} 2 \\ -1 \\ -1 \end{pmatrix}, \begin{pmatrix} 0 \\ 1 \\ -1 \end{pmatrix}$$

which, by normalization, becomes:

$$|2\rangle = \begin{pmatrix} 1 \\ -\frac{1}{2} \\ -\frac{1}{2} \end{pmatrix}, \quad |3\rangle = \begin{pmatrix} 0 \\ \frac{1}{\sqrt{2}} \\ -\frac{1}{\sqrt{2}} \end{pmatrix}$$

Problem 3.3.3. Find the eigenvalues of a unitary operator $\hat{U}$ and show that, assuming no degeneracy, the eigenfunctions form an orthogonal set.

(Answer): $|\lambda_i|^2 = 1$

3.4. Matrix Representations

We have seen, Eq.(3.44), that a state vector from the wave function space is completely defined by its components, that is by its scalar products with the basis vectors $\langle i|$ from the dual space. Similarly, a Hermitian operator is completely specified if the result of applying it to an arbitrary state vector Ψ, Eq.(3.42), is always known, and therefore the operator can be defined by its matrix elements (3.45) in a complete vector basis $|i\rangle$. If the eigenfunctions ψ_i of a Hermitian operator $\hat{A}$, which form a complete orthonormal set, are chosen as basis vectors, it is said that the state vectors and Hermitian operators are specified in the *representation* of the associated observable A. In other words, all the state vectors and Hermitian operators obeying the partial differential equations of wave mechanics may be so replaced by their matrix representations:

$$\Psi = \sum_i a_i \psi_i = \sum_i \langle \psi_i | \Psi \rangle \psi_i$$

$$(\hat{B})_{ik} = B_{ik} = \langle \psi_i | \hat{B} | \psi_k \rangle \tag{3.55}$$

which are systems of discrete numerical quantities a_i, B_{ik} that are defined by algebraic equations. Hence, the observable representations constitute a generalization of the familiar concept of a reference frame, where ordinary vectors and linear transformations are defined in terms of their components. The expectation value of the observable B in a dynamical state Ψ takes the form:

$$\langle B \rangle = \langle \Psi | \hat{B} | \Psi \rangle = \sum_{i,k} \langle \Psi | \psi_i \rangle \langle \psi_i | \hat{B} | \psi_k \rangle \langle \psi_k | \Psi \rangle = \sum_{i,k} a_i^* B_{ik} a_k \tag{3.56}$$

When written for the associated operator $\hat{A}$, the second equation (3.55) reduces to Eq.(3.51) and the matrix representative is diagonal. Therefore Eq.(3.56) reads in this case $\langle A \rangle = \sum_i |\langle \psi_i | \Psi \rangle|^2 \lambda_i$ as found for the discrete portion of Eq.(3.32). In the case of a change of basis, Eq.(3.53), induced by the unitary operator $\hat{U}$, the matrix elements of $\hat{B}$ transform as:

$$B'_{ik} = \langle i|\hat{B}|k\rangle = \sum_{p,q} U_{ip}\langle\psi_p|\hat{B}|\psi_q\rangle U^*_{kq} = \sum_{p,q} U_{ip} B_{pq} (\hat{U}^+)_{qk} \qquad (3.57)$$

which is the matrix equation:

$$(\hat{B}') = (\hat{U})(\hat{B})(\hat{U}^+) \qquad (3.58)$$

Assuming for the A-representation a purely continuous set of basis vectors $\psi(\lambda)$, Eqs.(3.20) and (3.21), the matrix representatives for state vectors and Hermitian operators have indices which take on a continuum of values, rather than being denumerable. Hence such matrices can no longer be written down as a column or a square array. However, in view of the expansion (3.18), which reads:

$$\Psi = \int \langle\psi(\lambda)|\Psi\rangle \psi(\lambda) d\lambda \qquad (3.59)$$

any state vector Ψ will be specified by the coefficients $\langle\psi(\lambda)|\Psi\rangle$, which may be regarded as elements of a column matrix representative. The scalar product between two state vectors, which is obtained from Eq.(3.59) as:

$$(\Phi|\Psi) = \int \langle\psi(\lambda)|\Psi\rangle\langle\Phi|\psi(\lambda)\rangle d\lambda \qquad (3.60)$$

is formed by the ordinary rule of matrix multiplication, see Eq.(1.71). If a Hermitian operator $\hat{B}$ is defined by:

$$\Phi = \hat{B}\Psi \qquad (3.61)$$

we have from Eqs.(3.59) and (3.60):

$$\langle\psi(\lambda)|\Phi\rangle = \langle\psi(\lambda)|\hat{B}|\Psi\rangle = \int \langle\psi(\lambda)|\hat{B}|\psi(\lambda')\rangle\langle\psi(\lambda')|\Psi\rangle d\lambda' \qquad (3.62)$$

This is the analogous of Eq.(3.41) for the case of a continuous vector basis, and therefore we may treat the quantities:

$$B_{\lambda\lambda'} = \langle\psi(\lambda)|\hat{B}|\psi(\lambda')\rangle \qquad (3.63)$$

as the matrix elements of $\hat{B}$, although the familiar square array representation is no possible for continuous indices. It is also appropriate to say that the matrix elements:

3. Eigenstates and the Third Postulate

$$A_{\lambda\lambda'} = \langle \psi(\lambda) | \hat{A} | \psi(\lambda') \rangle = \lambda \delta(\lambda - \lambda') \tag{3.64}$$

where use has been made of Eqs.(3.17) and (3.20), form the diagonal matrix representation of $\hat{A}$, in analogy with Eq.(3.51). The expectation value of the observable B, in a state Ψ described by Eq.(3.59), takes a form similar to Eq.(3.56):

$$\langle B \rangle = \langle \Psi | \hat{B} | \Psi \rangle = \int\int_{\lambda\lambda'} \langle \Psi | \psi(\lambda) \rangle \langle \psi(\lambda) | \hat{B} | \psi(\lambda') \rangle \langle \psi(\lambda') | \Psi \rangle d\lambda \, d\lambda' \tag{3.65}$$

which reduces for the observable A, in view of Eq.(3.64), to:

$$\langle A \rangle = \int |\langle \psi(\lambda) | \Psi \rangle|^2 \lambda \, d\lambda \tag{3.66}$$

The matrix representation in the eigenvector basis (3.23) of the momentum operator p_x is called the *momentum representation*. The expansion (3.59) of a state vector in this basis reads:

$$\Psi(x) = \frac{1}{\sqrt{2\pi\hbar}} \int \langle \psi(x, p_x) | \Psi \rangle e^{ip_x x/\hbar} dp_x$$

where, in view of Eq.(3.19), the coefficients:

$$\Phi(p_x) = \langle \psi(x, p_x) | \Psi \rangle = \frac{1}{\sqrt{2\pi\hbar}} \int \Psi(x) e^{-ip_x x/\hbar} dx$$

consist of all values $\Phi(p_x)$ of the Fourier transform of $\Psi(x)$, Eq.(2.39). In other words, the Fourier transform of the wave function $\Psi(x)$ is the wave function in the momentum representation. The matrix elements (3.63) of a Hermitian operator $\hat{B}$ read:

$$B_{p'_x p''_x} = \int_{-\infty}^{\infty} \psi^*(x, p'_x) \hat{B} \psi(x, p''_x) dx = \frac{1}{2\pi\hbar} \int_{-\infty}^{\infty} e^{-ip'_x x/\hbar} \hat{B} e^{ip''_x x/\hbar} dx$$

It follows, as expected, that the matrix representative of $\hat{p}_x$ is diagonal in its own eigenbasis:

$$(\hat{p}_x)_{p'_x p''_x} = \frac{1}{2\pi\hbar} \int_{-\infty}^{\infty} e^{-ip'_x x/\hbar} (\hat{p}_x e^{ip''_x x/\hbar}) dx = \frac{p''_x}{2\pi\hbar} \int_{-\infty}^{\infty} e^{i(p''_x - p'_x)x/\hbar} dx = p''_x \delta(p''_x - p'_x) \tag{3.67}$$

where p''_x is the eigenvalue of $\hat{p}_x$ in the eigenstate $\psi(x, p''_x)$. The matrix elements of the position operator $\hat{x}$:

$$x_{p'_x p''_x} = \frac{1}{2\pi\hbar} \int_{-\infty}^{\infty} e^{-ip'_x x/\hbar} (xe^{ip''_x x/\hbar}) dx = \frac{i\hbar}{2\pi\hbar} \int_{-\infty}^{\infty} e^{-ip'_x x/\hbar} \frac{\partial}{\partial p''_x} (e^{ip''_x x/\hbar}) dx$$

$$= i\hbar \frac{\partial}{\partial p''_x} \left[\frac{1}{2\pi\hbar} \int_{-\infty}^{\infty} e^{i(p''_x - p'_x)x/\hbar} dx \right] = i\hbar \frac{\partial}{\partial p''_x} \delta(p''_x - p'_x) = i\hbar \frac{\partial}{\partial p'_x} \delta(p'_x - p''_x) \qquad (3.68)$$

are those of a non-diagonal matrix, because the derivative $\delta'(p'_x - p''_x)$ is an odd function, see Eqs.(1.47). The expectation value of momentum p_x in an arbitrary state Ψ, is given by Eq.(3.66) in the form:

$$\langle p_x \rangle = \int |\Phi(p_x)|^2 p_x dp_x$$

The expectation value of the position x in this representation, Eq.(3.65), reads:

$$\langle x \rangle = \iint_{p_x p'_x} \Phi^*(p_x) x_{p_x p'_x} \Phi(p'_x) dp_x dp'_x = -i\hbar \iint_{p_x p'_x} \Phi^*(p_x) \left[\frac{\partial}{\partial p'_x} \delta(p_x - p'_x) \right] \Phi(p'_x) dp_x dp'_x$$

$$= i\hbar \iint_{p_x p'_x} \Phi^*(p_x) \delta(p_x - p'_x) \frac{\partial \Phi(p'_x)}{\partial p'_x} dp_x dp'_x = \int \Phi^*(p_x) \left(i\hbar \frac{\partial}{\partial p_x} \right) \Phi(p_x) dp_x$$

where we made use of the property:

$$\int_{-\infty}^{\infty} \frac{\partial}{\partial p'_x} \delta(p_x - p'_x) \Phi(p'_x) dp'_x = [\delta(p_x - p'_x) \Phi(p'_x)]_{-\infty}^{\infty} - \int_{-\infty}^{\infty} \delta(p_x - p'_x) \frac{\partial \Phi(p'_x)}{\partial p'_x} dp'_x$$

$$= -\int_{-\infty}^{\infty} \delta(p_x - p'_x) \frac{\partial \Phi(p'_x)}{\partial p'_x} dp'_x$$

which holds because $\Phi(p'_x)$ is vanishingly small for $p'_x \to \pm\infty$.

Since $\langle p_x \rangle$ and $\langle x \rangle$ are the same as derived in Eqs.(2.40) and (2.44), the same rules (2.49) for representation of the position and momentum operators in the momentum space follow. In other words, the matrix momentum representation for quantum states and quantum operators in the eigenbasis of the momentum operator is equivalent to the representation of wave functions and differential operators in momentum space, as described in Chapter 2.

In the matrix representation spanned by the eigenvectors (3.26) of the position operator x, called the *coordinate representation*, any wave vector can be specified by the expansion (3.59) which reads:

3. Eigenstates and the Third Postulate 141

$$\Psi(x) = \int_{-\infty}^{\infty} \langle \psi(x')|\Psi \rangle \psi(x')dx' = \int_{-\infty}^{\infty} \Psi(x')\delta(x-x')dx'$$

and shows that the coefficients of representation consists of all values of the function $\Psi(x')$, for any real value of the position x. Any Hermitian operator acting on the wave vectors $\Psi(x)$ is represented by the matrix elements (3.63), namely:

$$B_{x'x''} = \langle \psi(x')|\hat{B}|\psi(x'') \rangle$$

$$= \int_{-\infty}^{\infty} \psi^*(x,x')\hat{B}\psi(x,x'')dx = \int_{-\infty}^{\infty} \delta(x-x')\hat{B}_x\delta(x-x'')dx = \hat{B}_{x'}\delta(x'-x'') \quad (3.69)$$

where the indices x or x' attached to $\hat{B}$ indicate the variable in terms of which the operator should be expressed. If $\hat{B}$ is a multiplicative operator, $\hat{B} = f(x)$, the matrix representative is diagonal:

$$f_{x'x''}(x) = f(x')\delta(x'-x'') \quad (3.70)$$

since the Dirac δ-function is even, see Eqs.(1.46). This diagonal representation is certainly expected for the particular case $f(x) = x$.

It is convenient to use the unitary transformation $\hat{U}$ with matrix elements (3.53), where the basis vectors $|i\rangle$ are replaced by the position eigenvectors (3.26) and the eigenvectors $|\psi_k\rangle$ are those of the momentum operators, Eq.(3.23):

$$U(x', p_x) = \frac{1}{\sqrt{2\pi\hbar}} \int \delta(x-x')e^{ip_x x/\hbar}dx = \frac{1}{\sqrt{2\pi\hbar}} e^{ip_x x'/\hbar}$$

The matrix so defined is unitary, as required by Eq.(2.112), because it can easily be shown, in view of Eqs.(1.48) and (1.43), that:

$$\int U^*(x, p_x)U(x', p_x)dp_x = \delta(x-x')$$

$$\int U^*(x, p_x)U(x, p'_x)dx = \delta(p_x - p'_x)$$

While the matrix $(\hat{U})$ defines the change from the coordinate to the momentum eigenbasis, the reciprocal change of representation is given by the unitary matrix $(\hat{V})$ with elements:

$$V(p_x, x) = \frac{1}{\sqrt{2\pi\hbar}} \int e^{-ip_x x'/\hbar} \delta(x'-x)dx' = \frac{1}{\sqrt{2\pi\hbar}} e^{-ip_x x/\hbar}$$

142   Quantum Physics

and it is obvious that $\hat{V} = \hat{U}^+$. We may now apply Eq.(3.58) to derive the coordinate representation of the momentum operator, starting from the matrix elements (3.67) written in its own eigenbasis:

$$(\hat{p}_x)_{xx'} = \int\int_{p_x p'_x} V^*(p_x, x)(\hat{p}_x)_{p_x p'_x} V(p'_x, x') dp_x dp'_x$$

$$= \frac{1}{2\pi\hbar} \int\int_{p_x p'_x} e^{ip_x x/\hbar} p'_x \delta(p'_x - p_x) e^{-ip'_x x'/\hbar} dp_x dp'_x$$

$$= \frac{1}{2\pi\hbar} \int e^{ip_x x/\hbar} (p_x e^{-ip_x x'/\hbar}) dp_x = \frac{1}{2\pi\hbar} \int e^{ip_x x/\hbar} \left(-\frac{\hbar}{i}\right) \frac{\partial}{\partial x'} (e^{-ip_x x'/\hbar}) dp_x$$

$$= i\hbar \frac{\partial}{\partial x'} \left[\frac{1}{2\pi\hbar} \int e^{ip_x(x-x')/\hbar} dp_x \right] = i\hbar \frac{\partial}{\partial x'} \delta(x - x') = -i\hbar \frac{\partial}{\partial x} \delta(x - x')$$

If compared to Eq.(3.69), this result shows that the momentum operator in the coordinate representation assumes the form previously found in Eq.(2.7). We can write:

$$(p_x)_{xx'} = -i\hbar\, \delta'(x - x')$$

and the matrix representative, analogous to that found in Eq.(3.68), is no longer diagonal. It follows that the matrix coordinate representation in the eigenbasis of the position operator leads to the same description of the quantum states and quantum operators as that obtained in Chapter 2, using the coordinate representation of the wave functions and the basic assumptions of wave mechanics.

Problem 3.4.1. Show that the expectation value of a physical observable is conserved through a unitary change of basis.

(Solution): Consider a unitary transformation $\hat{U}$, inducing the following change of the state vectors and operators:

$$\Phi = \hat{U} \Psi, \quad \Psi = \hat{U}^{-1} \Phi$$

$$\hat{A}' = \hat{U} \hat{A} \hat{U}^{-1}, \quad \hat{A} = \hat{U}^{-1} \hat{A}' \hat{U}$$

The expectation value of an observable A in the states Ψ and Φ of a system is given by Eqs.(2.11) and (2.12) that reads:

3. Eigenstates and the Third Postulate 143

$$\langle A \rangle = \int \Psi^* \hat{A} \Psi dx, \quad \langle A \rangle = \int \Phi^* \hat{A}' \Phi dx$$

respectively. In view of the basic property of unitary operators, Eq.(2.111), we successively have:

$$\int \Phi^* \hat{A}' \Phi dx = \int (\hat{U}\Psi)^* \hat{A}' \hat{U} \Psi dx = \int \Psi^* \hat{U}^{-1} \hat{A}' \hat{U} \Psi dx = \int \Psi^* \hat{A} \Psi dx$$

and hence, any expectation value is conserved. If $\hat{A}$ is the identity operator, $\langle A \rangle$ reduces to the norm of the state vector, which is also conserved by a unitary transformation.

Problem 3.4.2. The trace of a square matrix is defined to be sum of its diagonal elements:

$$Tr(\hat{A}) = \sum_{i=1}^{n} A_{ii}$$

Show that the trace is invariant with respect to unitary transformations of the vector basis.

(Solution): Let us denote by $A_{ik} = \langle i | \hat{A} | k \rangle$, $A'_{ik} = \langle \psi_i | \hat{A} | \psi_k \rangle$, the matrix elements of the same operator in two different bases related by the unitary transformation (3.53). We successively have:

$$Tr(\hat{A})' = \sum_{i=1}^{n} \langle \psi_i | \hat{A} | \psi_i \rangle = \sum_{i=1}^{n} \sum_{j=1}^{n} \sum_{k=1}^{n} U_{ki}^* \langle k | \hat{A} | j \rangle U_{ji} = \sum_{j=1}^{n} \sum_{k=1}^{n} \left(\sum_{i=1}^{n} U_{ki}^* U_{ji} \right) A_{kj}$$

It follows that:

$$Tr(\hat{A})' = \sum_{j=1}^{n} \sum_{k=1}^{n} \delta_{jk} A_{kj} = \sum_{k=1}^{n} A_{kk} = Tr(\hat{A})$$

Problem 3.4.3. Show that, if $(\hat{A})$ and $(\hat{B})$ are the matrix representatives of two Hermitian operators, we have:

$$Tr(\hat{A} + \hat{B}) = Tr(\hat{A}) + Tr(\hat{B})$$

$$Tr(\hat{A}\hat{B}) = Tr(\hat{B}\hat{A})$$

---

4. COMMUTATION RELATIONS AND THE FOURTH POSTULATE

4.1. Commutator Algebra

In view of the Heisenberg uncertainty principle, introduced in Chapter 1, the application of Postulate 3 to a dynamical system faces one problem never encountered in classical mechanics, namely to define the conditions which make possible the simultaneous measurement of different observables. Because each physical observable takes definite values in the eigenstates of its associated operator, the simultaneous measurements of different observable will only be possible if their associated operators have a common eigenbasis. If such a complete orthonormal set of vectors ψ_i are simultaneous eigenfunctions of two operators $\hat{A}$ and $\hat{B}$:

$$\hat{A}\psi_i = \lambda_i \psi_i, \quad \hat{B}\psi_i = \mu_i \psi_i \qquad (4.1)$$

we have for each basis vector:

$$\hat{A}\hat{B}\psi_i = \hat{A}\mu_i \psi_i = \lambda_i \mu_i \psi_i = \mu_i \lambda_i \psi_i = \hat{B}\hat{A}\psi_i$$

It follows that for any state vector Ψ, which can be expanded in the form (3.55), we have:

$$(\hat{A}\hat{B} - \hat{B}\hat{A})\Psi = 0 \quad \text{or} \quad \hat{A}\hat{B} - \hat{B}\hat{A} = 0 \qquad (4.2)$$

This means, in view of Eq.(2.80), that simultaneous measurements on a dynamical system are restricted to those observables for which the associated operators commute.

It is convenient to introduce the *commutator* of two operators, defined as a bracket:

$$[\hat{A}, \hat{B}] = \hat{A}\hat{B} - \hat{B}\hat{A} \qquad (4.3)$$

4. Commutation Relations and the Fourth Postulate

which is itself an operator playing an important role in the development of quantum formalism. If the operators $\hat{A}$ and $\hat{B}$ commute, their commutator reduces to the null operator. The adjoint of the commutator (4.3) reads:

$$[\hat{A},\hat{B}]^+ = (\hat{A}\hat{B})^+ - (\hat{B}\hat{A})^+ = \hat{B}^+\hat{A}^+ - \hat{A}^+\hat{B}^+$$

and it follows that the commutator of two Hermitian operators is not Hermitian. However, if both $\hat{A}$ and $\hat{B}$ are Hermitian, the operator $i[\hat{A},\hat{B}]$ is Hermitian, for:

$$\left(i[\hat{A},\hat{B}]\right)^+ = -i[\hat{A},\hat{B}]^+ = -i(\hat{B}\hat{A} - \hat{A}\hat{B}) = i(\hat{A}\hat{B} - \hat{B}\hat{A}) = i[\hat{A},\hat{B}] \qquad (4.4)$$

The commutators are antisymmetric and linear in both factors:

$$[\hat{A},\hat{B}] = -[\hat{B},\hat{A}] \qquad (4.5)$$

$$[\alpha_1\hat{A}_1 + \alpha_2\hat{A}_2, \hat{B}] = \alpha_1[\hat{A}_1,\hat{B}] + \alpha_2[\hat{A}_2,\hat{B}] \qquad (4.6)$$

$$[\hat{A},\alpha_1\hat{B}_1 + \alpha_2\hat{B}_2] = \alpha_1[\hat{A},\hat{B}_1] + \alpha_2[\hat{A},\hat{B}_2] \qquad (4.7)$$

where α_1,α_2 are complex numbers. It is a straightforward matter to prove, by explicitly writing all the commutators, the following rules of commutator algebra:

$$[\hat{A}\hat{B},\hat{C}] = \hat{A}[\hat{B},\hat{C}] + [\hat{A},\hat{C}]\hat{B} \qquad (4.8)$$

$$[\hat{A},\hat{B}\hat{C}] = [\hat{A},\hat{B}]\hat{C} + \hat{B}[\hat{A},\hat{C}] \qquad (4.9)$$

$$\left[[\hat{A},\hat{B}],\hat{C}\right] + \left[[\hat{B},\hat{C}],\hat{A}\right] + \left[[\hat{C},\hat{A}],\hat{B}\right] = 0 \qquad (4.10)$$

Equation (4.10) is known as the Jacobi identity.

EXAMPLE 4.1. Functions of non-commuting operators

The operators representing physical observable should be expressed, by Postulate 2, in terms of the position and momentum operators in one dimension, which do not commute, Eq.(2.82), and this might rise the problem of handling functions of non-commuting operators. Using the powers of an operator $\hat{A}$, which are defined by the familiar rule of repeated operator multiplication, we can introduce polynomials with complex coefficients $P(\hat{A}) = \sum_{k=1}^{n} c_k \hat{A}^k$ and also define a function $f(\hat{A})$ by a power series expansion:

$$f(\hat{A}) = \sum_{n=0}^{\infty} c_n \hat{A}^n \qquad (4.11)$$

which is the analogous of the Taylor expansion for an ordinary function. Both $P(\hat{A})$ and $f(\hat{A})$ are linear operators, the simplest example being provided by the exponential operator:

$$e^{\hat{A}} = \sum_{n=0}^{\infty} \frac{\hat{A}^n}{n!} \qquad (4.12)$$

The differentiation of $f(\hat{A})$ follows from Eq.(4.11) as:

$$\frac{df(\hat{A})}{d\hat{A}} = \sum_{n=1}^{\infty} nc_n \hat{A}^{n-1} \qquad (4.13)$$

and takes a compact form if $\hat{A}$ is replaced by $\hat{x}_i$ or $\hat{p}_i$. Let us evaluate $[x_i, p_i^n]$ using the rule (4.9), which successively gives, in view of Eq.(2.82):

$$[\hat{x}_i, \hat{p}_i^2] = \hat{p}_i[\hat{x}_i, \hat{p}_i] + [\hat{x}_i, \hat{p}_i]\hat{p}_i = 2i\hbar \hat{p}_i$$

$$[\hat{x}_i, \hat{p}_i^3] = [\hat{x}_i, \hat{p}_i^2 \hat{p}_i] = \hat{p}_i^2[\hat{x}_i, \hat{p}_i] + [\hat{x}_i, \hat{p}_i^2]\hat{p}_i = 3i\hbar \hat{p}_i^2$$

and finally $[\hat{x}_i, \hat{p}_i^n] = i\hbar n \hat{p}_i^{n-1}$. Inserting this result into Eq.(4.13) one obtains:

$$i\hbar \frac{df(\hat{p}_i)}{d\hat{p}_i} = \sum_{n=1}^{\infty} c_n [\hat{x}_i, \hat{p}_i^n] = \left[\hat{x}_i, \sum_{n=0}^{\infty} c_n \hat{p}_i^n\right] = [\hat{x}_i, f(\hat{p}_i)] \qquad (4.14)$$

In a similar way we can show that $[\hat{x}_i^n, \hat{p}_i] = i\hbar n \hat{x}_i^{n-1}$ and therefore Eq.(4.13) takes the form:

$$i\hbar \frac{df(\hat{x}_i)}{d\hat{x}_i} = [f(\hat{x}_i), \hat{p}_i] \qquad (4.15)$$

It was possible to reduce Eq.(4.13) to the forms (4.14) or (4.15) only because the commutator of the operators involved is simply a complex number, which obviously commute with each of them. If $\hat{A}$ and $\hat{B}$ are two arbitrary operators of this kind, which both commute with their commutator, $[\hat{A}, \hat{B}]$, one may use the same procedure as before to prove that:

$$[\hat{B}, \hat{A}^n] = n\hat{A}^{n-1}[\hat{B}, \hat{A}] \qquad (4.16)$$

$$[\hat{B}^n, \hat{A}] = n\hat{B}^{n-1}[\hat{B}, \hat{A}] \qquad (4.17)$$

It follows from Eq.(4.12) that, for a real parameter λ, we have:

4. Commutation Relations and the Fourth Postulate

$$[\hat{B}, e^{-\lambda \hat{A}}] = \sum_{n=0}^{\infty}(-1)^n \frac{\lambda^n}{n!}[\hat{B}, \hat{A}^n] = \sum_{n=1}^{\infty}(-1)^n \frac{\lambda^n}{(n-1)!}\hat{A}^{n-1}[\hat{B}, \hat{A}] = -\lambda e^{-\lambda \hat{A}}[\hat{B}, \hat{A}]$$

This means that:

$$\hat{B}e^{-\lambda\hat{A}} = e^{-\lambda\hat{A}}\{\hat{B} - \lambda[\hat{B}, \hat{A}]\} \quad \text{or} \quad e^{\lambda\hat{A}}\hat{B}e^{-\lambda\hat{A}} = \hat{B} + \lambda[\hat{A}, \hat{B}] \qquad (4.18)$$

It is now convenient to we introduce the function of λ:

$$F(\lambda) = e^{\lambda\hat{A}} e^{\lambda\hat{B}} \qquad (4.19)$$

such that, in view of Eq.(4.18) one obtains:

$$\frac{dF(\lambda)}{d\lambda} = \hat{A}e^{\lambda\hat{A}}e^{\lambda\hat{B}} + e^{\lambda\hat{A}}\hat{B}e^{\lambda\hat{B}} = (\hat{A} + e^{\lambda\hat{A}}\hat{B}e^{-\lambda\hat{A}})F(\lambda) \qquad (4.20)$$

or:

$$\frac{dF(\lambda)}{d\lambda} = \left(\hat{A} + \hat{B} + \lambda[\hat{A}, \hat{B}]\right)F(\lambda)$$

Because the operators are independent of λ, this differential equation for $F(\lambda)$ can be integrated by treating these operators as ordinary constants, which gives:

$$F(\lambda) = e^{(\hat{A}+\hat{B})\lambda + [\hat{A},\hat{B}]\lambda^2/2}$$

The normalization constant was chosen to be equal to unity because, from the definition (4.19), it follows that $F(0) = 1$. By combining Eqs.(4.19) and (4.20) and setting $\lambda = 1$, we obtain the identity:

$$e^{\hat{A}}e^{\hat{B}} = e^{\hat{A}+\hat{B}+[\hat{A},\hat{B}]/2} \qquad (4.21)$$

which shows that the simple rule $e^{\hat{A}}e^{\hat{B}} = e^{\hat{A}+\hat{B}}$ only holds provided $\hat{A}$ and $\hat{B}$ are commuting operators.

If $\hat{A}$ and $\hat{B}$ are no longer restricted to obey same commutation rule, Eq.(4.18) assumes a different form, which can be found by successive differentiation of:

$$\hat{B}(\lambda) = e^{\lambda\hat{A}}\hat{B}e^{-\lambda\hat{A}}$$

with respect to λ:

$$\frac{d\hat{B}(\lambda)}{d\lambda} = \hat{A}e^{\lambda\hat{A}}\hat{B}e^{-\lambda\hat{A}} - e^{\lambda\hat{A}}\hat{B}e^{-\lambda\hat{A}}\hat{A} = [\hat{A}, \hat{B}(\lambda)]$$

$$\frac{d^2\hat{B}(\lambda)}{d\lambda^2} = \left[\hat{A}, \frac{d\hat{B}(\lambda)}{d\lambda}\right] = [\hat{A}, [\hat{A}, \hat{B}(\lambda)]]$$

and so on, such that, because $\hat{B}(0) = \hat{B}$, we may write the Taylor expansion:

$$\hat{B}(\lambda) = \hat{B} + \frac{\lambda}{1!}\frac{d\hat{B}}{d\lambda} + \frac{\lambda^2}{2!}\frac{d^2\hat{B}}{d\lambda^2} + \ldots = \hat{B} + \frac{\lambda}{1!}[\hat{A},\hat{B}] + \frac{\lambda^2}{2!}\left[\hat{A},[\hat{A},\hat{B}]\right] + \ldots$$

Hence, we have the identity:

$$e^{\lambda\hat{A}}\hat{B}e^{-\lambda\hat{A}} = \hat{B} + \frac{\lambda}{1!}[\hat{A},\hat{B}] + \frac{\lambda^2}{2!}\left[\hat{A},[\hat{A},\hat{B}]\right] + \ldots \qquad (4.22)$$

which can be inserted into Eq.(4.20) to give:

$$\frac{dF(\lambda)}{d\lambda} = \left(\hat{A} + \hat{B} + \lambda[\hat{A},\hat{B}] + \frac{\lambda^2}{2}\left[\hat{A},[\hat{A},\hat{B}]\right] + \ldots\right)F(\lambda)$$

If we integrate this expression, taking into account that $F(0) = 1$, we have:

$$F(\lambda) = e^{(\hat{A}+\hat{B})\lambda + [\hat{A},\hat{B}]\lambda^2/2 + [\hat{A},[\hat{A},\hat{B}]]\lambda^3/6 + \ldots}$$

By combining this result and Eq.(4.19), and setting $\lambda = 1$, one obtains the identity:

$$e^{\hat{A}}e^{\hat{B}} = e^{\hat{A}+\hat{B}+[\hat{A},\hat{B}]/2+[\hat{A},[\hat{A},\hat{B}]]/6+\ldots} \qquad (4.23)$$

Hence, Eq.(4.21) can be regarded as a particular case of Eq.(4.23). ◊

Problem 4.1.1. If the derivative of an operator with explicit dependence on a parameter λ is defined by:

$$\frac{d\hat{A}}{d\lambda} = \lim_{\varepsilon \to 0}\frac{\hat{A}(\lambda+\varepsilon) - \hat{A}(\lambda)}{\varepsilon}$$

find $d\hat{A}^{-1}/d\lambda$.

(Solution): It is convenient to express first the derivative of the product of two operators $\hat{A}\hat{B}$, which reads:

$$\frac{d(\hat{A}\hat{B})}{d\lambda} = \lim_{\varepsilon \to 0}\frac{\hat{A}(\lambda+\varepsilon)\hat{B}(\lambda+\varepsilon) - \hat{A}(\lambda)\hat{B}(\lambda)}{\varepsilon}$$

4. Commutation Relations and the Fourth Postulate

$$= \lim_{\varepsilon \to 0} \hat{A}(\lambda+\varepsilon) \frac{\hat{B}(\lambda+\varepsilon) - \hat{B}(\lambda)}{\varepsilon} + \lim_{\varepsilon \to 0} \frac{\hat{A}(\lambda+\varepsilon) - \hat{A}(\lambda)}{\varepsilon} \hat{B}(\lambda)$$

$$= \hat{A}\frac{d\hat{B}}{d\lambda} + \frac{d\hat{A}}{d\lambda}\hat{B}$$

For $\hat{B} = \hat{A}^{-1}$, the left-hand side vanishes and we obtain:

$$0 = \hat{A}\frac{d\hat{A}^{-1}}{d\lambda} + \frac{d\hat{A}}{d\lambda}\hat{A}^{-1}$$

Multiplying by $\hat{A}^{-1}$ leads to:

$$\frac{d\hat{A}^{-1}}{d\lambda} = -\hat{A}^{-1}\frac{d\hat{A}}{d\lambda}\hat{A}^{-1}$$

Problem 4.1.2. Show that from $[\hat{A}, \hat{B}] = \hat{I}$ it follows that:

$$[f(\hat{A}), \hat{B}] = \frac{df(\hat{A})}{d\hat{A}}$$

(Solution): Using the power series expansion (4.11), the problem is reduced to that of calculating the commutators $[\hat{A}^n, \hat{B}]$, which for $n = 2$ reads:

$$[\hat{A}^2, \hat{B}] = \hat{A}[\hat{A}, \hat{B}] + [\hat{A}, \hat{B}]\hat{A} = 2\hat{A} = \frac{d(\hat{A}^2)}{d\hat{A}}$$

Assuming that this result holds for a given n, that is:

$$[\hat{A}^n, \hat{B}] = n\hat{A}^{n-1} = \frac{d(\hat{A}^n)}{d\hat{A}} \quad \text{or} \quad \hat{A}^n\hat{B} = \hat{B}\hat{A}^n + n\hat{A}^{n-1}$$

we obtain that it is also true for $n + 1$:

$$[\hat{A}^{n+1}, \hat{B}] = \hat{A}(\hat{A}^n\hat{B}) - \hat{B}\hat{A}^{n+1} = \hat{A}(\hat{B}\hat{A}^n + n\hat{A}^{n-1}) - \hat{B}\hat{A}^{n+1}$$

$$= n\hat{A}^n + (\hat{A}\hat{B} - \hat{B}\hat{A})\hat{A}^n = (n+1)\hat{A}^n$$

It follows that:

$$[f(\hat{A}), \hat{B}] = \sum_n c_n[\hat{A}^n, \hat{B}] = \sum_n n c_n \hat{A}^{n-1} = \frac{df(\hat{A})}{d\hat{A}}$$

Problem 4.1.3. Using the coordinate representation of operators in three dimensions, show that:

$$\left[f(\vec{r}), \hat{\vec{p}}\right] = i\hbar \nabla f(\vec{r}), \quad \left[\hat{\vec{A}}, \hat{\vec{p}}\right] = i\hbar \nabla \cdot \hat{\vec{A}}$$

4.2. The Fourth Postulate

In order to apply the first three Postulates, concerned with the description of states and physical observables and the interpretation of the results of measurements, it is necessary to define the mutual relations between the operators representing different observables of a dynamical system. If these are known, one can set up, in a given representation, the form of new operators and predict the possible result of measuring the corresponding observable. It is convenient to postulate relations between quantum operators which preserve the analogy with the classical mechanics formalism, where the state of a system is specified by the values of the position and momentum variables, x_i and p_i, and any other dynamical variable is expressed in terms of these variables.

EXAMPLE 4.2. The Poisson brackets

The Newtonian equations of motion for a single particle of mass m, formulated in terms of the position x_i and velocity $\dot{x}_i$ variables:

$$\frac{d}{dt}(m\dot{x}_i) = -\frac{\partial U}{\partial x_i}$$

where $U(x,y,z)$ is the potential energy, assume a more convenient form if the velocity components are replaced by those of momentum, $p_i = m\dot{x}_i$, such that we have:

$$\dot{x}_i = \frac{dx_i}{dt} = \frac{p_i}{m}, \quad \dot{p}_i = \frac{dp_i}{dt} = -\frac{\partial U}{\partial x_i}$$

If the total energy is also expressed as a function of x_i and p_i, called the Hamiltonian of the particle, Eq.(2.17):

$$H(x_i, p_i) = \sum_i \frac{p_i^2}{2m} + U(x_i)$$

4. Commutation Relations and the Fourth Postulate

the equations of motion can be written in the symmetric form:

$$\dot{x}_i = \frac{\partial H}{\partial p_i}, \quad \dot{p}_i = -\frac{\partial H}{\partial x_i} \tag{4.24}$$

known as the canonical equations of motion. These equations are consistent with the principle of least action introduced in the form (P.48), which applies to the special case of a conservative system, $H(x_i, p_i) = E$. If we write:

$$pdl = \vec{p} \cdot \vec{dl} = p_x dx + p_y dy + p_z dz = \sum_i p_i dx_i = \sum_i p_i \dot{x}_i dt$$

such that:

$$\delta S = \delta \int_{p_1}^{p_2} pdl = \delta \int_{t_1}^{t_2} \left(\sum_i p_i \dot{x}_i \right) dt = 0$$

we can formally subtract the constant energy from the integrand, with no effect on the result:

$$\delta S = \delta \int_{t_1}^{t_2} \left(\sum_i p_i \dot{x}_i - E \right) dt = \delta \int_{t_1}^{t_2} \left(\sum_i p_i \dot{x}_i \right) dt - \delta \left[E(t_2 - t_1) \right] = 0$$

Hence, if the Hamiltonian is not an explicit function of the time, the principle of least action is satisfied by an action integral defined as:

$$S(\vec{r}, t) = \int_{t_1}^{t_2} \left[\sum_i p_i \dot{x}_i - H(x_i, p_i) \right] dt \tag{4.25}$$

This implies that:

$$\delta S = \int_{t_1}^{t_2} \sum_i \left(\dot{x}_i \delta p_i + p_i \delta \dot{x}_i - \frac{\partial H}{\partial p_i} \delta p_i - \frac{\partial H}{\partial x_i} \delta x_i \right) dt$$

$$= \sum_i \left[p_i \delta x_i \right]_{t_1}^{t_2} - \int_{t_1}^{t_2} \sum_i \left[\left(\dot{x}_i - \frac{\partial H}{\partial p_i} \right) \delta p_i - \left(\dot{p}_i + \frac{\partial H}{\partial x_i} \right) \delta x_i \right] dt = 0$$

and, because any real or virtual variation δx_i should vanish at $t = t_1$ and $t = t_2$, the canonical equations (4.24) are retrieved. In view of the equations of motion (4.24) for the conjugate variables x_i and p_i, a convenient form of describing the temporal evolution of any dynamical variable can be obtained in terms of the so-called Poisson bracket with respect to x_i and p_i, which for two variables A and B is introduced as:

$$\{A,B\} = \sum_i \left(\frac{\partial A}{\partial x_i} \frac{\partial B}{\partial p_i} - \frac{\partial A}{\partial p_i} \frac{\partial B}{\partial x_i} \right) \quad (4.26)$$

The basic properties of the Poisson brackets, namely the antisymmetry, the linearity and the product rules:

$$\{A,B\} = -\{B,A\} \quad (4.27)$$

$$\{\alpha_1 A_1 + \alpha_2 A_2, B\} = \alpha_1 \{A_1, B\} + \alpha_2 \{A_2, B\} \quad (4.28)$$

$$\{A, \alpha_1 B_1 + \alpha_2 B_2\} = \alpha_1 \{A, B_1\} + \alpha_2 \{A, B_2\} \quad (4.29)$$

$$\{AB, C\} = A\{B, C\} + \{A, C\}B \quad (4.30)$$

$$\{A, BC\} = \{A, B\}C + B\{A, C\} \quad (4.31)$$

derive in a straightforward manner from those of the partial derivatives. The antisymmetry of Eq.(4.26) leads to the Jacobi identity:

$$\{A, \{B, C\}\} + \{B, \{C, A\}\} + \{C, \{A, B\}\} = 0 \quad (4.32)$$

whose terms are homogenous linear functions of the second derivatives of the three functions A, B and C respectively, so that their sum vanishes identically. It is obvious that the properties (4.27) to (4.32) are similar to those of the commutators of quantum operators, Eqs.(4.5) to (4.10). If we consider the special cases where $B = x_i$ or $B = p_i$ respectively, we obtain further properties previously derived for commutators, Eqs.(4.14) and (4.15):

$$\frac{\partial A}{\partial p_i} = \{x_i, A\}, \quad \frac{\partial A}{\partial q_i} = \{A, p_i\} \quad (4.33)$$

and then, by setting $A = x_j$ and $A = p_j$ respectively in Eq.(4.33), we arrive to the *fundamental Poisson bracket* relations:

$$\{x_i, x_j\} = 0 \quad \{x_i, p_j\} = \delta_{ij}, \quad \{p_i, p_j\} = 0 \quad (4.34)$$

The time evolution of any dynamical variable is usually formulated in terms of its Poisson bracket with the Hamiltonian H of the system:

$$\frac{dA}{dt} = \sum_i \left(\frac{\partial A}{\partial x_i} \dot{x}_i + \frac{\partial A}{\partial p_i} \dot{p}_i \right) + \frac{\partial A}{\partial t} = \sum_i \left(\frac{\partial A}{\partial x_i} \frac{\partial H}{\partial p_i} - \frac{\partial A}{\partial p_i} \frac{\partial H}{\partial x_i} \right) + \frac{\partial A}{\partial t} = \{A, H\} + \frac{\partial A}{\partial t} \quad (4.35)$$

where use has been made of the canonical equations (4.24). If t does not occur explicitly, that is $\partial A/\partial t = 0$, the evolution equation reduces to:

4. Commutation Relations and the Fourth Postulate 153

$$\frac{dA}{dt} = \{A, H\} \tag{4.36}$$

which for $A = x_i$ and $A = p_i$ simply represents the equations of motion (4.24) in the form:

$$\dot{x}_i = \{x_i, H\}, \quad \dot{p}_i = \{p_i, H\} \tag{4.37}$$

A dynamical variable is called a *constant of motion* if:

$$\frac{dA}{dt} = 0 \quad \text{or} \quad \{A, H\} = 0 \tag{4.38}$$

and the following identity holds, in view of Eq.(4.35):

$$\frac{d}{dt}\{A, B\} = \left\{\frac{dA}{dt}, B\right\} + \left\{A, \frac{dB}{dt}\right\} \tag{4.39}$$

Assuming $dA/dt = 0$ and $dB/dt = 0$, it follows that $d\{A, B\}/dt$ also vanishes, and this provides a way of finding new constants of motion: if A and B are known, their Poisson bracket represents another constant of motion. ✥

It is convenient to evaluate the Poisson bracket $\{AA', BB'\}$, involving four dynamical variables expressed in terms of x_i, p_i, using the rules (4.30) and (4.31), that give either:

$$\{AA', BB'\} = A\{A', BB'\} + \{A, BB'\}A'$$

$$= A\{A', B\}B' + AB\{A', B'\} + \{A, B\}B'A' + B\{A, B'\}A'$$

or:

$$\{AA', BB'\} = \{AA', B\}B' + B\{AA', B'\}$$

$$= A\{A', B\}B' + \{A, B\}A'B' + BA\{A', B'\} + B\{A, B'\}A'$$

which, by comparison, yield:

$$(AB - BA)\{A', B'\} - \{A, B\}(A'B' - B'A') = 0$$

This means that the ratio:

$$\frac{AB - BA}{\{A, B\}}$$

is a constant in classical mechanics, irrespective of the dynamical variable involved. Therefore, it will be a reasonable requirement for the quantum operators $\hat{A}$ and $\hat{B}$, associated to the observables A and B, to exhibit a similar property, namely the ratio between the commutator $[\hat{A},\hat{B}]$ and the operator representing $\{A,B\}$ should also be a constant. It is then sufficient to determine the appropriate constant for a particular case, by taking the commutators of $\hat{x}_i$ and $\hat{p}_i$, for which we know the corresponding Poisson brackets, Eq.(4.34). Because it is immediate from Eqs.(4.14) and (4.15) that in both the coordinate and momentum representations we have:

$$[\hat{x}_i,\hat{x}_j]=0, \quad [\hat{x}_i,\hat{p}_j]=i\hbar\delta_{ij}, \quad [\hat{p}_i,\hat{p}_j]=0 \qquad (4.40)$$

we can formally introduce the $\hat{\delta}_{ij}$ operator representing $\{x_i,p_j\}$:

$$[\hat{x}_i,\hat{p}_j]=i\hbar\hat{\delta}_{ij}$$

and this determines the constant to be $i\hbar$. This value is consistent with the correspondence principle, formulated in Section P.1, in the sense that the classical relations (the Poisson brackets) proceed from those postulated in quantum mechanics (the commutation relations) for $\hbar\to 0$.

Postulate 4. *The commutation relation between any pair of quantum operators is derived from the Poisson bracket of the corresponding pair of classical variables, according to the correspondence rule:*

$$\{A,B\}=C \to [\hat{A},\hat{B}]=i\hbar\hat{C} \qquad (4.41)$$

EXAMPLE 4.3. Commutation relations for angular momentum

The orbital angular momentum operators for a single particle, introduced by Eq.(2.18), according to Postulate 2:

$$\hat{L}_x=\hat{y}\hat{p}_z-\hat{z}\hat{p}_y, \quad \hat{L}_y=\hat{z}\hat{p}_x-\hat{x}\hat{p}_z, \quad \hat{L}_z=\hat{x}\hat{p}_y-\hat{y}\hat{p}_x$$

are specified in terms of products of commuting Hermitian operators only, see Eqs.(4.40), and therefore all are Hermitian. If $\vec{\hat{r}}$ and $\vec{\hat{p}}$ are the vector operators for position and momentum of Cartesian components $\hat{x},\hat{y},\hat{z}$ and $\hat{p}_x,\hat{p}_y,\hat{p}_z$ respectively, the correspondence with the classical vector observable $\vec{L}$ is ensured by defining a vector operator:

$$\vec{L}=\vec{r}\times\vec{p} \to \vec{\hat{L}}=\vec{\hat{r}}\times\vec{\hat{p}} \qquad (4.42)$$

4. Commutation Relations and the Fourth Postulate

where the order of the two operators in the vector product (4.42) should always be observed, because from $[\hat{x}_i, \hat{p}_j] = 0$ for $i \neq j$, Eq.(4.40), it follows that $\hat{r} \times \hat{p} = -\hat{p} \times \hat{r}$. The square of the orbital angular momentum observable is represented by the operator $\hat{L}^2$, defined by Eq.(2.19). The commutation relations between the Cartesian components of the vector operators $\hat{L}, \hat{r}$ and $\hat{p}$ derive in a straightforward manner using Eqs.(4.40), (4.14) and (4.15). For example:

$$[\hat{L}_x, \hat{x}] = 0, \quad [\hat{L}_x, \hat{y}] = -i\hbar \frac{\partial \hat{L}_x}{\partial \hat{p}_y} = i\hbar \hat{z}, \quad [\hat{L}_x, \hat{z}] = -i\hbar \frac{\partial \hat{L}_x}{\partial \hat{p}_z} = -i\hbar \hat{y} \quad (4.43)$$

$$[\hat{L}_x, \hat{p}_x] = 0, \quad [\hat{L}_x, \hat{p}_y] = i\hbar \frac{\partial \hat{L}_x}{\partial \hat{y}} = i\hbar \hat{p}_z, \quad [\hat{L}_x, \hat{p}_z] = i\hbar \frac{\partial \hat{L}_x}{\partial \hat{z}} = -i\hbar \hat{p}_y \quad (4.44)$$

and corresponding results hold for $\hat{L}_y$ and $\hat{L}_z$. In other words, the Cartesian components $\hat{L}_i, \hat{x}_i$ and $\hat{p}_i$ along the same axis only are commuting operators. The commutation relations between the components of orbital angular momentum follow as:

$$[\hat{L}_x, \hat{L}_y] = [\hat{y}\hat{p}_z - \hat{z}\hat{p}_y, \hat{z}\hat{p}_x - \hat{x}\hat{p}_z]$$

$$= [\hat{y}\hat{p}_z, \hat{z}\hat{p}_x] - [\hat{z}\hat{p}_y, \hat{z}\hat{p}_x] - [\hat{y}\hat{p}_z, \hat{x}\hat{p}_z] + [\hat{z}\hat{p}_y, \hat{x}\hat{p}_z]$$

$$= \hat{y}[\hat{p}_z, \hat{z}]\hat{p}_x + \hat{x}[\hat{z}, \hat{p}_z]\hat{p}_y = i\hbar(\hat{x}\hat{p}_y - \hat{y}\hat{p}_x) = i\hbar \hat{L}_z \quad (4.45)$$

and similarly:

$$[\hat{L}_y, \hat{L}_z] = i\hbar \hat{L}_x, \quad [\hat{L}_z, \hat{L}_x] = i\hbar \hat{L}_y \quad (4.46)$$

These results may be summarised in the form:

$$\hat{L} \times \hat{L} = i\hbar \hat{L} \quad (4.47)$$

where the vector product of the operator $\hat{L}$ with itself does not vanish, because the Cartesian components $\hat{L}_i$ do not commute with each other.

The operator $\hat{L}^2$, introduced by Eq.(2.19), commutes with each of its components, as for instance:

$$[\hat{L}_x, \hat{L}^2] = [\hat{L}_x, \hat{L}_y^2] + [\hat{L}_x, \hat{L}_z^2]$$

$$= [\hat{L}_x, \hat{L}_y]\hat{L}_y + \hat{L}_y[\hat{L}_x, \hat{L}_y] + [\hat{L}_x, \hat{L}_z]\hat{L}_z + \hat{L}_z[\hat{L}_x, \hat{L}_z]$$

$$= i\hbar \hat{L}_z \hat{L}_y + i\hbar \hat{L}_y \hat{L}_z - i\hbar \hat{L}_y \hat{L}_z - i\hbar \hat{L}_z \hat{L}_y = 0$$

where use has been made of the commutation relations (4.46) established for the Cartesian components $\hat{L}_i$. In other words, if Eq.(4.47) holds, it always follows that:

$$[\hat{L}_x, \hat{L}^2] = [\hat{L}_y, \hat{L}^2] = [\hat{L}_z, \hat{L}^2] = 0 \quad \text{or} \quad [\hat{\vec{L}}, \hat{L}^2] = 0 \tag{4.48}$$

Any component of $\hat{\vec{L}}$ also commutes with the operator $\hat{p}^2$, representing the square of the momentum, as for example:

$$[\hat{L}_x, \hat{p}^2] = [\hat{L}_x, \hat{p}_y^2 + \hat{p}_z^2] = \hat{p}_y[\hat{L}_x, \hat{p}_y] + [\hat{L}_x, \hat{p}_y]\hat{p}_y + \hat{p}_z[\hat{L}_x, \hat{p}_z] + [\hat{L}_x, \hat{p}_z]\hat{p}_z$$

$$= i\hbar \hat{p}_y \hat{p}_z + i\hbar \hat{p}_z \hat{p}_y - i\hbar \hat{p}_z \hat{p}_y - i\hbar \hat{p}_y \hat{p}_z = 0$$

and hence it will commute with the kinetic energy operator, Eq.(2.16):

$$[\hat{L}_i, \hat{p}^2] = [\hat{L}_i, \hat{T}] = 0 \tag{4.49}$$

In the same way, if $\hat{r}^2 = \hat{x}^2 + \hat{y}^2 + \hat{z}^2$ we have:

$$[\hat{L}_x, \hat{r}^2] = [\hat{L}_y, \hat{r}^2] = [\hat{L}_z, \hat{r}^2] = 0 \tag{4.50}$$

and therefore $\hat{L}_i$ will commute with any function $f(\hat{r}^2) = f(r^2)$, which might express the potential energy of a particle under particular fields of force, such as the central force field. ☙

EXAMPLE 4.4. Unitary transformations of physical observables

A correspondence between some classical coordinate transformations and appropriate quantum operators can be established by taking the unitary operator $\hat{U}$, that induces a change of basis (3.53), in the form:

$$\hat{U} = e^{i\varepsilon \hat{G}} = 1 + \frac{i\varepsilon \hat{G}}{1!} + \frac{(i\varepsilon \hat{G})^2}{2!} + \ldots \tag{4.51}$$

where $\hat{G}$ is a Hermitian operator and ε is a real parameter. In view of Eq.(4.21), the inverse operator is:

$$\hat{U}^{-1} = e^{-i\varepsilon \hat{G}} = \hat{U}^+ \tag{4.52}$$

and, if $\hat{G}$ is the *generator* of a transformation for which ε is infinitesimal, we may retain in the expansion (4.51) only terms to the first order in ε, which gives:

$$\hat{U} = 1 + i\varepsilon \hat{G}$$
$$\hat{U}^{-1} = 1 - i\varepsilon \hat{G} \tag{4.53}$$

4. Commutation Relations and the Fourth Postulate

Consider a physical observable B, specified in relation to a coordinate system which is defined by the basis vectors $|i\rangle$, $(i = 1,2,3)$. A change of coordinate axes corresponds to a change of basis (3.53) in the three-dimensional Hilbert space, characterized by a unitary operator $\hat{U}$. We have seen that the physical observable should be specified with respect to the new coordinate system, which corresponds to the representation of the basis vectors $|\psi_i\rangle$, $(i = 1,2,3)$, as given by Eq.(3.58):

$$\hat{B}' = \hat{U}\hat{B}\hat{U}^+ \qquad (4.54)$$

In the case of an infinitesimal transformation generated by $\hat{G} = \hat{p}_i/\hbar$, the position operator $\hat{B} = \hat{x}_i$ is changed, according to Eq.(4.54), in the form:

$$\hat{x}'_i = (1+i\varepsilon\hat{p}_i/\hbar)\hat{x}_i(1-i\varepsilon\hat{p}_i/\hbar) = \hat{x}_i - \frac{i\varepsilon}{\hbar}[\hat{p}_i,\hat{x}_i] = \hat{x}_i - \varepsilon \qquad (4.55)$$

and this corresponds to an infinitesimal translation of the coordinate system in the x_i-direction. As illustrated in Figure 4.1, the value of an arbitrary state vector at a physical point should be the same in the two representations, $\Psi'(x') = \Psi(x)$, or in view of Eq.(4.55):

$$\Psi'(x) = \Psi(x+\varepsilon) \approx \Psi(x) + \varepsilon\frac{\partial\Psi}{\partial x} = (1+i\varepsilon\hat{p}_x/\hbar)\Psi(x) = \hat{U}\Psi(x) \qquad (4.56)$$

We conclude that the operators $\hat{p}_i/\hbar$ are generators of infinitesimal translations along the corresponding axes of a coordinate system.

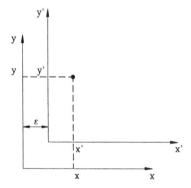

Figure 4.1. Infinitesimal translation of coordinate system

In a similar way, for an infinitesimal transformation generated by $\hat{G} = \hat{L}_z/\hbar$, the components $\hat{B} = \hat{p}_x, \hat{p}_y$ and $\hat{p}_z$ of the vector momentum operator along the same axis are transformed as:

$$\hat{p}'_x = (1+i\hat{L}_z\varepsilon/\hbar)\hat{p}_x(1-i\hat{L}_z\varepsilon/\hbar) = \hat{p}_x - \frac{i\varepsilon}{\hbar}[\hat{L}_z,\hat{p}_x] = \hat{p}_x + \varepsilon\hat{p}_y$$

158 Quantum Physics

$$\hat{p}'_y = \hat{p}_y - \frac{i\varepsilon}{\hbar}\left[\hat{L}_z, \hat{p}_y\right] = -\varepsilon\hat{p}_x + \hat{p}_y \tag{4.57}$$

$$\hat{p}'_z = \hat{p}_z - \frac{i\varepsilon}{\hbar}\left[\hat{L}_z, \hat{p}_z\right] = \hat{p}_z$$

This corresponds to the transformation of the $\vec{p}$ vector components induced by an infinitesimal rotation ε about the z-axis of the coordinate system (Figure 4.2), and can also be written in the matrix form:

$$\begin{pmatrix} p'_x \\ p'_y \\ p'_z \end{pmatrix} = \begin{pmatrix} 1 & \varepsilon & 0 \\ -\varepsilon & 1 & 0 \\ 0 & 0 & 1 \end{pmatrix}\begin{pmatrix} p_x \\ p_y \\ p_z \end{pmatrix} = \left(1 + \frac{i\varepsilon}{\hbar}\hat{L}_z\right)\begin{pmatrix} p_x \\ p_y \\ p_z \end{pmatrix} = \hat{U}\begin{pmatrix} p_x \\ p_y \\ p_z \end{pmatrix} \tag{4.58}$$

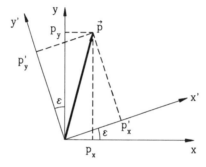

Figure 4.2. Infinitesimal rotation of coordinate system

It follows that $\hat{L}_z/\hbar$ can be regarded as generator of infinitesimal rotations about the z-axis, and has the matrix representation:

$$\left(\hat{L}_z\right) = \hbar\begin{pmatrix} 0 & -i & 0 \\ i & 0 & 0 \\ 0 & 0 & 0 \end{pmatrix}$$

For infinitesimal rotations about the x- and y-axis, generated by $\hat{L}_x/\hbar$ and $\hat{L}_y/\hbar$ respectively, we may apply the commutation relations (4.46) to obtain:

$$\left(\hat{L}_x\right) = \hbar\begin{pmatrix} 0 & 0 & 0 \\ 0 & 0 & -i \\ 0 & i & 0 \end{pmatrix}, \quad \left(\hat{L}_y\right) = \hbar\begin{pmatrix} 0 & 0 & i \\ 0 & 0 & 0 \\ -i & 0 & 0 \end{pmatrix}$$

4. Commutation Relations and the Fourth Postulate

Problem 4.2.1. Show that the operator:

$$\hat{S}(\lambda) = e^{i\lambda \hat{p}_i/\hbar}$$

is the generator of a finite translation of origin through λ along the x_i-axis, where x_i stands for x, y or z.

(Solution): A finite translation of origin transforms an arbitrary wave function in the form (see Figure 4.1):

$$\hat{S}(\lambda)\Psi(x_i) = \Psi(x_i - \lambda)$$

We may expand $\Psi(x_i - \lambda)$ as a power series:

$$\Psi(x_i - \lambda) = \Psi(x_i) - \frac{\lambda}{1!}\frac{d\Psi(x_i)}{dx_i} + \frac{\lambda^2}{2!}\frac{d^2\Psi(x_i)}{dx_i^2} + \ldots$$

$$= \Psi(x_i) + \sum_{n=1}^{\infty}(-1)^n \frac{\lambda^n}{n!}\frac{d^n\Psi(x_i)}{dx_i^n}$$

$$= \left[1 + \sum_{n=1}^{\infty}\frac{1}{n!}\left(\frac{i}{\hbar}\lambda\hat{p}_i\right)^n\right]\Psi(x_i) = e^{i\lambda\hat{p}_i/\hbar}\Psi(x_i)$$

which proves the property. An alternative proof is obtained by considering that a finite translation is made up of N infinitesimal translations, of the form (4.56), where $\varepsilon = \lambda/N$, so that:

$$\hat{S}(\lambda) = \lim_{N\to\infty}\hat{U}^N(\varepsilon) = \lim_{N\to\infty}(1 + i\lambda\hat{p}_i/N\hbar)^N = e^{i\lambda\hat{p}_i/\hbar}$$

The result also holds in three dimensions, where $x_i, \hat{p}_i, \lambda$ and $\lambda\hat{p}_i$ should be replaced by $\vec{r}, \hat{\vec{p}}, \vec{\lambda}$ and $\vec{\lambda}\cdot\hat{\vec{p}}$ in the form $\hat{S}(\vec{\lambda}) = e^{i\vec{\lambda}\cdot\hat{\vec{p}}/\hbar}$.

Problem 4.2.2. Show that the operator:

$$\hat{R}(\omega_i) = e^{i\omega_i \hat{L}_i/\hbar}$$

is the generator of a finite rotation through ω_i about the x_i-axis ($x_i = x, y, z$).

(Solution): The effect of a finite rotation through $\omega_z = \varphi$ about the z-axis (see Figure 4.2) can be written as:

$$\hat{R}(\varphi)\Psi(x, y, z) = \Psi(x\cos\varphi - y\sin\varphi, x\sin\varphi + y\cos\varphi, z)$$

and we will expand the latter function in powers of φ:

$$\Psi(x\cos\varphi - y\sin\varphi, x\sin\varphi + y\cos\varphi, z) = \Psi(x, y, z) + \sum_{n=1}^{\infty} \frac{\varphi^n}{n!}\left(\frac{\partial^n \Psi}{\partial \varphi^n}\right)_{\varphi=0}$$

where:

$$\left(\frac{\partial \Psi}{\partial \varphi}\right)_{\varphi=0} = \left(\frac{\partial \Psi}{\partial x}\frac{\partial x}{\partial \varphi} + \frac{\partial \Psi}{\partial y}\frac{\partial y}{\partial \varphi}\right)_{\varphi=0} = \left(-y\frac{\partial}{\partial x} + x\frac{\partial}{\partial y}\right)\Psi = \frac{i}{\hbar}\hat{L}_z\Psi$$

It follows that:

$$\hat{R}(\varphi)\Psi(x, y, z) = \left[1 + \sum_{n=1}^{\infty} \frac{1}{n!}\left(\frac{i}{\hbar}\varphi \hat{L}_z\right)^n\right]\Psi(x, y, z) = e^{i\varphi\hat{L}_z/\hbar}\Psi(x, y, z)$$

The same result also follows by combining N infinitesimal rotations, each through $\varepsilon = \varphi/N$, in the limit $N \to \infty$.

Problem 4.2.3. Show that the operators $\hat{L}_+$ and $\hat{L}_-$, defined in terms of $\hat{L}_x$ and $\hat{L}_y$ by Eqs.(2.30) and (2.31), satisfy the commutation relations:

$$[\hat{L}_z, \hat{L}_+] = \hbar \hat{L}_+, \quad [\hat{L}_z, \hat{L}_-] = -\hbar \hat{L}_-, \quad [\hat{L}_+, \hat{L}_-] = 2\hbar \hat{L}_z$$

4.3. Compatibility between Physical Observables

Physical observables are said to be *compatible* if they can be represented by operators which have a common set of eigenfunctions, Eq.(4.1), and hence commute. We now prove that two commuting Hermitian operators always represent observables that can be simultaneously measured.

Theorem. *If the Hermitian operators $\hat{A}$ and $\hat{B}$ commute, there exists a complete set of vectors which are simultaneously eigenfunctions of $\hat{A}$ and of $\hat{B}$.*

4. Commutation Relations and the Fourth Postulate

Proof. Let ψ_i be the basis vectors of the operator $\hat{A}$ with respective eigenvalues. Assuming that ψ_i is not necessarily an eigenfunction of the operator $\hat{B}$, the function $\hat{B}\psi_i$ can be expanded in terms of the eigenvectors of $\hat{A}$, according to Eq.(3.55):

$$\hat{B}\psi_i = \sum_k \langle \psi_k | \hat{B} | \psi_i \rangle \psi_k \tag{4.59}$$

On the other hand, the effect of the commutator $[\hat{A}, \hat{B}]$ on ψ_i can be written as:

$$[\hat{A}, \hat{B}]\psi_i = \hat{A}\hat{B}\psi_i - \hat{B}\hat{A}\psi_i = \sum_k \langle \psi_k | \hat{B} | \psi_i \rangle \hat{A}\psi_k - \lambda_i \hat{B}\psi_i$$

$$= \sum_k \langle \psi_k | \hat{B} | \psi_i \rangle \lambda_k \psi_k - \sum_k \langle \psi_k | \hat{B} | \psi_i \rangle \lambda_i \psi_k = \sum_k \langle \psi_k | \hat{B} | \psi_i \rangle (\lambda_k - \lambda_i) \psi_k$$

If $\hat{A}$ and $\hat{B}$ commute, the result must be set equal to zero, so that each coefficient of the eigenvectors ψ_k vanishes:

$$\langle \psi_k | \hat{B} | \psi_i \rangle (\lambda_k - \lambda_i) = 0 \tag{4.60}$$

Assuming distinct eigenvalues, it follows that every $\langle \psi_k | \hat{B} | \psi_i \rangle$ vanishes unless $k = i$, when its value may be denoted by μ_i:

$$\langle \psi_k | \hat{B} | \psi_i \rangle = \mu_i \delta_{ik} \tag{4.61}$$

If this result is into Eq.(4.59), we obtain the eigenvalue equation of the operator $\hat{B}$:

$$\hat{B}\psi_i = \mu_i \psi_i$$

which shows that the eigenvector ψ_i of $\hat{A}$ is also an eigenvector of $\hat{B}$.

If there is a k-fold degeneracy associated with λ_i, Eq.(3.10), from the commutivity of $\hat{A}$ with $\hat{B}$ it follows that any function $\hat{B}\psi_i^{(s)}, (s = 1, 2, \ldots, k)$ is also an eigenvector of $\hat{A}$ with eigenvalue λ_i:

$$\hat{A}(\hat{B}\psi_i^{(s)}) = \hat{B}\hat{A}\psi_i^{(s)} = \lambda_i (\hat{B}\psi_i^{(s)}) \tag{4.62}$$

and we have seen that a set of k orthonormal eigenvectors can be found such that:

$$\hat{B}\psi_i^{(s)} = \sum_{p=1}^{k} c_{sp}\psi_i^{(p)} \tag{4.63}$$

Although Eq.(4.63) shows that the $\psi_i^{(s)}$ are not eigenvectors of $\hat{B}$, a set of linear combinations $\psi_i'^{(s)}$ of the $\psi_i^{(s)}$ can always be constructed to be simultaneous eigenvectors of $\hat{A}$ and of $\hat{B}$:

$$\psi_i'^{(s)} = \sum_{r=1}^{k} \alpha_{sr}\psi_i^{(r)} \tag{4.64}$$

The condition that the $\psi_i'^{(s)}$ should be eigenvectors of $\hat{B}$, with eigenvalue μ_i, reads:

$$\hat{B}\psi_i'^{(s)} = \sum_{r=1}^{k} \alpha_{sr}\hat{B}\psi_i^{(r)} = \sum_{r=1}^{k}\sum_{p=1}^{k} \alpha_{sr}c_{rp}\psi_i^{(p)} = \mu_i\psi_i'^{(s)} = \sum_{r=1}^{k} \alpha_{sr}\mu_i\psi_i^{(r)} = \sum_{r=1}^{k}\sum_{p=1}^{k} \alpha_{sr}\mu_i\delta_{rp}\psi_i^{(p)}$$

where use has been made of Eq.(4.64). This leads, for each s, to k linear homogenous equations for the α_{sr}:

$$\sum_{p=1}^{k}(c_{rp} - \mu_i\delta_{rp})\alpha_{sr} = 0 \tag{4.65}$$

which is a result similar to the eigenvalue problem (3.7). The characteristic equation:

$$\det(c_{rp} - \mu_i\delta_{rp}) = 0$$

provides k eigenvalues $\mu_i^{(1)},\ldots,\mu_i^{(k)}$, not necessary distinct, and for each one set of coefficients α_{sr}, and hence an eigenvector $\psi_i'^{(s)}$ of $\hat{B}$ is determined. It is obvious from Eqs.(3.10) and (4.64) that all the $\psi_i'^{(s)}$ so found, are also eigenvectors of $\hat{A}$, with eigenvalue λ_i:

$$\hat{A}\psi_i'^{(s)} = \sum_{r=1}^{k} \alpha_{sr}\hat{A}\psi_i^{(r)} = \lambda_i\sum_{r=1}^{k} \alpha_{sr}\psi_i^{(r)} = \lambda_i\psi_i'^{(s)}$$

EXAMPLE 4.5. Removal of degeneracy

Consider the situation where the eigenfunction λ_i of $\hat{A}$ is doubly degenerate, having linearly independent orthonormal eigenfunctions $\psi_i^{(1)}$ and $\psi_i^{(2)}$, such that:

$$\psi_i' = \alpha_1\psi_i^{(1)} + \alpha_2\psi_i^{(2)}$$

4. Commutation Relations and the Fourth Postulate   163

is an eigenfunction of $\hat{A}$. The condition that ψ'_i is also an eigenfunction of $\hat{B}$ with eigenvalue μ_i reads:

$$\alpha_1 \hat{B}\psi_i^{(1)} + \alpha_2 \hat{B}\psi_i^{(2)} = \mu_i (\alpha_1 \psi_i^{(1)} + \alpha_2 \psi_i^{(2)})$$

Taking the scalar product of this equation with $\psi_i^{(1)}$ and $\psi_i^{(2)}$ respectively we obtain:

$$\alpha_1 \langle \psi_i^{(1)} | \hat{B} \psi_i^{(1)} \rangle + \alpha_2 \langle \psi_i^{(1)} | \hat{B} \psi_i^{(2)} \rangle = \mu_i \alpha_1$$

$$\alpha_1 \langle \psi_i^{(2)} | \hat{B} \psi_i^{(1)} \rangle + \alpha_2 \langle \psi_i^{(2)} | \hat{B} \psi_i^{(2)} \rangle = \mu_i \alpha_2$$

which are two simultaneous homogenous linear equations:

$$\alpha_1 (B_{11} - \mu_i) + \alpha_2 B_{12} = 0$$

$$\alpha_1 B_{21} + \alpha_2 (B_{22} - \mu_i) = 0$$

to be solved for α_1 and α_2. The characteristic equation:

$$\begin{vmatrix} B_{11} - \mu_i & B_{12} \\ B_{21} & B_{22} - \mu_i \end{vmatrix} = 0$$

has two roots $\mu_i^{(1)}$ and $\mu_i^{(2)}$. If $\mu_i^{(1)} \neq \mu_i^{(2)}$ the corresponding eigenvectors of $\hat{B}$ are orthogonal and hence linearly independent. It is then said that the degeneracy of λ_i was removed by the operator $\hat{B}$. This is because, for that eigenvalue of $\hat{A}$, a particular eigenvector can uniquely be identified by specifying a particular value μ_i of $\hat{B}$. If the operator $\hat{B}$ does not remove the degeneracy of every eigenvalue λ_i, we may use further operators $\hat{C}, \hat{D}, \ldots$ commuting with one another and with $\hat{A}$ and $\hat{B}$, and repeat the procedure until all their common eigenvector are uniquely identified. As a result, the degeneracy of the eigenvalues of $\hat{A}$ can be completely removed and in this case it is said that A, B, C, D, ... form a *complete set of compatible observables*. ↵

Therefore, the commutivity of two Hermitian operators $\hat{A}$ and $\hat{B}$ is a necessary and sufficient condition for the compatibility of the corresponding observables A and B. This means that measuring in turn the two observables, the probability of obtaining the results λ_i for A and μ_i for B is the same, irrespective of which observable is measured first:

$$P(\lambda_i, \mu_i) = P(\mu_i, \lambda_i) \tag{4.66}$$

Let us denote by $|\lambda\mu\rangle$ the simultaneous eigenvectors of $\hat{A}$ and $\hat{B}$:

$$\hat{A}|\lambda\mu\rangle = \lambda|\lambda\mu\rangle, \qquad \hat{B}|\lambda\mu\rangle = \mu|\lambda\mu\rangle$$

which are assumed to form a complete set such that the initial state of the system can be expanded in the form:

$$|\Psi\rangle = \sum_{\lambda,\mu} c_{\lambda\mu s}|\lambda\mu s\rangle \qquad (4.67)$$

where s runs over the eigenvectors corresponding to degenerate eigenvalues. If A is measured first, we obtain λ_i with a probability:

$$P(\lambda_i) = |\langle \lambda_i \mu s | \Psi \rangle|^2 = \sum_{\mu,s} |c_{\lambda_i \mu s}|^2 \qquad (4.68)$$

and, after the measurement, the system is in a state (3.40) which reads:

$$|\Psi_{\lambda_i}\rangle = \frac{\sum_{\mu,s} c_{\lambda_i \mu s}|\lambda_i \mu s\rangle}{|\langle \lambda_i \mu s | \Psi \rangle|} \qquad (4.69)$$

The measurement of B in this state yields μ_i with a probability:

$$P_{\lambda_i}(\mu_i) = |\langle \lambda_i \mu_i s | \Psi_{\lambda_i} \rangle|^2 = \frac{\sum_s |c_{\lambda_i \mu_i s}|^2}{|\langle \lambda_i \mu s | \Psi \rangle|^2} \qquad (4.70)$$

and the final state:

$$|\Psi_{\lambda_i,\mu_i}\rangle = \frac{\sum_s c_{\lambda_i \mu_i s}|\lambda_i \mu_i s\rangle}{|\langle \lambda_i \mu_i s | \Psi \rangle|} \qquad (4.71)$$

Using Eqs.(4.68) and (4.69), the probability of obtaining both λ_i and μ_i as a result of two independent measurements is given by the product:

$$P(\lambda_i,\mu_i) = P(\lambda_i) P_{\lambda_i}(\mu_i) = \sum_s |c_{\lambda_i \mu_i s}|^2 = |\langle \lambda_i \mu_i s | \Psi \rangle|^2 \qquad (4.72)$$

which is a result symmetrical in λ_i, μ_i, such that Eq.(4.66) follows. The final state $|\Psi_{\lambda_i,\mu_i}\rangle$, Eq.(4.71), is a common eigenvector of $\hat{A}$ and $\hat{B}$ corresponding to their eigenvalues λ_i and μ_i, which might be degenerate.

4. Commutation Relations and the Fourth Postulate

If the operators $\hat{A}$ and $\hat{B}$ do not commute and therefore have distinct eigenbases:

$$\hat{A}|\psi_i\rangle = \lambda_i|\psi_i\rangle, \quad \hat{B}|\varphi_i\rangle = \mu_i|\varphi_i\rangle$$

the corresponding observables are called *incompatible*, because both the probability of obtaining λ_i and μ_i respectively through a measurement, and the final state of the system, are determined by which observable is measured first:

$$P(\lambda_i, \mu_i) \neq P(\mu_i, \lambda_i)$$

Let us assume, for simplicity, that there is no degeneracy. Then the probability of obtaining the result λ_i, if A is measured first, reads:

$$P(\lambda_i) = |\langle\psi_i|\Psi\rangle|^2$$

and the state of the system, after the measurement, is $|\Psi_{\lambda_i}\rangle = |\psi_i\rangle$. The second measurement yields μ_i with a probability:

$$P_{\lambda_i}(\mu_i) = |\langle\varphi_i|\psi_i\rangle|^2$$

and the final state $|\Psi_{\lambda_i,\mu_i}\rangle = |\varphi_i\rangle$ is the eigenstate of the second operator with the measured eigenvalue. Hence we have:

$$P(\lambda_i, \mu_i) = |\langle\psi_i|\Psi\rangle|^2 |\langle\varphi_i|\psi_i\rangle|^2 \quad \text{and} \quad |\Psi_{\lambda_i\mu_i}\rangle = |\varphi_i\rangle$$

and, in the case of the measurement of B followed by that of A, one similarly obtains:

$$P(\mu_i, \lambda_i) = |\langle\varphi_i|\Psi\rangle|^2 |\langle\psi_i|\varphi_i\rangle|^2 \quad \text{and} \quad |\Psi_{\mu_i\lambda_i}\rangle = |\psi_i\rangle$$

The different results for $P(\lambda_i, \mu_i)$ and $|\Psi_{\lambda_i\mu_i}\rangle$ imply that the incompatible observables cannot be specified simultaneously. The commutator of their associated operators:

$$[\hat{A}, \hat{B}] = i\hat{C} \tag{4.73}$$

where $\hat{C}$ is a Hermitian operator, Eq.(4.4), can be introduced as a measure for the degree to which the precision loss should be estimated. If the deviation operators $\delta\hat{A}$ and $\delta\hat{B}$, Eq.(3.1), in terms of which we define the measurement dispersions $(\Delta A)^2 = \langle(\delta\hat{A})^2\rangle$ and

$(\Delta B)^2 = \langle (\delta \hat{B})^2 \rangle$, are both inserted into Eq.(4.73), one obtains in a straightforward manner that:

$$[\delta \hat{A}, \delta \hat{B}] = i\hat{C}$$

and the following relation, called the *Heisenberg inequality*, holds:

$$\langle (\delta \hat{A})^2 \rangle \langle (\delta \hat{B})^2 \rangle \geq \frac{1}{4} \langle \hat{C} \rangle^2 \qquad (4.74)$$

This can be shown by using a non-negative expression, written in terms of a real parameter α as follows:

$$\int |(\delta \hat{A} + \alpha \delta \hat{B}) \psi|^2 dx = \langle \psi | (\delta \hat{A} - i\alpha \delta \hat{B})(\delta \hat{A} + i\alpha \delta \hat{B}) | \psi \rangle$$

$$= \langle \psi | (\delta \hat{A})^2 + i\alpha [\delta \hat{A}, \delta \hat{B}] + \alpha^2 (\delta \hat{B})^2 | \psi \rangle$$

$$= \langle (\delta \hat{A})^2 \rangle - \alpha \langle \hat{C} \rangle + \alpha^2 \langle (\delta \hat{B})^2 \rangle \geq 0 \qquad (4.75)$$

The minimum value of the left hand side is attained when its derivative with respect to α is zero, that is for:

$$\alpha = \frac{\langle \hat{C} \rangle}{2 \langle (\delta \hat{B})^2 \rangle} = \frac{\langle \hat{C} \rangle}{2(\Delta B)^2}$$

Inserting this value of into the inequality, one obtains:

$$4 \langle (\delta \hat{A})^2 \rangle \langle (\delta \hat{B})^2 \rangle - \langle \hat{C} \rangle^2 \geq 0$$

which is the same as Eq.(4.74). It is common practice to also express the Heisenberg inequality in terms of the dispersion ΔA and ΔB of the two observables, in the form:

$$\Delta A \, \Delta B \geq \frac{1}{2} |\langle \hat{C} \rangle| \qquad (4.76)$$

If the operators $\hat{A}$ and $\hat{B}$ satisfy the commutation relation:

$$[\hat{A}, \hat{B}] = i\hbar \qquad (4.77)$$

it is said that A and B form a *canonically conjugate* pair of observable, for which Eq.(4.76) reduces to:

4. Commutation Relations and the Fourth Postulate 167

$$\Delta A \, \Delta B \geq \frac{1}{2} \hbar \qquad (4.78)$$

EXAMPLE 4.6. Minimum uncertainty states

Consider the case when both Eqs.(4.74) and (4.75) become equalities, which read:

$$\langle \Psi | (\delta \hat{A})^+ \delta \hat{A} | \Psi \rangle \langle \Psi | (\delta \hat{B})^+ \delta \hat{B} | \Psi \rangle = -\frac{1}{4} \langle \Psi | [\hat{A}, \hat{B}]^+ [\hat{A}, \hat{B}] | \Psi \rangle \qquad (4.79)$$

$$\left\langle \Psi \left| \left(\delta \hat{A} - \frac{i \langle \hat{C} \rangle}{2(\Delta B)^2} \delta \hat{B} \right)^+ \left(\delta \hat{A} - \frac{i \langle \hat{C} \rangle}{2(\Delta B)^2} \delta \hat{B} \right) \right| \Psi \right\rangle = 0 \qquad (4.80)$$

If the two observables A and B are compatible, the right-hand side of Eq.(4.79) vanishes, and therefore there might exist states Ψ for which both $\Delta A = 0$ and $\Delta B = 0$. These particular states are determined by:

$$\delta \hat{A} \Psi = 0 \quad \text{or} \quad \hat{A} \Psi = \langle A \rangle \Psi$$

$$\delta \hat{B} \Psi = 0 \quad \text{or} \quad \hat{B} \Psi = \langle B \rangle \Psi$$

and hence should be common eigenvectors of $\hat{A}$ and $\hat{B}$. The observables are said to be simultaneously well defined in these states, by the corresponding eigenvalues. If the two observables are not compatible, we can use Eq.(4.80), which is equivalent to:

$$\left(\delta \hat{A} - \frac{i \langle \hat{C} \rangle}{2(\Delta B)^2} \delta \hat{B} \right) \Psi = \left[\hat{A} - \langle A \rangle + \frac{1}{2} \frac{\langle [\hat{A}, \hat{B}] \rangle}{(\Delta B)^2} (\hat{B} - \langle B \rangle) \right] \Psi = 0 \qquad (4.81)$$

in order to determine the so-called *minimum uncertainty states*. Substituting $\hat{A} = \hat{p}_x = -i\hbar d/dx$ and $\hat{B} = x$, Eq.(4.81) reduces to the differential equation:

$$\left(-i\hbar \frac{d}{dx} - \langle p_x \rangle - \frac{i\hbar}{2} \frac{x - \langle x \rangle}{(\Delta x)^2} \right) \Psi(x) = 0$$

which is solved by the normalized wave function:

$$\Psi(x) = \frac{1}{\left[2\pi (\Delta x)^2 \right]^{\frac{1}{4}}} e^{-(x - \langle x \rangle)^2 / 4(\Delta x)^2} e^{i \langle \hat{p}_x \rangle x / \hbar} = A(x) e^{i \langle \hat{p}_x \rangle x / \hbar} \qquad (4.82)$$

where the normalization constant was determined from Eq.(1.37). The amplitude $A(x)$ of the plane wave function (4.82) corresponds to a Gaussian distribution of the position values about $\langle x \rangle$. If $\langle x \rangle = 0$ we obtain the wave packet introduced by Eq.(1.30), namely:

$$|\Psi(x)|^2 = \frac{1}{\sqrt{2\pi(\Delta x)^2}} e^{-x^2/2(\Delta x)^2} \tag{4.83}$$

This distribution has the dispersion $\Delta x = \hbar/2\Delta p_x$, as determined by the equality $\Delta x \Delta p_x = \hbar/2$ which holds in this case and ensures the highest precision for the simultaneous determination of position and momentum. It follows that this minimum uncertainty wave packet represents the best approximation, compatible with the uncertainty relations, for the description of a classical particle.

A similar wave packet for the momentum space can be obtained by substituting $\hat{A} = \hat{x} = i\hbar d/dp_x$ and $\hat{B} = p_x$ in Eq.(4.81), which becomes:

$$\left(i\hbar \frac{d}{dp_x} - \langle \hat{x} \rangle - \frac{i\hbar}{2} \frac{p_x - \langle p_x \rangle}{(\Delta p_x)^2} \right) \Phi(p_x) = 0 \tag{4.84}$$

The normalized solution:

$$\Phi(p_x) = \frac{1}{\left[2\pi(\Delta p_x)^2\right]^{\frac{1}{4}}} e^{-(p_x - \langle p_x \rangle)^2/4(\Delta p_x)^2} e^{-i\langle \hat{x} \rangle p_x/\hbar} \tag{4.85}$$

also leads to a Gaussian wave packet:

$$|\Phi(p_x)|^2 = \frac{1}{\sqrt{2\pi(\Delta p_x)^2}} e^{-(p_x - \langle p_x \rangle)^2/2(\Delta p_x)^2} \tag{4.86}$$

with the dispersion $\Delta p_x = \hbar/2\Delta x$. ◊

Problem 4.3.1. Show that the translation operator $\hat{S}(\lambda)$ defined in Problem 4.2.1 is compatible with the Hamiltonian in one dimension:

$$\hat{H} = \frac{\hat{p}_{x_i}^2}{2m} + U(x_i) = -\frac{\hbar^2}{2m} \frac{d^2}{dx_i^2} + U(x_i)$$

provided the potential energy is a periodic function with period λ, $U(x_i) = U(x_i - \lambda)$.

(Solution): The translation operator along the x_i-axis is expressed in terms of $\hat{p}_{x_i}$ (see Problem 4.2.1) and therefore commutes with kinetic energy. Alternatively, we can use the definition:

4. Commutation Relations and the Fourth Postulate

$$\hat{S}(\lambda)\Psi(x_i) = \Psi(x_i - \lambda)$$

and the coordinate representation of momentum, which yield the same result:

$$-\frac{\hbar^2}{2m}\left[\frac{d^2}{dx_i^2},\hat{S}(\lambda)\right]\Psi(x_i) = \frac{d^2}{dx_i^2}[\hat{S}(\lambda)\Psi(x_i)] - \hat{S}\frac{d^2\Psi(x_i)}{dx_i^2}$$

$$= \frac{d^2\Psi(x_i-\lambda)}{dx_i^2} - \frac{d^2\Psi(x_i-\lambda)}{d(x_i-\lambda)^2} = 0$$

Then we have:

$$[U(x_i),\hat{S}(\lambda)]\Psi(x_i) = U(x_i)\hat{S}(\lambda)\Psi(x_i) - \hat{S}(\lambda)U(x_i)\Psi(x_i)$$

$$= U(x_i)\Psi(x_i-\lambda) - U(x_i-\lambda)\Psi(x_i-\lambda)$$

and therefore, provided $U(x_i - \lambda) = U(x_i)$, we have:

$$[\hat{H},\hat{S}(\lambda)] = -\frac{\hbar^2}{2m}\left[\frac{d^2}{dx_i^2},\hat{S}(\lambda)\right] + [U(x_i),\hat{S}(\lambda)] = 0$$

Because $\hat{S}(\lambda)$ is unitary, its commutivity with $\hat{H}$ implies:

$$\hat{H}\hat{S}(\lambda) = \hat{S}(\lambda)\hat{H} \quad \text{or} \quad \hat{H} = \hat{S}(\lambda)\hat{H}\hat{S}^{-1}(\lambda)$$

and this means that the Hamiltonian is invariant under a finite translation λ of coordinates, provided the potential energy is a periodic function with period λ.

Problem 4.3.2. Show that the rotation operator defined in Problem 4.2.2 is compatible with the Hamiltonian of a particle in a central force field:

$$\hat{H} = \frac{1}{2m}(\hat{p}_x^2 + \hat{p}_y^2 + \hat{p}_z^2) + U(r) = -\frac{\hbar^2}{2m}\nabla^2 + U(r)$$

(Solution):

The central force is defined classically by the potential energy $U(r)$ which is a function of the radial distance only. We have seen, Eq.(4.49), that the operator $\hat{p}^2$, commutes with each $\hat{L}_i$, and therefore with any function $f(\hat{L}_i)$, so that it follows that:

$$[\hat{R}(\omega_i),\hat{p}^2] = [e^{i\omega_i\hat{L}_i/\hbar},\hat{p}^2] = 0$$

From Eq.(4.50) we may derive that $[\hat{L}_i,U(r)] = 0$, and also $[f(\hat{L}_i),U(r)] = 0$, hence:

$$[\hat{R}(\omega_i), U(r)] = [e^{i\omega_i \hat{L}_i/\hbar}, U(r)] = 0$$

This leads to $[\hat{R}(\omega_i), \hat{H}] = 0$, and also:

$$\hat{R}(\omega_i)\hat{H} = \hat{H}\hat{R}(\omega_i) \quad \text{or} \quad \hat{H} = \hat{R}(\omega_i)\hat{H}\hat{R}^{-1}(\omega_i)$$

Hence, the Hamiltonian of the motion in a central force field is left invariant by a finite rotation ω_i of coordinate, about each x_i-axis.

Problem 4.3.3. Show that, if the operators representing compatible observables commute with the parity operator, defined in Problem 3.1.2, all their common eigenvectors have a definite parity.

(Solution): Consider two commuting operators $\hat{B}$ and $\hat{C}$ with the common basis vectors $\psi_i(x)$:

$$\hat{B}\psi_i(x) = \mu_i \psi_i(x), \quad \hat{C}\psi_i(x) = v_i \psi_i(x)$$

Since $[\hat{\Pi}, \hat{B}] = [\hat{\Pi}, \hat{C}] = 0$ we have:

$$\hat{B}\hat{\Pi}\psi_i(x) = \hat{\Pi}\hat{B}\psi_i(x) = \mu_i \hat{\Pi}\psi_i(x)$$

$$\hat{C}\hat{\Pi}\psi_i(x) = \hat{\Pi}\hat{C}\psi_i(x) = v_i \hat{\Pi}\psi_i(x)$$

and this means that $\hat{\Pi}\psi_i(x)$ are also common eigenvectors of the two operators, corresponding to the same eigenvalues as the $\psi_i(x)$. These eigenvectors only might be different by a multiplicative constant, namely:

$$\hat{\Pi}\psi_i(x) = \lambda \psi_i(x)$$

In other words, all the $\psi_i(x)$ are also eigenvectors of $\hat{\Pi}$, and thus have a definite parity.

Problem 4.3.4. Show that the eigenstates of the commuting operators $\hat{L}_z, \hat{L}^2$ and $\hat{H}$, where $\hat{H}$ is the Hamiltonian of the motion in the central force field, introduced in Problem 4.3.2, have a definite parity.

(Answer): $[\hat{\Pi}, \hat{L}_z] = [\hat{\Pi}, \hat{L}^2] = [\hat{\Pi}, \hat{H}] = 0$

5. TEMPORAL EVOLUTION AND THE FIFTH POSTULATE

5.1. The Heisenberg Description

Quantum systems, described in terms of wave functions and Hermitian operators that are defined at any instant of time, should be capable of temporal evolution, and this is consistent with the first four Postulates. There is nothing in Postulate 1 to prevent us from considering state vectors $\Psi(t)$ which depend on time. A Hermitian operator $\hat{A}(t)$, representing a real observable A according to Postulate 2, depends on time either explicitly or through the position coordinate and conjugate momenta, following the correspondence rule introduced by Postulate 4.

The probabilistic interpretation given by Postulate 3 to the expectation value of any real observable remains valid if $\langle\Psi|\hat{A}|\Psi\rangle$ is a function of the time t. Although both the state functions and the operators might be considered dependent on time, it was found more convenient to develop the formalism of quantum mechanics in terms of either time dependent vectors $\Psi(t)$ or time dependent operators $\hat{A}(t)$.

In the *Schrödinger description*, the operator $\hat{A}$ representing a physical observable is taken to be independent of the time, and this implies that the position coordinates x_i and the time (in coordinate representation) or the momentum coordinates p_i and the time (in momentum representation) are independent variables. The temporal evolution is considered through the time dependence of the state vectors $\Psi(t)$, such that the expectation values should be written as $\langle\Psi(t)|\hat{A}|\Psi(t)\rangle$, with $\hat{A}$ independent of t. If, in contrast, the state vectors Ψ are taken to be independent of the time, we obtain the *Heisenberg description*, where the evolution with time is introduced through the time dependence of quantum operators $\hat{A}(t)$, and the expectation value reads $\langle\Psi|\hat{A}(t)|\Psi\rangle$. Since any expectation value represents the average result of a measurement on a physical observable, it should be independent of description, namely:

$$\langle\Psi(t)|\hat{A}|\Psi(t)\rangle = \langle\Psi|\hat{A}(t)|\Psi\rangle \tag{5.1}$$

172 Quantum Physics

The evolution of quantum systems can be described in terms of time-dependent representative operators $\hat{A}(t)$ by simply using the classical equation of motion for the observable A, in the Hamiltonian form (4.35) which, in view of the correspondence rule (4.41), becomes:

$$i\hbar \frac{d\hat{A}(t)}{dt} = \left[\hat{A}(t), \hat{H}\right] + i\hbar \frac{\partial \hat{A}(t)}{\partial t} \tag{5.2}$$

Here the Hamiltonian operator $\hat{H}$, Eq.(2.17), corresponds to the classical Hamiltonian H for the system and thus enters the evolution equation (5.2) as a time-independent operator, in the case of conservative systems. Provided the operator $\hat{A}(t)$ does not involve time explicitly, the evolution equation reduces to:

$$i\hbar \frac{d\hat{A}(t)}{dt} = \left[\hat{A}(t), \hat{H}\right] \tag{5.3}$$

The last two equations only contain the temporal evolution of representative operators, independent of how the state vector to which they might be applied changes with time. Hence, when calculating on their grounds the temporal evolution of the expectation values, if will be sufficient to consider time-independent state vectors Ψ.

EXAMPLE 5.1 Temporal evolution of position and momentum

Since the operators $\hat{x}_i$ and $\hat{p}_i$ do not depend explicity on time, their appropriate evolution equations have the form (5.3), which reads:

$$\frac{d\hat{x}_i}{dt} = \frac{i}{\hbar}\left[\hat{H}, \hat{x}_i\right], \quad \frac{d\hat{p}_i}{dt} = \frac{i}{\hbar}\left[\hat{H}, \hat{p}_i\right] \tag{5.4}$$

Taking $\hat{x}_i = \hat{x}$ we can then write the evolution equation of position along the x-axis as:

$$\frac{d\hat{x}}{dt} = \frac{i}{\hbar}\left[\hat{H}, \hat{x}\right] = \frac{i}{\hbar}(\hat{H}\hat{x} - \hat{x}\hat{H})$$

where $\hat{H}$ is the Hamiltonian operator in one dimension, as derived from Eq.(2.17). For an arbitrary time-independent state function $\Psi(x)$ we may write:

$$(\hat{H}\hat{x} - \hat{x}\hat{H})\Psi = -\frac{\hbar^2}{2m}\frac{d^2}{dx^2}(x\Psi) + Ux\Psi + x\frac{\hbar^2}{2m}\frac{d^2\Psi}{dx^2} - xU\Psi$$

$$= -\frac{\hbar^2}{2m}\left(x\frac{d^2\Psi}{dx^2} + 2\frac{d\Psi}{dx} - x\frac{d^2\Psi}{dx^2}\right) = \left(-\frac{\hbar^2}{m}\frac{d}{dx}\right)\Psi$$

5. Temporal Evolution and the Fifth Postulate

so that, independently of the particular state $\Psi(x)$, the evolution of the position operator is given by:

$$\frac{d\hat{x}}{dt} = \frac{i}{\hbar}\left[\hat{H}, \hat{x}\right] = -\frac{i\hbar}{m}\frac{d}{dx} = \frac{\hat{p}_x}{m} \tag{5.5}$$

In a similar way, taking $\hat{p}_i = \hat{p}_x$, the evolution equation of the conjugate momentum becomes:

$$\frac{d\hat{p}_x}{dt} = \frac{i}{\hbar}\left[\hat{H}, \hat{p}_x\right] = \frac{i}{\hbar}(\hat{H}\hat{p}_x - \hat{p}_x\hat{H})$$

The effect of the commutator $[\hat{H}, \hat{p}_x]$ on an arbitrary state function $\Psi(x)$ is:

$$(\hat{H}\hat{p}_x - \hat{p}_x\hat{H})\Psi = \left[\left(-\frac{\hbar^2}{2m}\frac{d^2}{dx^2} + U\right)\left(-i\hbar\frac{d}{dx}\right) - \left(-i\hbar\frac{d}{dx}\right)\left(-\frac{\hbar^2}{2m}\frac{d^2}{dx^2} + U\right)\right]\Psi$$

$$= -i\hbar\left(U\frac{d}{dx} - \frac{d}{dx}U\right)\Psi = -i\hbar\left(U\frac{d\Psi}{dx} - \frac{dU}{dx}\Psi - U\frac{d\Psi}{dx}\right) = i\hbar\frac{dU}{dx}\Psi$$

where we made use of the property that d/dx and d^2/dx^2 commute. Therefore we have the relation:

$$\frac{d\hat{p}_x}{dt} = -\frac{\partial U}{\partial x} = F(\hat{x}) \tag{5.6}$$

where the operator $F(\hat{x})$ must be associated with a conservative force. Upon insertion of Eqs.(5.5) and (5.6) into Eq.(2.11), the following relations between the expectation values of position coordinate, momentum and force are obtained:

$$\left\langle \frac{d\hat{x}}{dt}\right\rangle = \frac{1}{m}\langle \hat{p}_x\rangle \tag{5.7}$$

$$\left\langle \frac{d\hat{p}_x}{dt}\right\rangle = \langle F(\hat{x})\rangle \tag{5.8}$$

The first equation is almost identical to the relation between the classical momentum p_x and velocity dx/dt. It would be *exactly* identical, if and only if:

$$\left\langle \frac{d\hat{x}}{dt}\right\rangle = \frac{d}{dt}\langle \hat{x}\rangle \tag{5.9}$$

for in this case Eq.(5.7) would read:

$$\frac{d}{dt}\langle \hat{x}\rangle = \frac{1}{m}\langle \hat{p}_x\rangle \tag{5.10}$$

174 Quantum Physics

which is the same as the classical relation in one dimension, provided we identify $\langle \hat{x} \rangle$ with the position coordinate x. Assuming also that:

$$\left\langle \frac{d\hat{p}_x}{dt} \right\rangle = \frac{d}{dt} \langle \hat{p}_x \rangle \tag{5.11}$$

we can write Eq.(5.8) as:

$$\frac{d}{dt} \langle \hat{p}_x \rangle = \langle F(\hat{x}) \rangle \tag{5.12}$$

and, by means of Eq.(5.10), it follows that:

$$\frac{d}{dt}\left(m \frac{d}{dt} \langle \hat{x} \rangle \right) = \langle F(\hat{x}) \rangle \quad \text{or} \quad \frac{d^2}{dt^2} \langle \hat{x} \rangle = \frac{1}{m} \langle F(\hat{x}) \rangle \tag{5.13}$$

This equation would read:

$$\frac{d^2}{dt^2} \langle \hat{x} \rangle = \frac{1}{m} F(\langle \hat{x} \rangle) \tag{5.14}$$

hence would be identical to the Newton second law, provided that:

$$\langle F(\hat{x}) \rangle = F(\langle \hat{x} \rangle) \tag{5.15}$$

Equations (5.10) and (5.12) state that the evolution of $\langle \hat{x} \rangle$ and $\langle \hat{p}_x \rangle$ coincides with the evolution of x and p_x respectively, in other words that quantum mechanics corresponds to classical dynamics. However, this conclusion relies on the validity of Eqs.(5.9), (5.11) and (5.15). ☙

The requirement that the evolution of any expectation value $\langle \hat{A}(t) \rangle$ must be analogous to that of the corresponding classical function A is known as the Ehrenfest theorem:

$$\left\langle \frac{d\hat{A}(t)}{dt} \right\rangle = \frac{d}{dt} \langle \hat{A}(t) \rangle \tag{5.16}$$

which can be interpreted as a generalization of Eqs.(5.9) and (5.11). The Heisenberg description of the temporal evolution implies that Eq.(5.16) holds for any operator $\hat{A}(t)$, such that the evolution equation for a quantum operator will also be obeyed by its expectation value, in the form:

$$i\hbar \frac{d}{dt} \langle \hat{A}(t) \rangle = \langle [\hat{A}(t), \hat{H}] \rangle + i\hbar \left\langle \frac{\partial \hat{A}(t)}{\partial t} \right\rangle \tag{5.17}$$

5. Temporal Evolution and the Fifth Postulate

or, if $\hat{A}(t)$ has not an explicit dependence on the time:

$$i\hbar \frac{d}{dt}\langle \hat{A}(t)\rangle = \langle [\hat{A}(t),\hat{H}]\rangle \tag{5.18}$$

Assuming that $\hat{A}(t)$ also commutes with the Hamiltonian, Eq.(5.18) becomes:

$$\langle [\hat{A}(t),\hat{H}]\rangle = 0 \quad \text{or} \quad \frac{d}{dt}\langle \hat{A}(t)\rangle = 0 \tag{5.19}$$

and in this case the observable A is said to be conserved or to be a constant of motion.

Problem 5.1.1. Construct the position and momentum operators $\hat{x}(t)$ and $\hat{p}_x(t)$, for a linear harmonic motion.

(Solution): We use the evolution equation (5.3), which holds for both $\hat{x}(t)$ and $\hat{p}(t)$ with no explicit dependence on time:

$$i\hbar \frac{d\hat{x}(t)}{dt} = [\hat{x}(t),\hat{H}], \quad i\hbar \frac{d\hat{p}_x(t)}{dt} = [\hat{p}_x(t),\hat{H}]$$

The Hamiltonian operator should be written as:

$$\hat{H} = \frac{\hat{p}_x^2(t)}{2m} + \frac{m\omega^2}{2}\hat{x}^2(t)$$

which gives:

$$[\hat{x}(t),\hat{H}] = \frac{1}{2m}[\hat{x}(t),\hat{p}_x^2(t)] + \frac{m\omega^2}{2}[\hat{x}(t),\hat{x}^2(t)]$$

$$= \frac{1}{2m}\{\hat{p}_x(t)[\hat{x}(t),\hat{p}_x(t)] + [\hat{x}(t),\hat{p}_x(t)]\hat{p}_x(t)\} = \frac{i\hbar}{m}\hat{p}_x(t)$$

$$[\hat{p}_x(t),\hat{H}] = \frac{1}{2m}[\hat{p}_x(t),\hat{p}_x^2(t)] + \frac{m\omega^2}{2}[\hat{p}_x(t),\hat{x}^2(t)]$$

$$= \frac{m\omega^2}{2}\{\hat{x}(t)[\hat{p}_x(t),\hat{x}(t)] + [\hat{p}_x(t),\hat{x}(t)]\hat{x}(t)\} = -i\hbar m\omega^2 \hat{x}(t)$$

Quantum Physics

Hence, the evolution equations reduce to the linear differential forms:

$$\frac{d\hat{x}(t)}{dt} = \frac{\hat{p}_x(t)}{m}, \quad \frac{d\hat{p}_x(t)}{dt} = -m\omega^2 \hat{x}(t)$$

which, by further differentiation, take the familiar expression of the classical equations of motion:

$$\frac{d^2\hat{x}(t)}{dt^2} + \omega^2 \hat{x}(t) = 0, \quad \hat{p}_x(t) = m\frac{d\hat{x}(t)}{dt}$$

As a result, the solution may be written as:

$$\hat{x}(t) = a_1 \sin\omega t + a_2 \cos\omega t$$

which implies that:

$$\hat{p}(t) = m\omega a_1 \cos\omega t - m\omega a_2 \sin\omega t$$

The initial conditions $\hat{x}(0) = x$, $\hat{p}_x(0) = p_x$ determine the constants a_1 and a_2, such that:

$$\hat{x}(t) = x\cos\omega t + \frac{p_x}{m\omega}\sin\omega t$$

$$\hat{p}_x(t) = p_x \cos\omega t - m\omega x \sin\omega t$$

Problem 5.1.2. Find the commutation relations:

$$\left[\hat{x}(t_1), \hat{x}(t_2)\right], \quad \left[\hat{p}_x(t_1), \hat{p}_x(t_2)\right], \quad \left[\hat{x}(t_1), \hat{p}_x(t_2)\right]$$

for the operators determined in Problem 5.1.1.

(Answer):
$$\left[\hat{x}(t_1), \hat{x}(t_2)\right] = -\frac{i\hbar}{m\omega}\sin\omega(t_1 - t_2)$$

$$\left[\hat{p}_x(t_1), \hat{p}_x(t_2)\right] = -i\hbar m\omega \sin\omega(t_1 - t_2)$$

$$\left[\hat{x}(t_1), \hat{p}_x(t_2)\right] = i\hbar \cos\omega(t_1 - t_2)$$

Problem 5.1.3. Show that the position and momentum operators $\hat{x}(t,v)$ and $\hat{p}_x(t,v)$ for a linear harmonic oscillator whose centre of force is moving along the x-axis with constant velocity v, have the form:

5. Temporal Evolution and the Fifth Postulate

$$\hat{x}(t,v) = \hat{x}(t) + vt$$

$$\hat{p}_x(t,v) = \hat{p}_x(t) + mv$$

where $\hat{x}(t), \hat{p}_x(t)$ are the operators for the case considered in Problem 5.1.1, where the centre of the elastic force is fixed ($v = 0$).

5.2. The Fifth Postulate

Let us examine the temporal evolution of the expectation value of operators in the Schrödinger description where, in view of Eq.(5.1), Eq.(5.18) reads:

$$i\hbar \frac{d}{dt} \langle \Psi(t)| \hat{A}|\Psi(t)\rangle = \langle \Psi(t)|[\hat{A}, \hat{H}]|\Psi(t)\rangle \tag{5.20}$$

In this description all operators are assumed independent of time and we will consider time dependent state functions $\Psi(x, y, z, t)$ for which we have:

$$\frac{d\Psi(x,y,z,t)}{dt} = \frac{\partial \Psi(x,y,z,t)}{\partial t} \tag{5.21}$$

because the position variables x_i are time-independent. The left-hand side of Eq.(5.20) can be written as:

$$i\hbar \frac{d}{dt} \langle \hat{A} \rangle = i\hbar \frac{d}{dt} \int\int\int_{-\infty-\infty-\infty}^{\infty\;\infty\;\infty} \Psi^* \hat{A} \Psi dx dy dz = i\hbar \int_\infty \left(\frac{\partial \Psi^*}{\partial t} \hat{A} \Psi + \Psi^* \hat{A} \frac{\partial \Psi}{\partial t} \right) dV \tag{5.22}$$

while the right-hand side of the same equation reads:

$$\langle [\hat{A}, \hat{H}] \rangle = \int_\infty (-\Psi^* \hat{H}\hat{A}\Psi + \Psi^* \hat{A}\hat{H}\Psi) dV = \int_\infty \left[(-\hat{H}\Psi)^+ \hat{A}\Psi + \Psi^* \hat{A}\hat{H}\Psi \right] dV \tag{5.23}$$

By comparing Eqs.(5.22) and (5.23) it follows that the evolution equation (5.20) is true, in the Schrödinger description, only if:

$$i\hbar \frac{\partial \Psi(t)}{\partial t} = \hat{H}\Psi(t) \tag{5.24}$$

178 Quantum Physics

This basic result for describing the temporal evolution of quantum systems in terms of time-dependent state functions and time-independent operators is called the *time-dependent Schrödinger equation*. The equation is linear and homogenous, in the sense that if $\Psi_1(t)$ and $\Psi_2(t)$ are independent solutions:

$$i\hbar \frac{\partial \Psi_1(t)}{\partial t} = \hat{H}\Psi_1(t), \quad i\hbar \frac{\partial \Psi_2(t)}{\partial t} = \hat{H}\Psi_2(t) \qquad (5.25)$$

then any linear combination $a_1\Psi_1(t) + a_2\Psi_2(t)$ is also a solution (as $\hat{H}$ is a linear operator), and this is consistent with the principle of superposition of states, Eq.(1.63). The scalar product of the two solutions (5.25) is constant in time, because:

$$\frac{d}{dt}\langle \Psi_1(t)|\Psi_2(t)\rangle = \langle \frac{\partial \Psi_1}{\partial t}|\Psi_2\rangle + \langle \Psi_1|\frac{\partial \Psi_2}{\partial t}\rangle = \frac{i}{\hbar}\langle \Psi_1|\hat{H}|\Psi_2\rangle - \frac{i}{\hbar}\langle \Psi_1|\hat{H}|\Psi_2\rangle = 0 \qquad (5.26)$$

where use has been made of Eq.(5.21). In the special case where $\Psi_1(t) = \Psi_2(t)$, this means that the norm of the state vectors is conserved and therefore, if a state vector is normalized at $t = 0$, it will remain normalized at any later moment. The solution $\Psi(0)$ that describes the initial state of the system, always determines the state $\Psi(t)$ at any $t > 0$, by taking a series of infinitesimal translations $\varepsilon = dt$ in time:

$$\Psi(dt) = \Psi(0) + \left(\frac{\partial \Psi}{\partial t}\right)_0 dt = \Psi(0) - \frac{i\varepsilon}{\hbar}(\hat{H}\Psi)_0 = \left(1 - \frac{i\varepsilon}{\hbar}\hat{H}\right)\Psi(0) \qquad (5.27)$$

which are unitary transformations of the state functions, of the form (4.53), generated by the operator $\hat{H}/\hbar$. A finite translation in time, made up of N infinitesimal translations (5.27) with $\varepsilon = t/N$, will be given by the unitary operator:

$$\hat{T}(t) = \lim_{N\to\infty}\left(1 - \frac{it}{N\hbar}\hat{H}\right)^N = e^{-i\hat{H}t/\hbar} \qquad (5.28)$$

called the *evolution operator*, in terms of which the time dependence of the state function reads:

$$\Psi(t) = e^{-i\hat{H}t/\hbar}\Psi(0) \qquad (5.29)$$

This is a solution of Eq.(5.24):

$$\frac{\partial \Psi(t)}{\partial t} = -\frac{i}{\hbar}\hat{H}e^{-i\hat{H}t/\hbar}\Psi(0) = -\frac{i}{\hbar}\hat{H}\Psi(t) \qquad (5.30)$$

provided that the Hamiltonian does not depend explicitly of the time. The requirement that Eq.(5.24) must be satisfied is introduced as a postulate for the Schrödinger description of the temporal evolution.

Postulate 5. *The temporal evolution of the state function of a quantum system is given by the time-dependent Schrödinger equation:*

$$i\hbar \frac{\partial \Psi(t)}{\partial t} = \hat{H}\Psi(t)$$

where $\hat{H}$ is the Hamiltonian operator of the system.

If $\hat{H}$ is taken to be the Hamiltonian operator (2.17), introduced by Postulate 2 for a single particle in a force field $\vec{F} = -\nabla U(\vec{r},t)$, the time dependent states of the particle will be described by wave functions $\Psi(\vec{r},t)$ satisfying Eq.(5.24), which reads:

$$i\hbar \frac{\partial \Psi(\vec{r},t)}{\partial t} = -\frac{\hbar^2}{2m}\nabla^2 \Psi(\vec{r},t) + U(\vec{r},t)\Psi(\vec{r},t) \qquad (5.31)$$

The solutions are, in general, complex functions of the variables $\vec{r}$ and t, in terms of which one may express the probability density (1.26) of finding the particle at a given position, at time t:

$$P(\vec{r},t) = |\Psi(\vec{r},t)|^2 \qquad (5.32)$$

and also the scalar products from Eqs.(3.33) and (3.34), which represent the probability densities associated with the result of measurement on any physical observable of the particle (momentum, angular momentum, etc.). Hence, the time dependence of the state functions determines, in the Schrödinger description, the temporal evolution of the statistical predictions of quantum postulates.

EXAMPLE 5.2. The continuity equation for probability

The complex conjugate of Eq.(5.31) reads:

$$-i\hbar \frac{\partial \Psi^*}{\partial t} = -\frac{\hbar^2}{2m}\nabla^2 \Psi^* + U\Psi^*$$

Upon multiplication of Eq.(5.31) by Ψ^* and of its complex conjugate equation by Ψ, followed by subtraction, one obtains:

$$i\hbar \left(\Psi^* \frac{\partial \Psi}{\partial t} + \Psi \frac{\partial \Psi^*}{\partial t} \right) = -\frac{\hbar^2}{2m}(\Psi^*\nabla^2\Psi - \Psi\nabla^2\Psi^*) \qquad (5.33)$$

180   Quantum Physics

It is a straightforward matter to show that:

$$\Psi^* \frac{\partial^2 \Psi}{\partial x_i^2} - \Psi \frac{\partial^2 \Psi^*}{\partial x_i^2} = \frac{\partial}{\partial x_i}\left(\Psi^* \frac{\partial \Psi}{\partial x_i} - \Psi \frac{\partial \Psi^*}{\partial x_i}\right)$$

and therefore, Eq.(5.33) can be written in the equivalent form:

$$\frac{\partial}{\partial t}(\Psi^*\Psi) = \frac{i\hbar}{2m}\nabla\cdot(\Psi^*\nabla\Psi - \Psi\nabla\Psi^*) \qquad (5.34)$$

It is convenient to introduce a vector function called the *probability current density*, meaning the probability flow per unit time per unit area normal to $\vec{j}$, defined by:

$$\vec{j}(\vec{r},t) = \frac{i\hbar}{2m}(\Psi\nabla\Psi^* - \Psi^*\nabla\Psi) \qquad (5.35)$$

which is a real vector ($\vec{j}^* = \vec{j}$), that vanishes if Ψ is a real state function. In view of Eqs.(5.32) and (5.35), Eq.(5.34) can be written as:

$$\frac{\partial P}{\partial t} + \nabla\cdot\vec{j} = 0 \qquad (5.36)$$

and this form is analogous to the equation of continuity, found for classical quantities of spatial density ρ and current density $\vec{j}$:

$$\frac{\partial \rho}{\partial t} + \nabla\cdot\vec{j} = \frac{\partial \rho}{\partial t} + \nabla\cdot(\rho\vec{v}) = 0 \qquad (5.37)$$

Here, ρ might represent the mass per unit volume at a point in a fluid, and $\vec{j} = \rho\vec{v}$ the corresponding mass flow density, such that the equation of continuity shows the mass conservation for a volume element in a fluid. Similarly, ρ and $\vec{j} = \rho\vec{v}$ can be taken to be the charge density and the electric current density, respectively, so expressing the conservation of the electric charge in a volume element of a conducting medium. A factor proportional to the probability density $P = \Psi^*\Psi$ can also be formally separated in Eq.(5.35) which reads:

$$\vec{j} = \frac{i\hbar}{2m}\sum_{i=x,y,z}\vec{e}_i\left(\Psi\frac{\partial \Psi^*}{\partial x_i} - \Psi^*\frac{\partial \Psi}{\partial x_i}\right) = \frac{i\hbar}{2m}\sum_{i=x,y,z}\vec{e}_i\left(\frac{1}{\Psi^*}\frac{\partial \Psi^*}{\partial x_i} - \frac{1}{\Psi}\frac{\partial \Psi}{\partial x_i}\right)\Psi^*\Psi$$

$$= \frac{i\hbar}{2m}\sum_{i=x,y,z}\vec{e}_i\left[\frac{\partial}{\partial x_i}\left(\ln\frac{\Psi^*}{\Psi}\right)\right]\Psi^*\Psi = P\vec{v}$$

although the vector function:

$$\vec{v} = \frac{i\hbar}{2m}\sum_{i=x,y,z}\vec{e}_i\frac{\partial}{\partial x_i}\left(\ln\frac{\Psi^*}{\Psi}\right)$$

5. Temporal Evolution and the Fifth Postulate 181

cannot simply be interpreted as a quantum representative of velocity.

We may conclude that Eq.(5.36) shows that probability $P(\vec{r},t)$ is locally conserved. The integral form of this result, in a finite volume V enclosed by a surface S, reads:

$$\int_V \frac{\partial P}{\partial t} dV + \int_V (\nabla \cdot \vec{j}) dV = 0 \quad \text{or} \quad \frac{d}{dt} \int_V P dV + \oint_S \vec{j} \cdot d\vec{s} = 0$$

where the first integral is a function of time only, so that $\partial/\partial t$ was replaced by d/dt, and the Gauss theorem was used to reduce the second volume integral of $\nabla \cdot \vec{j}$ to the integral of the normal component of $\vec{j}$ over the surface bounding the volume. This equation states that any decrease in probability in the volume V is accounted for by a probability current, directed out of it. When the volume integral is taken over all space, S becomes the surface at infinity S_∞, where Ψ and $\nabla\Psi$ vanish for typical wave functions, which are normalizable to unity. One then obtains the conservation law for the total probability in the form:

$$\frac{d}{dt} \int_\infty P dV = 0 \quad \text{or} \quad \int_\infty P dV = \text{const}$$

In other words the normalization condition for state functions is constant in time, as already indicated by Eq.(5.26). ♦

EXAMPLE 5.3. The classical limit of the time-dependent Schrödinger equation

Basic classical results can be formally regained from the time-dependent Schrödinger equation, in the appropriate domain. If we consider a three-dimensional solution of Eq.(5.31), in the form:

$$\Psi(\vec{r},t) = A(\vec{r},t) e^{iS(\vec{r},t)/\hbar} \tag{5.38}$$

where $A(\vec{r},t)$ is a real amplitude, such that:

$$P(\vec{r},t) = |\Psi(\vec{r},t)|^2 = A^2(\vec{r},t) \tag{5.39}$$

and the phase function $S(\vec{r},t)$ has the dimensions of action (energy × time), it is a straightforward matter to calculate:

$$\frac{\partial \Psi}{\partial t} = \frac{\partial A}{\partial t} e^{iS/\hbar} + \frac{i}{\hbar} \frac{\partial S}{\partial t} A e^{iS/\hbar}, \quad \frac{\partial \Psi}{\partial x_i} = \frac{\partial A}{\partial x_i} e^{iS/\hbar} + \frac{i}{\hbar} \frac{\partial S}{\partial x_i} A e^{iS/\hbar}$$

$$\frac{\partial^2 \Psi}{\partial x_i^2} = \frac{\partial^2 A}{\partial x_i^2} e^{iS/\hbar} + 2 \frac{i}{\hbar} \frac{\partial S}{\partial x_i} \frac{\partial A}{\partial x_i} e^{iS/\hbar} + \frac{i}{\hbar} \frac{\partial^2 S}{\partial x_i^2} A e^{iS/\hbar} - \frac{1}{\hbar^2} \left(\frac{\partial S}{\partial x_i}\right)^2 A e^{iS/\hbar}$$

It follows that Eq.(5.31) takes the form:

182 Quantum Physics

$$i\hbar\left[\frac{\partial A}{\partial t}+\frac{i}{\hbar}\frac{\partial S}{\partial t}A\right]=-\frac{\hbar^2}{2m}\sum_{i=x,y,z}\left[\frac{\partial^2 A}{\partial x_i^2}+2\frac{i}{\hbar}\frac{\partial S}{\partial x_i}\frac{\partial A}{\partial x_i}+\frac{i}{\hbar}\frac{\partial^2 S}{\partial x_i^2}A-\frac{1}{\hbar^2}\left(\frac{\partial S}{\partial x_i}\right)^2 A\right]+UA$$

where the exponential common factor has been simplified. If we limit ourselves to the first-order terms in $\hbar$, in the so-called *quasi-classical approximation*, the first term on the right hand side should be neglected, and this equation reduces to:

$$i\hbar\left[\frac{\partial A}{\partial t}+\frac{1}{2m}\sum_{i=x,y,z}\left(2\frac{\partial A}{\partial x_i}\frac{\partial S}{\partial x_i}+A\frac{\partial^2 S}{\partial x_i^2}\right)\right]=\left[\frac{1}{2m}\sum_{i=x,y,z}\left(\frac{\partial S}{\partial x_i}\right)^2+U+\frac{\partial S}{\partial t}\right]A$$

The real and imaginary part should each be equal to zero, which provides one equation for the phase function:

$$\frac{1}{2m}\sum_{i=x,y,z}\left(\frac{\partial S}{\partial x_i}\right)^2+U+\frac{\partial S}{\partial t}=0 \qquad (5.40)$$

and a second equation involving the amplitude function:

$$\frac{\partial A}{\partial t}+\frac{1}{2m}\sum_{i=x,y,z}\left(2\frac{\partial A}{\partial x}\frac{\partial S}{\partial x_i}+A\frac{\partial^2 S}{\partial x_i^2}\right)=0 \qquad (5.41)$$

The equation for S is the so-called *Hamilton-Jacobi equation*, which is obtained in classical mechanics if the equations of motion are expressed in terms of the action integral (4.25). This equation is easily derived from Eq.(4.25), written in the equivalent form:

$$\frac{dS}{dt}-\sum_{i=x,y,x}p_i\dot{x}_i+H(x_i,p_i)=0 \quad \text{or} \quad \frac{dS}{dt}-\frac{1}{2}\sum_{i=x,y,z}m\dot{x}_i^2+U=0$$

if the time derivative of $S(\vec{r},t)$ is replaced by the expansion:

$$\frac{dS(\vec{r},t)}{dt}=\frac{\partial S}{\partial t}+\sum_{i=x,y,z}\frac{\partial S}{\partial x_i}\dot{x}_i$$

It follows that:

$$\frac{\partial S}{\partial t}-\frac{1}{2m}\sum_{i=x,y,z}\left(m\dot{x}_i-\frac{\partial S}{\partial x_i}\right)^2+\frac{1}{2m}\sum_{i=x,y,z}\left(\frac{\partial S}{\partial x_i}\right)^2+U=0$$

and this reduces to the Hamilton-Jacobi equation (5.40) by making the choice:

$$p_i=m\dot{x}_i=\frac{\partial S}{\partial x_i} \qquad (5.42)$$

The amplitude equation (5.41) takes a convenient form if written in terms of the probability amplitude A^2, which can be introduced through multiplication by $2A$, giving:

5. Temporal Evolution and the Fifth Postulate

$$2A\frac{\partial A}{\partial t} + \frac{1}{m}\sum_{i=x,y,z}\left(2A\frac{\partial A}{\partial x_i}\frac{\partial S}{\partial x_i} + A^2\frac{\partial^2 S}{\partial x_i^2}\right) = \frac{\partial A^2}{\partial t} + \frac{1}{m}\sum_{i=x,y,z}\left(\frac{\partial A^2}{\partial x_i}\frac{\partial S}{\partial x_i} + A^2\frac{\partial^2 S}{\partial x_i^2}\right)$$

$$= \frac{\partial A^2}{\partial t} + \sum_{i=x,y,z}\frac{\partial}{\partial x_i}\left(A^2\frac{1}{m}\frac{\partial S}{\partial x_i}\right) = 0 \quad (5.43)$$

In view of Eqs.(5.39) and (5.42), we may write:

$$A^2\frac{1}{m}\frac{\partial S}{\partial x_i} = P\frac{p_i}{m} = P\dot{x}_i = Pv_i$$

where v_i are the velocity components corresponding to the classical motion. Hence, Eq.(5.43) reduces to the continuity equation:

$$\frac{\partial P}{\partial t} + \sum_{i=x,y,z}\frac{\partial}{\partial x_i}(Pv_i) = 0 \quad \text{or} \quad \frac{\partial P}{\partial t} + \nabla\cdot(P\vec{v}) = 0 \quad (5.44)$$

which shows that, in the quasi-classical approximation, the probability current density $\vec{j} = P\vec{v}$ follows the classical trajectories. ♦

Problem 5.2.1. Show that the Schrödinger equation in one dimension is invariant under the Galilean transformation of coordinates:

$$x' = x - vt, \qquad t' = t$$

provided $U'(x',t') = U(x,t)$.

(Solution): The uniform translation of the reference frame should leave invariant the probability density:

$$|\Psi(x,t)|^2 = |\Psi'(x',t')|^2 \quad \text{or} \quad \Psi(x,t) = e^{i\varphi}\Psi'(x',t')$$

and hence, the wave functions $\Psi(x,t)$ and its Galilean transform should differ by a phase factor φ only, which is a real function of either x, t or x', t'. The invariance of the Schrödinger equation reads:

$$-\frac{\hbar^2}{2m}\frac{d^2\Psi'}{dx'^2} + U'\Psi' - i\hbar\frac{\partial\Psi'}{\partial t'} = -\frac{\hbar^2}{2m}\frac{d^2\Psi}{dx^2} + U\Psi - i\hbar\frac{\partial\Psi}{\partial t} = 0$$

184　Quantum Physics

If the left-hand side is expressed in terms of the variables x and t, using $\Psi'(x',t') = e^{-i\varphi(x,t)}\Psi(x,t)$, and then the exponential common factor is dropped, as each side is independently equal to zero, we obtain:

$$i\hbar\left(\frac{\hbar}{m}\frac{\partial\varphi}{\partial x} - v\right)\frac{\partial\Psi}{\partial x} + \left[i\frac{\hbar^2}{2m}\frac{\partial^2\varphi}{\partial x^2} + \frac{\hbar^2}{2m}\left(\frac{\partial\varphi}{\partial x}\right)^2 - \hbar v\frac{\partial\varphi}{\partial x} - \hbar\frac{\partial\varphi}{\partial t}\right]\Psi = 0$$

This leads to the simultaneous differential equations:

$$\frac{\partial\varphi}{\partial x} = \frac{mv}{\hbar}$$

$$\frac{\partial\varphi}{\partial t} = \frac{\hbar}{2m}\left(\frac{\partial\varphi}{\partial x}\right)^2 - v\frac{\partial\varphi}{\partial x} + i\frac{\hbar}{2m}\frac{\partial^2\varphi}{\partial x^2}$$

where v is a constant, so that the solution is immediately obtained as:

$$\varphi(x,t) = \frac{mv}{\hbar}x - \frac{mv^2}{2\hbar}t$$

Therefore, a solution $\varphi(x,t)$ always exists, such that the probability density remains unchanged:

$$\Psi(x,t) = e^{i(mvx/\hbar - mv^2t/2\hbar)}\Psi'(x-vt,t)$$

and this form ensures the invariance of the Schrödinger equation under the Galilean transformation.

Problem 5.2.2. Solve the Schrödinger equation for the motion of a mass m under a constant potential energy U_0.

(Solution):　Substituting a trial solution of the form:

$$\Psi(\vec{r},t) = \psi(\vec{r})f(t)$$

into the time-dependent Schrödinger equation:

$$i\hbar\frac{\partial\Psi(\vec{r},t)}{\partial t} = -\frac{\hbar^2}{2m}\nabla^2\Psi(\vec{r},t) + U_0\Psi(\vec{r},t)$$

we may separate two differential equations:

$$\frac{1}{f(t)}\frac{df(t)}{dt} = -\frac{i}{\hbar}C$$

5. Temporal Evolution and the Fifth Postulate

$$-\frac{\hbar^2}{2m}\nabla^2\psi(\vec{r})+U_0\psi(\vec{r})=C\psi(\vec{r})$$

that depend on the same arbitrary constant C and are solved by:

$$f(t)=f_0 e^{-iCt/\hbar}, \qquad \psi(\vec{r})=Ae^{i\vec{p}\cdot\vec{r}/\hbar}$$

Inserting $\psi(\vec{r})$ into the partial differential equation gives:

$$\nabla^2 e^{i\vec{p}\cdot\vec{r}/\hbar}=-\frac{1}{\hbar^2}p^2 e^{i\vec{p}\cdot\vec{r}/\hbar} \quad \text{and} \quad \frac{p^2}{2m}+U_0=C$$

This shows that C is the total energy of the particle, and hence will be denoted by E. Therefore, a particular solution to the Schrödinger equation for a single particle has the form:

$$\Psi(\vec{r},t)=Ae^{i(\vec{p}\cdot\vec{r}-Et)}$$

which is similar to the matter wave function, Eq.(P.52). Because the potential energy is defined up to an additive constant, we may always set $U_0=0$ and write the general solution in the form of the superposition:

$$\Psi(\vec{r},t)=\int_{-\infty}^{\infty}\int_{-\infty}^{\infty}\int_{-\infty}^{\infty} A(\vec{p})e^{i(\vec{p}\cdot\vec{r}-p^2 t/2m)/\hbar}dp_x dp_y dp_z$$

which describes a matter wave packet.

Problem 5.2.3. Find the probability density $P(x)$ and the probability current density $j(x)$ for a particle described at $t=0$ by the wave function:

$$\Psi(x,0)=Ae^{ip_x x/\hbar - x^2/2a}$$

where the amplitude A is a constant.

(Answer): $\quad P(x)=\dfrac{1}{a\sqrt{\pi}}e^{-x^2/a^2}, \quad j(x)=\dfrac{p_x}{m}P(x)$

5.3. The Equivalence between the Schrödinger and Heisenberg Descriptions

The condition of invariance for the expectation value, Eq.(5.1), when either the Schrödinger or the Heisenberg description of the temporal evolution is used, can be regarded as a change from the time-dependent $\Psi(t)$-representation to the time-independent Ψ-representation of the physical observable A. This implies that $\Psi(t)$ and Ψ must be related by the unitary transformation:

$$\Psi(t) = \hat{T}(t)\Psi \qquad (5.45)$$

which is equivalent to a finite translation in time in the Schrödinger description. Substituting Eq.(5.45) into the time-dependent Schrödinger equation (5.24) gives:

$$i\hbar \frac{\partial}{\partial t}\left[\hat{T}(t)\Psi\right] = \hat{H}\hat{T}(t)\Psi \quad \text{or} \quad i\hbar \frac{\partial \hat{T}(t)}{\partial t}\Psi = \hat{H}\hat{T}(t)\Psi$$

and this must be valid for any time independent wave function Ψ. Therefore we obtain the operator equation:

$$i\hbar \frac{\partial \hat{T}(t)}{\partial t} = \hat{H}\hat{T}(t) \qquad (5.46)$$

It is immediately apparent that the expression (5.28), found for the evolution operator $\hat{T}(t)$, is a particular solution to Eq.(5.46), when the Hamiltonian has no explicit time-dependence, as assumed in the Schrödinger description.

In view of Eq.(5.45), Eq.(5.1) becomes:

$$\langle \Psi | \hat{T}^+ \hat{A}\hat{T} | \Psi \rangle = \langle \Psi | \hat{A}(t) | \Psi \rangle$$

which means that, when the wave function Ψ_S and Ψ_H in the Schrödinger and Heisenberg descriptions are transformed according to Eq.(5.45), which reads:

$$\Psi_S = \hat{T}\Psi_H \quad \text{or} \quad \Psi_H = \hat{T}^{-1}\Psi_S \qquad (5.47)$$

the representative operators $\hat{A}_H$ and $\hat{A}_S$ in the two description are related by:

$$\hat{A}(t) = \hat{A}_H = \hat{T}^+ \hat{A}_S \hat{T} \qquad (5.48)$$

It is important to show that $\hat{A}_S$ and $\hat{A}_H$ have the same eigenvalue spectrum in the two descriptions. Assuming that λ is an arbitrary eigenvalue such that:

5. Temporal Evolution and the Fifth Postulate

$$\hat{A}_S \Psi_S = \lambda \Psi_S \tag{5.49}$$

we can apply $\hat{T}^{-1}$ to both sides, and insert the identity operator $\hat{I} = \hat{T}\hat{T}^{-1}$ in the left hand side, which gives:

$$\hat{T}^{-1} \hat{A}_S (\hat{T}\hat{T}^{-1}) \Psi_S = \lambda \hat{T}^{-1} \Psi_S$$

Using Eq.(5.47), this can be written as:

$$\hat{T}^{-1} \hat{A}_S \hat{T} \Psi_H = \lambda \Psi_H$$

which means, in view of Eq.(5.48), that:

$$\hat{A}_H \Psi_H = \lambda \Psi_H \tag{5.50}$$

In other words, any eigenvalue λ of a physical observable, obtained by solving the eigenvalue problem in the Schrödinger description, Eq.(5.49), is also a solution of the eigenvalue equation in the Heisenberg description, Eq.(5.50). The proof of the reciprocal statement is also immediate.

The temporal evolution of the representative operators $\hat{A}_H$ is obtained from Eq.(5.48) in the form:

$$\frac{d\hat{A}_H}{dt} = \frac{\partial \hat{T}^+}{\partial t} \hat{A}_S \hat{T} + \hat{T}^+ \hat{A}_S \frac{\partial \hat{T}}{\partial t} + \hat{T}^+ \frac{\partial \hat{A}_S}{\partial t} \hat{T} \tag{5.51}$$

with $\hat{T}, \hat{T}^+$ obeying Eq.(5.46) and its complex conjugate:

$$i\hbar \frac{\partial \hat{T}}{\partial t} = \hat{H}_S \hat{T}, \qquad -i\hbar \frac{\partial \hat{T}^+}{\partial t} = (\hat{H}_S \hat{T})^+ = \hat{T}^+ \hat{H}_S$$

where use has been made of Eq.(2.97). It follows that:

$$i\hbar \frac{d\hat{A}_H}{dt} = -\hat{T}^+ \hat{H}_S \hat{A}_S \hat{T} + \hat{T}^+ \hat{A}_S \hat{H}_S \hat{T} + i\hbar \hat{T}^+ \frac{\partial \hat{A}_S}{\partial t} \hat{T}$$

and it is convenient to insert the identity operator $\hat{T}\hat{T}^+$ where appropriate, to obtain:

$$i\hbar \frac{d\hat{A}_H}{dt} = -\hat{T}^+ \hat{H}_S (\hat{T}\hat{T}^+) \hat{A}_S \hat{T} + \hat{T}^+ \hat{A}_S (\hat{T}\hat{T}^+) \hat{H}_S \hat{T} + i\hbar \hat{T}^+ \frac{\partial \hat{A}_S}{\partial t} \hat{T}$$

$$= -\hat{H}_H \hat{A}_H + \hat{A}_H \hat{H}_H + i\hbar \frac{\partial \hat{A}_H}{\partial t} = \left[\hat{A}_H, \hat{H}_H\right] + i\hbar \frac{\partial \hat{A}_H}{\partial t} \tag{5.52}$$

This equation plays, in the Heisenberg description, the role of the time-dependent Schrödinger equation (5.24) in the Schrödinger description. The involved operators $\hat{H}_H, \hat{A}_H$ and also:

$$\frac{\partial \hat{A}_H}{\partial t} = \hat{T}^+ \frac{\partial \hat{A}_S}{\partial t} \hat{T} \quad (5.53)$$

have been introduced according to Eq.(5.48). The term (5.53) appears in Eq.(5.52) only if the observable A_S has a explicit dependence on the time. For the basic observables we always find $\partial \hat{A}_S/\partial t = 0$, and therefore $\partial \hat{A}_H/\partial t = 0$.

In the special case of a conservative system, where $\hat{T}(t)$ is given by Eq.(5.28), the Schrödinger and Heisenberg descriptions of the temporal evolution are related by:

$$\Psi_H = e^{i\hat{H}t/\hbar}\Psi_S$$
$$\hat{A}_H = e^{i\hat{H}t/\hbar}\hat{A}_S e^{-i\hat{H}t/\hbar} \quad (5.54)$$

The choice between the two descriptions is one of convenience, as it is for instance the option of describing a body at rest in a rotating set of axes or the body rotating relative to a fixed coordinate system.

EXAMPLE 5.4. The energy-time uncertainty relation

In the Schrödinger description, Eq.(5.24) shows that the energy operator $\hat{H}$ has the same effect on a time-dependent state vector $\Psi(t)$ as that of the differential operator $i\hbar\partial/\partial t$. This implies that energy and time form a pair of conjugate observables:

$$[\hat{H}, t] = i\hbar \quad (5.55)$$

because, for any state $\Psi(t)$, we have:

$$[\hat{H}, t]\Psi(t) = \left[i\hbar\frac{\partial}{\partial t}, t\right]\Psi(t) = i\hbar\left[\frac{\partial}{\partial t}(t\Psi(t)) - t\frac{\partial \Psi(t)}{\partial t}\right] = i\hbar\Psi(t)$$

It follows that the energy eigenvalues E and the time will be related by Eq.(4.78) which takes the form:

$$\Delta E \Delta t \geq \frac{1}{2}\hbar \quad (5.56)$$

which is the same as the energy-time uncertainty relation (1.36) between the minimum time involved in measurements of the energy E of the system and the uncertainties associated with these measurements.

5. Temporal Evolution and the Fifth Postulate

In the Heisenberg description, a similar result follows from the evolution equation (5.18), derived for the expectation value of a representative operator with no explicit dependence on the time. If $\langle[\hat{A},\hat{H}]\rangle = i\hbar d\langle\hat{A}\rangle/dt$ is substituted into the uncertainty relation (4.76), written for ΔE and ΔA, we have:

$$\Delta A \Delta E \geq \frac{1}{2}\hbar\left|\frac{d\langle\hat{A}\rangle}{dt}\right| \quad \text{or} \quad \Delta E \frac{\Delta A}{|d\langle\hat{A}\rangle/dt|} \geq \frac{1}{2}\hbar \qquad (5.57)$$

Here ΔE is independent of the time, but the standard deviation ΔA is time-dependent. The relation (5.56) can be obtained by taking:

$$\frac{d\langle\hat{A}\rangle}{dt} = \frac{\Delta\langle\hat{A}\rangle}{\Delta t} \quad \text{and} \quad \Delta\langle\hat{A}\rangle = \Delta A$$

In other words, we are introducing a characteristic time Δt for the measurement of a given observable A such that, during Δt, the expectation value of the representative operator $\hat{A}$ has a standard deviation $\Delta\langle\hat{A}\rangle$ equal to that of the observable. ✥

Problem 5.3.1. Find the time dependence of the position and momentum operators $\hat{x}(t)$ and $\hat{p}(t)$, in the Heisenberg description of the free particle motion in one dimension.

(Solution): The temporal evolution of position is described by Eq.(5.54) which reads:

$$\hat{x}(t) = e^{i\hat{H}t/\hbar}\hat{x}e^{-i\hat{H}t/\hbar}$$

where $\hat{H}$ is given by:

$$\hat{H} = -\frac{\hbar^2}{2m}\frac{\partial^2}{\partial x^2}, \quad \hat{H} = \frac{p_x^2}{2m}$$

in the coordinate and momentum representations, respectively. We have in the coordinate representation:

$$[\hat{H},\hat{x}]\Psi(x) = -\frac{\hbar^2}{2m}\left[\frac{\partial^2}{\partial x^2}(x\Psi(x)) - x\frac{\partial^2\Psi(x)}{\partial x^2}\right] = -\frac{\hbar^2}{m}\frac{\partial\Psi(x)}{\partial x}$$

and then:

$$[\hat{H},[\hat{H},x]]\Psi(x) = 0, \quad [\hat{H},[\hat{H},[\hat{H},x]]]\Psi(x) = 0,\ldots$$

and so on, which implies, in view of Eq.(4.22), that:

$$\hat{x}(t) = \hat{x} + \frac{it}{\hbar}[\hat{H}, x] = \hat{x} - \frac{i\hbar t}{m}\frac{\partial}{\partial x} = \hat{x} + \frac{\hat{p}_x}{m}t$$

The same result is obtained by using the momentum representation for $\hat{x} = i\hbar \partial/\partial p_x$, such that:

$$\hat{x}(t)\Phi(p_x) = e^{ip_x^2 t/2m\hbar}\left(i\hbar\frac{\partial}{\partial p_x}\right)e^{-ip_x^2 t/2m\hbar}\Phi(p_x) = i\hbar\frac{\partial \Phi(p_x)}{\partial p_x} + \frac{p_x t}{m}\Phi(p_x)$$

or:

$$\hat{x}(t) = i\hbar\frac{\partial}{\partial p_x} + \frac{p_x}{m}t = \hat{x} + \frac{\hat{p}_x}{m}t$$

Similarly we have for momentum:

$$\hat{p}_H(t) = e^{i\hat{H}t/\hbar}\hat{p}_S e^{-i\hat{H}t/\hbar} = \hat{p}_S$$

because $[\hat{H}, \hat{p}_S] = 0$, and hence the momentum of a free particle is a constant of motion.

Problem 5.3.2. Assuming that the commutation relation $[\hat{A}_S, \hat{B}_S] = i\hat{C}_S$ holds in the Schrödinger description, show that in the Heisenberg description we also have $[\hat{A}_H, \hat{B}_H] = i\hat{C}_H$.

(Solution): We use the relation between the representative operators in the two descriptions, Eq. (5.48), to evaluate the commutation relation:

$$[\hat{A}_H, \hat{B}_H] = \hat{A}_H \hat{B}_H - \hat{B}_H \hat{A}_H = \hat{T}^+ \hat{A}_S \hat{T} \hat{T}^+ \hat{B}_S \hat{T} - \hat{T}^+ \hat{B}_S \hat{T} \hat{T}^+ \hat{A}_S \hat{T}$$

$$= \hat{T}^+ \hat{A}_S \hat{B}_S \hat{T} - \hat{T}^+ \hat{B}_S \hat{A}_S \hat{T} = \hat{T}^+[\hat{A}_S, \hat{B}_S]\hat{T} = \hat{T}^+(i\hat{C}_S)\hat{T} = i\hat{C}_H$$

Problem 5.3.3. Show that, for a conservative system where $\hat{A}_S \Psi(t) = \lambda \Psi(t)$ at $t = 0$, we have at any $t > 0$:

$$\hat{A}_H(-t)\Psi(t) = \lambda \Psi(t)$$

5.4. The Schrödinger Equation for Stationary States

If the initial state of a system is the eigenfunction of a Hamiltonian $\hat{H}$, which has not an explicit time dependence in the Schrödinger description, its energy is given by the corresponding eigenvalue E, according to:

$$\hat{H}\psi(\vec{r}) = E\psi(\vec{r}) \tag{5.58}$$

From the series expansion of the evolution operator $\hat{T}(t)$, defined by Eq.(5.28):

$$e^{-i\hat{H}t/\hbar} = \sum_n \frac{1}{n!}\left(-\frac{i}{\hbar}\hat{H}t\right)^n \tag{5.59}$$

it follows that $\hat{T}(t)$ and $\hat{H}$ commute, and hence the eigenfunction $\psi(\vec{r}) = \Psi(\vec{r},0)$ of $\hat{H}$ is also an eigenfunction of $\hat{T}(t)$ with an eigenvalue $e^{-iEt/\hbar}$. The temporal evolution (5.29) of the state function thus takes the form:

$$\Psi(\vec{r},t) = e^{-iEt/\hbar}\psi(\vec{r}) \tag{5.60}$$

which implies that:

$$|\Psi(\vec{r},t)|^2 = \psi^*(\vec{r})e^{iEt/\hbar}e^{-iEt/\hbar}\psi(\vec{r}) = |\psi(\vec{r})|^2 = |\Psi(\vec{r},0)|^2$$

This shows that the state described by $\Psi(\vec{r},t)$ is physically equivalent to the initial state $\Psi(\vec{r},0)$, since the probability density does not vary with time and no evolution occurs. It is said that $\Psi(\vec{r},t)$ describes a *stationary state* of the system in which none of the observables changes with time. Actually, for any observable A, with eigenstates φ_n given by:

$$\hat{A}\varphi_n = \lambda_n \varphi_n$$

we may consider the expansion:

$$\Psi(\vec{r},t) = e^{-iEt/\hbar}\psi(r) = \sum_n c_n \varphi_n$$

and, according to Postulate 3, Eq.(3.33), the probability to obtain an eigenvalue λ_n as a result of measurements on A is time-independent:

$$P(\lambda_n) = |c_n|^2 = |\langle \varphi_n | \Psi(\vec{r},t) \rangle|^2 = |\langle \varphi_n | \psi(\vec{r}) \rangle|^2$$

In particular, the expectation value of any observable which has not an explicit time dependence is a constant of motion in a stationary state, because $\hat{H}\Psi(\vec{r},t) = e^{-iEt/\hbar}\hat{H}\psi(\vec{r}) = Ee^{-iEt/\hbar}\psi(\vec{r}) = E\Psi(\vec{r},t)$, and hence:

$$\langle\Psi(\vec{r},t)|\hat{A}\hat{H} - \hat{H}\hat{A}|\Psi(\vec{r},t)\rangle = \langle\Psi(\vec{r},t)|\hat{A}E|\Psi(\vec{r},t)\rangle - \langle\Psi(\vec{r},t)|E\hat{A}|\Psi(\vec{r},t)\rangle$$

$$= E\sum_n \lambda_n |c_n|^2 - E\sum_n \lambda_n |c_n|^2 = 0$$

From Eq.(5.18) it follows that $d\langle\hat{A}\rangle/dt = 0$, and this leads to the following rule:

If a system is initially in an eigenstate of the Hamiltonian, it remains in that state indefinitely and all the measured values of its physical observables are stationary.

This rule substantiates the Bohr postulate of stationary states for one-electron atoms, introduced in Preliminaries. It emphasizes that the Hamiltonian of a quantum system, which is governing its temporal evolution according to Postulate 5, has stationary eigenstates whose energies are definite eigenvalues of Eq.(5.58). Using the Hamiltonian operator (2.17), this energy eigenvalue equation takes the form:

$$\hat{H}\psi(\vec{r}) = -\frac{\hbar^2}{2m}\nabla^2\psi(\vec{r}) + U(\vec{r})\psi(\vec{r}) = E\psi(\vec{r}) \tag{5.61}$$

which is called the *time-independent Schrödinger equation*.

The temporal evolution of stationary states obeys Eq.(5.60), as each is an eigenfunction of the Hamiltonian in Eq.(5.58) for a given eigenvalue E. The principle of superposition of states implies that a general solution to the time-dependent Schrödinger equation (5.24) should be written as:

$$\Psi(\vec{r},t) = \sum_n a_n e^{-iE_n t/\hbar}\psi_n(\vec{r}) + \int a(E)e^{-iEt/\hbar}\psi(\vec{r},E)dE \tag{5.62}$$

provided $\psi_n(\vec{r})$ and $\psi(\vec{r},E)$ are solutions to Eq.(5.61), corresponding to the discrete and continuous eigenvalue spectrum respectively. The coefficients of the expansion are given by the scalar products of the form (3.14) or (3.19):

$$a_n = \langle\psi_n|\Psi(0)\rangle = \int \psi_n^*(\vec{r})\Psi(\vec{r},0)dV \tag{5.63}$$

$$a(E) = \langle\psi(E)|\Psi(0)\rangle = \int \psi^*(\vec{r},E)\Psi(\vec{r},0)dV \tag{5.64}$$

EXAMPLE 5.5. The classical limit of the time-independent Schrödinger equation

In a stationary state of the system, the solution (5.38) of the time-dependent Schrödinger equation should assume the form (5.60), namely:

$$\Psi(\vec{r},t) = A(\vec{r},t)e^{iS(\vec{r},t)/\hbar} = \psi(\vec{r})e^{-iEt/\hbar}$$

and therefore, we may take:

$$\psi(\vec{r}) = \Psi(\vec{r},0) = A(\vec{r},0)e^{iS(\vec{r},0)/\hbar} = A_0(\vec{r})e^{iS_0(\vec{r})/\hbar} \tag{5.65}$$

which implies that the action function will be expressed as:

$$S(\vec{r},t) = S_0(\vec{r}) - Et \tag{5.66}$$

Substituting the solution (5.65) into the time-independent Schrödinger equation (5.61), and following the same steps as in Example 5.3, we obtain:

$$\frac{i\hbar}{2m}\left(2\sum_i \frac{\partial A_0}{\partial x_i}\frac{\partial S_0}{\partial x_i} + A_0 \sum_i \frac{\partial^2 S_0}{\partial x_i^2}\right) - \frac{1}{2m}\sum_i A_0\left(\frac{\partial S_0}{\partial x_i}\right)^2 + (E-U)A_0 = 0$$

Each of the real and imaginary parts must be set equal to zero separately, which gives:

$$\frac{1}{2m}\sum_i \left(\frac{\partial S_0}{\partial x_i}\right)^2 + U = E \tag{5.67}$$

$$\frac{1}{m}\sum_i \frac{\partial A_0}{\partial x_i}\frac{\partial S_0}{\partial x_i} + \frac{A_0}{2m}\sum_i \frac{\partial S_0}{\partial x_i^2} = 0 \tag{5.68}$$

Equation (5.67) can simply be regarded as a special case of the Hamilton-Jacobi equation (5.40), corresponding to conservative systems, where the action integral assumes the form (5.66), such that:

$$p_i = \frac{\partial S}{\partial x_i} = \frac{\partial S_0}{\partial x_i}, \quad \frac{\partial S}{\partial t} = -E$$

and therefore the Hamilton-Jacobi equation reduces to:

$$H\left(x_i, \frac{\partial S_0}{\partial x_i}\right) = E \tag{5.69}$$

The amplitude equation (5.68) can be written (see Example 5.3) as:

$$\sum_i \frac{\partial}{\partial x_i}\left(A_0^2 \frac{1}{m}\frac{\partial S_0}{\partial x_i}\right) = \sum_i \frac{\partial}{\partial x_i}\left(A_0^2 v_i\right) = \nabla \cdot (P\vec{v}) = 0 \tag{5.70}$$

and this is a special case of the conservation law (5.44) when the probability density P is time-independent.

However, another classical correspondence for the time-independent Schrödinger equation can be evidenced, if Eqs.(5.67) and (5.68) are written as:

$$\sum_i \left(\frac{\partial S_0}{\partial x_i}\right)^2 = 2m(E-U) \quad \text{or} \quad (\nabla S_0)^2 = 2m(E-U) \tag{5.71}$$

$$2(\nabla A_0)\cdot(\nabla S_0) + A_0 \nabla^2 S_0 = 0 \quad \text{or} \quad \nabla\cdot(A_0^2 \nabla S_0) = 0 \tag{5.72}$$

Similar equations can be derived for classical waves, using the reduced or time-independent wave equation (P.50) which, for an inhomogenous medium, of index of refraction $n(\vec{r})$, reads:

$$\nabla^2 \psi(\vec{r}) + n^2(\vec{r}) k_0^2 \psi(\vec{r}) = 0 \tag{5.73}$$

Here n is the ratio of the velocity c of the wave in free space to the wave velocity u in the medium, so that the wave number was introduced in the wave equation (P.50) in the form $k^2 = \omega^2/u^2 = (\omega^2/c^2)(c^2/u^2) = n^2 k_0^2$. Substituting into Eq.(5.73) a solution of the form:

$$\psi(\vec{r}) = A_0(\vec{r}, k_0) e^{ik_0 \varphi_0(\vec{r})} \tag{5.74}$$

where $\varphi_0(\vec{r})$ is a function called *eikonal* (from the Greek word for image), which has the dimensions of length, one obtains:

$$\nabla^2 A_0 + k_0^2 A_0 [n^2 - \nabla\varphi_0 \cdot \nabla\varphi_0] + 2ik_0 (\nabla A_0)\cdot(\nabla\varphi_0) + ik_0 A_0 \nabla^2 \varphi_0 = 0 \tag{5.75}$$

If we further assume, in Eq.(5.75), that $A_0(\vec{r}, k_0)$ is a slowly varying amplitude, the second derivative term $\nabla^2 A_0$ may be neglected. This assumption is strictly true if the amplitude A_0 is independent of k_0 and implies that we are using the geometrical optics approximation of vanishingly small wavelengths ($k_0 \to \infty$ or $\lambda \to 0$). Consequently, equating to zero the real and the imaginary parts of Eq.(5.75) gives:

$$(\nabla\varphi_0)^2 = n^2 \quad \text{or} \quad |\nabla\varphi_0| = n \tag{5.76}$$

$$2(\nabla A_0)\cdot(\nabla\varphi_0) + A_0 \nabla^2 \varphi_0 = 0 \quad \text{or} \quad \nabla\cdot(A_0^2 \nabla\varphi_0) = 0 \tag{5.77}$$

Equation (5.76) is known as the *eikonal* equation, and enables us to visualize the wave propagation through the surfaces of constant phase $\varphi_0(\vec{r}) = $ const. For instance, in the case of spherical symmetry, $\varphi_0(\vec{r}) = \varphi_0(r)$, we have:

$$\frac{\partial \varphi_0}{\partial x_i} = \frac{\partial \varphi_0}{\partial r} \frac{\partial r}{\partial x_i} = \frac{\partial \varphi_0}{\partial r} \frac{x_i}{r} \quad \text{or} \quad \left(\frac{\partial \varphi_0}{\partial x_i}\right)^2 = \left(\frac{\partial \varphi_0}{\partial r}\right)^2 \frac{x_i^2}{r^2}$$

where $r = (x^2 + y^2 + z^2)^{1/2}$. Hence, Eq.(5.76) reads:

$$(\nabla\varphi_0)^2 = \left(\frac{\partial\varphi_0}{\partial x}\right)^2 + \left(\frac{\partial\varphi_0}{\partial y}\right)^2 + \left(\frac{\partial\varphi_0}{\partial z}\right)^2 = \left(\frac{\partial\varphi_0}{\partial r}\right)^2 \frac{x^2 + y^2 + z^2}{r^2} = \left(\frac{\partial\varphi_0}{\partial r}\right)^2 = n^2$$

or $\partial\varphi_0/\partial r = n$, which is satisfied by spherical surfaces $\varphi_0 = nr$ of constant phase. The gradient $\nabla\varphi_0 = n\vec{e}_r$ is perpendicular to the surface φ_0 = const and hence continuous lines called *rays* may be constructed, that are everywhere along the gradient of φ_0. If we set $\nabla\varphi_0 = n\vec{e}_r$ into Eq.(5.77), and then integrate over a volume that contains a given ray of a spherical wave (see Figure 5.1), one obtains:

$$\int_V \nabla\cdot(A_0^2 n\vec{e}_r)dV = \int_{s_1} A_0^2 n\vec{e}_r \cdot d\vec{s}_1 + \int_{s_2} A_0^2 n\vec{e}_r \cdot d\vec{s}_2 = 0$$

Assuming a uniform medium, where n is constant, we have:

$$A_0^2(s_1)ds_1 = A_0^2(s_2)ds_2 \quad \text{or} \quad \frac{A_0^2(s_1)}{A_0^2(s_2)} = \frac{R_2^2}{R_1^2} \tag{5.78}$$

where R_1 and R_2 are the radii of the two elementary spherical surfaces. Since the square of the amplitude is proportional to the energy of the wave, Eq.(5.78) is a formulation of the energy conservation for spherical waves, namely the energy crossing unit area is inversely proportional to the square of the distance from the source.

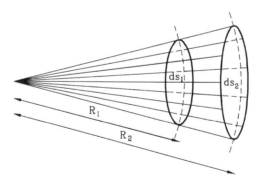

Figure 5.1. The equation of continuity for spherical waves

The eikonal equation (5.76) is equivalent to the principle of least time (P.46), since $\nabla\varphi_0$ is a conservative vector field:

$$\int_\Gamma \nabla\varphi_0 \cdot d\vec{l} = \int_S (\nabla\times\nabla\varphi_0)\cdot d\vec{s} = 0$$

where the Stokes theorem has been used and $\nabla\times\nabla\varphi_0 = 0$ is a vector identity. It follows that, along an arbitrary path, we have:

$$\int_{P_1}^{P_2} \nabla\varphi_0 \cdot d\vec{l} = \int_{P_1}^{P_2} |\nabla\varphi_0| dl \cos\alpha = \varphi_0(P_2) - \varphi_0(P_1)$$

On the other hand, the integral (P.45) can be written in terms of φ_0 as:

$$\int_{P_1}^{P_2} n \, dl = \int_{P_1}^{P_2} |\nabla\varphi_0| dl \geq \int_{P_1}^{P_2} |\nabla\varphi_0| dl \cos\alpha = \varphi_0(P_2) - \varphi_0(P_1)$$

where the equality holds for $\cos\alpha = 1$, when the path is tangent to $\nabla\varphi_0$ at any point. In other words, the path determined from the eikonal equation (5.76) is the same as that obtained from the principle of least time (P.46).

The classical limit of the time-independent Schrödinger equation, Eq.(5.71), becomes identical to the eikonal equation (5.76) by taking $S_0 = C\varphi_0$ which gives $(\nabla\varphi_0)^2 = 2m(E-U)/C^2 = n^2$. This is consistent with the correspondence rule (P.47) between the principles of least time and least action. Since the two integrands in Eqs.(P.46) and (P.48) have the same spatial dependence, they should be equal except a multiplicative constant C which has the dimensions of momentum. Because of this possible interpretation in terms of wave quantities, the time-independent Schrödinger equation is often called the Schrödinger wave equation. ♂

The temporal evolution of the system can be evidenced by taking the expectation value of an arbitrary observable in a state $\Psi(\vec{r},t)$, of the form (5.62), which is no longer a constant of motion, provided the sum has more than one term. Assuming, for simplicity, a discrete set of eigenvalues, we have in the Schrödinger description:

$$\langle \Psi(\vec{r},t)|\hat{A}|\Psi(\vec{r},t)\rangle = \langle \sum_n a_n e^{-iE_n t/\hbar} \psi_n |\hat{A}| \sum_m a_m e^{-iE_m t/\hbar} \psi_m \rangle$$

$$= \sum_n \sum_m a_n^* a_m e^{-i(E_m - E_n)t/\hbar} \langle \psi_n |\hat{A}| \psi_m \rangle$$

$$= \sum_n \sum_m a_n^* a_m e^{-i(E_m - E_n)t/\hbar} A_{nm} \tag{5.79}$$

where A_{nm} are the matrix elements of the operator $\hat{A}$ in the so-called *energy representation*. This representation is given with respect to the basis vectors $\psi_n(\vec{r})$, solutions to the time-independent Schrödinger equation (5.61) and hence eigenfunctions of the Hamiltonian operator. It is obvious that the expectation value (5.79) is time-dependent, through exponential factors which can be written in terms of frequencies similar to those introduced by the Bohr postulate (P.27):

$$\omega_{mn} = (E_m - E_n)/\hbar \tag{5.80}$$

5. Temporal Evolution and the Fifth Postulate 197

In the Heisenberg description, the expectation value (5.79) should be expressed in terms of the matrix elements of the time dependent operator $\hat{A}(t)$, which can be written, using Eq.(5.54), as:

$$A_{nm}(t) = \langle \psi_n | \hat{A}(t) | \psi_m \rangle = \langle \Psi_n | e^{i\hat{H}t/\hbar} \hat{A} e^{-i\hat{H}t/\hbar} | \psi_m \rangle = \langle \psi_n | \hat{A} | \psi_m \rangle e^{-i(E_m - E_n)t/\hbar}$$

and this gives the relation between the form of the energy representation of operators in the two descriptions, in the form:

$$A_{nm}(t) = A_{nm} e^{-i(E_m - E_n)t/\hbar} = A_{nm} e^{-i\omega_{mn}t} \tag{5.81}$$

or, if the energy eigenfunctions $\psi_n(\vec{r})$ are denoted by the corresponding eigenvalues $|E_n\rangle$:

$$\langle E_n | \hat{A}(t) | E_m \rangle = \langle E_n | \hat{A} | E_m \rangle e^{-i(E_m - E_n)t/\hbar} = \langle E_n | \hat{A} | E_m \rangle e^{-i\omega_{mn}t} \tag{5.82}$$

EXAMPLE 5.6. The energy representation of oscillatory motion

For a linear harmonic oscillator, with the Hamiltonian operator corresponding to the classical energy (P.5), the evolution equations (5.3) for the position and momentum operators can be rewritten in the familiar form (see Problem 5.1.1):

$$\hat{p}_x(t) = m \frac{d\hat{x}(t)}{dt} \tag{5.83}$$

$$\frac{d^2 \hat{x}(t)}{dt^2} + \omega^2 \hat{x}(t) = 0 \tag{5.84}$$

Using Eq.(5.82) for the matrix elements in the energy representation, it follows that:

$$\langle E_n | \frac{d^2 \hat{x}(t)}{dt^2} | E_m \rangle = \frac{d^2}{dt^2} \langle E_n | \hat{x}(t) | E_m \rangle = \langle E_n | \hat{x} | E_m \rangle \frac{d^2}{dt^2} \left[e^{-i(E_m - E_n)t/\hbar} \right]$$

$$= -\frac{1}{\hbar^2} (E_m - E_n)^2 \langle E_n | \hat{x}(t) | E_m \rangle$$

and Eqs.(5.84) becomes:

$$\left[\hbar^2 \omega^2 - (E_m - E_n)^2 \right] \langle E_n | \hat{x} | E_m \rangle = 0$$

In other words, all matrix elements $x_{nm} = \langle E_n | \hat{x} | E_m \rangle$ vanish, except those for which:

$$E_m = E_n \pm \hbar \omega \tag{5.85}$$

Quantum Physics

The non-vanishing matrix elements can be evaluated from the commutation relation (4.40), which is independent of representation:

$$\hat{x}(t)\hat{p}_x(t) - \hat{p}_x(t)\hat{x}(t) = i\hbar \quad \text{or} \quad \hat{x}(t)\frac{d\hat{x}(t)}{dt} - \frac{d\hat{x}(t)}{dt}\hat{p}_x(t) = \frac{i\hbar}{m}$$

where use has been made of Eq.(5.83). We can write:

$$\frac{i\hbar}{m} = \langle E_n|\hat{x}(t)\frac{d\hat{x}(t)}{dt}|E_n\rangle - \langle E_n|\frac{d\hat{x}(t)}{dt}\hat{x}(t)|E_n\rangle$$

$$= \sum_m \left[\langle E_n|\hat{x}(t)|E_m\rangle\langle E_m|\frac{d\hat{x}(t)}{dt}|E_n\rangle - \langle E_n|\frac{d\hat{x}(t)}{dt}|E_m\rangle\langle E_m|\hat{x}(t)|E_n\rangle \right]$$

$$= \sum_m \frac{2i}{\hbar}(E_m - E_n)|\langle E_n|\hat{x}(t)|E_m\rangle|^2 = \sum_m \frac{2i}{\hbar}(E_m - E_n)|\langle E_n|\hat{x}|E_m\rangle|^2 \quad (5.86)$$

because, from Eq.(5.82), we have:

$$\langle E_m|\frac{d\hat{x}(t)}{dt}|E_n\rangle = \frac{i}{\hbar}(E_m - E_n)\langle E_m|\hat{x}(t)|E_n\rangle \quad (5.87)$$

In view of the restriction (5.85), the sum in the right hand side of Eq.(5.86) is reduced to two terms only, corresponding to indices $m = n \pm 1$, and we have:

$$\frac{\hbar}{2m\omega} = |\langle E_n|\hat{x}|E_{n+1}\rangle|^2 - |\langle E_n|\hat{x}|E_{n-1}\rangle|^2 \quad (5.88)$$

If E_0 is the lowest energy eigenvalue, and thus $|E_{-1}\rangle$ vanishes, we obtain for $n = 0$ that:

$$\langle E_0|\hat{x}|E_1\rangle = \sqrt{\frac{\hbar}{2m\omega}}$$

Equation (5.88) is evidently a recurrence relation which, by iteration, yields:

$$\langle E_{n-1}|\hat{x}|E_n\rangle = \sqrt{\frac{n\hbar}{2m\omega}} = \langle E_n|\hat{x}|E_{n-1}\rangle \quad (5.89)$$

The matrix elements of momentum also result from Eq.(5.88), by using Eqs.(5.83) and (5.84), in the form:

$$\langle E_{n-1}|\hat{p}_x|E_n\rangle = -i\sqrt{n\frac{m\omega\hbar}{2}}, \quad \langle E_n|\hat{p}_x|E_{n-1}\rangle = i\sqrt{n\frac{m\omega\hbar}{2}}$$

and therefore we have the energy matrix representation:

5. Temporal Evolution and the Fifth Postulate

$$(\hat{x}) = \sqrt{\frac{\hbar}{2m\omega}} \begin{pmatrix} 0 & \sqrt{1} & 0 & \cdots \\ \sqrt{1} & 0 & \sqrt{2} & \cdots \\ 0 & \sqrt{2} & 0 & \cdots \\ \cdots & \cdots & \cdots & \cdots \end{pmatrix}, \quad (\hat{p}_x) = i\sqrt{\frac{m\omega\hbar}{2}} \begin{pmatrix} 0 & -\sqrt{1} & 0 & \cdots \\ \sqrt{1} & 0 & -\sqrt{2} & \cdots \\ 0 & \sqrt{2} & 0 & \cdots \\ \cdots & \cdots & \cdots & \cdots \end{pmatrix}$$

The energy E_n of a stationary state $|E_n\rangle$ is given by:

$$E_n = \langle E_n|\hat{H}|E_n\rangle = \frac{1}{2m}\sum_k \langle E_n|\hat{p}_x|E_k\rangle\langle E_k|\hat{p}_x|E_n\rangle + \frac{m\omega^2}{2}\sum_k \langle E_n|\hat{x}|E_k\rangle\langle E_k|\hat{x}|E_n\rangle$$

$$= \frac{1}{2m}\frac{m\omega\hbar}{2}(n+n+1) + \frac{m\omega^2}{2}\frac{\hbar}{2m\omega}(n+n+1) = \hbar\omega\left(n+\frac{1}{2}\right) \tag{5.90}$$

and the Hamiltonian:

$$(\hat{H}) = \hbar\omega \begin{pmatrix} \frac{1}{2} & 0 & 0 & \cdots \\ 0 & \frac{3}{2} & 0 & \cdots \\ 0 & 0 & \frac{5}{2} & \cdots \\ \cdots & \cdots & \cdots & \cdots \end{pmatrix}$$

has diagonal representation. ♦

Problem 5.4.1. Show that the energy eigenvalues of a particle of mass m satisfy the relation:

$$\sum_n (E_n - E_m)|x_{nm}|^2 = \frac{\hbar^2}{2m}$$

where the sum is taken over all the stationary states of the system.

(Solution): We use the energy representation (5.81) for position, which reads:

$$x_{nm}(t) = x_{nm}e^{-i(E_n-E_m)t/\hbar}$$

so that we can write:

$$\sum_n (E_n - E_m)|x_{nm}|^2 = \sum_n (E_n - E_m)|\langle E_n|\hat{x}|E_m\rangle|^2$$

$$= \sum_n (E_n - E_m)|\langle E_n|\hat{x}(t)|E_m\rangle|^2$$

$$= \frac{1}{2}\sum_n (E_n - E_m)\langle E_n|\hat{x}(t)|E_m\rangle\langle E_m|\hat{x}(t)|E_n\rangle$$

$$-\frac{1}{2}\sum_n (E_m - E_n)\langle E_m|\hat{x}(t)|E_n\rangle\langle E_n|\hat{x}(t)|E_m\rangle$$

$$= \frac{1}{2}\frac{\hbar}{i}\sum_n \langle E_m|\hat{x}(t)|E_n\rangle\langle E_n|\frac{d\hat{x}(t)}{dt}|E_m\rangle$$

$$-\frac{1}{2}\frac{\hbar}{i}\sum_n \langle E_m|\frac{d\hat{x}(t)}{dt}|E_n\rangle\langle E_n|\hat{x}(t)|E_m\rangle$$

where use has been made of Eq.(5.87). In view of Eq.(5.5) we further obtain:

$$\sum_n (E_n - E_m)|x_{nm}|^2 = \frac{\hbar}{2im}\sum_n \langle E_m|\hat{x}(t)|E_n\rangle\langle E_n|\hat{p}_x(t)|E_m\rangle$$

$$-\frac{\hbar}{2im}\sum_n \langle E_m|\hat{p}_x(t)|E_n\rangle\langle E_n|\hat{x}(t)|E_m\rangle$$

and it follows that:

$$\sum_n (E_n - E_m)|x_{nm}|^2 = \frac{\hbar}{2im}\langle E_m|\hat{x}(t)\hat{p}_x(t) - \hat{p}_x(t)\hat{x}(t)|E_m\rangle = \frac{\hbar^2}{2m}$$

because the commutation relations (4.40) are valid in the Heisenberg description (see Problem 5.3.2).

Problem 5.4.2. Show that, if the potential energy is a sum $U(\vec{r}) = \sum_{i=1}^{3} U_i(x_i)$, the Schrödinger wave equation (5.61) can be separated, in Cartesian coordinates, into one-dimensional eigenvalue equations of the form:

$$\frac{d^2\psi_i(x_i)}{dx_i^2} + \frac{2m}{\hbar^2}[E_i - U_i(x_i)]\psi_i(x_i) = 0$$

where $\psi(\vec{r}) = \psi_1(x)\psi_2(y)\psi_3(z)$ and $E = E_1 + E_2 + E_3$.

(Solution): Setting $\psi = \psi_1\psi_2\psi_3$, we have in Cartesian coordinates:

$$\nabla^2\psi = \sum_{i=1}^{3}\frac{\partial^2\psi}{\partial x_i^2} = \psi_2\psi_3\frac{\partial^2\psi_1}{\partial x^2} + \psi_1\psi_3\frac{\partial^2\psi_2}{\partial y^2} + \psi_1\psi_2\frac{\partial^2\psi_3}{\partial z^2}$$

5. Temporal Evolution and the Fifth Postulate   201

so that, dividing Eq.(5.61) by $\psi_1\psi_2\psi_3$, we obtain:

$$\sum_{i=1}^{3}\left(-\frac{\hbar^2}{2m}\frac{1}{\psi_i}\frac{d^2\psi_i}{dx_i^2}+U_i\right)=E$$

Because x_i are independent variables, each bracket on the left hand side should be equal to a constant, which is denoted by E_i:

$$-\frac{\hbar^2}{2m}\frac{1}{\psi_i}\frac{d^2\psi_i}{dx_i^2}+U_i=E_i \quad \text{or} \quad -\frac{\hbar^2}{2m}\frac{d^2\psi_i}{dx_i^2}+U_i\psi_i=E_i\psi_i$$

and it follows that $E_1+E_2+E_3=E$. Each function $\psi_i(x_i)$ is a solution to the Schrödinger wave equation in one dimension.

Problem 5.4.3. Assuming that the Hamiltonian $\hat{H}$ of a system, and hence the energy eigenvalues E_n, are dependent on a parameter α, show that:

$$\langle E_n|\frac{\partial \hat{H}}{\partial \alpha}|E_n\rangle = \frac{\partial E_n}{\partial \alpha}$$

---

6. ONE-DIMENSIONAL MOTION

6.1. Energy Eigenstates in One Dimension

Although one-dimensional problems are only an idealisation of naturally occurring physical situations, and hence may seem somewhat artificial, motion of a single particle in one dimension yields exact solutions of the Schrödinger equation which highlight most of the features of the three-dimensional quantum approach, avoiding, however, its complexity. Because the time-dependent Schrödinger equation in one dimension:

$$i\hbar \frac{\partial \Psi(x,t)}{\partial t} = -\frac{\hbar^2}{2m}\frac{\partial^2 \Psi(x,t)}{\partial x^2} + U(x)\Psi(x,t) \tag{6.1}$$

has the general solution, Eq.(5.62), which is expressed in terms of the eigenvalues and eigenfunctions of the Hamiltonian operator only:

$$\Psi(x,t) = \sum_n a_n e^{-iE_n t/\hbar} \psi_n(x) + \int a(E) e^{-iEt/\hbar} \psi(x,E) dE \tag{6.2}$$

the problem is always reduced to solving the Schrödinger wave equation:

$$-\frac{\hbar^2}{2m}\frac{d^2\psi(x)}{dx^2} + U(x)\psi(x) = E\psi(x) \tag{6.3}$$

This is the energy eigenvalue equation, which can also be written as:

$$\frac{d^2\psi(x)}{dx^2} + \frac{2m}{\hbar^2}[E - U(x)]\psi(x) = \frac{d^2\psi(x)}{dx^2} + k^2(x)\psi(x) = 0 \tag{6.4}$$

by setting:

$$k^2(x) = \frac{2m}{\hbar^2}[E - U(x)] \tag{6.5}$$

It follows that the energy eigenvalue problem is similar to the calculation of monochromatic waves in a medium with non-uniform refractive index. If $E < U(x)$, it is clear that $k(x)$ becomes imaginary, namely $k(x) = i\gamma(x)$, where:

$$\gamma^2(x) = \frac{2m}{\hbar^2}[U(x) - E], \qquad E < U(x) \tag{6.6}$$

For any given eigenvalue E, the second-order differential equation (6.4) yields two linearly independent solutions $\psi_1(x)$ and $\psi_2(x)$, such that:

$$\psi(x) = a_1\psi_1(x) + a_2\psi_2(x) \tag{6.7}$$

It follows that any eigenvalue might be at most two-fold degenerate. The solutions of interest are those which satisfy certain regularity conditions which make them admissible eigenfunctions:

(i) $\psi(x)$ *and its first derivative should be continuous functions of x, for all x*. This implies that we allow for finite discontinuities of the potential energy $U(x)$ at some points, where $d^2\psi/dx^2$ should exhibit an opposite finite discontinuity, so preventing the departure from continuity of the right-hand side of Eq.(6.1), whose left-hand side $\partial\Psi/\partial t$ is a continuous function. A step-like behaviour of $d^2\psi/dx^2$ requires the continuity of $d\psi/dx$ at the same points, otherwise the discontinuity becomes infinite.

(ii) $\psi(x)$ *should be finite or single-valued for all x*. This ensures the possibility of normalizing the wave functions, according to Eq.(1.37), and having a uniquely defined probability density $|\psi(x)|^2$.

The eigenfunctions $\psi(x)$ can always be chosen to be real, and this becomes apparent by conjugation of Eq.(6.4), which gives:

$$\frac{d^2\psi^*(x)}{dx^2} + k^2(x)\psi^*(x) = 0$$

where $k^2(x)$ is always real. Because $\psi(x)$ and $\psi^*(x)$ are eigenfunctions with the same eigenvalue E, their real and imaginary parts, namely:

$$\psi_r(x) = \frac{\psi(x) + \psi^*(x)}{2}, \qquad \psi_i(x) = \frac{\psi(x) - \psi^*(x)}{2i} \tag{6.8}$$

are real eigenfunctions corresponding to E. Hence, we will limit ourselves to the real solutions $\psi(x)$ to Eq.(6.4).

The nature of the eigenfunctions $\psi(x)$ can be determined by considering the classical counterpart of the energy problem. From Eq.(2.17) it follows that:

$$E - U(x) = \frac{p^2}{2m} \geq 0 \quad \text{or} \quad E - U(x) \geq 0 \tag{6.9}$$

and we will assume the potential energy function $U(x)$ illustrated in Figure 6.1, where $U_+ = \lim_{x \to +\infty} U(x), U_- = \lim_{x \to -\infty} U(x)$. It may describe, for instance, the motion of a small ring, of mass m, in a uniform gravitational field, in the absence of friction, along a rigid wire of given shape $z(x)$, situated in the vertical plane xOz, for which $U(x) = mgz(x)$.

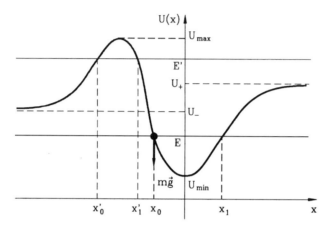

Figure 6.1. Classical turning points of one-dimensional motion

The classical motion is restricted to the regions where the condition (6.9) holds, called *classically allowed regions*. Assuming that the particle is given an energy E (for instance it is released at x_0 with $v = 0$), it moves in the allowed region (x_0, x_1), which represents a *potential energy well*. At x_0 and x_1, called *classical turning points*, the velocity of the particle vanishes, since $U(x) = E$, and the particle is turned back. For another given energy E', Figure 6.1 shows that the motion is allowed in the regions $(-\infty, x_0')$ and $(x_1', +\infty)$, on the two sides of the *potential energy barrier* rised between the turning points x_0', x_1', which are defined by $U(x) = E'$. The region (x_0', x_1'), where $E' < U(x)$, represents a *classically disallowed region*.

From Eq.(6.4) it follows that, for $E = U(x)$, $d^2\psi/dx^2$ vanishes and this implies that the classical turning points are either *inflexion points* of the eigenfunctions or *nodes*, at which $\psi(x) = 0$. In the classically allowed regions, where $E > U(x)$, Eq.(6.4) shows that $\psi(x)$ and $d^2\psi/dx^2$ must be of opposite sign. If the real solution $\psi(x)$ is assumed positive, $d^2\psi/dx^2$ is negative and so is the curvature of $\psi(x)$, which eventually might cross the x-axis (Figure 6.2 (a)). For $\psi(x) < 0$, $d^2\psi/dx^2$ becomes positive and the curvature of $\psi(x)$ turns also positive. This oscillatory behaviour is indicative for *sinusoidal* solutions to the energy eigenvalue problem in the classically allowed regions. Typical for the sinusoidal behaviour of eigenfunctions is that the nodes of $\psi(x)$ alternate with those of $d\psi/dx$.

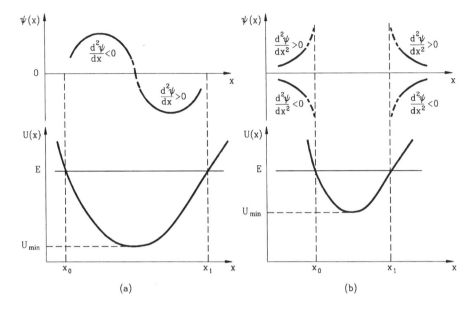

Figure 6.2. Energy eigenfunction behaviour in the classically allowed (a) and disallowed (b) regions

For $E < U(x)$, $\psi(x)$ and $d^2\psi/dx^2$ have the same sign, and therefore $\psi(x)$ can either increase indefinitely or vanish at infinity, as illustrated in Figure 6.2 (b). This means that the energy eigenfunctions are expected to exhibit an *exponential* behaviour in the classically disallowed regions. This behaviour consists of a monotonous increase or decrease of $\psi(x)$, which might have at most one node along the x-axis, because $d\psi/dx$ is not changing sign.

EXAMPLE 6.1. Motion in a constant potential energy

Assuming $U(x) = U_0$ in a given region, if $E > U_0$ one may set:

$$k^2(x) = \frac{2m}{\hbar^2}(E - U_0) = k^2$$

where k is real, and Eq.(6.4) reads:

$$\frac{d^2\psi(x)}{dx^2} + k^2\psi(x) = 0$$

The complex solutions are:

206 Quantum Physics

$$\psi_1(x) = e^{ikx}, \quad \psi_2(x) = e^{-ikx} \quad \text{or} \quad \psi(x) = a_1 e^{ikx} + a_2 e^{-ikx} \tag{6.10}$$

and the real solutions (6.8) can be written in the sinusoidal form:

$$\psi_r(x) = \cos kx, \quad \psi_i(x) = \sin kx \quad \text{or} \quad \psi(x) = a_1 \cos kx + a_2 \sin kx \tag{6.11}$$

If $E < U_0$, we introduce a real constant γ in Eq.(6.6):

$$\gamma^2(x) = \frac{2m}{\hbar^2}(U_0 - E) = \gamma^2$$

such that Eq.(6.4) reduces to:

$$\frac{d^2\psi(x)}{dx^2} - \gamma^2 \psi(x) = 0$$

and has two real independent solutions:

$$\psi_1(x) = e^{\gamma x}, \quad \psi_2(x) = e^{-\gamma x} \quad \text{or} \quad \psi(x) = a_1 e^{\gamma x} + a_2 e^{-\gamma x} \tag{6.12}$$

with the expected exponential behaviour.

The wave functions found for a constant potential energy can be adjusted to describe the eigenstates of a particle moving in a slowly varying potential energy. In this case, it is expected that $k(x)$ and $\gamma(x)$, defined by Eqs.(6.5) and (6.6), vary slowly as functions of x. In view of the somewhat similarly physical situation where the geometrical optics can be used as a good approximation for light propagation in media with slowly varying index of refraction, the quasi-classical approximation was applied by Wentzel, Kramers and Brillouin to the Schrödinger equation, assuming a slow change of the potential energy.

EXAMPLE 6.2. The WKB approximation

If we allow for the slow variation of $k(x)$ with x in Eq.(6.4), it is convenient to express the wave function solutions in the form:

$$\psi(x) = A e^{\pm i u(x)/\hbar} \tag{6.13}$$

where $u(x)$ is no longer linear in x, as found in Eq.(6.10) for a constant potential energy. We have in this case:

$$\frac{d\psi(x)}{dx} = \frac{i}{\hbar}\frac{du}{dx}\psi(x)$$

$$\frac{d^2\psi(x)}{dx^2} = \frac{i}{\hbar}\frac{d}{dx}\left(\frac{du}{dx}\right)\psi(x) + \frac{i}{\hbar}\frac{du}{dx}\frac{d\psi(x)}{dx} = \left[\frac{i}{\hbar}\frac{d}{dx}\left(\frac{du}{dx}\right) - \frac{1}{\hbar^2}\left(\frac{du}{dx}\right)^2\right]\psi(x)$$

which leads to an equation for du/dx:

$$i\hbar\frac{d}{dx}\left(\frac{du}{dx}\right) - \left(\frac{du}{dx}\right)^2 + 2m[E - U(x)] = 0 \qquad (6.14)$$

It is convenient to use the quasi-classical approximation (see Example 5.3), namely to consider an expansion of du/dx in terms of the powers of $i\hbar$:

$$\frac{du}{dx} = \left(\frac{du}{dx}\right)_0 + i\hbar\left(\frac{d^2u}{dx^2}\right)_0 + \ldots$$

which, by substitution into Eq.(6.14) gives, by equating the real and imaginary terms up to the first-order in $\hbar$:

$$\left(\frac{du}{dx}\right)_0^2 = 2m[E - U(x)], \qquad \frac{d}{dx}\left(\frac{du}{dx}\right)_0 - 2\left(\frac{d^2u}{dx^2}\right)_0\left(\frac{du}{dx}\right)_0 = 0$$

It follows that:

$$\left(\frac{d^2u}{dx^2}\right)_0 = \frac{1}{2}\frac{d}{dx}\ln\left(\frac{du}{dx}\right)_0 = \frac{1}{2}\frac{d}{dx}\ln\sqrt{2m[E-U(x)]}$$

and therefore:

$$\frac{du}{dx} = \hbar k(x) + i\hbar\frac{d}{dx}\ln\sqrt{k(x)}$$

where the function $k(x)$ is defined by Eq.(6.5). This determines $u(x)$, up to a constant of integration, as:

$$u(x) = \hbar\int_{x_0}^{x} k(x)dx + i\hbar\ln\sqrt{k(x)} \qquad (6.15)$$

and leads to the so-called *WKB approximation* for the function (6.13):

$$\psi(x) = \frac{A}{\sqrt{k(x)}}e^{\pm i\int_{x_0}^{x} k(x)dx} \qquad (6.16)$$

In a classically allowed region, where $k(x)$ is a real function, the solutions $\psi(x)$ have a sinusoidal behaviour, with modulated amplitude and frequency. If $U(x) > E$, $k(x) = i\gamma(x)$ and each $\psi(x)$ is

208 Quantum Physics

a real exponential function with amplitude modulated by $1/\sqrt{k(x)}$. However, the solutions (6.16) are not admissible wave functions, because of their divergent behaviour at the classical turning points, where $U(x) = E$ and hence $k(x) = 0$. The WKB approximation can be used nevertheless, if $U(x)$ is approximately linear through the neighbourhood of the turning points. Assuming that, for simplicity, such a turning point is taken as the origin, say $x_1 = 0$ in Figure 6.1, and taking the linear dependence $U(x) - E \sim x$, which means $k^2(x) = -\alpha^2 x$ about the origin, the wave function in the classically allowed region $(x < 0)$, can be written in terms of the WKB solutions (6.16) as:

$$\psi_-(x) = \frac{A_1}{\alpha\sqrt{-x}} e^{i\alpha\int_x^0 \sqrt{-x}\,dx} + \frac{A_2}{\alpha\sqrt{-x}} e^{-i\alpha\int_x^0 \sqrt{-x}\,dx} = \frac{2A_-}{\sqrt{k(x)}} \cos\left(\alpha\int_x^0 \sqrt{-x}\,dx - \varphi\right) \quad (6.17)$$

For $x > 0$, where $E - U(x)$ is negative and $k(x) \to i\gamma(x)$, Eq.(6.6), we only choose the WKB solution (6.16) vanishing at $x \to +\infty$, that is:

$$\psi_+(x) = \frac{A_+}{\sqrt{\gamma(x)}} e^{-\alpha\int_0^x \sqrt{x}\,dx} \quad (6.18)$$

The phase factor φ in Eq.(6.17) is determined from the continuity conditions at the turning point $x = 0$ for each $\psi_\pm(x)$ and for their first derivatives, which give:

$$\psi_-(0) = \psi_+(0) \quad : \quad 2A_-\cos\varphi = A_+$$

$$\left(\frac{d\psi_-}{dx}\right)_{x=0} = \left(\frac{d\psi_+}{dx}\right)_{x=0} \quad : \quad 2A_-\sin\varphi = A_+$$

and therefore $\tan\varphi = 1$ or $\varphi = \pi/4$. Substituting this value into Eqs.(6.17) we are allowed to connect the WKB solutions to the left and right of a turning point situated at the right of the classically allowed region, given by Eqs.(6.17) and (6.18). The general connection formula at the turning point $x = x_1$ of the classical region (x_0, x_1) is:

$$\frac{2}{\sqrt{k(x)}}\cos\left(\int_x^{x_1} k(x)\,dx - \frac{\pi}{4}\right) \leftarrow \frac{1}{\sqrt{\gamma(x)}} e^{-\int_{x_1}^x \gamma(x)\,dx} \quad (6.19)$$

A similar result is obtained if the turning point is to the left of the classically allowed region. In this case we take the origin at $x_0 = 0$, where the potential energy can be approximated by $U(x) - E \sim -x$, or $k^2(x) = \alpha^2 x$ (see Figure 6.1). We have:

$$\psi_-(x) = \frac{A_-}{\sqrt{\gamma(x)}} e^{-\alpha\int_x^0 \sqrt{-x}\,dx} \quad (x < 0), \quad \psi_+(x) = \frac{2A_+}{\sqrt{k(x)}}\cos\left(\alpha\int_0^x \sqrt{x}\,dx - \varphi\right) \quad (x > 0)$$

and the continuity conditions for $\psi_\pm(x)$ and their first derivatives, at $x = 0$, lead again to $\varphi = \pi/4$. Hence, the general connection formula at $x = x_0$ bears a form analogous to that given by Eq.(6.19):

$$\frac{1}{\sqrt{\gamma(x)}} e^{-\int_x^{x_0} \gamma(x) dx} \rightarrow \frac{2}{\sqrt{k(x)}} \cos\left(\int_{x_0}^x k(x) dx - \frac{\pi}{4}\right) \tag{6.20}$$

If the wave function to the right of x_0, in the classically allowed region (x_0, x_1), given by Eq.(6.20), is connected to that determined to the left of the other turning point x_1, Eq.(6.19), we obtain, for a particle in the potential energy well of Figure 6.1, that:

$$\frac{2A_0}{\sqrt{k(x)}} \cos\left(\int_{x_0}^x k(x) dx - \frac{\pi}{4}\right) = \frac{2A_1}{\sqrt{k(x)}} \cos\left(\int_x^{x_1} k(x) dx - \frac{\pi}{4}\right) = \frac{2A_1}{\sqrt{k(x)}} \cos\left(\frac{\pi}{4} - \int_x^{x_1} k(x) dx\right)$$

It follows that the arguments of the cosine functions can only differ by an integer multiple of π, and $A_1 = (-1)^n A_0$, which gives:

$$\int_{x_0}^x k(x) dx + \int_x^{x_1} k(x) dx = \frac{\pi}{2} + n\pi \quad \text{or} \quad \int_{x_0}^{x_1} k(x) dx = \left(n + \frac{1}{2}\right)\pi$$

In view of Eq.(6.5), this leads to:

$$2\int_{x_0}^{x_1} \sqrt{2m[E-U(x)]}\, dx = \left(n + \frac{1}{2}\right)h$$

which, written in terms of the momentum $p^2 = E - U(x)$, reduces to the Sommerfeld quantization rule (P.28) in one dimension, namely:

$$2\int_{x_0}^{x_1} p\,dx = \int_{x_0}^{x_1} pdx + \int_{x_0}^{x_1} |p|dx = \int_{x_0}^{x_1} pdx - \int_{x_1}^{x_0} |p|dx = \int_{x_0}^{x_1} pdx + \int_{x_1}^{x_0} pdx = \oint pdx = \left(n + \frac{1}{2}\right)h$$

The integral of the momentum of the classical particle is taken over one complete cycle of its motion, from x_0 to x_1 and then back to x_0 (where $p = -|p|$). The difference of $h/2$ between the two results is determined by the phase factor $\pi/4$ introduced two times in each period by the classical turning points at x_0 and x_1. ✥

Problem 6.1.1. Show that the energy eigenfunctions can always be chosen to have a definite parity for the one-dimensional motion in a symmetric potential energy $U(x) = U(-x)$.

(Solution): If $\psi(x)$ is a solution to Eq.(6.3), by changing the sign of x we obtain:

$$\frac{d^2\psi(-x)}{dx^2} + k^2(-x)\psi(-x) = \frac{d^2\psi(-x)}{dx^2} + k^2(x)\psi(-x) = 0$$

This shows that, provided the potential energy is an even function, $\psi(-x)$ is also an eigenfunction corresponding to the same eigenvalue as $\psi(x)$, and so are the linear combinations:

$$\psi_+(x) = \frac{1}{2}[\psi(x) + \psi(-x)], \quad \psi_-(x) = \frac{1}{2}[\psi(x) - \psi(-x)]$$

These functions can always be chosen as the even and odd eigenfunction corresponding to a given E value. If one of the two functions is equal to zero for all x, the other defines a non-degenerated state, of given parity.

Problem 6.1.2. Find the potential energy $U(x)$ and the energy eigenvalue E, given the corresponding eigenfunction of the form:

$$\psi(x) = Ax^n e^{-\gamma x} \quad x \geq 0$$

$$\psi(x) = 0 \quad x < 0$$

(Solution): It is obvious that the definition of the wave function satisfies the condition of continuity for $\psi(x)$ and $d\psi/dx$ at $x = 0$. Then we have:

$$\frac{d\psi(x)}{dx} = x^n e^{-\gamma x}\left(\frac{n}{x} - \gamma\right) = \left(\frac{n}{x} - \gamma\right)\psi(x)$$

$$\frac{d^2\psi(x)}{dx^2} = \left[\frac{n(n-1)}{x^2} - \frac{2n\gamma}{x} + \gamma^2\right]\psi(x)$$

and substituting into Eq.(6.4) gives:

$$\left[\frac{n(n-1)}{x^2} - \frac{2n\gamma}{x} - \frac{2m}{\hbar^2}U(x)\right] + \left(\gamma^2 + \frac{2m}{\hbar^2}E\right) = 0$$

It follows that:

6. One-Dimensional Motion 211

$$U(x) = \frac{\hbar^2}{2m}\frac{n}{x}\left(\frac{n-1}{x} - 2\gamma\right), \qquad E = -\frac{\hbar^2}{2m}\gamma^2$$

The potential energy vanishes for $x_0 = (n-1)/2\gamma$ and, for $x \to +\infty$, has a minimum for:

$$\frac{dU(x)}{dx} = \frac{\hbar^2}{m}\frac{n}{x^2}\left(\gamma - \frac{n-1}{x}\right) = 0 \qquad \text{or} \qquad x = 2x_0$$

namely $U_{min} = -\hbar^2 n\gamma/4mx_0$. There is an inflexion point for:

$$\frac{d^2 U(x)}{dx^2} = \frac{3\hbar^2}{m}\frac{n}{x^3}\left(\frac{n-1}{x} - \frac{2\gamma}{3}\right) = 0 \qquad \text{or} \qquad x = 3x_0$$

It is a straightforward matter to check that we always have $E > U_{min}$.

Problem 6.1.3. Find the eigenstates of a particle of mass m, that rotates on a circle of radius r_0 in a spherically symmetric potential energy $U(r)$.

(Solution): The problem is one-dimensional, because the equation of the planar motion can be expressed in terms of polar coordinates in the form:

$$x = r\cos\varphi, \qquad y = r\sin\varphi$$

where $r = r_0$. We have:

$$\nabla^2 = \frac{\partial^2}{\partial x^2} + \frac{\partial^2}{\partial y^2} = \frac{1}{r}\frac{\partial}{\partial r}\left(r\frac{\partial}{\partial r}\right) + \frac{1}{r^2}\frac{\partial^2}{\partial \varphi^2}$$

and the Schrödinger wave equation (6.4) reads:

$$\frac{1}{r}\frac{\partial}{\partial r}\left(r\frac{\partial\psi}{\partial r}\right) + \frac{1}{r^2}\frac{\partial^2\psi}{\partial\varphi^2} + \frac{2m}{\hbar^2}[E - U(r)]\psi = 0$$

Since $r = r_0$, it is apparent that $\partial\psi/\partial r = 0$ and $U(r_0) = U_0$ is a constant. Hence, we are left with an ordinary differential equation for the motion in a constant potential energy:

$$\frac{\partial^2\psi}{\partial\varphi^2} + k^2\psi = 0, \qquad k^2 = \frac{2m}{\hbar^2}r_0^2(E - U_0)$$

where $\psi = \psi(r_0,\varphi)$, so that we look for a solution of the form (6.11):

$$\psi(r_0,\varphi) = A\sin(k\varphi+\delta)$$

The appropriate admissibility condition for wave functions, given in terms of angular coordinate variables, consists of the single-valuedness requirement:

$$\sin(k\varphi+\delta) = \sin[k(\varphi+2\pi)+\delta]$$

which implies that $k = n$, where n is a positive integer $n = 1,2,3,\ldots$. The normalization condition:

$$A^2 \int_0^{2\pi} \sin^2(n\varphi+\delta)d\varphi = 1$$

gives $A = 1/\sqrt{\pi}$, and hence the eigenstates are described by:

$$\psi_n(r_0,\varphi) = \frac{1}{\sqrt{\pi}}\sin(n\varphi+\delta)$$

The corresponding eigenvalues form a discrete spectrum:

$$E_n = U_0 + \frac{n^2\hbar^2}{2mr_0^2}$$

Problem 6.1.4. Find the potential energy function $U(x)$ and the energy E of the eigenstate described by the wave function:

$$\psi(x) = Ae^{-\gamma x^2}$$

(Answer): $\quad U(x) = \dfrac{\hbar^2}{m}2\gamma^2 x^2, \quad E = \dfrac{\hbar^2}{m}\gamma$

Problem 6.1.5. Show that the WKB quantization rule, for a particle in the potential energy well of Figure 6.1, takes the form:

$$2\int_0^{x_1} \sqrt{2m[E-U(x)]} = \left(n+\frac{3}{4}\right)h$$

if $U(x)$ becomes infinite for $x \le 0$.

6.2. The Energy Spectrum for Discontinuous Potentials

Given the potential energy function $U(x)$, the admissibility conditions for the eigenfunctions will not be satisfied for any value of E in Eq.(6.4), and hence restrictions are imposed on the energy of stationary states. The nature of the energy eigenvalue spectrum is determined by the position of E with respect to the potential energy function $U(x)$ (see Figure 6.1). Assuming that $U(x)$ has a minimum U_{min}, we will always have:

$$E > U_{min} \qquad (6.21)$$

because otherwise the entire x-axis will correspond to a classically disallowed region, and this implies an exponential behaviour of eigenfunctions for all x, which cannot be finite, as required, in both asymptotic regions. Indeed, if there is a node of $\psi(x)$ at $x = \pm\infty$, then $\psi(x)$ becomes infinite at $x = \mp\infty$, and if the node is situated at a finite x, $\psi(x)$ becomes infinite at both $x \to +\infty$ and $x \to -\infty$. From Eq.(6.21) it then follows that for any eigenvalue E both classically allowed and disallowed regions will be found along the x-axis.

The energy eigenvalue spectrum is *discrete* for the range of E where both asymptotic regions $x \to \pm\infty$ are classically disallowed regions, namely:

$$\underset{x \to \pm\infty}{E - U(x) < 0} \qquad (6.22)$$

In this case we make use of Eq.(6.6), where it may be assumed, for simplicity, that $\underset{x \to \pm\infty}{U(x) \to 0}$, which implies $E < 0$. Then, in the asymptotic regions, we have:

$$\underset{x \to \pm\infty}{\gamma^2(x) = \gamma^2} = -\frac{2m}{\hbar^2}E > 0$$

Consider first the asymptotic behaviour at $x \to +\infty$, where the eigenfunctions will have the exponential form (6.12), namely:

$$\underset{x \to +\infty}{\psi(x) \sim a_1 e^{\gamma x} + a_2 e^{-\gamma x}} \qquad (6.23)$$

Normalization is only possible provided $a_1 = 0$, because otherwise the function will be divergent for $x \to +\infty$, and this condition determines $\psi(x)$ for all x, up to multiplicative constant a_2. It is thus expected that the eigenfunction also behaves like $e^{-\gamma x}$ at $x \to -\infty$, where:

$$\underset{x \to -\infty}{\psi(x) \sim a_1(E)e^{\gamma x} + a_2(E)e^{-\gamma x}} \qquad (6.24)$$

with coefficients $a_1(E), a_2(E)$ which are no longer arbitrary, but completely defined in terms of E. The restriction imposed by normalization requirements at $x \to -\infty$, namely:

$$a_2(E) = 0 \tag{6.25}$$

will be satisfied by eigenfunctions $\psi(x)$ corresponding to the solutions of Eq.(6.25), which represent a discrete set of eigenvalues E. This conclusion remains valid for $U(x) \underset{x \to \pm\infty}{\to} U_\pm \neq 0$ and even if $U(x) \underset{x \to \pm\infty}{\to} \infty$, because $\psi(x)$ is always a linear combination of two independent solutions with exponential behaviour in the classically disallowed regions. The admissible $\psi(x)$ should vanish at both ends, and this implies that the coefficient of the solution vanishing at $x \to +\infty$, which diverges as $x \to -\infty$, should be zero. This equation is satisfied by the energy eigenvalues. Therefore, the discrete energy eigenvalue spectrum is always associated with eigenfunctions for which the probability density vanishes at infinity:

$$\lim_{x \to \pm\infty} |\psi(x)|^2 = 0 \quad \text{or} \quad \lim_{x \to \pm\infty} \psi(x) = 0 \tag{6.26}$$

Such stationary states are called *bound states* and correspond to the bound orbits defined in classical mechanics. If $U(x) \underset{x \to \pm\infty}{\to} \infty$ the classical orbits are always bound, because the asymptotic regions are disallowed. If $U(x) \underset{x \to \pm\infty}{\to} 0$, the bound orbits have negative values of the total energy E, because otherwise the kinetic energy would be negative at infinity. Nevertheless, contrary to the bound classical orbits, where any E value is admissible, the bound quantum states correspond to a discrete energy spectrum.

For bound states in one dimension, there is no degeneracy, since, assuming the contrary, Eq.(6.4) gives:

$$\frac{d^2\psi_1(x)}{dx^2} + k^2(x)\psi_1(x) = 0, \quad \frac{d^2\psi_2(x)}{dx^2} + k^2(x)\psi_2(x) = 0$$

Multiplication by $\psi_2(x)$ and $\psi_1(x)$ respectively, followed by subtraction, yields:

$$\psi_2(x)\frac{d^2\psi_1(x)}{dx^2} - \psi_1(x)\frac{d^2\psi_2(x)}{dx^2} = \frac{d}{dx}\left[\psi_2(x)\frac{d\psi_1(x)}{dx} - \psi_1(x)\frac{d\psi_2(x)}{dx}\right] = 0$$

and hence:

$$\psi_2(x)\frac{d\psi_1(x)}{dx} - \psi_1(x)\frac{d\psi_2(x)}{dx} = \text{const}$$

where the constant should be equal to zero, because of the asymptotic behaviour (6.26) required for bound states. It follows that:

$$\frac{1}{\psi_1(x)}\frac{d\psi_1(x)}{dx} = \frac{1}{\psi_2(x)}\frac{d\psi_2(x)}{dx} \quad \text{or} \quad \ln\psi_1(x) = \ln\psi_2(x) + \varphi$$

where φ is a constant. This leads to:

$$\psi_1(x) = e^{i\varphi}\psi_2(x)$$

and therefore the two eigenfunctions should represent the same bound state, as they differ by a phase factor only.

EXAMPLE 6.3. Bound states in the square well potential energy

Consider the schematic representation of the potential energy $U(x)$ given in Figure 6.1, which consists of a function having the shape of a well with vertical sides, defined by:

$$\begin{aligned} U(x) &= U_- & \text{for} \quad & x < -\alpha \\ &= 0 & \text{for} \quad & -\alpha < x < \alpha \\ &= U_+ & \text{for} \quad & x > \alpha \end{aligned} \quad (6.27)$$

and illustrated in Figure 6.3.

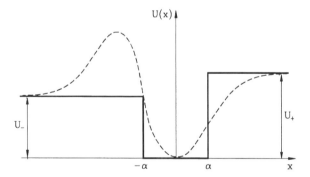

Figure 6.3. The square well approximation to the potential energy

The classical particle is bound by this potential energy in the allowed region $(-\alpha, \alpha)$, provided $0 < E < U_-$. This condition implies that $E - U(x)$ will be negative for $x < -\alpha$ and $x > \alpha$, and because the kinetic energy can never be negative, the particle will have to stay within the potential energy well. For a quantum particle, using the notations (6.5) and (6.6), which read:

$$k = \frac{1}{\hbar}\sqrt{2mE}, \quad \gamma_- = \frac{1}{\hbar}\sqrt{2m(U_- - E)}, \quad \gamma_+ = \frac{1}{\hbar}\sqrt{2m(U_+ - E)}$$

we must consider distinct forms of the Schrödinger wave equation (6.4) in the different regions:

(i) $$\frac{d^2\psi_-(x)}{dx^2} - \gamma_-^2 \psi_-(x) = 0 \quad \text{for} \quad x < -\alpha$$

with the admissible solution $\psi_-(x) = A_- e^{\gamma_- x}$ that vanishes as $x \to -\infty$.

(ii) $$\frac{d\psi(x)}{dx^2} + k^2 \psi(x) = 0 \quad \text{for} \quad -\alpha < x < \alpha$$

with the general solution (6.11), which is conveniently written as $\psi(x) = A \sin(kx + \varphi)$.

(iii) $$\frac{d^2\psi_+(x)}{dx^2} - \gamma_+^2 \psi_+(x) = 0 \quad \text{for} \quad x > \alpha$$

where we retain the solution vanishing at $x \to \infty$, $\psi_+(x) = A_+ e^{-\gamma_+ x}$.

The energy eigenvalue spectrum results from the condition that $\psi(x)$ and its first derivative are continuous everywhere, and in particular at the classical turning points $x = -\alpha$ and $x = \alpha$. At the border to the left of the classical region we then require that:

$$A_- e^{-\gamma_- \alpha} = A \sin(-k\alpha + \varphi), \quad \gamma_- A_- e^{-\gamma_- \alpha} = Ak \cos(-k\alpha + \varphi) \quad \text{or} \quad \tan(-k\alpha + \varphi) = \frac{k}{\gamma_-} \quad (6.28)$$

and at the border $x = \alpha$:

$$A_+ e^{-\gamma_+ \alpha} = A \sin(k\alpha + \varphi), \quad -\gamma_+ A_+ e^{-\gamma_+ \alpha} = Ak \cos(k\alpha + \varphi) \quad \text{or} \quad \tan(k\alpha + \varphi) = -\frac{k}{\gamma_+} \quad (6.29)$$

From Eq.(6.28) it follows that:

$$\sin(-k\alpha + \varphi) = \frac{\tan(-k\alpha + \varphi)}{\sqrt{1 + \tan^2(-k\alpha + \varphi)}} = \frac{k}{\sqrt{k^2 + \gamma_-^2}} = \sqrt{\frac{E}{U_-}} \quad \text{or} \quad \varphi = k\alpha + \arcsin\sqrt{\frac{E}{U_-}} \quad (6.30)$$

and similarly, from Eq.(6.29) one obtains:

$$\sin(k\alpha + \varphi) = -\frac{k}{\sqrt{k^2 + \gamma_+^2}} = -\sqrt{\frac{E}{U_+}} \quad \text{or} \quad \varphi = -k\alpha - \arcsin\sqrt{\frac{E}{U_+}} \quad (6.31)$$

The two values derived for φ might differ by an integer multiple of π only, that is:

$$2k\alpha + \arcsin\sqrt{\frac{E}{U_+}} + \arcsin\sqrt{\frac{E}{U_-}} = \frac{2\alpha}{\hbar}\sqrt{2mE} + \arcsin\sqrt{\frac{E}{U_+}} + \arcsin\sqrt{\frac{E}{U_-}} = n\pi \quad (6.32)$$

where $n = 1, 2, 3, \ldots$. Equation (6.32) is a quantization condition that leads to a discrete energy eigenvalue spectrum. There are no solutions to this equation in terms of elementary functions, so that it is common practice to use graphical techniques. Because $0 < E < U_-$ in the classical region, we may set $r^2 = E/U_-$, $\cos^2 R = U_-/U_+$ and $\chi = \sqrt{2mU_-}/\hbar$, such that Eq.(6.32) becomes:

$$\arcsin r + \arcsin(r \cos R) = n\pi - 2\alpha\chi r$$

If we represent, as in Figure 6.4, the curve $f(r) = \arcsin r + \arcsin(r \cos R)$, it increases from 0 to $(\pi/2) + \arcsin(\cos R) = \pi - R$ as E increases from 0 to U_-, and hence r increases from 0 to 1.

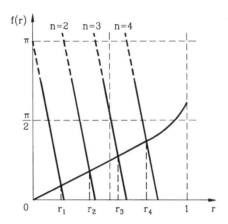

Figure 6.4. Graphical solution for the allowed values $r_n = E_n/U_-$

For each $n = 1, 2, 3, \ldots$, the right-hand side of the equation decreases linearly from $n\pi$ to $n\pi - 2\alpha\chi r$ and the solutions are determined by the intersection of these straight lines with the curve $f(r)$. The condition that at least one solution exists is $n\pi - 2\alpha\chi \leq \pi - R$, or $2\alpha\chi \geq (n-1)\pi + R$. If the intersections occur at $r = r_n$, the corresponding values of the energy are $E_n = U_- r_n$, and their number is finite. There will be no bound state for $2\alpha\chi < R$, one bound state for $R \leq 2\alpha\chi < R + \pi$, and in general n discrete energy levels E_n for $R + (n-1)\pi < 2\alpha\chi < R + n\pi$. A solution always exists if we assume, for simplicity $U_- = U_+ = U_0$, because $\cos R = 1$, hence $R = 0$, and $2\alpha\chi$ cannot be negative.

In the case when $\gamma_- = \gamma_+ = \gamma$, Eqs.(6.31) and (6.30) can be combined to give $2\varphi = n\pi$ and the discrete stationary states will be described by simpler wave functions $\psi(x) = A \sin(kx + n\pi/2)$, which are alternatively even or odd as n (and the energy eigenvalue) increases. For $n = 1, 3, 5, \ldots$, Eqs.(6.28) and (6.29) give $A_- = A_+ = Ae^{\gamma\alpha} \cos k\alpha$. Therefore we have:

$$\psi_-(x) = A \cos k\alpha \, e^{\gamma(x+\alpha)}, \quad \psi(x) = A \cos kx, \quad \psi_+(x) = A \cos k\alpha \, e^{-\gamma(x-\alpha)} \quad (6.33)$$

and it follows that the bound states are described by functions of even parity $\psi(x) = \psi(-x)$. The normalization constant results from:

$$\int_{-\infty}^{\infty} |\psi(x)|^2 dx = |A|^2 e^{2\gamma\alpha} \cos^2 k\alpha \int_{-\infty}^{-\alpha} e^{2\gamma x} dx + |A|^2 \int_{-\alpha}^{\alpha} \cos^2 kx \, dx + |A|^2 e^{2\gamma\alpha} \cos^2 k\alpha \int_{\alpha}^{\infty} e^{-2\gamma x} dx = 1$$

218 Quantum Physics

which gives the amplitude:

$$|A| = \left(\alpha + \frac{\cos^2 k\alpha}{\gamma} + \frac{\sin 2k\alpha}{2k}\right)^{-\frac{1}{2}} = \left(\alpha + \frac{1}{\gamma}\right)^{-\frac{1}{2}} \tag{6.34}$$

where use has been made of Eq.(6.28) to set $\tan k\alpha = \gamma/k$, if φ is an odd multiple of $\pi/2$. Therefore, the bound states of even parity are described by:

$$\psi_n(x) = \frac{1}{\sqrt{\alpha + 1/\gamma}} \cos kx \qquad (n = 1,3,5,\ldots) \tag{6.35}$$

Contrary to the classical situation, there is a nonvanishing probability of finding the particle in the classically disallowed region, given by:

$$\int_{-\infty}^{-\alpha} |\psi_-(x)|^2 dx + \int_{\alpha}^{\infty} |\psi_+(x)|^2 dx = 2\int_{\alpha}^{\infty} |\psi_+(x)|^2 dx = |A|^2 \frac{\cos^2 k\alpha}{\gamma} = (1+\alpha\gamma)\cos^2 k\alpha$$

This explains the behaviour of the wave functions (6.35) at the classical turning points, illustrated in Figure 6.5.

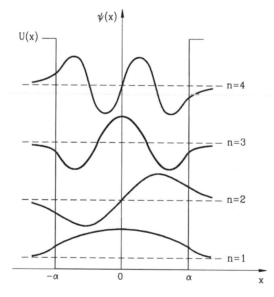

Figure 6.5. Bound state functions of even and odd parity for the lowest energy levels

Similar considerations are valid for $n = 2,4,6,\ldots$, in which case the bound states are described by wave functions of odd parity:

$$\psi_n(x) = \frac{1}{\sqrt{\alpha + 1/\gamma}} \sin kx \qquad (n = 2,4,6,\ldots) \tag{6.36}$$

as depicted in Figure 6.5.

6. One-Dimensional Motion   219

The energy eigenvalue spectrum is *continuous* for that range of E which makes at least one of the asymptotic regions to be a classically allowed region:

$$E - U(x) \underset{x \to +\infty}{> 0} \quad \text{or/and} \quad E - U(x) \underset{x \to -\infty}{> 0} \tag{6.37}$$

If we assume that $U(x)$ vanishes for both $x \to +\infty$ and $x \to -\infty$, the wave functions have sinusoidal asymptotic behaviour:

$$\psi(x) \underset{x \to +\infty}{\sim} a_1 e^{ikx} + a_2 e^{-ikx} \tag{6.38}$$

$$\psi(x) \underset{x \to -\infty}{\sim} b_1 e^{ikx} + b_2 e^{-ikx} \tag{6.39}$$

and hence, there is no longer need for normalization conditions which might determine the coefficients. The Schrödinger equation (6.4) has admissible solutions for any eigenvalue $E > 0$, which is two-fold degenerate. If only one asymptotic region is a classical region, say $U(x) \underset{x \to +\infty}{\to} 0$, and $U(x) \underset{x \to -\infty}{\to} U_-$, such that $0 < E < U_-$, Eq.(6.38) remains valid, but we obtain an exponential asymptotic behaviour as $x \to -\infty$, where one of the two linearly independent solutions diverges, and we must choose:

$$\psi(x) \underset{x \to -\infty}{\sim} b e^{\gamma x}, \quad \gamma = \frac{1}{\hbar} \sqrt{2m(U_- - E)} \tag{6.40}$$

This solution is admissible for all x, as it has a sinusoidal behaviour for $x \to \infty$. There is no longer degeneracy for $0 < E < U_-$, although the energy eigenvalue spectrum remains continuous.

The energy eigenstates corresponding to a continuous energy spectrum are called *unbound states* and correspond to the classical unbound orbits. Their physical interpretation does not rely on the form of the eigenfunctions but on the probability current density $\vec{j}$, Eq.(5.35), which can be used to find the probability that a particle in an unbound state will cross a given surface in unit time.

EXAMPLE 6.4. The tunnel effect

A continuous energy spectrum occurs in the motion of particles past a *potential energy barrier*, which may be qualitatively characterized by its height U_0 and width 2α, as in Figure 6.6. For instance, in terms of the potential energy $U(x)$ represented in Figure 6.1, we may choose $U_0 = U_{max} - U_-$, and the barrier width corresponding to the classically disallowed region (x_0', x_1'). The admissible solution of the Schrödinger equation (6.4) in the region $x < -\alpha$, where $k(x) = \sqrt{2mE}/\hbar = p_x/\hbar = k$, has the form:

$$\psi_-(x) = a_i e^{ikx} + a_r e^{-ikx} \tag{6.41}$$

and represents free particles with positive momentum $p_x = \hbar k$ incident on the barrier from the left and reflected free particles, moving away from the barrier with momentum $-\hbar k$.

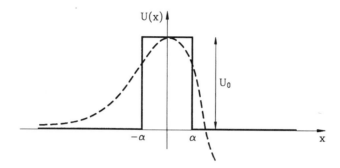

Figure 6.6. The square potential energy barrier

Assuming $0 < E < U_0$, the classical particles are all reflected by the barrier, because their kinetic energy cannot be negative, as implied by the barrier penetration. However, the Schrödinger equation has an admissible solution in the region $-\alpha < x < \alpha$:

$$\psi(x) = A_+ e^{\gamma x} + A_- e^{-\gamma x}, \qquad \gamma = \frac{1}{\hbar}\sqrt{2m(U_0 - E)} \qquad (6.42)$$

and also for $x > \alpha$:

$$\psi_+(x) = u_t e^{ikx} \qquad (6.43)$$

where it was assumed that, on the other side of the barrier, there are transmitted particles only, and no particles approaching the barrier. The probability current through the barrier is given by Eq.(5.35) for the incident particles:

$$j_i = \frac{i\hbar}{2m}\left[a_i e^{ikx}\frac{d}{dx}\left(a_i^* e^{-ikx}\right) - a_i^* e^{-ikx}\frac{d}{dx}\left(a_i e^{ikx}\right)\right] = \frac{\hbar k}{m}|a_i|^2 \qquad (6.44)$$

and similarly for the reflected and transmitted particles:

$$j_r = \frac{\hbar k}{m}|a_r|^2, \qquad j_t = \frac{\hbar k}{m}|a_t|^2 \qquad (6.45)$$

Hence we may define the transmission probability τ, which is the probability for tunnelling the barrier, and the probability for reflection ρ:

$$\tau = \frac{j_t}{j_i} = \left|\frac{a_t}{a_i}\right|^2, \qquad \rho = \frac{j_r}{j_i} = \left|\frac{a_r}{a_i}\right|^2 \qquad (6.46)$$

where the amplitudes a_i, a_r and a_t are related by the continuity conditions for wave functions and their first derivatives at $x = -\alpha$ and $x = \alpha$:

6. One-Dimensional Motion

$$a_i e^{-ik\alpha} + a_r e^{ik\alpha} = A_+ e^{-\gamma\alpha} + A_- e^{\gamma\alpha}$$

$$a_i e^{-ik\alpha} - a_r e^{ik\alpha} = \frac{\gamma}{ik}\left(A_+ e^{-\gamma\alpha} - A_- e^{\gamma\alpha}\right)$$

$$A_+ e^{\gamma\alpha} + A_- e^{-\gamma\alpha} = a_t e^{ik\alpha}$$

$$A_+ e^{\gamma\alpha} - A_- e^{-\gamma\alpha} = \frac{ik}{\gamma} a_t e^{ik\alpha}$$

The last two equations yield:

$$A_+ = \frac{a_t}{2} e^{-\gamma\alpha}\left(1 + \frac{ik}{\gamma}\right)e^{ik\alpha}, \qquad A_- = \frac{a_t}{2} e^{\gamma\alpha}\left(1 - \frac{ik}{\gamma}\right)e^{ik\alpha}$$

and then, substituting into the other two equations, and solving for a_i and a_r, we obtain:

$$a_i = a_t e^{2k\alpha}\left[\cosh(2\gamma\alpha) + \frac{i}{2}\left(\frac{\gamma}{k} - \frac{k}{\gamma}\right)\sinh(2\gamma\alpha)\right], \qquad a_r = \frac{a_t}{2i}\left(\frac{\gamma}{k} + \frac{k}{\gamma}\right)\sinh(2\gamma\alpha)$$

As a result, the probabilities (6.46) are given by:

$$\tau = \frac{1}{1 + \left(\frac{\gamma^2 + k^2}{2\gamma k}\right)^2 \sinh^2(2\gamma\alpha)} \qquad \text{and} \qquad \rho = 1 - \tau \tag{6.47}$$

It is convenient to rewrite τ in terms of the involved energies, in the form:

$$\tau = \frac{1}{1 + \frac{U_0^2}{4E(U_0 - E)}\sinh^2\left(\frac{2\alpha}{\hbar}\sqrt{2m(U_0 - E)}\right)} \tag{6.48}$$

where we can approximate $\sinh(2\gamma\alpha) = (e^{\gamma\alpha} - e^{-\gamma\alpha})/2 \approx e^{\gamma\alpha}/2$, because $\gamma\alpha \gg 1$, which gives the probability for tunnelling as:

$$\tau = 16\frac{E}{U_0}\left(1 - \frac{E}{U_0}\right)e^{-4\alpha\sqrt{2m(U_0 - E)}/\hbar} \tag{6.49}$$

It is obvious that $\tau = 0$ for $E = 0$ or $U_0 = \infty$, and also that $\tau \to 0$ for a very large barrier, of width $2\alpha \to \infty$. The classical result $\tau \to 0$ (and hence $\rho \to 1$) is obtained by taking the classical limit $\hbar \to 0$. For an arbitrary shape of the potential barrier function $U(x)$ we obtain a similar result, by dividing the classically disallowed region (x_0', x_1') into a series of small regions Δx_i, as in Figure 6.7, and using the transmission probabilities of the square potential barriers of height $U(x_i)$.

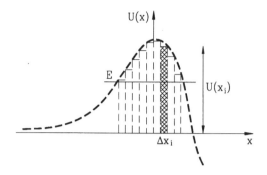

Figure 6.7. Potential energy barrier of arbitrary shape

The probability for tunnelling, in the case of a particle with energy E is:

$$\tau = \prod_{i=1}^{n} \tau_i = \tau_0 e^{-2\sum \sqrt{2m[U(x_i)-E]}\,\Delta x_i/\hbar}$$

and, at the limit $\Delta x_i \to 0$, we have:

$$\frac{2}{\hbar}\sum \sqrt{2m[U(x_i)-E]}\,\Delta x_i \to \frac{2}{\hbar}\int_{x_0'}^{x_1'} \sqrt{2m[U(x)-E]}\,dx$$

where x_0' and x_1' are the classical turning points. Therefore:

$$\tau = \tau_0\, e^{-2\int_{x_0'}^{x_1'} \sqrt{2m[U(x)-E]}\,dx/\hbar} \tag{6.50}$$

This relation allows us to estimate the tunnel effect probability if particular functions $U(x)$, appropriate for the physical situation, are assumed. ◊

If the potential energy is a periodic function:

$$U(x+L) = U(x) \tag{6.51}$$

which has an oscillatory asymptotic behaviour, rather than a well defined limit at $x \to \pm\infty$, the energy eigenvalue spectrum has a *band structure*, which consists of allowed energy bands and forbidden gaps. The occurrence of such a structure is determined by the general form of the eigenfunctions, imposed by the periodicity of the potential energy. If Eq.(6.4) is written for $\psi(x+L)$, we obtain, in view of Eq.(6.51):

$$\frac{d^2\psi(x+L)}{dx^2} + k^2(x)\psi(x+L) = 0$$

and hence the wave functions $\psi(x)$ and $\psi(x+L)$, that satisfy the same equation, must be constant multiples of each other:

$$\psi(x+L) = \lambda \psi(x) \tag{6.52}$$

Here λ is a complex number whose absolute value is equal to 1, because otherwise the wave function would be infinite at either $x \to +\infty$, if $|\lambda| > 1$, or $x \to -\infty$, if $|\lambda| < 1$. This condition allows us to determine the energy eigenvalue spectrum.

EXAMPLE 6.5. Energy bands in a one-dimensional array

Consider an infinite array of barriers in one dimension, Figure 6.8 (a), described by the periodic potential function:

$$U(x) = V_0 \sum_{n=-\infty}^{\infty} \delta(x - nL) \tag{6.53}$$

which, in view of the Dirac δ-function behaviour, consists of delta potential energy barriers, of height $U_0 \to \infty$ and width $2\alpha \to 0$, such that $2\alpha U_0$ remains constant, in our case equal to V_0. This function provides a simple approximation for the motion of an electron in a metal and is known as the *Kronig-Penney* model. For $x \ne nL$, the solution can be written in the form (6.11), which reads:

$$\psi(x) = a_n \sin k(x - nL) + b_n \cos k(x - nL) \quad (n-1)L \le x \le nL \tag{6.54}$$

$$\psi(x) = a_{n+1} \sin k[x - (n+1)L] + b_{n+1} \cos k[x - (n+1)L] \quad nL \le x \le (n+1)L \tag{6.55}$$

such that the continuity condition for $\psi(x)$ at $x = nL$ gives:

$$-a_{n+1} \sin kL + b_{n+1} \cos kL = b_n \tag{6.56}$$

The continuity condition for $d\psi/dx$ can be obtained by integrating Eq.(6.4), where $U(x)$ is given by Eq.(6.53), between $nL - \varepsilon$ and $nL + \varepsilon$:

$$\int_{nL-\varepsilon}^{nL+\varepsilon} \frac{d^2\psi(x)}{dx^2} dx - \frac{2m}{\hbar^2} V_0 \int_{nL-\varepsilon}^{nL+\varepsilon} \delta(x-nL)\psi(x)dx + \frac{2m}{\hbar^2} E \int_{nL-\varepsilon}^{nL+\varepsilon} \psi(x)dx = 0$$

As $\varepsilon \to 0$, the second term reduces to $-2mV_0\psi(nL)/\hbar^2$, according to the definition (1.44) of the Dirac δ-function, and the last term vanishes, so that we have:

$$\left(\frac{d\psi}{dx}\right)_{nL+\varepsilon} - \left(\frac{d\psi}{dx}\right)_{nL-\varepsilon} = \frac{2m}{\hbar^2} V_0 \psi(nL) \tag{6.57}$$

It is convenient to set $2mV_0/\hbar^2 = \rho$, such that Eq.(6.57) reads:

$$ka_{n+1} \cos kL + kb_{n+1} \sin kL - ka_n = \rho b_n \qquad (6.58)$$

Solving Eqs.(6.56) and (6.58) for a_{n+1} and b_{n+1} yields:

$$a_{n+1} = a_n \cos kL + b_n \left(\frac{\rho}{k} \cos kL - \sin kL\right)$$

$$b_{n+1} = a_n \sin kL + b_n \left(\frac{\rho}{k} \sin kL + \cos kL\right) \qquad (6.59)$$

The condition (6.52), when applied to the functions (6.54) and (6.55), implies that:

$$a_{n+1} = \lambda a_n = e^{i\varphi} a_n, \quad b_{n+1} = \lambda b_n = e^{i\varphi} b_n$$

and substituting into Eqs.(6.59) leads to:

$$e^{2i\varphi} - e^{i\varphi}\left(2\cos kL + \frac{\rho}{k}\sin kL\right) + 1 = 0$$

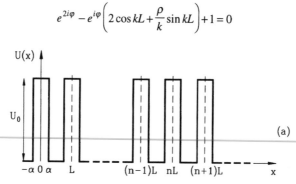

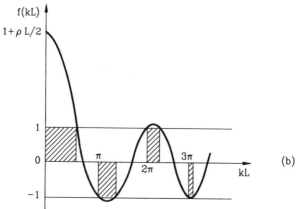

Figure 6.8. (a) Periodic potential energy in one dimension; (b) Allowed k values (energy bands) and forbidden gaps (the dashed regions)

Multiplication by $e^{-i\varphi}$ finally gives:

$$\cos\varphi = \cos kL + \frac{p}{2k}\sin kL \qquad (6.60)$$

Since $|\cos\varphi|\le 1$, we may represent the function $f(kL) = \cos kL + pL(\sin kL)/2kL$ in terms of kL, for a certain value $pL/2 = mLV_0/\hbar^2 = mL(2\alpha U_0)/\hbar^2$, and whenever $|f(kL)|>1$, the problem has no solution for kL, and hence no allowed energy eigenvalues exist. It is visible from Figure 6.8 (b) that the allowed energy regions, called *energy bands*, where the energy eigenvalue spectrum is continuous, are separated by regions that contain no eigenvalues, known as *energy gaps*. ♦

Problem 6.2.1. Find the transmission probability τ for the one-dimensional delta well:

$$U(x) = -V_0\delta(x) \qquad (V_0 > 0)$$

where $\delta(x)$ is the Dirac δ-function, considering unbound states $E > 0$.

(Solution): We use the wave functions given by Eqs.(6.41) and (6.42), namely:

$$\psi_-(x) = a_i e^{ikx} + a_r e^{-ikx} \qquad x < 0$$

$$\psi_+(x) = a_t e^{ikx} \qquad x > 0$$

which lead to the continuity condition for the wave function through $x = 0$:

$$\psi(0) = a_i + a_r = a_t$$

For $d\psi/dx$ we have the condition in the form (6.57) which reads:

$$\left(\frac{d\psi}{dx}\right)_\varepsilon - \left(\frac{d\psi}{dx}\right)_{-\varepsilon} = -\frac{2mV_0}{\hbar^2}\psi(0) \quad \text{or} \quad a_t + a_r - a_i = \frac{2imV_0}{k\hbar^2}a_t$$

Solving the two equations for a_t gives:

$$a_t = \left(1 - \frac{imV_0}{k\hbar^2}\right)a_i$$

and the transmission probability results from Eq.(6.46) as:

$$\tau = \left|\frac{a_t}{a_i}\right|^2 = \frac{1}{1 + m^2V_0^2/k^2\hbar^4} = \frac{1}{1 + mV_0^2/2\hbar^2 E}$$

226   Quantum Physics

Problem 6.2.2. Find the transmission probability through a potential energy barrier:

$$U(x) = 0 \qquad x < 0$$

$$U(x) = U_0\left(1 - \frac{x}{\alpha}\right) \qquad x \geq 0$$

for an electron of mass m and energy $E < U_0$. (U_0 may represent the extraction work function for metal electrons, and U_0/α the strength of an applied field).

(Solution):

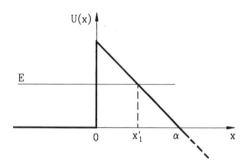

We may use the expression (6.50) for τ, where x_1' is given by:

$$U_0\left(1 - \frac{x_1'}{\alpha}\right) = E \qquad \text{or} \qquad x_1' = \alpha\left(1 - \frac{E}{U_0}\right)$$

Hence, we have:

$$\int_0^{x_1'} \sqrt{U(x) - E}\, dx = \int_0^{x_1'} \sqrt{(U_0 - E) - U_0 x/\alpha}\, dx = \frac{2}{3}\frac{\alpha}{U_0}(U_0 - E)^{\frac{3}{2}}$$

where use has been made of the standard integral:

$$\int \sqrt{a + bx}\, dx = \frac{2}{3b}(a + bx)^{\frac{3}{2}}$$

It follows from Eq.(6.50) that $\tau = \tau_0 e^{-4\alpha\sqrt{2m}\,(U_0 - E)^{3/2}/3\hbar U_0}$.

6. One-Dimensional Motion

Problem 6.2.3. Find the transmission probability τ for a square well potential energy:

$$U(x) = -U_0 \quad , \quad |x| \leq \alpha$$

$$U(x) = 0 \quad , \quad |x| > \alpha$$

assuming energies $E > 0$. Calculate the square well width 2α for which complete transmission ($\tau = 1$) occurs.

(Answer): $\quad \tau = \left[1 + \dfrac{U_0^2}{4E(E+U_0)} \sin^2\left(\dfrac{2\alpha}{\hbar} \sqrt{2m(E+U_0)} \right) \right]^{-1}, \quad 2\alpha = n\pi\hbar / \sqrt{2m(E+U_0)}$

Problem 6.2.4. Find the tunnelling probability through a potential energy barrier:

$$U(x) = U_0 \left(1 - \dfrac{x^2}{\alpha^2} \right) \quad , \quad |x| < \alpha$$

$$= 0 \quad , \quad |x| > \alpha$$

for a particle of mass m and energy $E < U_0$.

(Answer): $\quad \tau = \tau_0 e^{-\pi\alpha\sqrt{2m}(U_0 - E)/\hbar\sqrt{U_0}}$

Problem 6.2.5. Show that, for the motion in a periodic potential energy $U(x) = U(x+L)$, the solutions of the Schrödinger wave equation can be written as $\psi(x) = e^{ikx} u_k(x)$ where $u_k(x) = u_k(x+L)$.

Problem 6.2.6. Find the eigenfunctions and eigenvalues of a particle of mass m confined to a two-dimensional square well, represented by the potential energy function $U(x,y,z) = U_1(x) + U_2(y) + U_3(z)$, where:

$$U_1(x) = 0 \quad \text{for} \quad 0 < x < \alpha; \quad U_1(x) = \infty \quad \text{otherwise}$$

$$U_2(y) = 0 \quad \text{for} \quad 0 < y < \beta; \quad U_2(x) = \infty \quad \text{otherwise}$$

$$U_3(z) = 0 \quad \text{for all } z.$$

(Answer): $\quad \psi_{n_1 n_2}(x,y) = \dfrac{2}{\sqrt{\alpha\beta}} \sin\left(\dfrac{n_1 \pi}{\alpha} x\right) \sin\left(\dfrac{n_2 \pi}{\beta} y\right), \quad E = \dfrac{\pi^2 \hbar^2}{2m} \left(\dfrac{n_1^2}{\alpha^2} + \dfrac{n_2^2}{\beta^2} \right)$

6.3. The Linear Harmonic Motion

There are only a few exactly soluble eigenvalue problems in one dimension, which rely on certain types of second order differential equations that often occur in quantum physics and have been studied in detail. The physical problem consists of finding solutions $\psi(x)$ that satisfy the admissibility conditions described in Section 6.1. If potential energy functions with discontinuities, like the square wells and barriers, are not involved, the continuity conditions for $\psi(x)$ and $d\psi/dx$ need no longer be considered, such that we are left with the finiteness requirement for all x. The admissible solutions $\psi(x)$ can be found in this case by a *polynomial method* in three stages:

(i) Determination of the approximate solution $\psi_\infty(x)$ valid in the regions $x \to \pm\infty$.

(ii) Factorization of $\psi_\infty(x)$ out of $\psi(x)$, which is set as $\psi(x) = u(x)\psi_\infty(x)$, and so $u(x)$ satisfies a second order differential equation of the general form:

$$\frac{d^2 u}{dx^2} + f(x)\frac{du}{dx} + g(x)u = 0 \tag{6.61}$$

If $f(x)$ and $g(x)$ are finite and continuous, it can be shown that such equations are solved by substituting a power series expansion of $u(x)$.

(iii) Enforcement of the finiteness condition at infinity, which makes it necessary to terminate the infinite series. This requirement determines the energy eigenvalue spectrum and the corresponding eigenfunctions.

Three-dimensional problems, involving partial differential equations that can be separated into a set of ordinary differential equations, will also be solved by the polynomial method, which provides in each case a specific class of *special functions*, appropriate to describe the bound eigenstates. We follow this procedure to describe the simple harmonic oscillator, which is an example of bound motion in one dimension, with quantum eigenstates defined in terms of specific special functions.

Substituting the governing potential energy $U(x) = m\omega^2 x^2/2$ into Eq.(6.4), which becomes:

$$\frac{d^2\psi(x)}{dx^2} + \frac{2m}{\hbar^2}\left(E - \frac{m\omega^2}{2}x^2\right)\psi(x) = 0 \tag{6.62}$$

it is convenient to make the equation dimensionless, by introducing the parameters:

$$\lambda = \frac{2E}{\hbar\omega}, \quad \xi = \sqrt{\frac{m\omega}{\hbar}}x \tag{6.63}$$

so that Eq.(6.62) reduces to:

$$\frac{d^2\psi(\xi)}{d\xi^2} + (\lambda - \xi^2)\psi(\xi) = 0 \tag{6.64}$$

Stage (i). For $x \to +\infty$ the term involving the constant λ becomes negligible, for any eigenvalue E, such that $\psi_\infty(\xi)$ is a solution to:

$$\frac{d^2\psi_\infty(\xi)}{d\xi^2} - \xi^2 \psi_\infty(\xi) = 0$$

On multiplication by $2d\psi_\infty/d\xi$ this becomes:

$$\frac{d}{d\xi}\left(\frac{d\psi_\infty}{d\xi}\right)^2 - \xi^2 \frac{d}{d\xi}(\psi_\infty^2) = 0 \quad \text{or} \quad \frac{d}{d\xi}\left[\left(\frac{d\psi_\infty}{d\xi}\right)^2 - \xi^2 \psi_\infty^2\right] = -2\xi\psi_\infty^2$$

We may drop the right-hand side, which is negligible for large values of ξ, as $2\xi\psi_\infty^2 \ll d(\xi^2\psi_\infty^2)/d\xi$, and the last equation reduces to:

$$\left(\frac{d\psi_\infty}{d\xi}\right)^2 = \text{const} + \xi^2 \psi_\infty^2 \quad \text{or} \quad \frac{d\psi_\infty}{d\xi} = \pm\xi\psi_\infty$$

where the constant of integration was set equal to zero, in order for $\psi_\infty(\xi)$ and its derivative to vanish at infinity. It follows that:

$$\psi_\infty(\xi) = e^{-\xi^2/2} \tag{6.65}$$

and the asymptotic inequality stated above is satisfied.

Stage (ii). We look for solutions of Eq.(6.64) where $\psi_\infty(\xi)$ factors out of $\psi(\xi)$:

$$\psi(\xi) = u(\xi) e^{-\xi^2/2} \tag{6.66}$$

and this leads to a differential equation for $u(\xi)$ of the form (6.61):

$$\frac{d^2 u(\xi)}{d\xi^2} - 2\xi \frac{du(\xi)}{d\xi} + (\lambda - 1)u(\xi) = 0 \tag{6.67}$$

This equation is solved by the power series:

$$u(\xi) = \sum_{k=0}^{\infty} a_k \xi^k \tag{6.68}$$

which gives:

$$\frac{du(\xi)}{d\xi} = \sum_{k=1}^{\infty} a_k k \xi^{k-1} \quad \text{and} \quad \frac{d^2 u(\xi)}{d\xi^2} = \sum_{k=2}^{\infty} a_k k(k-1) \xi^{k-2} = \sum_{k=0}^{\infty} a_{k+2}(k+2)(k+1)\xi^k$$

where we have redefined every index k for convenience. Substituting into Eq.(6.67) we obtain:

$$\sum_{k=0}^{\infty} \left[(k+1)(k+2) a_{k+2} - (2k+1-\lambda) a_k \right] \xi^k = 0$$

This can be true only if the coefficient of each power in ξ vanishes and so we obtain a recurrence relation:

$$a_{k+2} = \frac{2k+1-\lambda}{(k+1)(k+2)} a_k \quad \text{or} \quad a_{k+2} \cong \frac{2}{k} a_k \qquad (6.69)$$

The last equality only holds as $k \to \infty$ and is identical to the recurrence relation between successive terms of the power series:

$$e^{\xi^2} = \sum_{k=0,2,4,\ldots} \frac{1}{(k/2)!} \xi^k$$

and hence, $u(\xi)$ will tend to infinity with ξ, like e^{ξ^2}. As a result, $\psi(\xi)$ given by Eq.(6.66) will diverge like $e^{\xi^2/2}$.

Stage (iii). According to the finiteness requirement for $\psi(\xi)$, it is necessary that the power series (6.68) terminates after a finite number of terms. This can be accomplished by choosing λ such that the numerator of Eq.(6.69) vanishes for some finite value $k = n$, that is by choosing $\lambda = 2n + 1$. If this condition is combined with the definition of λ, Eq.(6.63), it follows that the total energy of the system is quantized according to:

$$E_n = \left(n + \frac{1}{2} \right) \hbar \omega \qquad n = 0,1,2,\ldots \qquad (6.70)$$

The term $n\hbar\omega$ gives the Planck series of energy levels, as postulated by Eq.(P.15) for the normal modes of thermal radiation. However, the minimum energy of the quantum oscillator is not zero, but takes the value $\hbar\omega/2$, known as the zero point energy, which is a manifestation of the uncertainty principle. To show this, consider a classical harmonic oscillator, bound to the origin, and thus described by:

$$x = A\cos(\omega t + \varphi), \quad p_x = m\frac{dx}{dt} = -m\omega A \sin(\omega t + \varphi)$$

which has average position and momentum equal to zero. This must also be true for the expectation values:

$$\langle \hat{x} \rangle_n = \langle \psi_n | \hat{x} | \psi_n \rangle = 0, \quad \langle \hat{p}_x \rangle_n = \langle \psi_n | \hat{p}_x | \psi_n \rangle = 0$$

in any oscillator eigenstate $\psi_n(x)$, as stated by the Ehernfest theorem, Eqs.(5.9) and (5.11). The quantum equation corresponding to the classical virial theorem (see Problem 2.1.1) shows that, for a potential energy $U(x)$ which is a second order homogenous function of position, we have:

$$\frac{1}{2m}\langle \hat{p}_x^2 \rangle_n = \frac{1}{2}m\omega^2 \langle \hat{x}^2 \rangle_n = \frac{1}{2}E_n = \frac{1}{2}\hbar\omega\left(n + \frac{1}{2}\right)$$

As a result, the uncertainties of the position and momentum for a quantum oscillator can be expressed as:

$$(\Delta p_x)_n^2 = \langle (\delta \hat{p}_x)^2 \rangle_n = \langle \hat{p}_x^2 \rangle_n - \langle \hat{p}_x \rangle_n^2 = m\hbar\omega\left(n + \frac{1}{2}\right)$$

$$(\Delta x)_n^2 = \langle (\delta \hat{x})^2 \rangle_n = \langle \hat{x}^2 \rangle_n - \langle \hat{x} \rangle_n^2 = \frac{\hbar}{m\omega}\left(n + \frac{1}{2}\right)$$

so that:

$$(\Delta x)_n (\Delta p_x)_n = \left(n + \frac{1}{2}\right)\hbar$$

As stated by the uncertainty principle, Eq.(4.78), this product must never be smaller that $\hbar/2$, even in the lowest energy state $n = 0$, where a zero point energy $\hbar\omega/2$ is thus required to exist.

Taking $\lambda = 2n + 1$, the eigenfunctions (6.66) can be expressed as:

$$\psi_n(\xi) = N_n e^{-\xi^2/2} H_n(\xi) \tag{6.71}$$

where N_n are normalization constants and $H_n(\xi)$ are the so-called *Hermite polynomial* solutions to Eq.(6.67), which reads:

$$\frac{d^2 H_n(\xi)}{d\xi^2} - 2\xi \frac{dH_n(\xi)}{d\xi} + 2nH_n(\xi) = 0 \tag{6.72}$$

232 Quantum Physics

EXAMPLE 6.6. The Hermite polynomials

The principle of superposition of the stationary states, Eq.(5.62), shows that, for a quantum oscillator, we have:

$$\Psi(\xi,t) = \sum_n a_n N_n e^{-i\omega t(n+1/2)} H_n(\xi) e^{-\xi^2/2} \tag{6.73}$$

where $\Psi(\xi,t)$ is a solution to the time-dependent Schrödinger equation (5.31), which in terms of ξ, introduced by Eq.(6.63), reads:

$$\frac{2i}{\omega}\frac{\partial \Psi(\xi,t)}{\partial t} = -\frac{\partial^2 \Psi(\xi,t)}{\partial \xi^2} + \xi^2 \Psi(\xi,t) \tag{6.74}$$

Because ξ is a dimensionless parameter, we look for an exponential solution:

$$\Psi(\xi,t) = e^{S(\xi,t)} = e^{\alpha\xi^2 + 2\beta\xi + \gamma} \tag{6.75}$$

where $\beta = \beta(t)$ and $\gamma = \gamma(t)$ are assumed to be time-dependent coefficients, which gives:

$$\frac{\partial \Psi}{\partial t} = \left(2\frac{d\beta}{dt}\xi + \frac{d\gamma}{dt}\right)\Psi \qquad \frac{\partial^2 \Psi}{\partial \xi^2} = \left[2\alpha + 4(\alpha\xi + \beta)^2\right]\Psi$$

Substituting these expressions into Eq.(6.74), we obtain:

$$\frac{2i}{\omega}\left(2\frac{d\beta}{dt}\xi + \frac{d\gamma}{dt}\right) + 2\alpha + 4(\alpha\xi + \beta)^2 - \xi^2 = 0$$

where the coefficients of the same powers of ξ should vanish. Hence, we obtain a set of equations:

$$4\alpha^2 - 1 = 0, \qquad \frac{i}{\omega}\frac{d\beta}{dt} + 2\alpha\beta = 0, \qquad \frac{i}{\omega}\frac{d\gamma}{dt} + \alpha + 2\beta^2 = 0$$

to be solved for α, β and γ. We first choose the negative solution $\alpha = -1/2$, because otherwise $\Psi(\xi,t)$ diverges for large ξ values. Then we determine β from:

$$\frac{d\beta}{dt} = -i\omega\beta \qquad \text{or} \qquad \beta = be^{-i\omega t} \tag{6.76}$$

and finally we obtain γ from:

$$\frac{d\gamma}{dt} = 2i\omega b^2 e^{-2i\omega t} - \frac{i\omega}{2} \qquad \text{or} \qquad \gamma = -\left(be^{-i\omega t}\right)^2 - \frac{i\omega t}{2} = -\beta^2 - \frac{i\omega t}{2}$$

where the constants of integration, corresponding to some phase factors, have been neglected. Substituting the calculated coefficients into Eq.(6.75), we obtain $\Psi(\xi,t)$ in the form:

6. One-Dimensional Motion

$$\Psi(\xi,t) = e^{-\xi^2/2 - i\omega t/2 + 2\beta\xi - \beta^2} \tag{6.77}$$

Comparison with Eq.(6.73) suggests that the factor containing the variable β should be expanded in a power series, with coefficients that are polynomials in ξ of order n:

$$e^{2\beta\xi - \beta^2} = \sum_{n=0}^{\infty} \frac{\beta^n}{n!} H_n(\xi) \tag{6.78}$$

We can rewrite this equation in a more useful form, leading to explicit expressions for $H_n(\xi)$, by observing that:

$$e^{\xi^2} e^{-(\xi-\beta)^2} = \sum_{n=0}^{\infty} \frac{\beta^n}{n!} H_n(\xi)$$

and therefore:

$$H_n(\xi) = e^{\xi^2} \left[\frac{\partial^n}{\partial \beta^n} e^{-(\xi-\beta)^2} \right]_{\beta=0} = e^{\xi^2} \left[(-1)^n \frac{\partial^n}{\partial \xi^n} e^{-(\xi-\beta)^2} \right]_{\beta=0} = (-1)^n e^{\xi^2} \frac{d^n}{d\xi^n} e^{-\xi^2} \tag{6.79}$$

The first few polynomials are obtained from Eq.(6.79) as:

$$H_0(\xi) = 1 \qquad H_1(\xi) = 2\xi$$
$$\tag{6.80}$$
$$H_2(\xi) = 4\xi^2 - 2 \qquad H_3(\xi) = 8\xi^3 - 12\xi$$

and it is clear that $H_n(\xi)$ are either odd or even functions. Because the function $v = (-1)^n e^{-\xi^2}$ satisfies the differential equation $dv/d\xi + 2\xi v = 0$ we obtain, by differentiating this expression $(n+1)$ times:

$$\frac{d^2}{d\xi^2}\left(\frac{d^n v}{d\xi^n}\right) + 2\xi \frac{d}{d\xi}\left(\frac{d^n v}{d\xi^n}\right) + 2(n+1)\frac{d^n v}{d\xi^n} = 0 \tag{6.81}$$

and, from Eq.(6.79), we have:

$$H_n(\xi) = e^{\xi^2} \frac{d^n v}{d\xi^n} \quad \text{or} \quad \frac{d^n v}{d\xi^n} = e^{-\xi^2} H_n(\xi)$$

Substituting this expression into Eq.(6.81) shows that the polynomials $H_n(\xi)$, defined by the *generating function* introduced in Eq.(6.78), satisfy the differential equation (6.72), and hence are the Hermite polynomials.

Now substituting Eqs.(6.78) and (6.76) into Eq.(6.77) gives:

$$\Psi(\xi,t) = \sum_{n=0}^{\infty} \frac{\beta^n}{n!} e^{-\xi^2/2-i\omega t/2} H_n(\xi) = \sum_{n=0}^{\infty} \frac{b^n}{n!} e^{-i\omega t(n+1/2)} H_n(\xi) e^{-\xi^2/2} \quad (6.82)$$

and, by comparing this expression with Eq.(6.73), it follows that the time-dependent stationary states of the quantum oscillator are expressed in terms of the Hermite polynomials as:

$$\Psi_n(\xi,t) = N_n e^{-i\omega t(n+1/2)} H_n(\xi) e^{-\xi^2/2} \quad (6.83)$$

which corresponds to the stationary state eigenfunctions (6.71), obtained by solving the Schrödinger wave equation using the polynomial method. ♣

The eigenfunctions of the harmonic oscillator form an orthonormal set, as it can be shown by using the generating function of the Hermite polynomials, rewritten as:

$$N_n e^{-\xi^2/2+2\beta\xi-\beta^2} = \sum_{n=0}^{\infty} \frac{\beta^n}{n!} \psi_n(\xi)$$

which reduces the orthonormality condition to a Gaussian integral:

$$\sum_{n=0}^{\infty} \sum_{m=0}^{\infty} \frac{\beta^n}{n!} \frac{\delta^m}{m!} \int_{-\infty}^{\infty} \psi_n(\xi) \psi_m(\xi) d\xi = N_n^2 \int_{-\infty}^{\infty} e^{-(\xi^2-2\beta\xi-2\delta\xi+\beta^2+\delta^2)} d\xi$$

$$= N_n^2 e^{2\beta\delta} \int_{-\infty}^{\infty} e^{-(\xi-\beta-\delta)^2} d(\xi-\beta-\delta) = N_n^2 e^{2\beta\delta} \sqrt{\pi}$$

If $e^{2\beta\delta}$ is also expanded into a power series, we obtain an identity in β and δ:

$$\sum_{n=0}^{\infty} \sum_{m=0}^{\infty} \frac{\beta^n}{n!} \frac{\delta^m}{m!} \int_{-\infty}^{\infty} \psi_n(\xi) \psi_m(\xi) d\xi = N_n^2 \sqrt{\pi} \sum_{n=0}^{\infty} \frac{(2\beta\delta)^n}{n!}$$

and it follows that the wave functions are orthogonal for $n \neq m$, and, for $m = n$, we have $2^n n! \sqrt{\pi} N_n^2 = 1$. This determines the orthonormal eigenfunctions to be:

$$\psi_n(\xi) = \left(2^n n! \sqrt{\pi}\right)^{-\frac{1}{2}} H_n(\xi) e^{-\xi^2/2} \quad (6.84)$$

The probability densities $|\psi_n(\xi)|^2$, which describe the probability of finding the quantum oscillator at various distances $\xi = \sqrt{m\omega/\hbar}\, x$ from the origin, are plotted in Figure 6.9, using the Hermite polynomials given by Eq.(6.80). It is clear that the classical restriction (6.9) which reads:

$$E - \frac{\hbar\omega}{2}\xi^2 \geq 0 \quad \text{or} \quad -\sqrt{2n+1} \leq \xi \leq \sqrt{2n+1}$$

no longer exists.

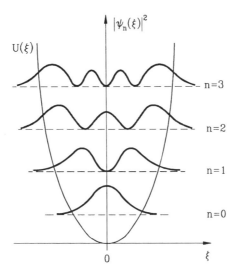

Figure 6.9. Probability densities of the harmonic oscillator for the lowest stationary states

The quantum oscillator can be found outside the parabolic potential energy barrier $U(\xi)$, in a region that is forbidden for the classical oscillator. However, the probability to find ξ in the classically allowed region is close to unity.

EXAMPLE 6.7. The harmonic oscillator in the Heisenberg description

The eigenvalues and eigenfunctions of the quantum oscillator, obtained as a result of the finiteness condition imposed to the stationary state functions in the Schrödinger description, can also be derived using the operator formalism of the Heisenberg description. As the Hamiltonian of a classical oscillator may be factorized into:

$$H = \frac{p_x^2}{2m} + \frac{m\omega^2}{2}x^2 = \omega\left(\sqrt{\frac{m\omega}{2}}x - i\frac{p_x}{\sqrt{2m\omega}}\right)\left(\sqrt{\frac{m\omega}{2}}x + i\frac{p_x}{\sqrt{2m\omega}}\right) \tag{6.85}$$

it follows that the representative operators $\hat{x}$ and $\hat{p}_x$ must be related by:

$$\omega\left(\sqrt{\frac{m\omega}{2}}\hat{x} - i\frac{\hat{p}_x}{\sqrt{2m\omega}}\right)\left(\sqrt{\frac{m\omega}{2}}\hat{x} + i\frac{\hat{p}_x}{\sqrt{2m\omega}}\right) = \frac{\hat{p}_x^2}{2m} + \frac{m\omega}{2}\hat{x}^2 - \frac{i\omega}{2}(\hat{p}_x\hat{x} - \hat{x}\hat{p}_x) = \hat{H} - \frac{1}{2}\hbar\omega$$

where we made use of the commutation relations (4.40). It is convenient to introduce the operators:

$$\hat{a} = \sqrt{\frac{m\omega}{2}}\hat{x} + i\frac{\hat{p}_x}{\sqrt{2m\omega}}$$

$$\hat{a}^+ = \sqrt{\frac{m\omega}{2}}\hat{x} - i\frac{\hat{p}_x}{\sqrt{2m\omega}}$$

(6.86)

where $\hat{a}^+$ is the adjoint of $\hat{a}$. Thus the Hamiltonian takes the form:

$$\hat{H} = \omega \hat{a}^+ \hat{a} + \frac{1}{2}\hbar\omega \qquad (6.87)$$

and the following commutation relations are immediately apparent:

$$[\hat{a},\hat{a}^+] = \left[\sqrt{\frac{m\omega}{2}}\hat{x}, -i\frac{\hat{p}_x}{\sqrt{2m\omega}}\right] + \left[i\frac{\hat{p}_x}{\sqrt{2m\omega}}, \sqrt{\frac{m\omega}{2}}\hat{x}\right] = \hbar$$

$$[\hat{H},\hat{a}] = [\omega \hat{a}^+ \hat{a}, \hat{a}] = \omega[\hat{a}^+,\hat{a}]\hat{a} = -\hbar\omega\hat{a} \qquad (6.88)$$

$$[\hat{H},\hat{a}^+] = [\omega \hat{a}^+ \hat{a}, \hat{a}^+] = \omega \hat{a}^+[\hat{a},\hat{a}^+] = \hbar\omega\hat{a}^+$$

If ψ_n is an eigenfunction of $\hat{H}$ corresponding to the eigenvalue E_n:

$$\hat{H}\psi_n = E_n \psi_n$$

by applying the operator $\hat{a}$ to this eigenvalue equation, we obtain:

$$\hat{a}\hat{H}\psi_n = (\hat{H}\hat{a} + \hbar\omega\hat{a})\psi_n = (\hat{H} + \hbar\omega)\hat{a}\psi_n = E_n \hat{a}\psi_n$$

where use has been made of Eq.(6.88). It follows that:

$$\hat{H}(\hat{a}\psi_n) = (E_n - \hbar\omega)(\hat{a}\psi_n)$$

which means that, if ψ_n is an eigenfunction of $\hat{H}$, then $\hat{a}\psi_n$ is also eigenfunction of $\hat{H}$ but corresponds to an eigenvalue lowered by $\hbar\omega$. In the same way we can show that:

$$\hat{H}(\hat{a}^+\psi_n) = (E_n + \hbar\omega)(\hat{a}^+\psi_n)$$

that is $\hat{a}^+\psi_n$ is also an eigenfunction of $\hat{H}$ with the eigenvalue raised by $\hbar\omega$. The operators $\hat{a}$ and $\hat{a}^+$ are accordingly known as *lowering* and *raising* operators. Denoting by ψ_0 the ground state function, corresponding to the minimum eigenvalue E_0, the condition:

$$\hat{a}\psi_0 = 0$$

6. One-Dimensional Motion

must be true. If this condition is substituted into the eigenvalue equation, the energy of the ground state E_0 is obtained as:

$$\hat{H}\psi_0 = \left(\omega \hat{a}^+\hat{a} + \frac{1}{2}\hbar\omega\right)\psi_0 = \frac{1}{2}\hbar\omega\psi_0 \quad \text{or} \quad E_0 = \frac{1}{2}\hbar\omega$$

In view of Eq.(6.86), where the momentum operator is substituted in coordinate representation, the same condition gives the form of the ground state function:

$$\left(m\omega x + \hbar\frac{d}{dx}\right)\psi_0 = 0 \quad \text{or} \quad \psi_0(x) = N_0 e^{-m\omega x^2/2\hbar}$$

which includes N_0 as a normalization constant. The complete set of eigenfunctions can be generated by successive raising operations on ψ_0:

$$\hat{H}(\hat{a}^+)^n\psi_0 = (E_0 + n\hbar\omega)(\hat{a}^+)^n\psi_0$$

It follows that the energy eigenvalues are given by:

$$E_n = E_0 + n\hbar\omega = \left(n + \frac{1}{2}\right)\hbar\omega, \quad n = 0, 1, 2, \ldots \quad (6.89)$$

and the eigenfunctions have the form:

$$\psi_n(x) = N_n(\hat{a}^+)^n\psi_0 = N_n'\left(m\omega x - \hbar\frac{d}{dx}\right)^n e^{-m\omega x^2/2\hbar} \quad (6.90)$$

where N_n, N_n' denote normalization constants. A comparison of Eqs.(6.89) and (6.90) with Eqs.(6.70) and (6.71) shows that these results are identical to those derived in the Schrödinger description of the linear harmonic motion. This can be made apparent by using the expression (6.79) of the Hermite polynomials, in view of the auxiliary identity, valid for any function $f(\xi)$:

$$\frac{d}{d\xi}[e^{-\xi^2/2}f(\xi)] = -e^{-\xi^2/2}\left(\xi - \frac{d}{d\xi}\right)f(\xi) \quad (6.91)$$

The operators $\hat{a}^+$ and $\hat{a}$, which raise and respectively lower the energy of eigenstates by quanta $\hbar\omega$ are often called *ladder operators*, since they raise and respectively lower the eigenfunctions up and down the ladder of energy eigenvalues. On substitution of Eqs.(6.87) and (6.89) into the eigenvalue equation, we obtain:

$$\hat{a}^+\hat{a}\psi_n = n\hbar\psi_n, \quad n = 0, 1, 2, \ldots \quad (6.92)$$

If the eigenfunctions $\psi_n(x)$ are assumed to be normalized, the lowering operation reads:

$$\hat{a}\psi_n = C_n\psi_{n-1} \quad (6.93)$$

where C_n is a complex number. When multiplying Eq.(6.93) by $\hat{a}^+$ we obtain, in view of Eq.(6.92):

$$\hat{a}^+\hat{a}\psi_n = C_n\hat{a}^+\psi_{n-1} = n\hbar\,\psi_n \quad \text{or} \quad \hat{a}^+\psi_{n-1} = \frac{n\hbar}{C_n}\psi_n \qquad (6.94)$$

Since $\hat{a}^+$ is the adjoint of $\hat{a}$, we have:

$$\langle\psi_{n-1}|\hat{a}\psi_n\rangle = \langle\hat{a}^+\psi_{n-1}|\psi_n\rangle \quad \text{or} \quad C_n\langle\psi_{n-1}|\psi_{n-1}\rangle = \frac{n\hbar}{C_n^*}\langle\psi_n|\psi_n\rangle$$

which gives $|C_n|^2 = n\hbar$. It is convenient to choose C_n to be a real positive number, up to phase factor, such that the recurrence relations (6.93) and (6.94) become:

$$\hat{a}\psi_n = \sqrt{n\hbar}\,\psi_{n-1}, \quad \hat{a}^+\psi_{n-1} = \sqrt{n\hbar}\,\psi_n \qquad (6.95)$$

As $\hat{x}$ and $\hat{p}_x$ can be expressed in terms of $\hat{a}$ and $\hat{a}^+$ from the definition (6.86), Eqs.(6.95) can be used to calculate the effect of any function of position or momentum on the energy eigenfunctions of the harmonic oscillator. ✣

Problem 6.3.1. Show that, if a simple harmonic oscillator of mass m and charge q is subjected to a uniform electric field of intensity F along the x-axis, its energy levels are shifted by $q^2F^2/2m\omega^2$.

(Solution): The potential energy of the harmonic oscillator can be written in this case as:

$$U(x) = \frac{m\omega^2}{2}x^2 + qFx = \frac{m\omega^2}{2}\left(x + \frac{qF}{m\omega^2}\right)^2 - \frac{q^2F^2}{2m\omega^2}$$

such that the Schrödinger wave equation reads:

$$\frac{d^2\psi}{dx^2} + \frac{2m}{\hbar^2}\left[E + \frac{q^2F^2}{2m\omega^2} - \frac{m\omega^2}{2}\left(x + \frac{qF}{m\omega^2}\right)^2\right]\psi = 0$$

Hence, it is convenient to set:

$$E_q = E + \frac{q^2F^2}{2m\omega^2}, \quad x_q = x + \frac{qF}{m\omega^2}$$

so reducing the eigenvalue problem to that of a simple harmonic oscillator, Eq.(6.62):

$$\frac{d^2\psi}{dx_q^2} + \frac{2m}{\hbar^2}\left(E_q - \frac{m\omega^2}{2}x_q^2\right)\psi = 0$$

It follows that the eigenvalues E_q are given by Eq.(6.70), and thus:

$$E = \left(n+\frac{1}{2}\right)\hbar\omega - \frac{q^2F^2}{2m\omega^2}$$

Problem 6.3.2. Use the WKB approximation to calculate the energy eigenvalues of a harmonic oscillator.

(Solution): We use the quantization rule derived in Example 6.2 from the WKB connection relations, with $U(x) = m\omega^2 x^2/2$:

$$2\int_{x_0}^{x_1} \sqrt{2m(E - m\omega^2 x^2/2)}\, dx = \left(n+\frac{1}{2}\right)h$$

where the integration limits are given by:

$$E - \frac{m\omega^2}{2}x_0^2 = E - \frac{m\omega^2}{2}x_1^2 = 0 \quad \text{or} \quad x_1 = -x_0 = \sqrt{\frac{2E}{m\omega^2}}$$

We may set for simplicity $x = x_1 u$, such that the integral becomes:

$$2x_1\sqrt{2mE}\int_{-1}^{1}\sqrt{1-u^2}\,du = 4\frac{E}{\omega}\frac{\pi}{2} = 2\pi\frac{E}{\omega}$$

Substituting into the quantization rule gives:

$$E = \left(n+\frac{1}{2}\right)\hbar\omega$$

In other words the WKB approximation gives, in this special case, the exact solution for the energy levels.

We notice that, if $U(x) = \infty$ for $x \leq 0$, the WKB quantization rule should be that given in Problem 6.1.5. In this case the previous integral is evaluated for $x_0 = 0$ and one obtains:

$$E = \left(2n + \frac{3}{2}\right)\hbar\omega = \left[(2n+1) + \frac{1}{2}\right]\hbar\omega$$

This has a straightforward interpretation, because the eigenfunctions of the harmonic oscillator must, in this case, vanish for $x < 0$, where $U(x) = \infty$, and coincide with those of the simple harmonic motion for $x \geq 0$. These means that the admissible eigenstates will be defined by the odd eigenfunctions of the simple harmonic oscillator only, which correspond to odd quantum numbers, $n \to 2n + 1$.

Problem 6.3.3. Find the energy levels of the harmonic oscillator in three dimensions described by the Hamiltonian:

$$\hat{H} = \frac{\hat{p}^2}{2m} + \frac{m}{2}\left(\omega_1^2 x^2 + \omega_2^2 y^2 + \omega_3^2 z^2\right)$$

and show that there is degeneracy when the oscillator is isotropic ($\omega_1 = \omega_2 = \omega_3$).

(Solution): The Schrödinger wave equation can be separated into three ordinary differential equations (see Problem 5.4.2):

$$\frac{d^2\psi_i(x_i)}{dx_i^2} + \frac{2m}{\hbar^2}\left(E_i - \frac{m\omega_i^2}{2}x_i^2\right)\psi_i(x_i) = 0$$

by taking:

$$\psi(x, y, z) = \psi_1(x_1)\psi_2(x_2)\psi_3(x_3), \qquad E = E_1 + E_2 + E_3$$

such that, from Eq.(6.70), we obtain:

$$E_{n_1,n_2,n_3} = \left(n_1 + \frac{1}{2}\right)\hbar\omega_1 + \left(n_2 + \frac{1}{2}\right)\hbar\omega_2 + \left(n_3 + \frac{1}{2}\right)\hbar\omega_3$$

For an isotropic oscillator the eigenvalues will be given by:

$$E_n = \left(n + \frac{3}{2}\right)\hbar\omega \qquad n = n_1 + n_2 + n_3$$

and the levels are degenerate. Assuming that n_3 is fixed, the sum $n_1 + n_2$ takes the definite value $n - n_3$, and as n_1 is given values from zero to $n - n_3$, we obtain $n - n_3 + 1$ values of the sum, which completely determine n_1 and n_2. Hence, for each n_3 value there will be $n - n_3 + 1$ distinct values for n, and thus for E_n. Summing over all the allowed n_3 values, from 0 to n we obtain:

$$\sum_{n_3=0}^{n}(n-n_3+1) = \sum_{n_3=0}^{n}(n+1) - \sum_{n_3=0}^{n}n_3 = (n+1)^2 - \frac{n(n+1)}{2} = \frac{1}{2}(n+1)(n+2)$$

which is the degree of degeneracy of E_n.

Problem 6.3.4. Prove the recurrence relation of the Hermite polynomials:

$$2xH_n(x) = H_{n+1}(x) + 2nH_{n-1}(x)$$

and use it to show that the matrix elements of the operator $\hat{x}$, relative to a basis defined by the energy eigenfunctions $\psi_n(x)$, are:

$$x_{jk} = \int_{-\infty}^{\infty} \psi_j^* x \psi_k \, dx = \left(\frac{\hbar}{2m\omega}\right)^{\frac{1}{2}} \left[\sqrt{j}\,\delta_{j,k+1} + \sqrt{j+1}\,\delta_{j,k-1}\right]$$

Problem 6.3.5. Prove the recurrence relation of the Hermite polynomials:

$$\frac{dH_n(x)}{dx} = 2nH_{n-1}(x)$$

and then show that the matrix elements of momentum are given for the harmonic oscillator by:

$$p_{jk} = -i\hbar \int_{-\infty}^{\infty} \psi_j^* \frac{d\psi_k}{dx} dx = i\left(\frac{m\hbar\omega}{2}\right)^{\frac{1}{2}} \left[\sqrt{j}\,\delta_{j,k+1} - \sqrt{j+1}\,\delta_{j,k-1}\right]$$

Problem 6.3.6. Show that the energy of a harmonic oscillator can be expressed as:

$$E_n = m\omega^2 \langle x^2 \rangle_n = \frac{1}{m}\langle p_x^2 \rangle_n$$

Problem 6.3.7. Show that, if $\hat{A}$ is an operator which contains p factors $\hat{a}^+$ and q factors $\hat{a}$, in any order, its expectation value $\langle \psi_n | \hat{A} | \psi_n \rangle$ vanishes unless $p = q$.

7. ELECTRON MOTION IN THE ATOM

7.1. Rotational Motion for a Single Particle

Spherically symmetric systems, where the potential energy $U(r)$ is independent of the direction of $\vec{r}$, are usually treated using spherical polar coordinates (r,θ,φ), related to the Cartesian coordinates as shown by Eqs.(2.20) and (2.21). Therefore, in view of the appropriate form (2.36) of the Hamiltonian, the Schrödinger wave equation for $\Psi(r,\theta,\varphi)$ reads:

$$\hat{H}\Psi = -\frac{\hbar^2}{2m}\left[\frac{1}{r^2}\frac{\partial}{\partial r}\left(r^2\frac{\partial \Psi}{\partial r}\right) + \frac{1}{r^2\sin\theta}\frac{\partial}{\partial \theta}\left(\sin\theta\frac{\partial \Psi}{\partial \theta}\right) + \frac{1}{r^2\sin^2\theta}\frac{\partial^2 \Psi}{\partial \varphi^2}\right] + U(r)\Psi = E\Psi \quad (7.1)$$

In contrast to the motion of a particle whose potential energy depends on Cartesian coordinates, where the Schrödinger equation can be separated into three equations pertaining to motions along the axes x, y, z, which are independent of one another (see Problem 5.4.2), Eq.(7.1) also separates into equations pertaining to motions along the coordinates r, θ and φ, but the three types of motion are no longer independent. The separation of variables is made in two stages, first substituting into the energy eigenvalue equation the trial solution:

$$\Psi(r,\theta,\varphi) = R(r)Y(\theta,\varphi) \quad (7.2)$$

This transforms Eq.(7.1) in the form:

$$\frac{\hbar^2}{2m}\left[\frac{1}{R}\frac{\partial}{\partial r}\left(r^2\frac{\partial R}{\partial r}\right)\right] + r^2[E - U(r)] = -\frac{\hbar^2}{2m}\frac{1}{Y}\left[\frac{1}{\sin\theta}\frac{\partial}{\partial \theta}\left(\sin\theta\frac{\partial Y}{\partial \theta}\right) + \frac{1}{\sin^2\theta}\frac{\partial^2 Y}{\partial \varphi^2}\right] \quad (7.3)$$

where one side is independent of θ and φ, and the other side is independent of r, so that they must be separately equal to some constant, which is conveniently chosen as $L^2/2m$. In this way we separate the *radial motion*, described by an ordinary differential equation where $\partial/\partial r$ is replaced by d/dr:

$$\left[-\frac{\hbar^2}{2m}\frac{1}{r^2}\frac{d}{dr}\left(r^2\frac{d}{dr}\right)+U(r)+\frac{L^2}{2mr^2}\right]R(r)=ER(r) \tag{7.4}$$

from the *rotational motion*, described by the angular equation:

$$-\hbar^2\left[\frac{1}{\sin\theta}\frac{\partial}{\partial\theta}\left(\sin\theta\frac{\partial}{\partial\theta}\right)+\frac{1}{\sin^2\theta}\frac{\partial^2}{\partial\varphi^2}\right]Y=L^2Y \quad\text{or}\quad \hat{L}^2Y=L^2Y \tag{7.5}$$

which is the eigenvalue problem for orbital angular momentum, represented by the operator (2.33). This separation corresponds to the classical splitting of kinetic energy among the two types of motion. The second stage of the separation process assumes that:

$$Y(\theta,\varphi)=P(\theta)\Phi(\varphi) \tag{7.6}$$

such that Eq.(7.5) becomes:

$$\hbar^2\left[\frac{\sin\theta}{P(\theta)}\frac{\partial}{\partial\theta}\left(\sin\theta\frac{\partial P(\theta)}{\partial\theta}\right)+\frac{L^2}{\hbar^2}\sin^2\theta\right]=-\hbar^2\frac{1}{\Phi(\varphi)}\frac{\partial^2\Phi(\varphi)}{\partial\varphi^2} \tag{7.7}$$

Since one side is independent of φ, while the other is independent of θ, they must each be equal to a constant, which we call L_z^2, and this yields two other ordinary differential equations:

$$-\hbar^2\frac{d^2}{d\varphi^2}\Phi(\varphi)=L_z^2\Phi(\varphi) \quad\text{or}\quad \hat{L}_z^2\Phi(\varphi)=L_z^2\Phi(\varphi) \tag{7.8}$$

where $\hat{L}_z$ is defined by Eq.(2.25), and:

$$\frac{1}{\sin\theta}\frac{d}{d\theta}\left(\sin\theta\frac{dP(\theta)}{d\theta}\right)+\left(\frac{L^2}{\hbar^2}-\frac{L_z^2}{\hbar^2\sin^2\theta}\right)P(\theta)=0 \tag{7.9}$$

In other words the rotational motion is studied by splitting the kinetic energy of rotation into components of motion along parallel and meridian lines, described by Eqs.(7.8) and (7.9) respectively. Each of the equations (7.1), (7.5) and (7.8) is an eigenvalue equation for the operators $\hat{H}, \hat{L}^2$ and $\hat{L}_z$ respectively, which represent compatible observables in the central force field. This is implied by their commutation relations (see Example 4.2), which also show that L^2 and L_z are constants of motion, as defined by Eq.(5.19). The eigenvalue problem for $\hat{L}_z$, Eq.(7.8), actually depends on the φ coordinate only and should be solved first. Then, we can solve the equation of motion (7.9) along θ, which involves as a parameter the eigenvalue L_z^2 of the motion along φ, and provides the

244 Quantum Physics

magnitude L^2 of the orbital angular momentum, that is the kinetic energy of rotation. The energy eigenvalues E are finally obtained from the radial equation (7.4), where L^2 is a parameter.

Eigenstates of motion along a parallel line are solutions to the eigenvalue equation for $\hat{L}_z$, which reads:

$$\hat{L}_z \Phi(\varphi) = -i\hbar \frac{d}{d\varphi} \Phi(\varphi) = L_z \Phi(\varphi) \tag{7.10}$$

and hence the eigenfunctions have the form:

$$\Phi(\varphi) = e^{iL_z \varphi / \hbar} \tag{7.11}$$

which also satisfies Eq.(7.8). Because all the values of the angular coordinate φ, differing by a multiple of 2π radians, represent the same point, $\Phi(\varphi)$ is subjected to the single-valuedness condition:

$$e^{iL_z(\varphi+2\pi)/\hbar} = e^{iL_z \varphi / \hbar} \quad \text{or} \quad e^{iL_z 2\pi/\hbar} = 1$$

which determines a discrete spectrum of the eigenvalue L_z:

$$L_z = m\hbar \qquad m = 0, \pm 1, \pm 2 \tag{7.12}$$

This quantum condition for the z-component of the orbital angular momentum is formulated in terms of the integer m, known as the *magnetic quantum number*, and consequently, the eigenfunctions (7.11) have the form:

$$\Phi_m(\varphi) = \frac{1}{\sqrt{2\pi}} e^{im\varphi} \tag{7.13}$$

which has been normalized by integration over the variable φ, between 0 and 2π. In other words, there is a discrete set of eigenstates of motion along φ.

Substituting L_z given by Eq.(7.12), the equation of motion along meridian lines, Eq.(7.9), becomes:

$$\frac{1}{\sin\theta} \frac{d}{d\theta} \left(\sin\theta \frac{dP(\theta)}{d\theta} \right) + \left(\frac{L^2}{\hbar^2} - \frac{m^2}{\sin^2\theta} \right) P(\theta) = 0 \tag{7.14}$$

and is made simpler if we set:

$$u = \cos\theta \quad \text{or} \quad \frac{d}{d\theta} = -\sin\theta \frac{d}{du}$$

This immediately gives:

7. Electron Motion in the Atom

$$\frac{d}{du}\left[(1-u^2)\frac{dP(u)}{du}\right] + \left(\frac{L^2}{\hbar^2} - \frac{m^2}{1-u^2}\right)P(u) = 0 \tag{7.15}$$

and it is convenient to find first the particular solutions of this equation for $m = 0$, which describe the eigenstates of no motion along φ. In this case we have:

$$\frac{d}{du}\left[(1-u^2)\frac{dP(u)}{du}\right] + \frac{L^2}{\hbar^2}P(u) = 0 \tag{7.16}$$

and the admissible solutions $P(u) = P(\cos\theta)$ can be obtained by the polynomial method described in Section 6.3. We assume that:

$$P(u) = \sum_{k=0}^{\infty} a_k u^k, \quad \frac{dP(u)}{du} = \sum_{k=1}^{\infty} k a_k u^{k-1} \tag{7.17}$$

and substitution into Eq.(7.16) yields:

$$\sum_{k=0}^{\infty}(k+1)(k+2)a_{k+2}u^k - \sum_{k=0}^{\infty}k(k+1)a_k u^k + \frac{L^2}{\hbar^2}\sum_{k=0}^{\infty}a_k u^k = 0$$

The condition that the coefficient of each power should be zero, leads to a recurrence relation:

$$a_{k+2} = \frac{k(k+1) - L^2/\hbar^2}{(k+1)(k+2)} a_k \quad \text{or} \quad \lim_{k\to\infty}\frac{a_{k+2}}{a_k} = 1 \tag{7.18}$$

For large k values, this equation becomes identical to the recurrence relation between successive terms of the power expansion of the function $(1-u^2)^{-1}$, which is divergent for $u = 1$. The finiteness requirement implies that the series must terminate for a finite value $k = l$, for instance by taking $a_l \neq 0$ and $a_{l+2} = 0$ in Eq.(7.18), which gives:

$$L^2 = l(l+1)\hbar^2 \quad l = 0, 1, 2, \ldots \tag{7.19}$$

This is a second quantum condition for the orbital angular momentum, which introduces the integer l called the *orbital quantum number*. The series (7.17) is so reduced to a polynomial of order l which satisfies the equation:

$$\frac{d}{du}\left[(1-u^2)\frac{dP_l(u)}{du}\right] + l(l+1)P_l(u) = (1-u^2)\frac{d^2P_l(u)}{du^2} - 2u\frac{dP_l(u)}{du} + l(l+1)P_l(u) = 0 \tag{7.20}$$

called the *Legendre equation*.

EXAMPLE 7.1. Legendre polynomials

A set of polynomials that are solutions to Eq.(7.20), called the *Legendre polynomials*, are given by the Rodrigue formula:

$$P_l(u) = \frac{1}{2^l l!} \frac{d^l}{du^l}(u^2-1)^l \qquad (7.21)$$

from which the Legendre equation is easily derived. To this end we consider the function $f(u) = (-1)^l (1-u^2)^l$ which satisfies the equation:

$$(1-u^2)\frac{df(u)}{du} + 2lu\, f(u) = 0$$

where $(l+1)$ times differentiation gives:

$$(1-u^2)\frac{d^{l+2}f(u)}{du^{l+2}} - 2u\frac{d^{l+1}f(u)}{du^{l+1}} + l(l+1)\frac{d^l f(u)}{du^l} = 0$$

This is the same as Eq.(7.20), because $P_l(u) = (d^l f/du^l)/2^l l!$. The Legendre polynomials (7.21) form an orthogonal set of functions for real values of u in the interval $-1 \leq u \leq 1$, corresponding to the motion along meridian lines ($u = \cos\theta$), as it can be shown, through l times integration by parts, that:

$$\int_{-1}^{1} P_l(u) P_{l'}(u)\, du = \frac{2}{2l+1}\delta_{ll'} \qquad (7.22)$$

The first few Legendre polynomials have the form:

$$P_0(u) = 1 \qquad\qquad P_1(u) = u$$

$$P_2(u) = \frac{1}{2}(3u^2 - 1) \qquad P_3(u) = \frac{1}{2}(5u^3 - 3u) \qquad (7.23)$$

$$P_4(u) = \frac{1}{8}(35u^4 - 30u^2 + 3) \qquad P_5(u) = \frac{1}{8}(63u^5 - 70u^3 + 15u)$$

and exhibit the following properties that are often used:

$$P_l(1) = 1, \qquad P_l(-u) = (-1)^l P_l(u) \qquad (7.24)$$

The power expansion formula:

$$(u^2 - 1)^l = \sum_{s=0}^{l} \frac{(-1)^s l!}{s!(l-s)!} u^{2l-2s}$$

can be differentiated k times to obtain, in view of Eq.(7.21):

$$P_l(u) = \frac{1}{2^l l!} \frac{d^l}{du^l}(u^2-1)^l = \frac{1}{2^l} \sum_{s=0}^{k} \frac{(-1)^s (2l-2s)! \, u^{l-2s}}{s!(l-s)!(l-2s)!}$$

where $k = l/2$, when l is even, and $k = (l-1)/2$, when l is odd. The same result can be derived, if the Legendre polynomials are defined as the coefficients of the power series:

$$G(u,\beta) = (1 - 2\beta u + \beta^2)^{-\frac{1}{2}} = \sum_{l=0}^{\infty} P_l(u)\beta^l \tag{7.25}$$

in other words, if we use a *generating function* $G(u,\beta)$ for $P_l(u)$, as usual in the study of special functions. Some recurrence relations are immediate from Eq.(7.25), and can also be derived from the definition (7.21). For instance, the generating function satisfies the equation:

$$(1 - 2\beta u + \beta^2)\frac{\partial G}{\partial \beta} = (u - \beta)G$$

where, substituting the expansion (7.25), we find:

$$(2l+1)u \, P_l(u) = l P_{l-1}(u) + (l+1) P_{l+1}(u) \tag{7.26}$$

Similarly, using the property:

$$\beta \frac{\partial G}{\partial \beta} = (u - \beta)\frac{\partial G}{\partial u}$$

we obtain:

$$l \, P_l(u) = u \frac{dP_l(u)}{du} - \frac{dP_{l-1}(u)}{du} \tag{7.27}$$

and additional recurrence relations can be further derived by combining Eqs.(7.26) and (7.27).

The eigenfunctions of Eq.(7.15), for non-zero values of m, are given by the so-called *associated Legendre equation*, which is obtained, after inserting the quantum condition (7.19), in the form:

$$\frac{d}{du}\left[(1-u^2)\frac{dP(u)}{du}\right] + \left[l(l+1) - \frac{m^2}{1-u^2}\right]P(u) = 0 \tag{7.28}$$

A trial solution is:

$$P(u) = (1 - u^2)^{m/2} F(u) \tag{7.29}$$

where the function $F(u)$ will be related to the Legendre polynomials. Substituting $P(u)$ and its derivatives $dP(u)/du$, $d^2P(u)/du^2$ into Eq.(7.28), the equation obeyed by $F(u)$ is obtained as:

$$(1-u^2)\frac{d^2F(u)}{du^2} - (m+1)2u\frac{dF(u)}{du} + [l(l+1) - m(m+1)]F(u) = 0 \qquad (7.30)$$

An analogous equation is obtained through m-fold differentiation of the Legendre equation (7.20), which gives:

$$(1-u^2)\frac{d^2}{du^2}\left(\frac{d^m P_l}{du^m}\right) - (m+1)2u\frac{d}{du}\left(\frac{d^m P_l}{du^m}\right) + [l(l+1) - m(m+1)]\frac{d^m P_l}{du^m} = 0 \qquad (7.31)$$

Comparison between Eq.(7.30) and (7.31) shows that we must choose:

$$F(u) = \frac{d^m}{du^m} P_l(u)$$

which is defined for positive integers m. It follows that the solutions (7.29) have the form:

$$P_l^{|m|}(u) = (1-u^2)^{|m|/2} \frac{d^{|m|}}{du^{|m|}} P_l(u) \qquad (7.32)$$

called the *associated Legendre functions*, which are well characterized by $|m|$ because Eq.(7.28) is independent of the sign of m. Since $P_l(u)$ are polynomials of degree l, their $|m|$th derivative will be zero if $|m|$ is greater than l, which implies that:

$$|m| \leq l \qquad (7.33)$$

and this provides a basic relation between the magnetic and orbital quantum numbers. It can be shown that the associated Legendre functions (7.32) form an orthogonal set in the physically relevant interval $-1 \leq u < 1$, according to:

$$\int_{-1}^{1} P_l^{|m|}(u) P_{l'}^{|m|}(u) du = \frac{(l+|m|)!}{(l-|m|)!} \frac{2}{2l+1} \delta_{ll'} \qquad (7.34)$$

such that $P_l^{|m|}(u) = P_l^{|m|}(\cos\theta)$ are admissible eigenfunctions for the motion along θ. We may now combine the solutions (7.13) and (7.32) to express the common eigenfunctions of $\hat{L}^2$ and $\hat{L}_z$, Eq.(7.6), as:

$$Y_{lm}(\theta,\varphi) = N_{lm} P_l^{|m|}(\cos\theta) e^{im\varphi}$$

7. Electron Motion in the Atom

with the normalization constants given by the condition:

$$\int_0^{2\pi}\int_0^\pi Y_{lm}^*(\theta,\varphi)Y_{l'm'}(\theta,\varphi)\sin\theta\, d\theta\, d\varphi = N_{lm}^2 \int_{-1}^1 P_l^{|m|}(u) P_{l'}^{|m'|} du \int_0^{2\pi} e^{i(m-m')\varphi} d\varphi = \delta_{ll'}\delta_{mm'}$$

where the integral over φ is zero if $m \neq m'$, and, if $m = m'$, the integral over θ is zero unless $l = l'$. In view of Eq.(7.34), we obtain:

$$\int_0^{2\pi} d\varphi \int_0^\pi Y_{lm}^*(\theta,\varphi) Y_{l'm}(\theta,\varphi)\sin\theta\, d\theta = N_{lm}^2 \frac{4\pi}{2l+1} \frac{(l+|m|)!}{(l-|m|)!} \delta_{ll'}$$

and therefore the orthonormal set of eigenfunctions of the orbital angular momentum consists of the so-called *spherical harmonics*:

$$Y_{lm}(\theta,\varphi) = \left[\frac{2l+1}{4\pi}\frac{(l-|m|)!}{(l+|m|)!}\right]^{\frac{1}{2}} P_l^{|m|}(\cos\theta) e^{im\varphi} \tag{7.35}$$

defined on the spherical surface of unit radius. The following properties are immediate and of current use:

(i) $$Y_{lm}(\pi-\theta,\varphi+\pi) = (-1)^l Y_{lm}(\theta,\varphi)$$

which shows that the spherical harmonics have a definite parity under reflection at the origin, which changes θ into $\theta - \pi$ and φ into $\varphi + \pi$, and:

(ii) $$Y_{l0}(\theta,\varphi) = \sqrt{(2l+1)/4\pi}\, P_l(\cos\theta)$$

The spherical harmonic eigenstates of the combined rotational motion along θ and φ correspond to the eigenvalues (7.19) of the squared angular momentum, which depend on l but are often called by standard code letters, originating from spectroscopy, rather than by their l values, as follows:

s-states: $l = 0, m = 0$; $L = L_z = 0$; $Y_{00} = \dfrac{1}{\sqrt{4\pi}}$

p-states: $l = 1, m = 0, \pm 1$; $L = \sqrt{2}\,\hbar$, $L_z = 0, \pm\hbar$

$$Y_{10} = \sqrt{\frac{3}{4\pi}}\cos\theta,\quad Y_{1,\pm 1} = \sqrt{\frac{3}{8\pi}}\sin\theta\, e^{\pm i\varphi} \tag{7.36}$$

d-states: $l = 2, m = 0, \pm 1, \pm 2$; $L = \sqrt{6}\,\hbar$, $L_z = 0, \pm\hbar, \pm 2\hbar$

$$Y_{20} = \sqrt{\frac{5}{16\pi}}(3\cos^2\theta - 1), \quad Y_{2,\pm 1} = \sqrt{\frac{15}{8\pi}}\cos\theta\sin\theta\, e^{\pm i\varphi}, \quad Y_{2,\pm 2} = \sqrt{\frac{15}{32\pi}}\sin^2\theta\, e^{\pm 2i\varphi}$$

(7.37)

Because the spherical harmonics form a complete set of functions, it is always possible to expand an arbitrary wave function $\Psi(\theta,\varphi)$ in terms of $Y_{lm}(\theta,\varphi)$ as follows:

$$\Psi(\theta,\varphi) = \sum_{l=0}^{\infty}\sum_{m=-l}^{l} a_{lm} Y_{lm}(\theta,\varphi) \qquad (7.38)$$

where the coefficients are given by:

$$a_{lm} = \int Y_{lm}^*(\theta,\varphi)\Psi(\theta,\varphi)d\Omega = \int_0^\pi\int_0^{2\pi} Y_{lm}^*(\theta,\varphi)\Psi(\theta,\varphi)\sin\theta\, d\theta\, d\varphi \qquad (7.39)$$

The orthonormality of the spherical harmonics, defined by Eq.(7.35), implies that:

$$\sum_{l=0}^{\infty}\sum_{m=-l}^{l} |a_{lm}|^2 = 1 \qquad (7.40)$$

and this means that the coefficients a_{lm} can be interpreted in terms of the probability amplitudes for the eigenstates of the orbital angular momentum.

Problem 7.1.1. Find the eigenvalues and the eigenfunctions of a rigid rotator, which is an idealization of a diatomic molecule, consisting of two atoms of masses M_1 and M_2 rigidly joined to each other at a distance R.

(Solution): The total kinetic energy of the two particles can always be separated as:

$$\frac{1}{2}(M_1 v_1^2 + M_2 v_2^2) = \frac{M_1+M_2}{2}\left(\frac{M_1\vec{v}_1 + M_2\vec{v}_2}{M_1+M_2}\right)^2 + \frac{1}{2}\frac{M_1 M_2}{M_1+M_2}(\vec{v}_1 - \vec{v}_2)^2$$

where the first term represents the energy associated with the centre of mass and the second is the energy of the relative motion. Only the second term is relevant to the internal energy of the system. It is thus convenient to choose the origin of the coordinate system in the centre of mass of the molecule, such that $M_1 r_1 = M_2 r_2$ and, because $r_1 + r_2 = R$, we obtain:

$$r_1 = \frac{M_2}{M_1+M_2}R, \qquad r_2 = -\frac{M_1}{M_1+M_2}R$$

7. Electron Motion in the Atom

The spherical polar coordinates of the relative motion of the two atoms are (r_1, θ, φ) and $(r_2, \pi - \theta, \pi + \varphi)$ respectively, and their kinetic energies have the form:

$$\frac{M_i}{2}\left[\left(\frac{dx_i}{dt}\right)^2 + \left(\frac{dy_i}{dt}\right)^2 + \left(\frac{dz_i}{dt}\right)^2\right] = \frac{M_i}{2}\left[r_i^2\left(\frac{d\theta}{dt}\right)^2 + r_i^2 \sin^2\theta\left(\frac{d\varphi}{dt}\right)^2\right] \quad (i = 1,2)$$

which yield the total kinetic energy:

$$\frac{1}{2}(M_1 r_1^2 + M_2 r_2^2)\left[\left(\frac{d\theta}{dt}\right)^2 + \sin^2\theta\left(\frac{d\varphi}{dt}\right)^2\right] = \frac{\mu R^2}{2}\left[\left(\frac{d\theta}{dt}\right)^2 + \sin^2\theta\left(\frac{d\varphi}{dt}\right)^2\right]$$

$$= \frac{\mu}{2}\left[\left(R\frac{d\theta}{dt}\right)^2 + \left(R\sin\theta\frac{d\varphi}{dt}\right)^2\right] = \frac{\mu}{2}(v_\theta^2 + v_\varphi^2)$$

where μ is the reduced mass of the system:

$$\frac{1}{\mu} = \frac{1}{M_1} + \frac{1}{M_2} \quad \text{or} \quad \mu = \frac{M_1 M_2}{M_1 + M_2}$$

and $\vec{v}_\theta$ and $\vec{v}_\varphi$ are the velocities of the motion along the meridian and parallel lines respectively. There is no relative radial motion, so that the classical Hamiltonian of the rigid rotator reduces to the kinetic energy of rotation. This can be rewritten in terms of the orbital angular momentum:

$$\vec{L} = \vec{R} \times \vec{p} = \mu \vec{R} \times \vec{v} = \mu \vec{R} \times (\vec{v}_\theta + \vec{v}_\varphi)$$

whose magnitude is given by:

$$L^2 = |\vec{L}|^2 = \mu^2 R^2 (v_\theta^2 + v_\varphi^2)$$

If we denote by $I = \mu R^2$ the moment of inertia of the molecule about the centre of mass, the Hamiltonian reads:

$$H = \frac{L^2}{2\mu R^2} = \frac{L^2}{2I} \quad \rightarrow \quad \hat{H} = \frac{1}{2I}\hat{L}^2$$

and the eigenvalue problem of the rigid rotator takes the form:

$$\frac{1}{2I}\hat{L}^2 \Psi = E\Psi \quad \text{or} \quad \hat{L}^2 \Psi = 2IE\Psi$$

This equation is identical to Eq.(7.5), provided we take $L^2 = 2IE$, and hence has the same solutions, namely the eigenvalue spectrum is given by:

$$E_l = l(l+1)\frac{\hbar^2}{2I} \qquad l = 0, 1, 2, 3, \ldots$$

and the corresponding eigenfunctions are the spherical harmonics (7.35). There is a $(2l + 1)$-fold degeneracy of each energy level, except of the ground level $l = 0$, due to the allowed m values which define $2l + 1$ linearly independent eigenfunctions $Y_{lm}(\theta, \varphi)$ for a given l.

Problem 7.1.2. A state is characterized by the wave function:

$$\Psi(\theta, \varphi) = \sqrt{\frac{3}{8\pi}}(\sin\theta \sin\varphi + i\cos\theta)$$

Find the probability of obtaining for a measurement of L^2 the result $2\hbar^2$, and for L_z the results $\hbar, 0$ and $-\hbar$.

(Solution)   If we write the expansion (7.38) as:

$$\Psi(\theta, \varphi) = a_{11}Y_{11}(\theta, \varphi) + a_{10}Y_{10}(\theta, \varphi) + a_{1,-1}Y_{1,-1}(\theta, \varphi)$$

the coefficients are obtained by inserting $\Psi(\theta,\varphi)$ and the spherical harmonics (7.36) into Eq.(7.39) which gives:

$$a_{11} = \frac{i}{2}, \quad a_{10} = \frac{i}{\sqrt{2}}, \quad a_{1,-1} = -\frac{i}{2}$$

The expansion becomes:

$$\Psi(\theta, \varphi) = i\left(\frac{1}{2}Y_{11} + \frac{1}{\sqrt{2}}Y_{10} - \frac{1}{2}Y_{1,-1}\right)$$

and shows that, by measuring L_z, we obtain the eigenvalues $m\hbar = \hbar, 0, -\hbar$ with probabilities $1/4, 1/2$ and $1/4$ respectively. The expectation value of L_z is:

$$\sum_{m=-l}^{l} |a_{lm}|^2 m\hbar = \frac{1}{4}\hbar - \frac{1}{4}\hbar = 0$$

On the other hand, $\Psi(\theta,\varphi)$ is an eigenfunction of $\hat{L}^2$, and hence the eigenvalue $l(l+1)\hbar^2 = 2\hbar^2$ is for certainty obtained.

Problem 7.1.3. Show that the expectation values of $\hat{L}_x$ and $\hat{L}_y$ are zero in the eigenstates of $\hat{L}_z$.

Problem 7.1.4. Show that the highest precision of a measurement on L_x and L_y, in a state described by the spherical harmonics Y_{lm}, is obtained for $|m| = l$.

(Answer): $(\Delta L_x)^2 + (\Delta L_y)^2 = (l^2 + l - m^2)\hbar^2$

Problem 7.1.5. Show that the quantum treatment of angular momentum implies that L_z can never point in the direction of $\vec{L}$, and find the minimum angle made by the vector $\vec{L}$ with the z direction.

(Answer): $\theta_{\min} = \arcsin(l+1)^{-\frac{1}{2}}$

Problem 7.1.6. Find the expression of the spherical harmonics Y_{lm} in Cartesian coordinates, for $l = 1$, and use it to show that at a measurement of L_z in a state described by the eigenfunction:

$$\psi(x, y, z) = \frac{1}{\sqrt{4\pi}}(x + y + z)e^{-(x^2 + y^2 + z^2)}$$

the results $\hbar, 0$ and $-\hbar$ are obtained with equal probabilities.

7.2. Radial Motion in Central Force Field

The radial wave equation (7.4), which upon substitution of the quantum condition (7.19) becomes:

$$-\frac{\hbar^2}{2m}\left(\frac{d^2}{dr^2} + \frac{2}{r}\frac{d}{dr}\right)R(r) + \left[U(r) + \frac{l(l+1)\hbar^2}{2mr^2}\right]R(r) = ER(r) \qquad (7.41)$$

can be conveniently simplified if we set:

$$R(r) = \frac{\xi(r)}{r} \qquad (7.42)$$

which gives:

$$\left(\frac{d^2}{dr^2} + \frac{2}{r}\frac{d}{dr}\right)\frac{\xi(r)}{r} = \frac{1}{r}\frac{d^2\xi(r)}{dr^2}$$

Hence, Eq.(7.41) reduces to:

$$-\frac{\hbar^2}{2m}\frac{d^2\xi(r)}{dr^2} + \left[U(r) + \frac{l(l+1)\hbar^2}{2mr^2}\right]\xi(r) = E\xi(r) \qquad (7.43)$$

and this is formally identical to the Schrödinger wave equation in one dimension, Eq.(6.3), provided the central force potential $U(r)$ is replaced by an effective potential energy:

$$U_{eff}(r) = U(r) + \frac{l(l+1)\hbar^2}{2mr^2} \qquad (7.44)$$

The differences from the one-dimensional motion are as follows:

(i) There is a repulsive potential energy, associated with the magnitude of the orbital angular momentum, and hence with the rotational motion, where the centrifugal force keeps the particle away from the centre of force placed at the origin. This term is called *centrifugal potential energy*.

(ii) The variable r takes values from 0 to ∞, in contrast with the variable x for the motion in one dimension.

(iii) The admissible solution $\xi(r)$ must vanish at $r = 0$:

$$\xi(0) = 0 \qquad (7.45)$$

as required by the finiteness condition on $R(r)$, defined by Eq.(7.42).

The nature of eigenfunctions and the energy spectrum of the radial motion can be discussed assuming the following restrictions on the potential energy:

$$U(r) \to 0 \quad \text{for} \quad r \to \infty \qquad (7.46)$$

$$U(r) \sim Ar^\alpha \quad \text{with} \quad \alpha > -2, \quad \text{for} \quad r \to 0 \qquad (7.47)$$

where Eq.(7.46) implies that the central force vanishes at large distances from the centre of force, and Eq.(7.47) is imposed by the finiteness requirement at the origin. The two assumptions (7.46) and (7.47) correspond to a large class of central force potentials, such as the electron-proton atomic interaction, the neutron-proton nuclear interaction and the atom-atom interaction in a diatomic molecule. By comparing the admissible form of the solutions, near the origin and in the asymptotic region, we may infer the nature of the eigenvalue spectrum.

Near the origin, where the condition (7.47) holds, and the centrifugal potential energy diverges rapidly and dominates over the other terms, the radial equation (7.43) can be approximated by:

$$\frac{d^2\xi_0(r)}{dr^2} - \frac{l(l+1)}{r^2}\xi_0(r) \approx 0 \tag{7.48}$$

and has a solution r^α, where α is determined by the characteristic equation:

$$\alpha(\alpha-1) - l(l+1) = 0$$

The two solutions are $\alpha_1 = l+1$ and $\alpha_2 = -l$, but the second root leads to a function $\xi(r)$ which is infinite for $r = 0$ and hence is inadmissible as an eigenfunction, such that we choose:

$$\alpha = l+1 \tag{7.49}$$

This implies that the radial function (7.42), near the origin, has the form:

$$R(r) \sim r^l \tag{7.50}$$

and although the exact solution $R(r)$ can only be derived by integrating the radial equation with given $U(r)$, any acceptable solution must behave like r^l near the origin, for the angular momentum l.

For large r values, both the potential energy $U(r)$ and the centrifugal potential can be neglected and the radial equation becomes:

$$\frac{d^2\xi_\infty(r)}{dr^2} + \frac{2m}{\hbar^2}E\xi_\infty(r) \approx 0 \tag{7.51}$$

and the form of the solution depends on the sign of the energy eigenvalues, as discussed before in the case of one-dimensional motion. We set:

$$k^2 = \frac{2mE}{\hbar^2} \quad (E>0), \qquad \gamma^2 = -\frac{2mE}{\hbar^2} \quad (E<0) \tag{7.52}$$

and hence the asymptotic equation (7.51) has for $E > 0$ the solution:

$$\xi_\infty(r) = a_1(E,l)e^{ikr} + a_2(E,l)e^{-ikr} \tag{7.53}$$

where the coefficients $a_1(E,l)$ and $a_2(E,l)$ are determined in terms of E and l by the behaviour (7.50) of the radial function near the origin. The solution (7.53) remains finite as $r \to \infty$, irrespective of $a_1(E,l)$ and $a_2(E,l)$. For $E < 0$ the asymptotic solution is:

$$\xi_\infty(r) = b_1(E,l) e^{\gamma r} + b_2(E,l) e^{-\gamma r} \tag{7.54}$$

and this is a finite solution for large r values only if:

$$b_1(E,l) = 0 \tag{7.55}$$

In other words, admissible solutions of the radial equation can be obtained for any $E > 0$ and given l, and hence, for $E > 0$, the energy spectrum is continuous. There will be an infinity of solutions $R_{El}(r)$ with $E > 0$ and l taking all the integer non-negative values. For $E < 0$, the admissible solutions only corresponds to the energy eigenvalues that satisfy Eq.(7.55). There will be a finite number of solutions of this equation for each given l value, denoted by $E_{l1}, E_{l2}, \ldots, E_{ln}$ and the set of roots corresponding to all l values ($l = 0, 1, 2, \ldots$) form a discrete eigenvalue spectrum. The asymptotic behaviour of the type $r^{l+1} e^{-\gamma r}$ ensures that $\xi(r)$ are square integrable functions and therefore the radial integral $\int_0^\infty r^2 R_{El}(r) dr$ is not divergent for these eigenstates. It also follows from the asymptotic behaviour that the probability to find the particle at large distances from the centre of force is small, and therefore the eigenstates corresponding to discrete negative energy eigenvalues are *bound states* of the motion.

Assuming that the centre of force is the atomic nucleus, of pozitive charge Ze, and considering the motion of an electron of mass m_e and charge $-e$, the potential energy is conveniently written, if we set $e^2 = 4\pi\varepsilon_0 e_0^2$, as:

$$U(r) = -\frac{Ze^2}{4\pi\varepsilon_0 r} = -\frac{Ze_0^2}{r} \tag{7.56}$$

This expression describes the one-electron atoms, with atomic number Z and all but one of its electrons removed (H, He$^+$, Li^{2+}, etc.) and satisfies the conditions (7.47), such that we expect to obtain an energy eigenvalue spectrum which is discrete for $E < 0$ and continuous for $E > 0$. Substituting Eq.(7.56), the radial equation (7.43) becomes:

$$\frac{d^2\xi(r)}{dr^2} + \frac{2m_e}{\hbar^2}\left[E + \frac{Ze_0^2}{r} - \frac{l(l+1)\hbar^2}{2m_e r^2}\right]\xi(r) = 0 \tag{7.57}$$

We first discuss the bound states, which are the most significant for the atomic electron, and correspond to negative eigenvalues $E = -|E|$. Introducing a new variable ρ defined as:

$$\rho = 2\eta r \tag{7.58}$$

the radial equation takes the form:

$$\frac{d^2\xi(\rho)}{d\rho^2} + \left\{ -\frac{m_e|E|}{2\hbar^2\eta^2} + \frac{m_e Ze_0^2}{\hbar^2\eta}\frac{1}{\rho} - \frac{l(l+1)}{\rho^2} \right\}\xi(\rho) = 0 \tag{7.59}$$

It is common practice to choose the constant η such as:

$$\frac{m_e|E|}{2\hbar^2\eta^2} = \frac{1}{4} \quad \text{or} \quad \eta = \sqrt{\frac{2m_e|E|}{\hbar^2}} \tag{7.60}$$

and to set:

$$\lambda = \frac{m_e Ze_0^2}{\hbar^2\eta} = \frac{Ze_0^2}{\hbar}\sqrt{\frac{m_e}{2|E|}} \tag{7.61}$$

in order to simplify the coefficients of Eq.(7.59), which so reduces to:

$$\frac{d^2\xi(\rho)}{d\rho^2} + \left\{ -\frac{1}{4} + \frac{\lambda}{\rho} - \frac{l(l+1)}{\rho^2} \right\}\xi(\rho) = 0 \tag{7.62}$$

For large r values, this equation is approximated by:

$$\frac{d^2\xi_\infty(\rho)}{d\rho^2} - \frac{1}{4}\xi_\infty(\rho) = 0$$

and has solutions of the form (7.54), where $\gamma = 1/2$, subjected to the finiteness requirement (7.55). This leads to a trial solution:

$$\xi(\rho) = e^{-\rho/2} S(\rho) \tag{7.63}$$

to solve Eq.(7.62), which upon substitution yields:

$$\frac{d^2 S(\rho)}{d\rho^2} - \frac{dS(\rho)}{d\rho} + \left[\frac{\lambda}{\rho} - \frac{l(l+1)}{\rho^2} \right] S(\rho) = 0 \tag{7.64}$$

The behaviour near the origin, Eq.(7.49), implies that $S(\rho)$ has the form:

$$S(\rho) = \rho^{l+1} F(\rho) \tag{7.65}$$

where $F(\rho)$ must be a finite solution to:

Quantum Physics

$$\rho\frac{d^2F(\rho)}{d\rho^2}+(2l+2-\rho)\frac{dF(\rho)}{d\rho}+(\lambda-l-1)F(\rho)=0 \tag{7.66}$$

which is called the *confluent hypergeometric equation*.

EXAMPLE 7.2. Confluent hypergeometric functions

Equation (7.66) can be reduced to the standard form:

$$\rho\frac{d^2F(\rho)}{d\rho^2}+(b-\rho)\frac{dF(\rho)}{d\rho}-aF(\rho)=0 \tag{7.67}$$

if we substitute:

$$a=l+1-\lambda, \qquad b=2l+2 \tag{7.68}$$

Equation (7.67) is solved by a power series expansion for $F(\rho)$:

$$F(\rho)=\rho^\alpha\sum_{k=0}^{\infty}c_k\rho^k \tag{7.69}$$

where the admissibility condition (7.45) implies that α should be chosen such as $c_0 \neq 0$. Substitution of Eq.(7.69) into Eq.(7.67) gives:

$$\sum_{k=0}^{\infty}(k+\alpha)(k+\alpha+b-1)c_k\rho^{k+\alpha-1}-\sum_{k=0}^{\infty}(k+\alpha+a)c_k\rho^{k+\alpha}=0$$

which is conveniently rewritten as:

$$\alpha(\alpha+b-1)c_0\rho^{\alpha-1}+\sum_{k=0}^{\infty}\left[(k+\alpha+1)(k+\alpha+b)c_{k+1}-(k+\alpha+a)c_k\right]\rho^{k+\alpha}=0 \tag{7.70}$$

All coefficients should be zero, and this yields the equation:

$$\alpha(\alpha+b-1)c_0=0 \tag{7.71}$$

and the recurrence relation:

$$(k+\alpha+1)(k+\alpha+b)c_{k+1}=(k+\alpha+a)c_k, \quad k=0,1,2,\ldots \tag{7.72}$$

Since $c_0 \neq 0$, Eq.(7.71) has two solutions, $\alpha=0$ and $\alpha=1-b$, that define two functions:

$$F(\rho)=\sum_{k=0}^{\infty}c_k\rho^k \qquad \text{and} \qquad F'(\rho)=\rho^{1-b}\sum_{k=0}^{\infty}c'_k\rho^k$$

which coincide if $b = 1$. It is shown that $F(\rho)$ and $F'(\rho)$ are linearly independent only if b is not an integer. In view of Eq.(7.68), it is then sufficient to consider the value $\alpha = 0$, and the recurrence relation (7.72) successively yields:

$$b c_1 = a c_0 \quad \text{or} \quad c_1 = \frac{a}{b} c_0 = \frac{a}{b}$$

where c_0 is arbitrary and can be chosen $c_0 = 1$ for simplicity, and also:

$$2(b+1) c_2 = (a+1) c_1 \quad \text{or} \quad c_2 = \frac{a(a+1)}{2b(b+1)}$$

$$3(b+2) c_3 = (a+2) c_2 \quad \text{or} \quad c_3 = \frac{a(a+1)(a+2)}{3! \, b(b+1)(b+2)}$$

and so on. It follows that the solution (7.69) can be written in the form:

$$F(a,b;\rho) = 1 + \frac{a}{b} \frac{\rho}{1!} + \frac{a(a+1)}{b(b+1)} \frac{\rho^2}{2!} + \ldots = \sum_{k=0}^{\infty} \frac{a(a+1)\cdots(a+k-1)}{b(b+1)\cdots(b+k-1)} \frac{\rho^k}{k!} \quad (7.73)$$

known as the *confluent hypergeometric function*. Since the recurrence relation (7.72) gives:

$$c_{k+1} = \frac{a+k}{b+k} \frac{1}{k+1} c_k \quad (7.74)$$

the function (7.73) has the important property that, if α is taken equal to a negative integer $-\nu$, it reduces to a polynomial of order ν, because $k = \nu$ and $c_{\nu+1} = 0$ along with all subsequent coefficients.

Because the solution $F(l+1-\lambda, 2l+2; \rho)$ behaves asymptotically as e^ρ for large ρ, the radial function $R(\rho)$, which is given, in view of Eqs.(7.42), (7.63) and (7.65), by:

$$R(\rho) = A \rho^l e^{-\rho/2} F(l+1-\lambda, 2l+2; \rho) \quad (7.75)$$

will diverge exponentially as $e^{\rho/2}$ at infinity, unless the infinite series (7.73) is reduced to a finite polynomial. This can be achieved by taking:

$$l + 1 - \lambda = -\nu, \quad \nu = 0, 1, 2, 3, \ldots \quad (7.76)$$

where the integer ν is called the *radial quantum number*. Eq.(7.76) shows that the parameter λ must always be an integer, which is usually denoted by n:

$$\lambda = \nu + l + 1 = n, \quad n = 1, 2, 3, \ldots \quad (7.77)$$

This quantum condition introduces the so-called *principal quantum number* n and implies the inequality:

$$l \leq n - 1 \tag{7.78}$$

In other words, for a given n, the quantum number l can only take the values:

$$l = 0, 1, 2, \ldots, n-1 \tag{7.79}$$

and the corresponding stationary states will be described by the spectroscopic notation: $1s$ ($n=1$; $l=0$); $2s, 2p$ ($n=2$; $l=0,1$); $3s, 3p, 3d$ ($n=3$; $l=0,1,2$); $4s, 4p, 4d, 4f$ ($n=4$; $l=0,1,2,3$), etc. Because λ was defined in terms of the energy eigenvalues for bound states, Eq.(7.61), it follows that there is an infinite set of discrete energy levels of one electron atoms, defined by:

$$E_n = -\frac{Z^2}{n^2}\frac{m_e e_0^4}{2\hbar^2} = -\frac{Z^2}{n^2}\frac{e_0^2}{2a_0} \tag{7.80}$$

where a is the so-called *Bohr radius* of the atom:

$$a_0 = \frac{\hbar^2}{m_e e_0^2} = \frac{4\pi\varepsilon_0 \hbar^2}{m_e e^2} = 0.53 \times 10^{-10} \, m \tag{7.81}$$

The same energy levels were derived before in Eq.(P.26) from the Bohr model of one-electron atoms.

Equation (7.80) indicates that all eigenfunctions corresponding to a given n value, with $0 \leq l \leq n-1$, have the same energy. Consequently each radial function (7.75) should be specified in term of two quantum numbers, n and l, respectively related to the energy and the orbital angular momentum eigenvalues:

$$R_{nl}(r) = A_{nl}\, \rho^l e^{-\rho/2} F(-n+l+1, 2l+2; \rho), \quad \rho = \frac{2Z}{na_0} r \tag{7.82}$$

where the confluent hypergeometric function is reduced to a polynomial of order $n - l - 1$. The total number of states belonging to E_n represents the order of degeneracy, which can be derived by observing that each stationary state of one-electron atoms is determined by the quantum numbers n, l and m and there are $2l + 1$ possible values of the quantum number m for each value of l, Eq.(7.33), and hence:

$$\sum_{l=0}^{n-1}(2l+1) = n^2 \tag{7.83}$$

The complete set of eigenfunctions corresponding to the eigenvalue spectrum (7.80) is now given by Eq.(7.2) in the form:

$$\Psi_{nlm}(r,\theta,\varphi) = R_{nl}(r)Y_{lm}(\theta,\varphi) \tag{7.84}$$

subjected to the normalization condition:

$$\int_{\infty}^{\infty} R_{nl}^2(r)|Y_{lm}(\theta,\varphi)|^2 dV = \int_0^{\infty} R_{nl}^2(r)r^2 dr \int_0^{\pi}\int_0^{2\pi}|Y_{lm}(\theta,\varphi)|^2 \sin\theta\, d\theta\, d\varphi = \int_0^{\infty} R_{nl}^2(r)r^2 dr = 1 \tag{7.85}$$

where the spherical harmonics (7.35) are normalized on the spherical surface of unit radius. The constants A_{nl} can be derived from Eq.(7.85) as:

$$A_{nl} = \frac{1}{(2l+1)!}\sqrt{\frac{(n+l)!}{2n(n-l-1)!}}\left(\frac{2Z}{na_0}\right)^3 \tag{7.86}$$

such that the first few radial functions for one-electron atoms read:

$$R_{1s}(r) = 2\left(\frac{Z}{a_0}\right)^{\frac{3}{2}} e^{-Zr/a_0}$$

$$R_{2s}(r) = \left(\frac{Z}{2a_0}\right)^{\frac{3}{2}}\left(2 - \frac{Zr}{a_0}\right)e^{-Zr/2a_0}, \quad R_{2p}(r) = \frac{2}{\sqrt{3}}\left(\frac{Z}{2a_0}\right)^{\frac{5}{2}} r\, e^{-Zr/2a_0}$$

$$\tag{7.87}$$

$$R_{3s}(r) = 2\left(\frac{Z}{3a_0}\right)^{\frac{3}{2}}\left(1 - \frac{2Zr}{3a_0} + \frac{2Z^2 r^2}{27a_0^2}\right)e^{-Zr/3a_0}$$

$$R_{3p}(r) = \frac{4\sqrt{2}}{3}\left(\frac{Z}{3a_0}\right)^{\frac{5}{2}} r\left(1 - \frac{Zr}{6a_0}\right)e^{-Zr/3a_0}, \quad R_{3d}(r) = \frac{4}{3\sqrt{10}}\left(\frac{Z}{3a_0}\right)^{\frac{7}{2}} r^2 e^{-Zr/3a_0}$$

If we now consider *unbound states* of positive energy $E > 0$, the parameters η and λ of the radial equation, defined by Eqs.(7.60) and (7.61), become imaginary and it is convenient to set:

$$\eta = -ik \quad \text{and} \quad k = \sqrt{\frac{2m_e E}{\hbar^2}}, \quad \lambda = i\ell \quad \text{and} \quad \ell = \frac{Ze^2}{\hbar}\sqrt{\frac{m_e}{2E}} \tag{7.88}$$

which implies that the variable ρ from Eq.(7.58) becomes:

$$\rho = -2ikr \tag{7.89}$$

and that Eq.(7.62) reads:

$$\frac{d^2\xi(\rho)}{d\rho^2} + \left\{-\frac{1}{4} + \frac{i\ell}{\rho} - \frac{l(l+1)}{\rho^2}\right\}\xi(\rho) = 0 \tag{7.90}$$

Following the same further steps as before, we obtain the confluent hypergeometric equation solved by the radial functions of the type (7.75):

$$R_l(r) = A_l r^l e^{ikr} F(l+1-i\ell, 2l+2; -2ikr) \tag{7.91}$$

These solutions remain finite for large r values, as shown by their asymptotic behaviour, which is of the form (7.53). In this region the potential energy can be neglected, so that the energy E, which from Eq.(7.88) results as $(\hbar k)^2/2m_e$, can be interpreted as the kinetic energy of the electron far for the nucleus. Hence the positive energy states, having a continuous eigenvalue spectrum, correspond to the motion of an unbound electron in the neibourhood of the nucleus, with enough energy to escape to infinity.

EXAMPLE 7.3. Radial motion of a free particle

In the special case of a free particle, where the potential energy $U(r) = 0$ has spherical symmetry, the radial equation (7.41) reads:

$$\frac{d^2R(r)}{dr^2} + \frac{2}{r}\frac{dR(r)}{dr} + \left[k^2 - \frac{l(l+1)}{r^2}\right]R(r) = 0 \tag{7.92}$$

where it was assumed that $E > 0$ and $E = (\hbar k)^2/2m_e$. For $l = 0$ we obtain the solution:

$$R_{k0}(r) = \frac{A}{r}\sin kr \tag{7.93}$$

Since the eigenvalue spectrum is continuous and R_{k0} depends on the continuous parameter $k = p/\hbar$, where p is interpreted as the radial momentum of the free particle, the appropriate normalization condition is of the type (3.20), which reads:

$$\int_0^\infty R_{k0}^*(r) R_{k'0}(r) r^2 dr = \frac{A^2}{4} \int_0^\infty \left(e^{i(k-k')r} + e^{-i(k-k')r} - e^{i(k+k')r} - e^{-i(k+k')r}\right) dr$$

$$= \frac{A^2}{4}\left[\int_{-\infty}^\infty e^{i(k-k')r} dr - \int_{-\infty}^\infty e^{i(k+k')r} dr\right] = \frac{A^2}{4} 2\pi\delta(k-k') = \delta(k-k')$$

where use has been made of Eq.(1.43) and of $\delta(k+k') = 0$, as k and k' are both positive parameters. It follows that:

$$A = \sqrt{\frac{2}{\pi}} \quad \text{and} \quad R_{k0}(r) = \sqrt{\frac{2}{\pi}}\frac{\sin kr}{r} \tag{7.94}$$

For $l \neq 0$, it is convenient to write:

$$R_{kl}(r) = r^l u_{kl}(r) \tag{7.95}$$

where the new function is a solution to the equation:

$$\frac{d^2 u_{kl}}{dr^2} + \frac{2(l+1)}{r}\frac{d u_{kl}}{dr} + k^2 u_{kl} = 0 \tag{7.96}$$

or, by differentiation:

$$\frac{d}{dr}\left(\frac{d^2 u_{kl}}{dr^2}\right) + \frac{2(l+1)}{r}\frac{d}{dr}\left(\frac{d u_{kl}}{dr}\right) + \left[k^2 - \frac{2(l+1)}{r^2}\right]\frac{d u_{kl}}{dr} = 0 \tag{7.97}$$

If we further set $du_{kl}/dr = rf(r)$, one obtains:

$$\frac{d^2 f(r)}{dr^2} + \frac{2(l+2)}{r}\frac{df(r)}{dr} + k^2 f(r) = 0 \tag{7.98}$$

which, by comparison with Eq.(7.96), shows that $f = u_{k,l+1}$, and this yields the recurrence relation:

$$u_{k,l+1} = \frac{1}{r}\frac{du_{kl}}{dr}$$

For $l = 0$, we obtain from Eq.(7.95) that $u_{k0}(r) = R_{k0}(r)$, and this successively leads to:

$$R_{kl}(r) = r^l u_{kl}(r) = \sqrt{\frac{2}{\pi}}(-1)^l r^l \left(\frac{1}{kr}\frac{d}{dr}\right)^l \frac{\sin kr}{r} \tag{7.99}$$

where use has been made of Eq.(7.94).

Problem 7.2.1. Solve the energy eigenvalue problem for the three-dimensional harmonic oscillator in spherical polar coordinates.

(Solution): The potential energy of the isotropic harmonic oscillator has spherical symmetry:

$$U(r) = \frac{m\omega^2}{2}(x^2 + y^2 + z^2) = \frac{m\omega^2}{2}r^2$$

and hence the Schrödinger equation can be solved in spherical polar coordinates. Separating the angular motion, the radial equation is of the type (7.41), which becomes:

$$-\frac{\hbar^2}{2m}\left(\frac{d^2}{dr^2}+\frac{2}{r}\frac{d}{dr}\right)R(r)+\left[\frac{m\omega^2}{2}r^2+\frac{l(l+1)\hbar^2}{2mr^2}\right]R(r)=ER(r)$$

Near the origin, we find that $R(r) \sim r^l$, as obtained before in Eq.(7.50), and hence, we may take:

$$R(r)=r^l K(r)$$

Substituting this expression gives:

$$-\frac{\hbar^2}{2m}\left[\frac{d^2 K(r)}{dr^2}+\frac{2(l+1)}{r}\frac{dK(r)}{dr}\right]+\frac{m\omega^2}{2}r^2 K(r)=EK(r)$$

and this equation is further simplified if we set:

$$r=\sqrt{\frac{\hbar}{m\omega}}\,\rho,\quad \lambda=\frac{2E}{\hbar\omega}$$

such that we obtain:

$$\frac{d^2 K(r)}{d\rho^2}+\frac{2(l+1)}{\rho}\frac{dK(\rho)}{d\rho}+(\lambda-\rho^2)K(\rho)=0$$

The asymptotic behaviour is of the type $K(r) \sim e^{-\rho^2/2}$, which leads to the trial solution:

$$K(\rho)=e^{-\rho^2/2}S(\rho)$$

where $S(\rho)$ is a solution to:

$$\frac{d^2 S(\rho)}{d\rho^2}+\left[\frac{2(l+1)}{\rho}-2\rho\right]\frac{dS(\rho)}{d\rho}+(\lambda-2l-3)S(\rho)=0$$

This reduces to the confluent hypergeometric equation (7.67), is we finally set $\rho^2 = x$, giving:

$$x\frac{d^2 S}{dx^2}+\left(l+\frac{3}{2}-x\right)\frac{dS}{dx}+\frac{1}{4}(\lambda-2l-3)\,S(x)=0$$

where we identify:

$$a = -\frac{1}{4}(\lambda - 2l - 3), \qquad b = l + \frac{3}{2}$$

As discussed in Example 7.3, admissible polynomial solutions only exist if a is taken equal to a negative integer $-v$, which implies:

$$\frac{1}{4}(\lambda - 2l - 3) = v \quad \text{or} \quad \lambda = 2(l + 2v) + 3 = 2n + 3$$

It follows that the eigenvalues can be written as:

$$E_n = \left(n + \frac{3}{2}\right)\hbar\omega \qquad n = 0, 1, 2, \ldots$$

as found before in Problem 6.3.3, provided the principal and orbital quantum numbers n and l are related by $n = l + 2v$. The corresponding eigenfunctions of the isotropic oscillator will have the form:

$$R_{nl}(\rho) = A_{nl}\rho^l e^{-\rho^2/2} F\left(\frac{l}{2} - \frac{n}{2}, l + \frac{3}{2}; \rho^2\right)$$

Problem 7.2.2. Use the radial equation for one-electron atoms to calculate the expectation value $\langle 1/r^2 \rangle_{nl}$ in a stationary state of the hydrogen atom.

(Solution): Because the spherical harmonic eigenfunctions of the rotational motion are normalized on the unit sphere, the expectation value of any power of r is obtained from:

$$\langle r^k \rangle_{nl} = \langle \Psi_{nlm} | r^k | \Psi_{nlm} \rangle = \int_0^\infty r^k R_{nl}^2(r) r^2 dr = \int_0^\infty r^k \xi_{nl}^2(r) dr$$

where the functions $\xi_{nl}(r)$ are solutions to the radial equation (7.57), and are normalized according to Eq.(7.85). Substituting the energy eigenvalues E_n from Eq.(7.80) gives:

$$\frac{d^2 \xi_{nl}(r)}{dr^2} + \frac{2m_e}{\hbar^2}\left[-\frac{e_0^2}{2a_0 n^2} + \frac{e_0^2}{r} - \frac{l(l+1)\hbar^2}{2m_e r^2}\right]\xi_{nl}(r) = 0$$

or:

$$\frac{d^2 \xi_{nl}(r)}{dr^2} + \left[-\frac{1}{a_0^2 n^2} + \frac{2}{a_0 r} - \frac{l(l+1)}{r^2}\right]\xi_{nl}(r) = 0$$

It is convenient to eliminate the Bohr radius a_0 by setting $r = a_0 \rho$:

$$\frac{d^2\xi_{nl}(\rho)}{d\rho^2} = \left[\frac{1}{n^2} - \frac{2}{\rho} + \frac{l(l+1)}{\rho^2}\right]\xi_{nl}(\rho)$$

The property of the radial functions:

$$\int_0^\infty \xi_{nl}^2 \frac{\partial}{\partial l}\left(\frac{\partial^2 \xi_{nl}/\partial\rho^2}{\xi_{nl}}\right)d\rho = \int_0^\infty \xi_{nl}\frac{\partial}{\partial l}\left(\frac{\partial^2 \xi_{nl}}{\partial\rho^2}\right)d\rho - \int_0^\infty \frac{\partial^2 \xi_{nl}}{\partial\rho^2}\frac{\partial\xi_{nl}}{\partial l}d\rho = 0$$

implies that:

$$\int_0^\infty \xi_{nl}^2 \frac{\partial}{\partial l}\left[\frac{1}{n^2} - \frac{2}{\rho} + \frac{l(l+1)}{\rho^2}\right]d\rho = -\frac{2}{n^3}\int_0^\infty \xi_{nl}^2\, d\rho + (2l+1)\int_0^\infty \frac{1}{\rho^2}\xi_{nl}^2\, d\rho$$

$$= -\frac{2}{a_0 n^3}\int_0^\infty \xi_{nl}^2\, dr + (2l+1)a_0 \int_0^\infty \frac{1}{r^2}\xi_{nl}^2\, dr = 0$$

which yields:

$$\left\langle \frac{1}{r^2} \right\rangle_{nl} = \frac{2}{(2l+1)n^3}\frac{1}{a_0^2}$$

Problem 7.2.3. Show that from the radial equation of the hydrogen atom one can derive the Kramers recurrence relation for the expectation values of the powers of r, in the form:

$$\frac{s+1}{n^2}\langle r^s\rangle_{nl} - (2s+1)a_0\langle r^{s-1}\rangle_{nl} + \frac{sa_0^2}{4}\left[(2l+1)^2 - s^2\right]\langle r^{s-2}\rangle_{nl} = 0$$

where $s > -2l - 3$.

Problem 7.2.4. Show that:

$$\left\langle \frac{1}{r} \right\rangle_{nl} = \frac{1}{n^2 a_0},\quad \langle r \rangle_{nl} = \frac{1}{2}\left[3n^2 - l(l+1)\right]a_0,\quad \langle r^2\rangle_{nl} = \frac{n^2}{2}\left[5n^2 + 1 - 3l(l+1)\right]a_0^2$$

using the Kramers recurrence relation and Problem 7.2.2.

7.3. One-Electron Atoms

The results (7.80) and (7.84), obtained for the motion of a particle in a Coulomb force field, are of fundamental importance for the description of eigenstates of one-electron atoms, which consist of a negatively charged electron moving in the electric field of a nucleus of mass M. However, because the nucleus is not infinitely massive, we have to solve a two-body problem, for which the Schrödinger wave equation should be written in terms of the coordinates $\vec{r}_1$ and $\vec{r}_2$ of the nucleus and electron, respectively:

$$\left(-\frac{\hbar^2}{2M}\nabla_1^2 - \frac{\hbar^2}{2m_e}\nabla_2^2 - \frac{Ze_0^2}{|\vec{r}_1 - \vec{r}_2|}\right)\Psi(\vec{r}_1,\vec{r}_2) = E_a \Psi(\vec{r}_1,\vec{r}_2) \qquad (7.100)$$

This equation can be conveniently separated into two independent motions, by introducing the coordinates $\vec{R}(X,Y,Z)$ of the centre of mass of the atom and the relative coordinates $\vec{r}(x,y,z)$, defined as:

$$\vec{R} = \frac{M\vec{r}_1 + m_e \vec{r}_2}{M + m_e}, \qquad \vec{r} = \vec{r}_2 - \vec{r}_1 \qquad (7.101)$$

such that we have:

$$\frac{\partial}{\partial x_1} = \frac{M}{M+m_e}\frac{\partial}{\partial X} - \frac{\partial}{\partial x}, \quad \frac{\partial^2}{\partial x_1^2} = \left(\frac{M}{M+m_e}\right)^2 \frac{\partial^2}{\partial X^2} + \frac{\partial^2}{\partial x^2} - \frac{2M}{M+m_e}\frac{\partial^2}{\partial X\,\partial x}$$

$$(7.102)$$

$$\frac{\partial}{\partial x_2} = \frac{m_e}{M+m_e}\frac{\partial}{\partial X} + \frac{\partial}{\partial x}, \quad \frac{\partial^2}{\partial x_2^2} = \left(\frac{m_e}{M+m_e}\right)^2 \frac{\partial^2}{\partial X^2} + \frac{\partial^2}{\partial x^2} + \frac{2m_e}{M+m_e}\frac{\partial^2}{\partial X\,\partial x}$$

which means that:

$$\frac{1}{M}\nabla_1^2 + \frac{1}{m_e}\nabla_2^2 = \frac{1}{M+m_e}\nabla_R^2 + \left(\frac{1}{M}+\frac{1}{m_e}\right)\nabla_{\vec{r}}^2 = \frac{1}{M+m_e}\nabla_R^2 + \frac{1}{\mu}\nabla_{\vec{r}}^2$$

where $\mu = Mm_e/(M+m_e)$ is the reduced mass of the atom. The Schrödinger equation (7.100) can thus be written in terms of $\vec{R}$ and $\vec{r}$ in the form:

$$\left[-\frac{\hbar^2}{2(M+m_e)}\nabla_R^2 - \frac{\hbar^2}{2\mu}\nabla_{\vec{r}}^2 - \frac{Ze_0^2}{r}\right]\Psi(\vec{r},\vec{R}) = E_a \Psi(\vec{r},\vec{R}) \qquad (7.103)$$

which is separable by taking:

$$\Psi(\vec{r},\vec{R}) = \Psi(\vec{r})\varphi(\vec{R}), \qquad E_a = E + E_R \qquad (7.104)$$

The motion of the centre of mass is obtained by solving:

$$-\frac{\hbar^2}{2(M+m_e)}\nabla_{\vec{R}}^2 \varphi(\vec{R}) = E_R \varphi(\vec{R}) \qquad (7.105)$$

which is a Schrödinger wave equation for a free particle, with a continuous energy eigenvalue spectrum and plane wave eigenfunctions:

$$\varphi(\vec{R}) = a_1 e^{i\vec{k}\vec{R}} + a_2 e^{-i\vec{k}\vec{R}}, \qquad E_R = \frac{\hbar^2 k^2}{2(M+m_e)} \qquad (7.106)$$

This solution represents the whole atom of mass $M+m_e$ and kinetic energy E_R moving as a plane wave in space. The total momentum, which can be derived, using Eq.(7.102), as:

$$\hat{P}_x = \hat{p}_{1x} + \hat{p}_{2x} = -i\hbar\left(\frac{\partial}{\partial x_1} + \frac{\partial}{\partial x_2}\right) = -i\hbar\frac{\partial}{\partial X} \quad \text{or} \quad \hat{\vec{P}} = -i\hbar\nabla_{\vec{R}} \qquad (7.107)$$

is a constant of this motion. The relative motion of the electron about the nucleus satisfies the equation:

$$\left(-\frac{\hbar^2}{2\mu}\nabla_{\vec{r}}^2 - \frac{Ze_0^2}{r}\right)\Psi(\vec{r}) = E\Psi(\vec{r}) \qquad (7.108)$$

which is the same as the Schrödinger equation with a central potential, Eq.(7.1), if the electronic mass is replaced by the reduced mass:

$$\mu = \frac{m_e}{1+m_e/M} \approx m_e\left(1-\frac{m_e}{M}\right) \qquad (7.109)$$

The relative momentum of this motion is defined in correspondence to the classical result:

$$\vec{p} = \mu\vec{v} = \mu\left(\frac{d\vec{r}_2}{dt} - \frac{d\vec{r}_1}{dt}\right) = \frac{m_e}{M+m_e}\vec{p}_2 - \frac{M}{M+m_e}\vec{p}_1$$

which, in our case, in view of Eq.(7.102), reads:

$$\hat{p}_x = -i\hbar\left(\frac{m_e}{M+m_e}\frac{\partial}{\partial x_2} - \frac{M}{M+m_e}\frac{\partial}{\partial x_1}\right) = -i\hbar\frac{\partial}{\partial x} \quad \text{or} \quad \hat{\vec{p}} = -i\hbar\nabla_{\vec{r}} \qquad (7.110)$$

7. Electron Motion in the Atom

We observe that the total orbital angular momentum of the atom can also be separated in the form:

$$\hat{\vec{L}} = \vec{r}_1 \times \hat{\vec{p}}_1 + \vec{r}_2 \times \hat{\vec{p}}_2 = \vec{R} \times \hat{\vec{P}} + \vec{r} \times \hat{\vec{p}} = \hat{\vec{L}}_R + \hat{\vec{L}}_r \qquad (7.111)$$

where $\hat{\vec{L}}_R$ and $\hat{\vec{L}}_r$ commute, and each represents a constant of motion. Separating the relative angular motion, whose eigenfunctions are the spherical harmonics (7.35), as discussed in Section 7.1, we are left with the one-dimensional problem, Eq.(7.57):

$$\frac{d^2\xi(r)}{dr^2} + \frac{2\mu}{\hbar^2}\left[E + \frac{Ze_0^2}{r} - \frac{l(l+1)\hbar^2}{2\mu r^2}\right]\xi(r) = 0 \qquad (7.112)$$

written for a particle of mass μ moving in the effective potential energy (7.44), which for one-electron atoms reads:

$$U_{eff}(r) = -\frac{Ze_0^2}{r} + \frac{l(l+1)\hbar^2}{2\mu r^2} \qquad (7.113)$$

This function is plotted in Figure 7.1 for a few lowest values of the orbital quantum number.

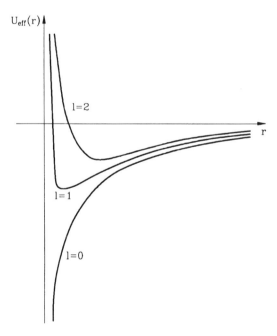

Figure 7.1. The effective potential energy curves for the lowest values of the orbital quantum number

Figure 7.1 indicates that radial motion with positive energies corresponds to the physical situation of scattering of an electron by a positive charge Ze, with which it is made to collide. The bound states of one-electron atoms result from radial motion with negative energies, and it is expected that the highest probability to find the electron will correspond to the classically allowed region, defined by:

$$E + \frac{Ze_0^2}{r} - \frac{l(l+1)\hbar^2}{2\mu r^2} \geq 0 \tag{7.114}$$

The exact solution of the radial equation yields the energy eigenvalues (7.80), where the reduced mass μ should take the place of the electronic mass:

$$E_n = -\frac{Z^2}{n^2} \frac{\mu e_0^4}{2\hbar^2} = -\frac{Z^2}{n^2} \frac{e_0^2}{2a} \tag{7.115}$$

and a new constant a, which approaches the Bohr radius a_0 when $M \to \infty$, is introduced:

$$a = \left(1 + \frac{m_e}{M}\right) a_0 \tag{7.116}$$

Substituting the energy level E_n into Eq.(7.114), the inequality:

$$-\frac{Z^2}{n^2} \frac{e_0^2}{2a} + \frac{Ze_0^2}{r} - \frac{l(l+1)\hbar^2}{2\mu r^2} \geq 0 \tag{7.117}$$

is satisfied by the r values situated between the two roots of the equation:

$$r^2 - 2\frac{an^2}{Z} r + \frac{a^2 n^2}{Z^2} l(l+1) = 0 \tag{7.118}$$

which are:

$$r_{1,2} = a\frac{n}{Z}\left[n \pm \sqrt{n^2 - l(l+1)}\right] \tag{7.119}$$

This means that, for the stationary state defined by the quantum numbers n and l, the electron will be found with highest probability in a region of width $an\sqrt{n^2 - l(l+1)}/Z$ about the value $r_n = an^2/Z$. For $n = 1$, $Z = 1$ we obtain $r_n = a$ and the electron moves in a spherical region defined by the Bohr radius. The width of the probability distribution decreases as l increases up to the maximum value $n-1$, in a stationary state with a defined n value. When $r \to 0$, the repulsive potential energy dominates the Coulomb

7. Electron Motion in the Atom

attraction, such that for small values l, there will be a significant probability for the electron to get near to the nucleus, but this probability will die out fast as l increases up to its highest value $l = n - 1$, and hence, the electron will be kept away from the nucleus. The details of the quantum description of radial motion result from the graphical representation of the *radial distribution function* for the atomic electron:

$$P_{nl}(r) = \xi_{nl}^2(r) = R_{nl}^2(r) r^2 \tag{7.120}$$

which gives the probability density to find the electron between the spheres of radii r and $r + dr$, in terms of the normalized solutions $R_{nl}(r)$ of the radial equation. In Figure 7.2 it is plotted $P_{nl}(r)$ in terms of r, expressed in a/Z units, using the radial functions from Eq.(7.87). It can be seen that for the 1s, 2p and 3d electron states, where $l = n - 1$, the radial distribution function exhibits a unique maximum situated at $r_n = n^2 a/Z$, whose width is increasing as $n^{3/2}$. For electron states characterized by l values lower than $n - 1$, the radial solution $R_{nl}(r)$ have $n - 1 - l$ nodes, and consequently the radial probability includes secondary maxima of lower intensity.

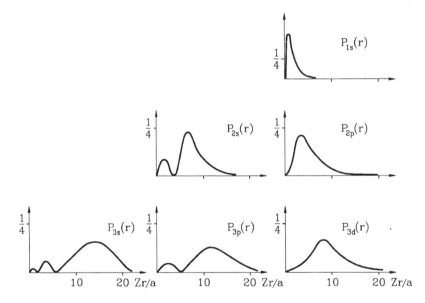

Figure 7.2. The radial distribution function for electron states of lowest energies

A stationary state, determined by the quantum numbers n, l and m, is completely described by the probability density $P_{nlm}(r, \theta, \varphi)$ that the polar coordinates of the electron, are between r and $r + dr$, θ and $\theta + d\theta$ and φ and $\varphi + d\varphi$, defined as:

$$P_{nlm}(r, \theta, \varphi) r^2 \sin\theta \, d\theta \, d\varphi \, dr = |\Psi_{nlm}|^2 r^2 dr \, d\Omega = \left(R_{nl}^2(r) r^2 dr \right) \left(|Y_{lm}(\theta, \varphi)|^2 d\Omega \right) \tag{7.121}$$

272 Quantum Physics

which shows that $P_{nlm}(r,\theta,\varphi)$ is a product between the radial distribution function (7.120) and the *angular distribution function*:

$$P_{lm}(\theta,\varphi) = |Y_{lm}(\theta,\varphi)|^2 \qquad (7.122)$$

The probability density $P_{lm}(\theta,\varphi)$ of finding the electron within a solid angle element $d\Omega$ about the origin, is plotted in Figure 7.3 in polar coordinates, using the spherical harmonics (7.35), which were normalized on a spherical surface of unit radius. Because the angular probability does not depend on φ, the polar diagrams are drawn by taking the origin at the nucleus, where $r = 0$, the z-axis along the direction from which the angle θ is measured, and the distance from the origin to the surface, measured at the angle θ, equal to the value of $P_{lm}(\theta,\varphi)$ for that angle. The three-dimensional surface is obtained by rotating each diagram about the z-axis, through the 2π range of the angle φ.

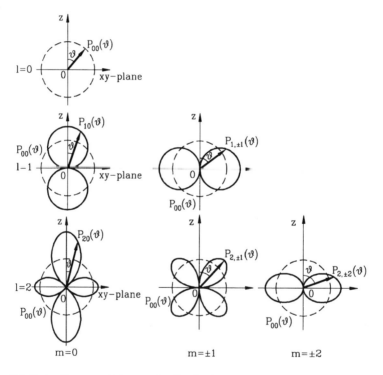

Figure 7.3. Polar diagrams of the angular distribution function for electron states of lowest l values

For s electron states, the angular distribution function is spherically symmetrical and $P_{nlm}(r,\theta,\varphi)$ reduces to the radial distribution function (7.120). Each p electron state is three-fold degenerate, and we can write, by combining Eqs.(7.36) and (7.87), the normalized 2p eigenfunctions as:

$$\Psi_{2p_0} = R_{2p}(r)Y_{1,0}(\theta,\varphi) = \frac{1}{4\sqrt{2\pi}}\left(\frac{Z}{a}\right)^{\frac{5}{2}} r\cos\theta\, e^{-Zr/2a}$$

(7.123)

$$\Psi_{2p_{\pm 1}} = R_{2p}(r)Y_{1,\pm 1}(\theta,\varphi) = \frac{1}{8\sqrt{\pi}}\left(\frac{Z}{a}\right)^{\frac{5}{2}} r\sin\theta\, e^{\pm i\varphi} e^{-Zr/2a}$$

As discussed in Chapter 3, it is often convenient to replace these eigenfunctions by a different set $\Psi_{2p_{x_i}}$, by taking linear combinations of Ψ_{2p_m}. In view of Eq.(2.20), we may introduce the new eigenfunctions:

$$\Psi_{2p_x} = \frac{1}{\sqrt{2}}\left(\Psi_{2p_1} + \Psi_{2p_{-1}}\right) = \frac{1}{4\sqrt{2\pi}}\left(\frac{Z}{a}\right)^{\frac{5}{2}} x e^{-Zr/2a}$$

$$\Psi_{2p_y} = \frac{i}{\sqrt{2}}\left(\Psi_{2p_1} - \Psi_{2p_{-1}}\right) = \frac{1}{4\sqrt{2\pi}}\left(\frac{Z}{a}\right)^{\frac{5}{2}} y e^{-Zr/2a}$$

(7.124)

$$\Psi_{2p_z} = \Psi_{2p_0} = \frac{1}{4\sqrt{2\pi}}\left(\frac{Z}{a}\right)^{\frac{5}{2}} z e^{-Zr/2a}$$

It is clear that Ψ_{2p_z} is directed along the z-axis and is zero in the xy-plane, and similarly Ψ_{2p_x} and Ψ_{2p_y} are directed along the x-axis and y-axis respectively, as shown by the polar diagram for $l = 1$ in Figure 7.3.

The d electron states are five-fold degenerate, as described for $n = 3$ by the eigenfunctions:

$$\Psi_{3d_0} = R_{3d}(r)Y_{2,0}(\theta,\varphi) = \frac{1}{81\sqrt{6\pi}}\left(\frac{Z}{a}\right)^{\frac{7}{2}} r^2 (3\cos^2\theta - 1) e^{-Zr/3a}$$

$$\Psi_{3d_{\pm 1}} = R_{3d}(r)Y_{2,\pm 1}(\theta,\varphi) = \frac{1}{81\sqrt{\pi}}\left(\frac{Z}{a}\right)^{\frac{7}{2}} r^2 \sin\theta\cos\theta\, e^{\pm i\varphi} e^{-Zr/3a}$$

(7.125)

$$\Psi_{3d_{\pm 2}} = R_{3d}(r)Y_{2,\pm 2}(\theta,\varphi) = \frac{1}{81\sqrt{4\pi}}\left(\frac{Z}{a}\right)^{\frac{7}{2}} r^2 \sin^2\theta\, e^{\pm 2i\varphi} e^{-Zr/3a}$$

where use has been made of Eqs.(7.37) and (7.87). By taking appropriate linear combinations of Ψ_{3d_m} we obtain the new set of degenerate eigenfunctions:

$$\Psi_{3d_{zz}} = \Psi_{3d_0} = \frac{1}{81\sqrt{6\pi}} \left(\frac{Z}{a}\right)^{\frac{7}{2}} (r^2 - 3z^2) e^{-Zr/3a}$$

$$\Psi_{3d_{xz}} = \frac{1}{81} \sqrt{\frac{2}{\pi}} \left(\frac{Z}{a}\right)^{\frac{7}{2}} xz\, e^{-Zr/3a}, \quad \Psi_{3d_{yz}} = \frac{1}{81} \sqrt{\frac{2}{\pi}} \left(\frac{Z}{a}\right)^{\frac{7}{2}} yz\, e^{-Zr/3a} \qquad (7.126)$$

$$\Psi_{3d_{xy}} = \frac{1}{81} \sqrt{\frac{2}{\pi}} \left(\frac{Z}{a}\right)^{\frac{7}{2}} xy\, e^{-Zr/3a}, \quad \Psi_{3d_{x^2-y^2}} = \frac{1}{81\sqrt{2\pi}} \left(\frac{Z}{a}\right)^{\frac{7}{2}} (x^2 - y^2) e^{-Zr/3a}$$

The function $\Psi_{3d_{zz}}$ has the relative maxima along the z-axis and in the xy-plane ($z = 0$) and is zero for $z = r/\sqrt{3}$, which defines a cone passing through the origin and making an angle $\theta = \arccos(1/\sqrt{3})$ with the z-axis. The function $\Psi_{3d_{xz}}$ is zero for $x = 0$ and $z = 0$ and its relative maxima are contained in the planes $x = \pm z$. A similar behaviour, corresponding to $l = 2$ and $m = \pm 1$ in Figure 7.3, has $\Psi_{3d_{yz}}$. Finally, $\Psi_{3d_{xy}}$ is zero along the z-axis ($x = 0$, $y = 0$) and the planes $x = \pm y$ contain the relative maxima, and $\Psi_{3d_{x^2-y^2}}$ behaves similarly if we rotate $\pi/4$ around the z-axis (see the angular distribution function for $l = 2$, $m = \pm 2$). It is apparent from Figure 7.3 that, as m increases, the angular probability density shifts from the z-axis towards the equatorial plane xy. When $|m| = l$ and also l takes large values, the classical limit of the Sommerfeld planar orbits, discussed in Example P.4, is approached.

The ground state of the one-electron atom is characterized by the lowest values of the quantum numbers: $n = 1$, $l = 0$, $m = 0$. Any state with $n > 1$ is called an *excited state* and, in view of Eq.(7.83), it is always *degenerate*. In other words there are several electron eigenstates, with different l and m, which have the same energy E_n. The degeneracy induced by the n values $l \leq n - 1$ corresponding to a given value of the principal quantum number, is called *accidental*. Furthermore, the $2l + 1$ values of the magnetic quantum number allowed for each l electron state gives rise to a *spatial degeneracy*. The correspondence between the electron states and the elliptical orbits, derived for the electron motion from the quantization rule (P.28), is convenient for illustrating the concept of degeneracy of the stationary states in one-electron atoms. The accidental degeneracy is comparable with the case where different elliptical orbits have the same energy, provided they have the same major axis. The spatial degeneracy corresponds to the property of the orbit to preserve its energy during a rotation of its plane, which is restricted by the spatial quantization of the orbital angular momentum.

The degeneracy of the excited states of one-electron atoms is removed by small interactions of the electron with external fields. A splitting of the degenerate energy level E_n will be expected to occur, because some of the corresponding eigenfunctions Ψ_{nlm} might be more affected than others.

Problem 7.3.1. Calculate the expectation value of the potential energy of the electron in the ground state of the hydrogen atom, and show that the virial theorem is satisfied.

(Solution): The ground state of the hydrogen atom is described by the normalized wave function $\Psi_{100}(r)$, which is obtained from Eqs.(7.35) and (7.87), for $Z = 1$, as:

$$\Psi_{100}(r) = R_{10}(r)Y_{10}(\theta,\varphi) = \frac{1}{\sqrt{\pi a^3}} e^{-r/a}$$

In view of the spherical symmetry, we only may define the radial probability density:

$$P(r) = |\Psi_{100}(r)|^2 4\pi r^2 = \frac{4}{a^3} r^2 e^{-2r/a}$$

of finding the electron in a spherical shell of thickness dr, at a distance r from the nucleus, which is taken at origin. The expectation value of the potential energy of the ground state is then obtained from:

$$\left\langle \frac{1}{r} \right\rangle_{10} = \left\langle \Psi_{100} \left| \frac{1}{r} \right| \Psi_{100} \right\rangle = \int_0^\infty \frac{1}{r} P(r)dr = \frac{4}{a^3} \int_0^\infty e^{-2r/a} r\, dr = \frac{1}{a}$$

where a is the Bohr radius. Substituting into Eq.(7.56) this implies:

$$\langle U(r) \rangle_{10} = -e_0^2 \left\langle \frac{1}{r} \right\rangle_{10} = -\frac{e_0^2}{a} = 2E_1$$

where the ground energy level E_1 is given by Eq.(7.115). It follows that the kinetic energy has the expectation value:

$$\langle \hat{T} \rangle_{10} = E_1 - \langle U \rangle_{10} = -E_1 = -\frac{1}{2} \langle U \rangle_{10}$$

which is in agreement with the virial theorem (see Problem 2.1.1) for a homogenous function $U(r)$ of the order -1.

Problem 7.3.2. Show that the Bohr radius a represents the radial position of maximum probability for an electron in the ground state of the hydrogen atom.

Problem 7.3.3. Calculate the expectation values $\langle r \rangle, \langle r^2 \rangle, \langle p_r \rangle$ and $\langle p_r^2 \rangle$ in the ground state of the hydrogen atom and show that the uncertainty principle is satisfied.

Problem 7.3.4. Calculate the expectation value of kinetic energy of the electron in the stationary state Ψ_{nlm} of hydrogen.

(Answer): $\langle T \rangle_{nlm} = \dfrac{e^2}{2n^2 a}$

7.4. The Magnetic Moment of the Electron

The orbital motion of the electron charge $-e$ around the nucleus is classically equivalent to a current flow of density $\vec{j}_e = \rho \vec{v}$, where ρ is the charge density in the atomic volume V:

$$-e = \int_V \rho dV \qquad (7.127)$$

The orbital magnetic moment $\vec{\mu}_l$, Eq.(P.34), which is assigned to each electron orbit of angular momentum $\vec{L}$, can be written as:

$$\vec{\mu}_l = -\dfrac{e}{2m_e}\vec{L} = \dfrac{1}{2m_e}\int_V \rho \vec{L} dV = \dfrac{1}{2m_e}\int_V \rho(\vec{r}\times\vec{p})dV = \dfrac{1}{2}\int_V (\vec{r}\times\rho\vec{v})dV = \dfrac{1}{2}\int_V (\vec{r}\times\vec{j}_l)dV$$

(7.128)

The quantum mechanical description of the electron, in terms of the allowed stationary states Ψ_{nlm}, leads to a statistical distribution of its position with probability density $P_{nlm} = |\Psi_{nlm}|^2$, given by Eq.(7.121), and hence the charge density will be given by:

$$\rho = -e \Psi^*_{nlm} \Psi_{nlm} \qquad (7.129)$$

which satisfies Eq.(7.127), because the eigenfunctions Ψ_{nlm} are normalized, according to Eq.(7.85). Consequently, the continuity equation for probability, Eq.(5.36), implies that the electron current density $\vec{j}_l$ should be taken as:

$$\hat{\vec{j}}_l = -e\hat{\vec{j}} = \dfrac{ie\hbar}{2m_e}(\Psi^*_{nlm}\nabla\Psi_{nlm} - \Psi_{nlm}\nabla\Psi^*_{nlm}) \qquad (7.130)$$

where the probability current density $\hat{\vec{j}}$ is given by Eq.(5.35). It follows that the magnitude of the orbital magnetic moment (7.128) will be given by:

$$\vec{\mu}_l = \frac{ie\hbar}{4m_e} \int_V \left[\Psi_{nlm}^* (\vec{r} \times \nabla) \Psi_{nlm} - \Psi_{nlm} (\vec{r} \times \nabla) \Psi_{nlm}^* \right] dV$$

$$= -\frac{e}{4m_e} \left[\int_\infty \Psi_{nlm}^* \hat{\vec{L}} \Psi_{nlm} dV + \int_\infty (\hat{\vec{L}} \Psi_{nlm})^* \Psi_{nlm} dV \right]$$

$$= -\frac{e}{4m_e} \left(\langle \Psi_{nlm} | \hat{\vec{L}} \Psi_{nlm} \rangle + \langle \hat{\vec{L}} \Psi_{nlm} | \Psi_{nlm} \rangle \right) = -\frac{e}{2m_e} \langle \Psi_{nlm} | \hat{\vec{L}} | \Psi_{nlm} \rangle \quad (7.131)$$

which is the expectation value of an *orbital magnetic moment* operator $\hat{\vec{\mu}}_l$, defined in correspondence to the classical observables:

$$\hat{\vec{\mu}}_l = -\frac{e}{2m_e} \hat{\vec{L}} \quad (7.132)$$

The orbital magnetic moment is associated with the angular distribution properties of the stationary states, and only vanishes for the s electron states of spherical symmetry, where $l = 0$. Because of the quantization of $\hat{L}^2$ and $\hat{L}_z$, Eqs.(7.19) and (7.12), the magnitude of the orbital magnetic moment and of its component along the z-axis have discrete sets of allowed values. Namely:

$$\mu_{lz} = -\frac{e}{2m_e} m\hbar = -m\mu_B \quad (7.133)$$

where the parameter μ_B, called the *Bohr magneton*, is defined in terms of the characteristic electron constants:

$$\mu_B = \frac{e\hbar}{2m_e} \quad (7.134)$$

EXAMPLE 7.4. Classical motion in electric and magnetic fields

If the canonical equations of motion (4.24):

$$\vec{v} = \frac{d\vec{r}}{dt} = \frac{\partial H}{\partial \vec{p}} \quad \text{and} \quad \frac{d\vec{p}}{dt} = -\frac{\partial H}{\partial \vec{r}} = -\nabla H$$

are applied to the motion of an electron in an electrostatic field, defined by the scalar potential V, with the Hamiltonian:

$$H = \frac{p^2}{2m_e} + U = \frac{p^2}{2m_e} - eV$$

we obtain $\vec{p} = m_e \vec{v}$ and:

$$\frac{d\vec{p}}{dt} = -\frac{\partial H}{\partial \vec{r}} = e \nabla V \quad \text{or} \quad \frac{d(m_e \vec{v})}{dt} = -e\vec{E} = \vec{F} \tag{7.135}$$

In the presence of an electromagnetic field, in which the electric and magnetic field intensities $\vec{E}$ and $\vec{B}$ are defined in terms of the scalar and vector potentials V and $\vec{A}$:

$$\vec{E} = -\nabla V - \frac{\partial \vec{A}}{\partial t}, \quad \vec{B} = \nabla \times \vec{A} \tag{7.136}$$

the equation of motion (7.135) should involve the Lorentz force:

$$\vec{F} = -e(\vec{E} + \vec{v} \times \vec{B}) = -e\left[-\nabla V - \frac{\partial \vec{A}}{\partial t} + \vec{v} \times (\nabla \times \vec{A})\right]$$

$$= -e\left[-\nabla V - \frac{\partial \vec{A}}{\partial t} + \nabla(\vec{v} \cdot \vec{A}) - (\vec{v} \cdot \nabla)\vec{A}\right] = e\left[\frac{\partial \vec{A}}{\partial t} + (\vec{v} \cdot \nabla)\vec{A} + \nabla(V - \vec{v} \cdot \vec{A})\right]$$

Because $\vec{A}$ is, in general, a function of the position vector $\vec{r}$ and of the time, we have:

$$\frac{d\vec{A}}{dt} = \frac{\partial \vec{A}}{\partial t} + (\vec{v} \cdot \nabla)\vec{A}$$

and therefore:

$$\vec{F} = e\left[\frac{d\vec{A}}{dt} + \nabla(V - \vec{v} \cdot \vec{A})\right]$$

such that Eq.(7.135) is replaced by:

$$\frac{d}{dt}(m_e \vec{v} - e\vec{A}) = -\nabla(-eV + e\vec{v} \cdot \vec{A}) \tag{7.137}$$

This equation implies that the appropriate Hamiltonian function in electromagnetic fields should be written as:

$$H = \frac{1}{2m_e}(\vec{p} + e\vec{A})^2 - eV \tag{7.138}$$

which gives:

$$\vec{v} = \frac{\partial H}{\partial \vec{p}} = \frac{1}{m_e}(\vec{p} + e\vec{A}) \quad \text{or} \quad \vec{p} = m_e \vec{v} - e\vec{A} \tag{7.139}$$

Moreover, we have:

$$\frac{dp_i}{dt} = -\frac{\partial H}{\partial x_i} = -\frac{1}{m_e}(\vec{p}+e\vec{A})\cdot\frac{\partial}{\partial x_i}(\vec{p}+e\vec{A})+e\frac{\partial V}{\partial x_i} = e\frac{\partial V}{\partial x_i}-\vec{v}\cdot\frac{\partial}{\partial x_i}(e\vec{A}) = e\frac{\partial V}{\partial x_i}-e\frac{\partial}{\partial x_i}(\vec{v}\cdot\vec{A})$$

which can be combined with Eq.(7.139) to yield the i component of Eq.(7.137):

$$\frac{d}{dt}(m_e v_i - eA_i) = e\frac{\partial V}{\partial x_i} - e\frac{\partial}{\partial x_i}(\vec{v}\cdot\vec{A})$$

The Hamiltonian operator for the motion of the atomic electron in the presence of an external magnetic field is obtained from Eq.(7.138), if $U(r)=-eV$ denotes the potential energy of the central electrostatic field of the nucleus and if $\vec{r}$ and $\vec{p}$ are replaced by their representative operators. ⚘

The Schrödinger equation (5.24), in an external magnetic field, becomes:

$$i\hbar\frac{\partial\Psi(\vec{r},t)}{\partial t} = \hat{H}\Psi(\vec{r},t) = \left[\frac{1}{2m_e}\left(\hat{\vec{p}}+e\vec{A}\right)^2 + U(r)\right]\Psi(\vec{r},t) \qquad (7.140)$$

where the operators $\hat{p}_i$ and A_i do not commute:

$$\hat{p}_i A_i \Psi = -i\hbar\frac{\partial}{\partial x_i}(A_i\Psi) = \left(A_i\hat{p}_i - i\hbar\frac{\partial A_i}{\partial x_i}\right)\Psi$$

so that we have:

$$(\hat{\vec{p}}+e\vec{A})^2 = (\hat{\vec{p}}+e\vec{A})\cdot(\hat{\vec{p}}+e\vec{A}) = \hat{\vec{p}}^2 + e(\vec{A}\cdot\hat{\vec{p}}+\hat{\vec{p}}\cdot\vec{A}) + e^2\vec{A}^2$$

It follows that Eq.(7.140) can be written in the form (see Problem 2.3.2):

$$i\hbar\frac{\partial\Psi}{\partial t} = -\frac{\hbar^2}{2m_e}\nabla^2\Psi - \frac{ie\hbar}{m_e}\vec{A}\cdot\nabla\Psi - \frac{ie\hbar}{2m_e}(\nabla\cdot\vec{A})\Psi + \frac{e^2}{2m_e}A^2\Psi + U(r)\Psi \quad (7.141)$$

which by complex conjugation becomes:

$$i\hbar\frac{\partial\Psi^*}{\partial t} = \frac{\hbar^2}{2m_e}\nabla^2\Psi^* - \frac{ie\hbar}{m_e}\vec{A}\cdot\nabla\Psi^* - \frac{ie\hbar}{2m_e}(\nabla\cdot\vec{A})\Psi^* - \frac{e^2}{2m_e}A^2\Psi^* - U(r)\Psi^*$$

Multiplying the last two equations by Ψ^* and Ψ respectively, and summing we obtain:

$$i\hbar\frac{\partial(\Psi^*\Psi)}{\partial t} = -\frac{\hbar^2}{2m_e}(\Psi^*\nabla^2\Psi - \Psi\nabla^2\Psi^*) - \frac{ie\hbar}{m_e}\vec{A}\cdot(\Psi^*\nabla\Psi + \Psi\nabla\Psi^*) - \frac{ie\hbar}{2m_e}(\nabla\cdot\vec{A})\Psi^*\Psi$$

which can be written in the form of the continuity equation:

$$\frac{\partial}{\partial t}(-e\Psi^*\Psi) + \nabla\cdot\left[\frac{ie\hbar}{2m_e}(\Psi^*\nabla\Psi - \Psi\nabla\Psi^*) - \frac{e^2}{m_e}\vec{A}\Psi^*\Psi\right] = 0 \quad (7.142)$$

In view of Eqs.(7.129) and (7.130), Eq.(7.142) reads:

$$\frac{\partial\rho}{\partial t} + \nabla\cdot\left(\vec{j}_l - \frac{e^2}{m_e}\vec{A}\Psi^*\Psi\right) = 0 \quad (7.143)$$

and it follows that the presence of an external magnetic field introduces a supplementary term to the probability current density. If $\vec{j}_l$ gives rise to the orbital magnetic moment of the electron, according to Eq.(7.128), which is not related to $\vec{B}$, there exists an additional magnetic moment, of magnitude given by Eq.(7.128) as:

$$\vec{\mu}_d = -\frac{e^2}{2m_e}\int_V \Psi^*(\vec{r}\times\vec{A})\Psi dV = -\frac{e^2}{2m_e}\langle\Psi|\vec{r}\times\vec{A}|\Psi\rangle \quad (7.144)$$

This is the expectation value of the operator:

$$\hat{\mu}_d = -\frac{e^2}{2m_e}(\vec{r}\times\vec{A}) \quad (7.145)$$

which is the quantum representative of the a magnetic moment induced by the presence of an external magnetic field.

EXAMPLE 7.5. One-electron atoms in a uniform magnetic field

If the external magnetic field $\vec{B}$ is uniform, it is convenient to take:

$$\vec{A} = \frac{1}{2}(\vec{B}\times\vec{r}) \quad (7.146)$$

which obeys the constraint (7.136) on the vector potential, for all the spatial derivatives of $\vec{B}$ vanish:

$$\nabla\times\vec{A} = \nabla\times\left(\frac{1}{2}\vec{B}\times\vec{r}\right) = \frac{1}{2}\left[(\nabla\cdot\vec{r})\vec{B} - (\vec{B}\cdot\nabla)\vec{r}\right] = \frac{1}{2}(3\vec{B} - \vec{B}) = \vec{B}$$

With this choice for $\vec{A}$, the induced magnetic moment (7.145) takes the form:

$$\hat{\vec{\mu}}_d = -\frac{e^2}{2m_e}\left[\vec{r}\times\left(\frac{1}{2}\vec{B}\times\vec{r}\right)\right] = -\frac{e^2}{4m_e}\left[(\vec{r}\cdot\vec{B})\vec{r} - r^2\vec{B}\right] \qquad (7.147)$$

which is linear in $\vec{B}$. For weak magnetic fields, $\vec{\mu}_d$ is significant to the s electrons only, because the orbital magnetic moment $\vec{\mu}_l$ vanishes in the stationary states with $l = 0$:

$$\vec{\mu}_d = -\frac{e^2}{2m_e}\langle\Psi_{n00}|(\vec{r}\cdot\vec{B})\vec{r} - r^2\vec{B}|\Psi_{n00}\rangle \qquad (7.148)$$

Otherwise, the interaction of the induced magnetic moment with the external magnetic field is negligible as compared with that of the orbital magnetic moment. The wave functions Ψ_{n00} are spherically symmetrical, and in spherical polar coordinates we have:

$$\int_\Omega x_i x_j d\Omega = 0 \quad (i \neq j), \quad \int_\Omega x_i^2 d\Omega = \frac{1}{3}\int_\Omega r^2 d\Omega$$

so that Eq.(7.148) gives:

$$\vec{\mu}_d = -\frac{e^2\vec{B}}{4m_e}\int_V (r^2 - z^2)|\Psi_{n00}|^2\,dV = -\frac{e^2\vec{B}}{6m_e}\int_V r^2|\Psi_{n00}|^2\,dV = -\alpha_n\vec{B} \qquad (7.149)$$

where α_n is a positive constant, called the *magnetic polarizability*. It follows that the induced magnetic moment has always an antiparallel orientation to the magnetic field, which accounts for the so-called *diamagnetic* effect of an external magnetic field $\vec{B}$ on a macroscopic solid. Substituting Eq.(7.146), the Hamiltonian (7.138) can be written as:

$$H = \frac{1}{2m_e}\left(\vec{p} + \frac{e}{2}\vec{B}\times\vec{r}\right)^2 + U(r) = \frac{\vec{p}^2}{2m_e} + U(r) + \frac{e}{2m_e}\vec{p}\cdot(\vec{B}\times\vec{r}) + \frac{e^2}{2m_e}\left(\frac{1}{2}\vec{B}\times\vec{r}\right)^2$$

$$= \frac{\vec{p}^2}{2m_e} + U(r) + \frac{e}{2m_e}(\vec{r}\times\vec{p})\cdot\vec{B} + \frac{e^2}{2m_e}\left(\frac{1}{2}\vec{B}\times\vec{r}\right)^2$$

and the corresponding Hamiltonian operator reads:

$$\hat{H} = \frac{\hat{\vec{p}}^2}{2m_e} + U(r) + \frac{e}{2m_e}(\vec{r}\times\hat{\vec{p}})\cdot\vec{B} + \frac{e^2}{2m_e}\left(\frac{1}{2}\vec{B}\times\vec{r}\right)^2 = \hat{H}_0 + \frac{e}{2m_e}\hat{\vec{L}}\cdot\vec{B} + \frac{e^2}{8m_e}\left[r^2B^2 - (\vec{r}\cdot\vec{B})^2\right]$$

$$(7.150)$$

where $\hat{H}_0$ describes the electron motion in one-electron atoms in the absence of external fields. In the stationary states with $l \neq 0$, the contribution from the quadratic term in $\vec{B}$ may be dropped out from Eq.(7.150). If we take the z-direction to coincide with that of $\vec{B}$, the Hamiltonian will commute with $\hat{H}_0$ and $\hat{L}_z$, and hence will have the common set of eigenfunctions Ψ_{nlm} given by Eq.(7.84). The energy eigenvalue spectrum in a uniform magnetic field is obtained from:

$$\hat{H}\Psi_{nlm} = \hat{H}_0\Psi_{nlm} + \frac{eB}{2m_e}\hat{L}_z\Psi_{nlm} = \left(E_n + \frac{eB}{2m_e}m\hbar\right)\Psi_{nlm} = (E_n + m\mu_B B)\Psi_{nlm} \quad (7.151)$$

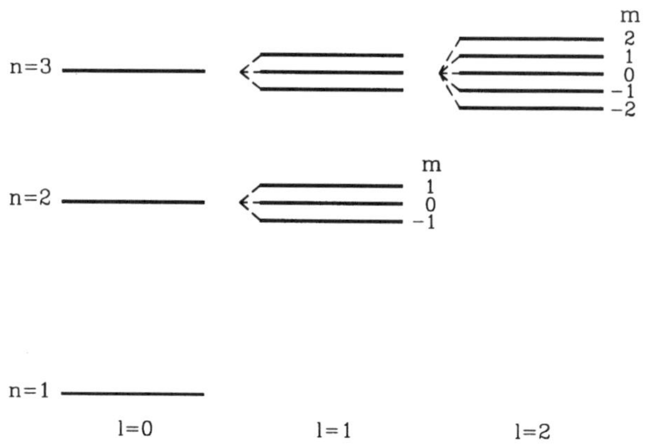

Figure 7.4. Normal Zeeman splitting of the lowest energy levels in one-electron atoms

Each energy level corresponding to degenerate l-states is splitted into $2l + 1$ distinct levels, the difference between the consecutive energy levels being $\mu_B B$, as illustrated in Figure 7.4. This coupling energy between the orbital magnetic moment and the external magnetic field:

$$\mu_B B = \frac{e\hbar}{2m_e}B = \hbar\omega_L$$

can be formally related to the classical picture of an angular momentum vector $\vec{L}$ precessing about the z-axis with the Larmor frequency ω_L, as discussed in Example P.4, provided that $\hbar\omega_L$ is interpreted as a quantum of energy for the precession motion. The splitting of the energy levels under the influence of magnetic field is called the *normal Zeeman effect* and removes the spatial degeneracy, with respect to m. However, the accidental degeneracy, with respect to l, is not lifted.

Problem 7.4.1. Find the energy eigenvalues and eigenfunctions of an electron moving in a uniform magnetic field $\vec{B}$.

(Solution): The eigenvalue problem reads:

7. Electron Motion in the Atom

$$\frac{1}{2m_e}(\hat{\vec{p}}+e\vec{A})^2 \Psi = E\Psi$$

and we may take the direction of $\vec{B}$ as the z-axis. Consequently, if we choose the vector potential of the uniform field of the form given in Eq.(7.146), we obtain:

$$\frac{1}{2}\vec{B}\times\vec{r} = \frac{1}{2}B(x\vec{e}_y - y\vec{e}_x) \quad \text{or} \quad A_x = -\frac{1}{2}By, \quad A_y = \frac{1}{2}Bx, \quad A_z = 0$$

There is, however, more convenient to take:

$$A_x = -By, \quad A_y = 0, \quad A_z = 0$$

which also satisfies the restriction $\vec{B} = \nabla \times \vec{A}$ when $\vec{B}$ is along the z-axis. With this choice, we have $\nabla \vec{A} = 0$ and the eigenvalue problem simplifies to:

$$-\frac{\hbar^2}{2m_e}\nabla^2\Psi + \frac{i\hbar}{m_e}eBy\frac{\partial\Psi}{\partial x} + \frac{e^2}{2m_e}B^2y^2\Psi = E\Psi$$

The Hamiltonian operator commutes with $\hat{p}_x$ and $\hat{p}_z$, which have eigenfunctions of the type (3.23), so that a trial solution has the form:

$$\Psi(x,y,z) = e^{i(p_x x + p_z z)/\hbar}u(y)$$

Substituting into the eigenvalue equation and dropping the exponential factors gives:

$$-\frac{\hbar^2}{2m_e}\frac{d^2u(y)}{dy^2} + \frac{e^2B^2}{2m_e}\left(y^2 - \frac{2yp_x}{eB} + \frac{p_x^2}{e^2B^2}\right)u(y) = \left(E - \frac{p_z^2}{2m_e}\right)u(y)$$

We introduce the new variable $\rho = y + p_x/eB$ such that:

$$-\frac{\hbar^2}{2m_e}\frac{d^2u(\rho)}{d\rho^2} + \frac{e^2B^2}{2m_e}\rho^2 u(\rho) = \left(E - \frac{p_z^2}{2m_e}\right)u(\rho)$$

This equation can be reduced to the eigenvalue equation (6.62) of the linear harmonic motion, if we set:

$$\varepsilon = E - \frac{p_z^2}{2m_e}, \quad \omega = \frac{eB}{m_e}$$

which gives:

$$\frac{d^2u(\rho)}{d\rho^2} + \frac{2m_e}{\hbar^2}\left(\varepsilon - \frac{m_e\omega^2}{2}\rho^2\right)u(\rho) = 0$$

It follows that the energy eigenvalue of an electron in a uniform magnetic field can be written as:

$$E_{n,p_z} = \frac{p_z^2}{2m_e} + \varepsilon_n = \frac{p_z^2}{2m_e} + \left(n+\frac{1}{2}\right)\hbar\omega = \frac{p_z^2}{2m_e} + \left(n+\frac{1}{2}\right)\hbar\frac{eB}{m_e}$$

where the first term has a continuous set of values, corresponding to the free motion along the z-axis, and the second term has a discrete set of values associated to the electron oscillation in a plane normal to $\vec{B}$. The corresponding eigenfunctions have the form (6.71), namely:

$$u_n(\xi) = N_n e^{-\xi^2/2} H_n(\xi)$$

where $H_n(\xi)$ are the Hermite polynomials and:

$$\xi = \sqrt{\frac{m_e\omega}{\hbar}}\rho = \sqrt{\frac{eB}{\hbar}}\left(y - \frac{p_x}{eB}\right)$$

Finally we obtain:

$$\Psi_{np_xp_z}(x, y, z) = N_n e^{i(p_x x + p_z z)/\hbar} e^{-\xi^2/2} H_n(\xi)$$

and there is infinite degeneracy, because for n and p_z fixed by E_{np_z}, p_x can take a continuous set of values.

Problem 7.4.2. Show that the total magnetic moment of the atomic electron in the presence of an external uniform magnetic field, can be expressed as:

$$\vec{\mu}_l + \vec{\mu}_d = -\frac{\partial \hat{H}}{\partial \vec{B}}$$

Problem 7.4.3. Calculate the magnetic polarizability α_n for s-electrons, in the presence of an external field B.

(Answer): $\alpha_n = \dfrac{e^2}{12m_e} n^2 (5n^2 + 1) a^2$

8. ANGULAR MOMENTUM

8.1. The Matrix Eigenvalue Problem for Angular Momentum

The orbital angular momentum operator has been introduced by Postulate 2 as a vector operator $\hat{\vec{L}} = \vec{r} \times \hat{\vec{p}}$ whose Cartesian components, Eq.(2.18), correspond to those of the classical observable for a single particle. As a consequence of Postulate 4, the components $\hat{L}_x, \hat{L}_y$ and $\hat{L}_z$ do not commute with one another, Eq.(4.47), and all commute with $\hat{L}^2$, and this implies that the eigenstates of $\hat{\vec{L}}$ are simultaneous eigenfunctions of $\hat{L}^2$ and of one component $\hat{L}_i$ at a time. Similar properties are exhibited by the angular momentum vector operator of a system of particles, which is defined as the vector sum over the individual particle operators:

$$\hat{\vec{J}} = \sum_p \hat{\vec{L}}(p), \quad \hat{J}^2 = \hat{J}_x^2 + \hat{J}_y^2 + \hat{J}_z^2$$

such that, because the coordinate and momentum operator for two different particles are commuting operators, we have:

$$[\hat{J}_x, \hat{J}_y] = \left[\sum_p \hat{L}_x(p), \sum_q \hat{L}_y(q) \right] = \sum_p \sum_q \left[\hat{L}_x(p), \hat{L}_y(q) \right]$$

$$= \sum_p \left[\hat{L}_x(p), \hat{L}_y(p) \right] = i\hbar \sum_p \hat{L}_z(p) = i\hbar \hat{J}_z \quad (8.1)$$

and similarly:

$$[\hat{J}_y, \hat{J}_z] = i\hbar \hat{J}_x, \quad [\hat{J}_z, \hat{J}_x] = i\hbar \hat{J}_y \quad (8.2)$$

Because these relations are the same as the commutation relations (4.45) and (4.46) for the single particle operators, we obtain, as shown in Example 4.2, that the operator $\hat{J}^2$ commutes with $\hat{J}_x, \hat{J}_y$ and $\hat{J}_z$:

286 Quantum Physics

$$[\hat{J}_x,\hat{J}^2]=[\hat{J}_y,\hat{J}^2]=[\hat{J}_z,\hat{J}^2]=0 \tag{8.3}$$

The commutation relations for the component operators $\hat{J}_i$, Eqs.(8.1) and (8.2), define essential properties of the angular momentum operators of any mechanical system, including the properties of the system with respect to rotations of coordinate axes.

EXAMPLE 8.1. Rotation properties of vector observables

In the case of a system of particles, the operator of an infinitesimal rotation of the coordinate system through $\vec{\varepsilon}=\vec{e}_x\varepsilon_x+\vec{e}_y\varepsilon_y+\vec{e}_z\varepsilon_z$ can be written, using Eq.(4.58), in the general form:

$$\hat{R}(\vec{\varepsilon})=1+\frac{i}{\hbar}\vec{\varepsilon}\cdot\sum_p \hat{\vec{L}}(p)=1+\frac{i}{\hbar}\vec{\varepsilon}\cdot\hat{\vec{J}} \tag{8.4}$$

It follows from Eq.(4.54) that the infinitesimal transformation under rotation of any observable represented by a vector operator $\hat{\vec{V}}$, of components $\hat{V}_x,\hat{V}_y,\hat{V}_z$, reads:

$$\hat{\vec{V}}'=\left(1+\frac{i}{\hbar}\vec{\varepsilon}\cdot\hat{\vec{J}}\right)\hat{\vec{V}}\left(1-\frac{i}{\hbar}\vec{\varepsilon}\cdot\hat{\vec{J}}\right)\approx\hat{\vec{V}}+\frac{i}{\hbar}[\vec{\varepsilon}\cdot\hat{\vec{J}},\hat{\vec{V}}]$$

where terms in ε^2 have been neglected. If this result is compared to the classical transformation of a vector $\vec{V}$ under the infinitesimal rotation $\vec{\varepsilon}$:

$$\vec{V}'=\vec{V}+\vec{\varepsilon}\times\vec{B}$$

we obtain a set of equations for the components of $\hat{\vec{V}}$:

$$\frac{i}{\hbar}[\varepsilon_x\hat{J}_x+\varepsilon_y\hat{J}_y+\varepsilon_z\hat{J}_z,\hat{V}_x]=\varepsilon_y\hat{V}_z-\varepsilon_z\hat{V}_y$$

$$\frac{i}{\hbar}[\varepsilon_x\hat{J}_x+\varepsilon_y\hat{J}_y+\varepsilon_z\hat{J}_z,\hat{V}_y]=\varepsilon_z\hat{V}_x-\varepsilon_x\hat{V}_z$$

$$\frac{i}{\hbar}[\varepsilon_x\hat{J}_x+\varepsilon_y\hat{J}_y+\varepsilon_z\hat{J}_z,\hat{V}_z]=\varepsilon_x\hat{V}_y-\varepsilon_y\hat{V}_x$$

Because $\vec{\varepsilon}$ is an arbitrary vector, it is convenient to equate the coefficients of each component of $\vec{\varepsilon}$ in these equations, so obtaining an equivalent set of nine commutation rules:

$$[\hat{J}_x,\hat{V}_x]=0, \quad [\hat{J}_y,\hat{V}_x]=-i\hbar\hat{V}_z, \quad [\hat{J}_z,\hat{V}_x]=i\hbar\hat{V}_y$$

$$[\hat{J}_x,\hat{V}_y]=i\hbar\hat{V}_z, \quad [\hat{J}_y,\hat{V}_y]=0, \quad [\hat{J}_z,\hat{V}_y]=-i\hbar\hat{V}_x \tag{8.5}$$

$$[\hat{J}_x,\hat{V}_z]=-i\hbar\hat{V}_y, \quad [\hat{J}_y,\hat{V}_z]=i\hbar\hat{V}_x, \quad [\hat{J}_z,\hat{V}_z]=0$$

Hence, the transformation of an arbitrary vector under infinitesimal rotations of the coordinate system can be defined in terms of commutation relations with the angular momentum components. For $\vec{\omega} = N\vec{\varepsilon}$, we obtain from Eq.(8.4) the operator:

$$\hat{R}(\vec{\omega}) = \lim_{N \to \infty}\left(1 + \frac{i}{\hbar}\frac{1}{N}\vec{\omega}\cdot\hat{\vec{J}}\right)^N = e^{i\vec{\omega}\cdot\hat{\vec{J}}/\hbar} \qquad (8.6)$$

corresponding to a finite rotation. ♦

Therefore we will postulate that a vector operator $\hat{\vec{J}}$ represents an angular momentum provided that:

(i) the operators $\hat{J}_x, \hat{J}_y, \hat{J}_z$ are Hermitian, and hence are associated with real observables;

(ii) the commutation relations (8.1) and (8.2) hold for the $\hat{J}_x, \hat{J}_y$ and $\hat{J}_z$, such that:

$$\hat{\vec{J}} \times \hat{\vec{J}} = i\hbar\hat{\vec{J}} \qquad (8.7)$$

The operators $\hat{J}^2$ and $\hat{J}_z$ represent compatible observables and will therefore have a common set of eigenfunctions $\psi_{\lambda m}$, corresponding to eigenfunctions $\lambda\hbar^2$ and $m\hbar$, according to the eigenvalue equations:

$$\hat{J}^2 \psi_{\lambda m} = \lambda\hbar^2 \psi_{\lambda m}$$
$$\hat{J}_z \psi_{\lambda m} = m\hbar \psi_{\lambda m} \qquad (8.8)$$

The obvious inequality for the expectation values:

$$\langle\psi_{\lambda m}|\hat{J}^2|\psi_{\lambda m}\rangle = \langle\psi_{\lambda m}|\hat{J}_x^2|\psi_{\lambda m}\rangle + \langle\psi_{\lambda m}|\hat{J}_y^2|\psi_{\lambda m}\rangle + \langle\psi_{\lambda m}|\hat{J}_z^2|\psi_{\lambda m}\rangle \geq \langle\psi_{\lambda m}|\hat{J}_z^2|\psi_{\lambda m}\rangle$$

yields the following restriction:

$$\lambda\hbar^2 \geq m^2\hbar^2 \qquad (8.9)$$

which implies that eigenfunctions $\psi_{\lambda m}$ with $m^2 \geq \lambda + 1$ must not exist. If the eigenvalue problem of angular momentum is solved entirely based on the rules of commutation (8.7), the results will be valid for any operators with similar commuting properties, whether a classical analogue exists or not. The eigenvalues and eigenfunctions obtained in Section 7.1 for the orbital angular momentum of a single particle, by solving the differential equations that were found by expressing the operators in terms of differential operators in coordinate space, will appear as a special case.

The procedure is somewhat similar to the operator formalism used for the linear harmonic motion in Example 6.7, and consists of introducing the ladder operators:

$$\hat{J}_+ = \hat{J}_x + i\hat{J}_y, \qquad \hat{J}_- = \hat{J}_x - i\hat{J}_y \tag{8.10}$$

in terms of which we can make the factorization:

$$\hat{J}^2 - \hat{J}_z^2 = (\hat{J}_x + i\hat{J}_y)(\hat{J}_x - i\hat{J}_y) - \hbar\hat{J}_z = (\hat{J}_x - i\hat{J}_y)(\hat{J}_x + i\hat{J}_y) + \hbar\hat{J}_z$$

where use has been made of Eq.(8.1). Hence, we have:

$$\hat{J}^2 - \hat{J}_z^2 = \hat{J}_+\hat{J}_- - \hbar\hat{J}_z = \hat{J}_-\hat{J}_+ + \hbar\hat{J}_z \tag{8.11}$$

The physical significance of $\hat{J}_+$ can be demonstrated by applying it to one of the eigenfunctions $\psi_{\lambda m}$, giving:

$$\hat{J}_z(\hat{J}_+\psi_{\lambda m}) = \hat{J}_z(\hat{J}_x + i\hat{J}_y)\psi_{\lambda m} = (\hat{J}_x + i\hat{J}_y)(\hat{J}_z + \hbar)\psi_{\lambda m} = (m+1)\hbar(\hat{J}_+\psi_{\lambda m}) \tag{8.12}$$

This means that if $\psi_{\lambda m}$ is an eigenfunction of $\hat{J}_z$ with the eigenvalue $m\hbar$, $\hat{J}_+\psi_{\lambda m}$ is an eigenfunction of $\hat{J}_z$ with the eigenvalue $(m+1)\hbar$. In other words, $\hat{J}_+$ is a raising operator in the sense that it raises the eigenvalue $m\hbar$ of any eigenstate of angular momentum by $\hbar$. Similarly, the lowering operator $\hat{J}_-$ lowers any eigenvalue of $\hat{J}_z$ by $\hbar$. We observe that, because of the commutation relations (8.3), we have:

$$\hat{J}^2(\hat{J}_+\psi_{\lambda m}) = \hat{J}_+(\hat{J}^2\psi_{\lambda m}) = \lambda\hbar^2(\hat{J}_+\psi_{\lambda m}) \tag{8.13}$$

such that $\hat{J}_+$, and similarly $\hat{J}_-$, leave the quantum number λ of $\psi_{\lambda m}$ unchanged. There must be a limit to the number of times $\hat{J}_+$ can be applied in succession to $\Psi_{\lambda m}$, and this will be given by the highest value of m, which can be set equal to an arbitrary integer, $m = j$, provided that:

$$\hat{J}_+\psi_{\lambda j} = 0 \tag{8.14}$$

From Eqs.(8.8), (8.11) and (8.14) we obtain:

$$(\hat{J}^2 - \hat{J}_z^2)\psi_{\lambda j} = (\hat{J}_-\hat{J}_+ + \hbar\hat{J}_z)\psi_{\lambda j} = j\hbar^2\psi_{\lambda j}$$

which gives:

$$\hat{J}^2\psi_{\lambda j} = j(j+1)\hbar^2\psi_{\lambda j} \tag{8.15}$$

In the same way, if $\hat{J}_-$ is successively applied to $\psi_{\lambda m}$, there will be a lowest integer value of m, say $m = k$, such that:

$$\hat{J}_- \psi_{\lambda k} = 0 \qquad (8.16)$$

and by combining this result with Eq.(8.11) we have:

$$(\hat{J}^2 - \hat{J}_z^2)\psi_{\lambda k} = (\hat{J}_+ \hat{J}_- - \hbar \hat{J}_z)\psi_{\lambda k} = -k\hbar^2 \psi_{\lambda k}$$

or:

$$\hat{J}^2 \psi_{\lambda k} = k(k-1)\hbar^2 \psi_{\lambda k}$$

When compared to Eq.(8.15), this results shows that the same eigenvalue $\lambda\hbar^2$ can be expressed in terms of either j or k, which implies:

$$j(j+1) = k(k-1)$$

Since, for a given eigenvalue $\lambda\hbar^2$, the eigenvalues of $\hat{J}_z$ are separated by integral multiples of $\hbar$, it follows that the difference between the highest and the lowest values of m must be a *positive* integer, say p, where $p = j - k$. With this choice, we obtain:

$$k = -j \quad \text{or} \quad p = 2j$$
$$k = j+1 \quad \text{or} \quad p = -1 \qquad (8.17)$$

and it is obvious that the second solution leads to an unacceptable p value. Hence, we must take $k = -j$, which determines the eigenvalue of $\hat{J}^2$ to be:

$$\langle \psi_{\lambda m} | \hat{J}^2 | \psi_{\lambda m} \rangle = j(j+1)\hbar^2, \quad j = 0, 1, 2, \ldots \qquad (8.18)$$

For a given magnitude of angular momentum, fixed by the quantum number j, there will be $2j + 1$ allowed eigenvalues for $\hat{J}_z$:

$$\langle \psi_{\lambda m} | \hat{J}_z | \psi_{\lambda m} \rangle = m\hbar \qquad (8.19)$$

$$-j \leq m \leq j \qquad (8.20)$$

The quantum conditions (8.18), (8.19) and (8.20) are analogous to those found for the orbital angular momentum operator, Eqs.(7.19), (7.12) and (7.33). However, Eq.(8.17) shows that, since p is an integer, j can be either an integer or a half-integer:

$$j = 0, \tfrac{1}{2}, 1, \tfrac{3}{2}, 2, \ldots \tag{8.21}$$

and it is obvious that if j assumes integer or half-integer values, so does m. The half-integer values cannot be found from the transcription of classical problems into the eigenvalue differential equations through operator representation in coordinate space, because angular momenta corresponding to classical observables yield integer quantum numbers j and m. Such values will be assigned to angular momentum operators having no classical counterpart.

The common set of eigenfunctions of $\hat{J}^2$ and $\hat{J}_z$ are usually labelled by the quantum number j and m_j, rather than by λ and m, so that we set:

$$\psi_{\lambda m} \equiv \psi_{jm_j} \tag{8.22}$$

and the eigenvalue equations read:

$$\hat{J}^2 \psi_{jm_j} = j(j+1)\hbar^2 \psi_{jm_j} \tag{8.23}$$

$$\hat{J}_z \psi_{jm_j} = m_j \hbar \psi_{jm_j} \tag{8.24}$$

where use has been made of Eqs.(8.18) and (8.19). The Hermitian operators $\hat{J}^2$ and $\hat{J}_z$ have no repeated eigenvalues, and hence their common set of eigenfunctions ψ_{jm_j} form a mutually orthonormal set, as discussed in Section 3.1:

$$\langle \psi_{jm_j} | \psi_{j'm'_j} \rangle = \delta_{jj'} \delta_{m_j m'_j} \tag{8.25}$$

The eigenfunctions corresponding to different magnitudes of angular momentum are orthogonal and, for each j value, there are $2j + 1$ eigenfunctions with definite properties with respect to the operators $\hat{J}_+$ and $\hat{J}_-$, and hence with respect to $\hat{J}_x$ and $\hat{J}_y$, as determined by the commutation properties (8.7). For the raising operator $\hat{J}_+$, in view of Eqs.(8.12) and (8.13), we can write:

$$\hat{J}_+ |\psi_{jm_j}\rangle = \alpha_{jm_j} \hbar |\psi_{j,m_j+1}\rangle \tag{8.26}$$

and the conjugate equation:

$$\langle \psi_{jm_j} | \hat{J}_- = \alpha^*_{jm_j} \hbar \langle \psi_{j,m_j+1} | \tag{8.27}$$

such that we may determine the constant α_{jm_j} from:

$$\langle\psi_{jm_j}|\hat{J}_-\hat{J}_+|\psi_{jm_j}\rangle = \langle\psi_{jm_j}|\hat{J}^2 - \hat{J}_z^2 - \hbar\hat{J}_z|\psi_{jm_j}\rangle$$

$$= \left[j(j+1) - m_j(m_j+1)\right]\hbar^2 = |\alpha_{jm_j}|^2\hbar^2 \qquad (8.28)$$

It is convenient to choose α_{jm_j} as a real and positive constant:

$$\alpha_{jm_j} = \sqrt{j(j+1) - m_j(m_j+1)} \qquad (8.29)$$

and hence the ψ_{jm_j} are transformed by $\hat{J}_+$ as:

$$\hat{J}_+\psi_{jm_j} = \alpha_{jm_j}\hbar\psi_{j,m_j+1} = \sqrt{j(j+1) - m_j(m_j+1)}\,\hbar\,\psi_{j,m_j+1} \qquad (8.30)$$

In the same way, if the lowering operator $\hat{J}_-$ is applied to ψ_{jm_j}, one obtains:

$$\hat{J}_-\psi_{jm_j} = \alpha_{j,m_j-1}\hbar\psi_{j,m_j-1} = \sqrt{j(j+1) - m_j(m_j-1)}\,\hbar\,\psi_{j,m_j-1} \qquad (8.31)$$

where the coefficient α_{j,m_j-1} has the form (8.29), where m_j is replaced by $m_j - 1$. The last two equations, in view of the definitions (8.10), immediately give:

$$\hat{J}_x\psi_{jm_j} = \frac{1}{2}\alpha_{jm_j}\hbar\psi_{j,m_j+1} + \frac{1}{2}\alpha_{j,m_j-1}\hbar\psi_{j,m_j-1}$$

$$\hat{J}_y\psi_{jm_j} = -\frac{i}{2}\alpha_{jm_j}\hbar\psi_{j,m_j+1} + \frac{i}{2}\alpha_{j,m_j-1}\hbar\psi_{j,m_j-1} \qquad (8.32)$$

The $2j + 1$ common eigenfunctions of $\hat{J}^2$ and $\hat{J}_z$ form a *canonical basis* in terms of which can be expressed all eigenstates corresponding to a given magnitude j of angular momentum. When j and m_j assume the integer values of the orbital and magnetic quantum numbers l and m, the canonical basis is generated by the equations:

$$\hat{L}^2\psi_{lm} = l(l+1)\hbar^2\psi_{lm}$$

$$\hat{L}_z\psi_{lm} = m\hbar\psi_{lm} \qquad (8.33)$$

$$\hat{L}_\pm\psi_{lm} = \sqrt{l(l+1) - m(m\pm 1)}\,\hbar\,\psi_{l,m\pm 1}$$

which are satisfied by the spherical harmonic eigenfunctions $Y_{lm}(\theta,\varphi)$, defined by Eq.(7.35).

EXAMPLE 8.2. Spherical harmonic eigenstates of angular momentum

The eigenvalue equation of $\hat{L}_z$ has been previously solved in spherical polar coordinates, Eq.(7.10), indicating that the angular momentum eigenfunctions have the φ dependence as:

$$\psi_{lm}(\theta,\varphi) = \frac{1}{\sqrt{2\pi}} e^{im\varphi} F_l^m(\theta) \tag{8.34}$$

For the lowest value of the magnetic quantum number $m = -l$, we obtain from Eqs.(8.33) that $\hat{L}_-\psi_{l,-l} = 0$, and substituting the representation (2.31) of $\hat{L}_-$ in spherical polar coordinates we obtain an equation for $F_l^{-l}(\theta)$:

$$e^{-i\varphi}\left(-\frac{\partial}{\partial\theta} + i\frac{\cos\theta}{\sin\theta}\frac{\partial}{\partial\varphi}\right) e^{-il\varphi} F_l^{-l}(\theta) = e^{-i(l+1)\varphi}\left[-\frac{dF_l^{-l}(\theta)}{d\theta} + l\frac{\cos\theta}{\sin\theta} F_l^{-l}(\theta)\right] = 0$$

which can be integrated in a straightforward manner:

$$\frac{dF_l^{-l}(\theta)}{F_l^{-l}(\theta)} = l\frac{\cos\theta}{\sin\theta} d\theta \quad \text{or} \quad \ln F_l^{-l}(\theta) = l\int\frac{\cos\theta d\theta}{\sin\theta} = l\ln(\sin\theta)$$

This gives the normalized function:

$$F_l^{-l}(\theta) = N_l \sin^l\theta = \sqrt{\frac{2l+1}{2}} \frac{\sqrt{(2l)!}}{2^l l!} \sin^l\theta \tag{8.35}$$

where the constant N_l was obtained from:

$$\int_0^\pi [F_l^{-l}(\theta)]^2 \sin\theta\, d\theta = N_l^2 \int_0^\pi \sin^{2l}\theta \sin\theta\, d\theta = 1$$

Integrating by parts provides a recurrence relation:

$$I_l = \int_0^\pi \sin^{2l}\theta \sin\theta\, d\theta = -\int_0^\pi \sin^{2l}\theta\, d(\cos\theta) = \left[-\sin^{2l}\theta \cos\theta\right]_0^\pi + 2l\int_0^\pi \sin^{2l-1}\theta \cos^2\theta\, d\theta$$

$$= 2l\int_0^\pi \sin^{2l-1}\theta(1-\sin^2\theta)\, d\theta = 2l\int_0^\pi \sin^{2(l-1)}\theta \sin\theta\, d\theta - 2l\int_0^\pi \sin^{2l}\theta \sin\theta\, d\theta = 2lI_{l-1} - 2lI_l$$

that successively gives:

$$I_l = \frac{2l}{2l+1} I_{l-1} = \frac{2l(2l-2)\ldots 2}{(2l+1)(2l-3)\ldots 3} I_0 = 2\frac{2^l l!}{(2l+1)(2l-3)\ldots 3} = 2\frac{(2^l l!)^2}{(2l+1)!}$$

because $I_0 = 2$. The constant N_l immediately follows as given in Eq.(8.35). The other functions ψ_{lm} of the canonical basis, corresponding to a given l, can be generated by using Eq.(8.33) for $\hat{L}_+$, which, by substituting Eq.(2.30), gives:

$$e^{i\varphi}\left(\frac{\partial}{\partial\theta} + i\frac{\cos\theta}{\sin\theta}\frac{\partial}{\partial\varphi}\right)e^{im\varphi}F_l^m(\theta) = \alpha_{lm}e^{i(m+1)\varphi}F_l^{m+1}(\theta)$$

so providing the recurrence relation:

$$F_l^{m+1}(\theta) = -\frac{1}{\alpha_{lm}}\left(\frac{d}{d\theta} - m\frac{\cos\theta}{\sin\theta}\right)F_l^m(\theta)$$

It is convenient to introduce the new variable $u = \cos\theta$, such that we obtain:

$$F_l^{m+1}(u) = -\frac{1}{\alpha_{lm}}\left(\sqrt{1-u^2}\frac{d}{du} + \frac{mu}{\sqrt{1-u^2}}\right)F_l^m(u)$$

If we set:

$$F_l^m(u) = (1-u^2)^{m/2} f_l^m(u)$$

and consider the following identity, which can be directly proved:

$$\left(\sqrt{1-u^2}\frac{d}{du} + \frac{mu}{\sqrt{1-u^2}}\right)\left[(1-u^2)^{m/2} f_l^m(u)\right] = (1-u^2)^{(m+1)/2}\frac{df_l^m(u)}{du}$$

the recurrence relation simplifies to:

$$f_l^{m+1}(u) = -\frac{1}{\alpha_{lm}}\frac{d}{du}f_l^m(u) \quad \text{or} \quad f_l^m(u) = -\frac{1}{\alpha_{l,m-1}}\frac{d}{du}f_l^{m-1}(u)$$

This immediately leads to:

$$f_l^m(u) = (-1)^{l+m}\frac{1}{\alpha_{l,m-1}\alpha_{l,m-2}\cdots\alpha_{l,-l}}\frac{d^{l+m}}{du^{l+m}}f_l^{-l}(u)$$

where:

$$\alpha_{l,m-1}\alpha_{l,m-2}\cdots\alpha_{l,-l} = \sqrt{\frac{(l+m)!}{(l-m)!}(2l)!} \quad \text{and} \quad f_l^{-l} = \sqrt{\frac{2l+1}{2}}\frac{\sqrt{(2l)!}}{2^l l!}(1-u^2)^l$$

It follows that:

$$F_l^m(u) = (-1)^m \sqrt{\frac{2l+1}{2}} \sqrt{\frac{(l-m)!}{(l+m)!}} \frac{1}{2^l l!} (1-u^2)^{m/2} \frac{d^{l+m}}{du^{l+m}} (u^2-1)^l$$

$$= (-1)^m \sqrt{\frac{2l+1}{2}} \sqrt{\frac{(l-m)!}{(l+m)!}} (1-u^2)^{m/2} \frac{d^m}{du^m} P_l(u) = (-1)^m \sqrt{\frac{2l+1}{2}} \sqrt{\frac{(l-m)!}{(l+m)!}} P_l^m(u)$$

where $P_l(u)$ are the Legendre polynomials, Eq.(7.21), and $P_l^m(u)$ are the associated Legendre functions, as defined by Eq.(7.32), if m is assumed a positive integer. Substituting into Eq.(8.34), we obtain the eigenfunctions in the form:

$$Y_l^m(\theta, \varphi) = (-1)^m \left[\frac{2l+1}{4\pi} \frac{(l-|m|)!}{(l+|m|)!} \right]^{\frac{1}{2}} P_l^{|m|}(\cos\theta) e^{im\varphi} \tag{8.36}$$

It is often convenient to introduce the phase convention:

$$\psi_{lm}(\theta, \varphi) = (-1)^m Y_l^m(\theta, \varphi) = Y_{lm}(\theta, \varphi) = \left[\frac{2l+1}{4\pi} \frac{(l-|m|)!}{(l+|m|)!} \right]^{\frac{1}{2}} P_l^{|m|}(\cos\theta) e^{im\varphi} \tag{8.37}$$

which allows us to express the canonical basis functions in terms of the spherical harmonics, as defined by Eq.(7.35).

We observe that the representation (8.37) of the canonical basis is only valid for integer quantum numbers l and m. If we choose, for example $l = \frac{1}{2}, m = \frac{1}{2}$, we obtain from Eq.(8.37):

$$\psi_{\frac{1}{2}, \pm\frac{1}{2}}(\theta, \varphi) = \frac{e^{\pm i\varphi/2}}{\sqrt{4\pi}} \sqrt{\sin\theta}$$

and it is immediate that Eqs.(8.33) for $\hat{L}_+$ and $\hat{L}_-$ are not satisfied, since:

$$\hat{L}_+ \psi_{\frac{1}{2}, \frac{1}{2}} = e^{i\varphi} \left(\frac{\partial}{\partial\theta} + i\frac{\cos\theta}{\sin\theta} \frac{\partial}{\partial\varphi} \right) e^{i\varphi/2} \sqrt{\sin\theta} = e^{i3\varphi/2} \left(\frac{\partial}{\partial\theta} - \frac{1}{2}\frac{\cos\theta}{\sin\theta} \right) \sqrt{\sin\theta} \neq 0$$

$$\hat{L}_- \psi_{\frac{1}{2}, -\frac{1}{2}} = e^{-i\varphi} \left(-\frac{\partial}{\partial\theta} + i\frac{\cos\theta}{\sin\theta} \frac{\partial}{\partial\varphi} \right) e^{-i\varphi/2} \sqrt{\sin\theta} = e^{-i3\varphi/2} \left(-\frac{\partial}{\partial\theta} + \frac{1}{2}\frac{\cos\theta}{\sin\theta} \right) \sqrt{\sin\theta} \neq 0$$

In other words the functions (8.37) corresponding to half-integer l values are not admissible eigenfunctions of the orbital angular momentum. ♂

Since there are no functions that describe the angular momentum properties of systems with half-integer values of the quantum numbers l and m, we introduce the alternative description, by means of the matrix representation of angular momentum operators in the canonical basis, as discussed in Section 3.4. Any state function Ψ_j with a definite magnitude of angular momentum is expanded in terms of the $2j+1$ eigenvectors ψ_{jm_j} of $\hat{J}^2$ and $\hat{J}_z$ according to Eq.(3.4):

$$\Psi_j = \sum_{m_j=-j}^{j} a_{m_j} \psi_{jm_j} = \sum_{m_j=-j}^{j} \langle \psi_{jm_j} | \Psi_j \rangle \psi_{jm_j} \tag{8.38}$$

and will be completely specified by its scalar products a_{m_j} with the canonical basis vectors, which in the Dirac notation, Eq.(1.69), can be represented by a $(2j+1) \times 1$ column matrix. The angular momentum operators are represented in the same basis by the matrix elements:

$$(\hat{J})_{m_j m'_j} = J_{m_j m'_j} = \langle \psi_{jm_j} | \hat{J} | \psi_{jm'_j} \rangle \tag{8.39}$$

such that the expectation value of the observable J in the state Ψ_j takes the form (3.56), which reads:

$$\langle J \rangle = \langle \Psi_j | \hat{J} | \Psi_j \rangle = \sum_{m_j, m'_j} \langle \Psi_j | \psi_{jm_j} \rangle \langle \psi_{jm_j} | \hat{J} | \psi_{jm'_j} \rangle \langle \psi_{jm'_j} | \Psi_j \rangle = \sum_{m_j, m'_j} a^*_{m_j} J_{m_j m'_j} a_{m'_j} \tag{8.40}$$

This algebraic formalism provides, for integer j and m_j values, the same results as the usual way of finding the expectation values of angular momentum observables:

$$\langle J \rangle = \int \Psi_j^* \hat{J} \Psi_j dV$$

where differential operators are applied to the functions describing angular momentum states, which are expressed in terms of spherical harmonics. But the matrix method for computing expectation values will also be valid in the case of half-integer values of the angular momenta. From the eigenvalue equations (8.23) and (8.24) it follows that the operators $\hat{J}^2$ and $\hat{J}_z$ have in this basis a diagonal representation of the form:

$$\langle \psi_{jm'_j} | \hat{J}^2 | \psi_{jm_j} \rangle = j(j+1)\hbar^2 \delta_{m_j m'_j} \tag{8.41}$$

$$\langle \psi_{jm'_j} | \hat{J}_z | \psi_{jm_j} \rangle = m_j \hbar \delta_{m_j m'_j} \tag{8.42}$$

The only nonvanishing matrix elements of the operators $\hat{J}_+$ and $\hat{J}_-$ in the canonical basis representation are obtained from Eqs.(8.30) and (8.31) as:

$$\langle \psi_{j,m_j+1} | \hat{J}_+ | \psi_{jm_j} \rangle = \alpha_{jm_j} \hbar = \sqrt{j(j+1) - m_j(m_j+1)} \, \hbar$$

$$\langle \psi_{j,m_j-1} | \hat{J}_- | \psi_{jm_j} \rangle = \alpha_{j,m_j-1} \hbar = \sqrt{j(j+1) - m_j(m_j-1)} \, \hbar \tag{8.43}$$

It is also a straightforward matter to show, using the definitions (8.10), that the nonvanishing matrix elements of $\hat{J}_x$ and $\hat{J}_y$ read:

$$\langle \psi_{j,m_j+1} | \hat{J}_x | \psi_{jm_j} \rangle = i \langle \psi_{j,m_j+1} | \hat{J}_y | \psi_{jm_j} \rangle = \frac{1}{2} \alpha_{jm_j} \hbar$$

$$\langle \psi_{j,m_j-1} | \hat{J}_x | \psi_{jm_j} \rangle = -i \langle \psi_{j,m_j-1} | \hat{J}_y | \psi_{jm_j} \rangle = \frac{1}{2} \alpha_{j,m_j-1} \hbar$$

(8.44)

The results (8.41) to (8.44) for the matrix representation do not depend on the properties of the angular momentum eigenfunctions, but only on the commutation properties of the angular momentum operators applied to the eigenfunctions of $\hat{J}^2$ and $\hat{J}_z$. Therefore these properties can be assumed to also apply to operators representing half-integer angular momenta, if defined by the commutation relations (8.7).

EXAMPLE 8.3. Matrix representation of orbital angular momentum

Consider the particular case of an orbital angular momentum operator in a p-state, where $j = l = 1$ and $m_j = m = -1, 0, 1$. The matrix elements of $\hat{L}_x, \hat{L}_y, \hat{L}_z$ and $\hat{L}^2$ in the canonical basis representation are given by Eqs.(8.41), (8.42), (8.43) and (8.44), on substitution of the appropriate quantum numbers, as follows:

$$\begin{array}{cccc} & 1 & 0 & -1 \quad m \end{array}$$

$$\hat{L}_z = \hbar \begin{pmatrix} 1 & 0 & 0 \\ 0 & 0 & 0 \\ 0 & 0 & -1 \end{pmatrix} \begin{matrix} 1 \\ 0, \\ -1 \end{matrix} \qquad \hat{L}^2 = \hbar^2 \begin{pmatrix} 2 & 0 & 0 \\ 0 & 2 & 0 \\ 0 & 0 & 2 \end{pmatrix}$$

$$\hat{L}_x = \frac{\hbar}{\sqrt{2}} \begin{pmatrix} 0 & 1 & 0 \\ 1 & 0 & 1 \\ 0 & 1 & 0 \end{pmatrix}, \qquad \hat{L}_y = \frac{\hbar}{\sqrt{2}} \begin{pmatrix} 0 & -i & 0 \\ i & 0 & -i \\ 0 & i & 0 \end{pmatrix}$$

(8.45)

It is easily confirmed that these matrices satisfy the commutation relations (4.45), (4.46) and (4.48). The canonical basis consists of the spherical harmonics $Y_{lm}(\theta, \varphi)$, so that the common eigenfunctions of $\hat{L}^2$ and $\hat{L}_z$ can be represented by the column vectors:

$$\psi_{11} = \begin{pmatrix} 1 \\ 0 \\ 0 \end{pmatrix}, \quad \psi_{10} = \begin{pmatrix} 0 \\ 1 \\ 0 \end{pmatrix}, \quad \psi_{1,-1} = \begin{pmatrix} 0 \\ 0 \\ 1 \end{pmatrix}$$

(8.46)

which obviously satisfy the matrix eigenvalue equations (8.33) for $l = 1$, if the matrix representation (8.45) for $\hat{L}^2$ and $\hat{L}_z$ is used. The eigenstates of another particular component of orbital angular momentum, say $\hat{L}_x$, can be determined without using the properties of spherical harmonics. This can be done by solving the matrix eigenvalue equation for $\hat{L}_x$, which reads:

$$\hat{L}_x \begin{pmatrix} a_1 \\ a_0 \\ a_{-1} \end{pmatrix} = \frac{\hbar}{\sqrt{2}} \begin{pmatrix} 0 & 1 & 0 \\ 1 & 0 & 1 \\ 0 & 1 & 0 \end{pmatrix} \begin{pmatrix} a_1 \\ a_0 \\ a_{-1} \end{pmatrix} = m_x \hbar \begin{pmatrix} a_1 \\ a_0 \\ a_{-1} \end{pmatrix}$$

where the column vectors represent a state of the system defined by the wave function:

$$\psi_{1m_x} = a_1 Y_{11} + a_0 Y_{10} + a_{-1} Y_{1,-1}$$

The eigenvalues are given by the characteristic equation (3.8), which reduces to:

$$\begin{vmatrix} -m_x & \frac{1}{\sqrt{2}} & 0 \\ \frac{1}{\sqrt{2}} & -m_x & \frac{1}{\sqrt{2}} \\ 0 & \frac{1}{\sqrt{2}} & -m_x \end{vmatrix} = -m_x^3 + m_x = 0$$

and leads to $m_x \hbar = 0, \pm \hbar$. The eigenvectors are obtained from the simultaneous equations:

$$\frac{1}{\sqrt{2}} a_0 = m_x a_1, \quad \frac{1}{\sqrt{2}} (a_1 + a_{-1}) = m_x a_0, \quad \frac{1}{\sqrt{2}} a_0 = m_x a_{-1}$$

where, by substitution of the m_x values, one obtains:

$$\psi_{11} = \frac{1}{2} \begin{pmatrix} 1 \\ \sqrt{2} \\ 1 \end{pmatrix}, \quad \psi_{10} = \frac{1}{\sqrt{2}} \begin{pmatrix} 1 \\ 0 \\ -1 \end{pmatrix}, \quad \psi_{1,-1} = \frac{1}{2} \begin{pmatrix} 1 \\ -\sqrt{2} \\ 1 \end{pmatrix}$$

which means that:

$$\psi_{11} = \frac{1}{2}(Y_{11} + \sqrt{2} Y_{10} + Y_{1,-1}), \quad \psi_{10} = \frac{1}{\sqrt{2}}(Y_{11} - Y_{1,-1}), \quad \psi_{1,-1} = \frac{1}{2}(Y_{11} - \sqrt{2} Y_{10} + Y_{1,-1})$$

If this set of functions is chosen as a basis for the matrix representation, the matrix $\hat{L}_x$ becomes diagonal. The expectation values of $\hat{L}_x$ in the states ψ_{1m_x} are given by Eq.(8.40), which reads:

$$\langle \psi_{1,\pm 1} | \hat{L}_x | \psi_{1,\pm 1} \rangle = \frac{1}{4} \frac{\hbar}{\sqrt{2}} (1, \pm \sqrt{2}, 1) \begin{pmatrix} 0 & 1 & 0 \\ 1 & 0 & 1 \\ 0 & 1 & 0 \end{pmatrix} \begin{pmatrix} 1 \\ \pm \sqrt{2} \\ 1 \end{pmatrix} = \pm \hbar$$

$$\langle \psi_{10} | \hat{L}_x | \psi_{10} \rangle = \frac{1}{2} \frac{\hbar}{\sqrt{2}} (1,0,-1) \begin{pmatrix} 0 & 1 & 0 \\ 1 & 0 & 1 \\ 0 & 1 & 0 \end{pmatrix} \begin{pmatrix} 1 \\ 0 \\ -1 \end{pmatrix} = 0$$

In the same way, we can use Eqs.(8.41) to (8.43) to construct the angular momentum matrices in the case of half-integer values of j and m_j and then solve the eigenvalue problem for systems that exhibit half-integer angular momenta. ♌

Problem 8.1.1. Show that spherical harmonics $Y_{lm}(\theta,\varphi)$ can also be represented in the form:

$$Y_{lm}(\theta,\varphi) = \frac{e^{im\varphi}}{2^l l!} \left[\frac{2l+1}{4\pi} \frac{(l+m)!}{(l-m)!} \right]^{\frac{1}{2}} (1-u^2)^{-m/2} \frac{d^{l-m}}{du^{l-m}} (u^2-1)^l$$

(Solution): We start from Eq.(8.34) which, for $m = l$, becomes:

$$Y_{ll}(\theta,\varphi) = \frac{e^{il\varphi}}{\sqrt{2\pi}} F_l^l(\theta)$$

to which we apply Eq.(8.33) in the form:

$$\hat{L}_+ Y_{ll} = \hbar e^{i\varphi} \left(\frac{\partial}{\partial \theta} + i \frac{\cos\theta}{\sin\theta} \frac{\partial}{\partial \varphi} \right) Y_{ll} = \hbar e^{i(l+1)\varphi} \left(\frac{dF_l^l}{d\theta} - l \frac{\cos\theta}{\sin\theta} F_l^l \right) = 0$$

This leads to:

$$F_l^l(\theta) = \frac{(-1)^l}{2^l l!} \sqrt{\frac{2l+1}{2}(2l)!} \, \sin^l \theta$$

where the normalization constants is the same as determined in Eq.(8.35). The function $Y_{lm}(\theta,\varphi)$ can be generated by applying $l-m$ times, the operator $\hat{L}_-$, according to Eq.(8.33), to $Y_{ll}(\theta,\varphi)$. Equation (8.33) provides, by substituting Eq.(2.31) for $\hat{L}_-$, a recurrence relation:

$$F_l^{m-1}(\theta) = -\frac{1}{\alpha_{l,m-1}} \left(\frac{d}{d\theta} + m \frac{\cos\theta}{\sin\theta} \right) F_l^m(\theta)$$

In terms of the new variable $u = \cos\theta$, this reads:

8. Angular Momentum

$$F_l^{m-1}(u) = \frac{1}{\alpha_{l,m-1}}\left(\sqrt{1-u^2}\,\frac{d}{du} - \frac{mu}{\sqrt{1-u^2}}\right)F_l^m(u)$$

By taking:

$$F_l^m(u) = (1-u^2)^{-m/2} f_l^m(u)$$

we have:

$$f_l^{m-1}(u) = \frac{1}{\alpha_{l,m-1}}\frac{df_l^m(u)}{du} \quad \text{or} \quad f_l^m(u) = \frac{1}{\alpha_{lm}}\frac{df_l^{m+1}(u)}{du}$$

Successive application of this relation gives:

$$f_l^m(u) = \frac{1}{\alpha_{lm}\,\alpha_{l,m+1}\cdots\alpha_{ll}}\frac{d^{l-m}}{du^{l-m}}f_l^l(u) = \left[\frac{1}{(2l)!}\frac{(l+m)!}{(l-m)!}\right]^{\frac{1}{2}}\frac{d^{l-m}}{du^{l-m}}f_l^l(u)$$

where:

$$f_l^l(u) = (1-u^2)^{1/2}F_l^l(u) = \frac{1}{2^l l!}\sqrt{\frac{2l+1}{2}(2l)!}\,(1-u^2)^l$$

Finally by substituting $F_l^m(\theta)$, so determined, into Eq.(8.34), the required result follows.

Problem 8.1.2. Solve the eigenvalue equation for the operator:

$$\hat{L}_r = \sin\theta\cos\varphi\,\hat{L}_x + \sin\theta\sin\varphi\,\hat{L}_y + \cos\theta\,\hat{L}_z$$

which represents the orbital angular momentum component in a direction $\vec{e}_r$ defined by the spherical polar coordinates θ and φ (see Figure 2.1).

(Solution): The eigenvalue problem reads:

$$\hat{L}_r \psi_{l\lambda}(\theta,\varphi) = \lambda\hbar\,\psi_{l\lambda}(\theta,\varphi)$$

where $\psi_{l\lambda}$ can be expanded in terms of the spherical harmonics according to Eq.(7.38). It is convenient to write the operator $\hat{L}_r$ in the form:

$$\hat{L}_r = \frac{1}{2}\sin\theta\,e^{-i\varphi}\hat{L}_+ + \frac{1}{2}\sin\theta\,e^{i\varphi}\hat{L}_- + \cos\theta\,\hat{L}_z$$

and to use Eq.(8.33) which yields, for a given l value:

$$\sum_{m=-l}^{l} a_m \left[\frac{1}{2}\sin\theta e^{-i\varphi}\alpha_{lm}Y_{l,m+1} + \frac{1}{2}\sin\theta e^{i\varphi}\alpha_{l,m-1}Y_{l,m-1} + (m\cos\theta - \lambda)Y_{lm}\right] = 0$$

where α_{lm} are given by Eq.(8.29). This can be rewritten as:

$$\sum_{m=-l}^{l}\left[\frac{a_{m+1}}{2}\sin\theta e^{i\varphi}\alpha_{lm} + a_m(m\cos\theta - \lambda) + \frac{a_{m-1}}{2}\sin\theta e^{-i\varphi}\alpha_{l,m-1}\right]Y_{lm} = 0$$

where all the coefficients of the spherical harmonics must vanish, and this provides $2l + 1$ equations to be solved for the a_m. In the special case $l = 1$, for example, we immediately obtain, by taking $m = 1, 0, -1$:

$$a_1(\cos\theta - \lambda) + a_0 \frac{1}{\sqrt{2}}\sin\theta e^{-i\varphi} = 0$$

$$a_1 \frac{1}{\sqrt{2}}\sin\theta e^{i\varphi} - a_0\lambda + a_{-1}\frac{1}{\sqrt{2}}\sin\theta e^{-i\varphi} = 0$$

$$a_0 \frac{1}{\sqrt{2}}\sin\theta e^{i\varphi} - a_{-1}(\cos\theta + \lambda) = 0$$

with the determinantal equation $-\lambda^3 + \lambda = 0$ which gives $\lambda = 0, \pm 1$. Hence the eigenvalues of $\hat{L}_r$ are $0, \pm\hbar$ and the corresponding normalized eigenfunctions are obtained by calculating a_1, a_0 and a_{-1} for each λ and then substituting into Eq.(7.38) which gives:

$$\psi_{11} = \cos^2\frac{\theta}{2}e^{-i\varphi}Y_{11} + \frac{1}{\sqrt{2}}\sin\theta Y_{10} + \sin^2\frac{\theta}{2}e^{i\varphi}Y_{1,-1}$$

$$\psi_{10} = -\frac{1}{\sqrt{2}}\sin\theta e^{-i\varphi}Y_{11} + \cos\theta Y_{10} + \frac{1}{\sqrt{2}}\sin\theta e^{i\varphi}Y_{1,-1}$$

$$\psi_{1,-1} = -\sin^2\frac{\theta}{2}e^{-i\varphi}Y_{11} + \frac{1}{\sqrt{2}}\sin\theta Y_{10} - \cos^2\frac{\theta}{2}e^{i\varphi}Y_{1,-1}$$

Problem 8.1.3. Show that the expectation values of $\hat{L}_z$ in the eigenstates of the operator $\hat{L}_r$, defined in Problem 8.1.2, are $m\hbar\cos\theta$.

Problem 8.1.4. Find the expectation values of the operator:

$$\hat{J}_r = \sin\theta\cos\varphi\,\hat{J}_x + \sin\theta\sin\varphi\,\hat{J}_y + \cos\theta\,\hat{J}_z$$

in the eigenstates ψ_{jm_j} of the canonical basis.

(Answer): $\langle \hat{J}_r \rangle = m\hbar \cos\theta$

8.2. Electron Spin

The spin of the electron, was first observed by applying a weak magnetic field of induction $\vec{B}$ in the direction of the z-axis, which had the effect of splitting the energy level E_{nl} into $2l + 1$ levels, according to Eq.(7.151). If a beam of one-electron atoms is directed through an inhomogenous field $\vec{B}$, there is a force on the magnetic moment, parallel to the field, of the form:

$$F_z = -\frac{d}{dz}(m\mu_B B) = -m\mu_B \frac{dB}{dz}$$

which results in splitting the beam into $2l + 1$ components. The atoms which are in the ground state, where $l = 0$, should not be affected by the field. However, it was observed experimentally that the beam is in fact split in two by the action of the inhomogenous field. Since the possible values of the component orbital angular momenta, and of the associated orbital magnetic moments, defined by Eq.(7.133), are integral multiples of $\hbar$ and μ_B respectively, the splitting into two components of equal intensity suggests that the atomic electron has an intrinsic magnetic moment with the same magnitude and two possible orientations: parallel or antiparallel to the magnetic field. The origin of this observable can be understood if it is assumed that an electron has an intrinsic *spin angular momentum*, in addition to any orbital angular momentum. If the electron spin is characterized by two quantum numbers s and m_s, which are the analogues of l and m, associated with orbital angular momentum, the experiment implies that the z-component of spin can take only two values $(2s + 1 = 2)$, so that the quantum numbers are restricted to $s = \frac{1}{2}$ and $m_s = \pm\frac{1}{2}$. Therefore the electron spin can be represented by an angular momentum operator $\hat{J} = \hat{S}$, defined in the special case $j = s = \frac{1}{2}$. Its description will be developed as a matrix representation which gives the transformation properties of the eigenfunctions, in complete agreement with the experiment, and with no concern about the analytical form of these functions.

As a manifestation of the spin as angular momentum, the spin component operators $\hat{S}_x, \hat{S}_y$ and $\hat{S}_z$ and the total spin momentum operator $\hat{S}^2$ must satisfy the commutation relations (8.7), which read:

$$\vec{\hat{S}} \times \vec{\hat{S}} = i\hbar \vec{\hat{S}} \qquad (8.47)$$

Because the spin observable is a supplementary degree of freedom, independent of the electron motion, the operators $\hat{S}_x, \hat{S}_y$ and $\hat{S}_z$ will not operate on the wave functions $\Psi(\vec{r})$ defined in terms of the electron position coordinates, but in a two-dimensional wave function space, where each eigenstate can be expanded in a canonical basis consisting of two common eigenvectors χ of $\hat{S}^2$ and $\hat{S}_z$. In the $\hat{S}^2, \hat{S}_z$ representation, the matrix eigenvalue equations to be solved are:

$$\hat{S}^2 \chi = s(s+1)\hbar^2 \chi$$
$$\hat{S}_z \chi = m_s \hbar \chi \tag{8.48}$$

where $s = \frac{1}{2}$ and $m_s = \pm \frac{1}{2}$. It is clear from Eq.(8.48) that the existence of spin is only possible in the frame of quantum mechanics, as it disappears when $\hbar \to 0$. It will be noticed that the eigenvalues of the orbital angular momentum $l(l+1)\hbar^2$ may tend to a finite value at the classical limit if, as $\hbar \to 0$, the quantum number l tends to infinity. The spin matrices are obtained from Eqs.(8.42) and (8.44), for $j = s = \frac{1}{2}$ and $m_j = m_s = \pm \frac{1}{2}$, as:

$$\hat{S}_z = \begin{pmatrix} \frac{1}{2} & 0 \\ 0 & -\frac{1}{2} \end{pmatrix}\hbar = \frac{1}{2}\hbar\begin{pmatrix} 1 & 0 \\ 0 & -1 \end{pmatrix} = \frac{1}{2}\hbar\hat{\sigma}_z$$

$$\hat{S}_x = \begin{pmatrix} 0 & \frac{1}{2} \\ \frac{1}{2} & 0 \end{pmatrix}\hbar = \frac{1}{2}\hbar\begin{pmatrix} 0 & 1 \\ 1 & 0 \end{pmatrix} = \frac{1}{2}\hbar\hat{\sigma}_x \tag{8.49}$$

$$\hat{S}_y = \begin{pmatrix} 0 & -\frac{i}{2} \\ \frac{i}{2} & 0 \end{pmatrix}\hbar = \frac{1}{2}\hbar\begin{pmatrix} 0 & -i \\ i & 0 \end{pmatrix} = \frac{1}{2}\hbar\hat{\sigma}_y$$

where $\hat{\sigma}_x, \hat{\sigma}_y$ and $\hat{\sigma}_z$ are known as the *Pauli matrices*. They obviously have the property (8.47), which reads:

$$[\hat{\sigma}_i, \hat{\sigma}_j] = 2i\hat{\sigma}_k \tag{8.50}$$

and also are *anticommuting*:

$$\hat{\sigma}_i \hat{\sigma}_j + \hat{\sigma}_j \hat{\sigma}_i = 0 \tag{8.51}$$

Because of the obvious property:

$$\hat{\sigma}_x^2 = \hat{\sigma}_y^2 = \hat{\sigma}_z^2 = 1 \tag{8.52}$$

by combining Eqs.(8.51) and (8.52) we obtain the more compact form:

$$\hat{\sigma}_i \hat{\sigma}_j + \hat{\sigma}_j \hat{\sigma}_i = 2\delta_{ij} \tag{8.53}$$

The total spin is defined as:

$$\hat{S}^2 = \hat{S}_x^2 + \hat{S}_y^2 + \hat{S}_z^2 = \frac{1}{4}\hbar^2(\hat{\sigma}_x^2 + \hat{\sigma}_y^2 + \hat{\sigma}_z^2) = \frac{3}{4}\hbar^2 \begin{pmatrix} 1 & 0 \\ 0 & 1 \end{pmatrix} = \frac{3}{4}\hbar^2 \tag{8.54}$$

and Eqs.(8.49) and (8.55) show that the matrix eigenvalue equations for $\hat{S}^2$ and $\hat{S}_z$ have the eigenvalues postulated by Eqs.(8.48), namely:

$$s(s+1)\hbar^2 = \frac{1}{2}\left(\frac{1}{2}+1\right)\hbar^2 = \frac{3}{4}\hbar^2, \quad m_s\hbar = \pm\frac{1}{2}\hbar \tag{8.55}$$

The common set of eigenfunctions consists of two component column vectors χ called *spinors*, which follow from Eq.(8.48) as:

$$\hat{S}_z \chi = \frac{1}{2}\hbar \begin{pmatrix} 1 & 0 \\ 0 & -1 \end{pmatrix} \begin{pmatrix} a_+ \\ a_- \end{pmatrix} = \pm \frac{1}{2}\hbar \begin{pmatrix} a_+ \\ a_- \end{pmatrix}$$

which gives:

$$\begin{pmatrix} a_+ \\ -a_- \end{pmatrix} = \pm \begin{pmatrix} a_+ \\ a_- \end{pmatrix}$$

Hence, in the $\hat{S}^2, \hat{S}_z$ representation, we obtain the following orthonormal spinors as a basis:

$$\chi_+ = \begin{pmatrix} 1 \\ 0 \end{pmatrix}, \quad \chi_- = \begin{pmatrix} 0 \\ 1 \end{pmatrix} \tag{8.56}$$

which define the spatial quantization for electron spin, although the analytical form of the eigenfunctions is not specified. An arbitrary spinor will be expanded in terms of the complete set (8.56) in the form:

$$\chi = \begin{pmatrix} a_+ \\ a_- \end{pmatrix} = a_+ \begin{pmatrix} 1 \\ 0 \end{pmatrix} + a_- \begin{pmatrix} 0 \\ 1 \end{pmatrix} = a_+ \chi_+ + a_- \chi_- = \sum_{m_s=-1/2}^{1/2} a_{m_s} \chi_{m_s} \tag{8.57}$$

where, as a consequence of Postulate 3, $|a_{m_s}|^2$ can be interpreted as the probability of obtaining the eigenvalue $m_s\hbar$ for $\hat{S}_z$ as a result of the measurement. It is understood that the normalization condition:

$$|a_+|^2 + |a_-|^2 = 1$$

must be satisfied. Any operator in the spin space will be represented by a 2×2 matrix with complex elements, using the orthonormal spinors (8.56) as a basis:

$$(\hat{A}) = \begin{pmatrix} \langle\chi_+|\hat{A}|\chi_+\rangle & \langle\chi_+|\hat{A}|\chi_-\rangle \\ \langle\chi_-|\hat{A}|\chi_+\rangle & \langle\chi_-|\hat{A}|\chi_-\rangle \end{pmatrix} \tag{8.58}$$

and this can be expressed as a linear combination of the unit matrix $\hat{I}$ and the Pauli matrices:

$$(\hat{A}) = a_0 \hat{I} + a_x \hat{\sigma}_x + a_y \hat{\sigma}_y + a_z \hat{\sigma}_z = \begin{pmatrix} a_0 + a_z & a_x - ia_y \\ a_x + ia_y & a_0 - a_z \end{pmatrix} \tag{8.59}$$

The eigenfunction of $\hat{S}_x$ with eigenvalues $\pm\frac{1}{2}\hbar$ are obtained from the matrix eigenvalue equation as:

$$\hat{S}_x \chi = \frac{1}{2}\hbar \begin{pmatrix} 0 & 1 \\ 1 & 0 \end{pmatrix} \begin{pmatrix} a_+ \\ a_- \end{pmatrix} = \pm\hbar \begin{pmatrix} a_+ \\ a_- \end{pmatrix} \quad \text{or} \quad \begin{pmatrix} a_- \\ a_+ \end{pmatrix} = \pm \begin{pmatrix} a_+ \\ a_- \end{pmatrix}$$

that is $a_+ = \pm a_-$. The normalization condition then gives $a_+ = a_- = 1/\sqrt{2}$ or $a_+ = -a_- = 1/\sqrt{2}$, and the eigenfunctions follow from Eq.(8.57) as:

$$\chi_+^x = \frac{1}{\sqrt{2}}(\chi_+ + \chi_-) = \frac{1}{\sqrt{2}}\begin{pmatrix} 1 \\ 1 \end{pmatrix}, \quad \chi_-^x = \frac{1}{\sqrt{2}}(\chi_+ - \chi_-) = \frac{1}{\sqrt{2}}\begin{pmatrix} 1 \\ -1 \end{pmatrix} \tag{8.60}$$

Similarly, the eigenfunctions of $\hat{S}_y$ with eigenvalues $\pm\frac{1}{2}\hbar$ are:

$$\chi_+^y = \frac{1}{\sqrt{2}}(\chi_+ + i\chi_-) = \frac{1}{\sqrt{2}}\begin{pmatrix} 1 \\ i \end{pmatrix}, \quad \chi_-^y = \frac{1}{\sqrt{2}}(\chi_+ - i\chi_-) = \frac{1}{\sqrt{2}}\begin{pmatrix} 1 \\ -i \end{pmatrix} \tag{8.61}$$

The fact that $\chi_\pm^x, \chi_\pm^y$ and $\chi_\pm$ are all different, means that no pair of S_x, S_y and S_z observables are compatible, in accordance with the commutation rules (8.47) of the spin component operators. The expectation value of $\hat{S}_z$ is zero:

$$\langle \hat{S}_z \rangle = |a_+|^2 \left(\frac{1}{2}\hbar\right) + |a_-|^2 \left(-\frac{1}{2}\hbar\right) = \frac{1}{2}\hbar\left(2|a_+|^2 - 1\right) = 0$$

in each of the eigenstates of $\hat{S}_x$ and $\hat{S}_y$ given above, which all have $a_+ = 1/\sqrt{2}$. This is consistent with the orientation of spin components along mutually perpendicular axes.

A rotation of spin coordinates through angle φ about the z-axis, and hence in the xy-plane, produces a state in which the spin components $\hat{S}_x, \hat{S}_y, \hat{S}_z$ are replaced by $\hat{S}'_x, \hat{S}'_y, \hat{S}'_z$, such that:

$$\hat{S}_x = \hat{S}'_x \cos\varphi - \hat{S}'_y \sin\varphi, \quad \hat{S}_y = \hat{S}'_x \sin\varphi + \hat{S}'_y \cos\varphi, \quad \hat{S}_z = \hat{S}'_z$$

The eigenfunctions of $\hat{S}_x$ in the new spin coordinates are obtained from the eigenvalue equation:

$$\hat{S}_x \chi = (\hat{S}'_x \cos\varphi - \hat{S}'_y \sin\varphi)\chi = \frac{1}{2}\hbar \begin{pmatrix} 0 & e^{i\varphi} \\ e^{-i\varphi} & 0 \end{pmatrix} \begin{pmatrix} a_+ \\ a_- \end{pmatrix} = \pm\frac{1}{2}\hbar \begin{pmatrix} a_+ \\ a_- \end{pmatrix}$$

which gives:

$$\begin{pmatrix} a_- e^{i\varphi} \\ a_+ e^{-i\varphi} \end{pmatrix} = \pm \begin{pmatrix} a_+ \\ a_- \end{pmatrix}$$

so that the normalized eigenfunctions are given by:

$$\chi_+^\varphi = \frac{1}{2}(e^{i\varphi/2}\chi_+ + e^{-i\varphi/2}\chi_-) = \sum_{m_s=-1/2}^{1/2} e^{im_s\varphi} a_{m_s} \chi_{m_s}$$

$$\chi_-^\varphi = \frac{1}{2}(e^{i\varphi/2}\chi_+ - e^{-i\varphi/2}\chi_-) = \sum_{m_s=-1/2}^{1/2} e^{im_s\varphi} a'_{m_s} \chi_{m_s}$$

(8.62)

It follows that the change of an arbitrary spinor, Eq.(8.57), by rotation of spin coordinates through φ about the z-axis, can be expressed in terms of $\hat{S}_z$ as:

$$\chi^\varphi = \sum_{m_s=-1/2}^{1/2} e^{im_s\varphi} a_{m_s} \chi_{m_s} = e^{i\hat{S}_z\varphi/\hbar} \chi \qquad (8.63)$$

In other words, the operator $e^{i\hat{S}_z\varphi/\hbar}$ is interpreted as a spin rotation generator about the z-axis. This property is similar to that of the operator $e^{i\hat{L}_z\varphi/\hbar}$, defined by Eq.(8.6), in the particular case of a single particle and $\vec{\omega} = \vec{e}_z \varphi$, which is used to express the change of

306   Quantum Physics

spatial state functions $\Psi(x,y,z) = \Psi(r,\theta,\varphi)$ upon rotation of space coordinates through φ about the z-axis. We observe that a rotation through 2π about the z-axis, which is physically equivalent to the identity transformation, leaves any single valued spatial eigenfunction of the orbital angular momentum $\Psi(r,\theta,\varphi)$ unchanged, whereas the rotation changes the sign of any spinor, as $e^{im_s 2\pi} = e^{\pm i\pi} = -1$. Although the spinors are double valued, no spin observable S_i ($i = x, y, z$) is changed by a rotation through 2π, since the spinors which can be used to define the matrix elements of $\hat{S}_i$ either do not change sign or both change sign.

Problem 8.2.1. Solve the eigenvalue problem for the operator representing the spin component in a direction defined by the spherical polar coordinates:

$$\hat{S}_r = \sin\theta\cos\varphi\,\hat{S}_x + \sin\theta\sin\varphi\,\hat{S}_y + \cos\theta\,\hat{S}_z$$

(Solution):   It is convenient to use the representation (8.49), such that the matrix eigenvalue equation reads:

$$\frac{\hbar}{2}\begin{pmatrix} \cos\theta & \sin\theta\,e^{-i\varphi} \\ \sin\theta\,e^{i\varphi} & -\cos\theta \end{pmatrix}\begin{pmatrix} a_+ \\ a_- \end{pmatrix} = \lambda\hbar\begin{pmatrix} a_+ \\ a_- \end{pmatrix}$$

which is equivalent to the simultaneous equations:

$$\left(\frac{\cos\theta}{2} - \lambda\right)a_+ + \frac{\sin\theta}{2}e^{-i\varphi}a_- = 0$$

$$\frac{\sin\theta}{2}e^{i\varphi}a_+ - \left(\frac{\cos\theta}{2} + \lambda\right)a_- = 0$$

The characteristic equation immediately gives $\lambda = \pm 1/2$, as expected, such that the eigenvalues are $\pm\frac{1}{2}\hbar$, with corresponding spinors:

$$\chi_+^r = \begin{pmatrix} \cos\frac{\theta}{2}\,e^{-i\varphi/2} \\ \sin\frac{\theta}{2}\,e^{i\varphi/2} \end{pmatrix}, \qquad \chi_-^r = \begin{pmatrix} -\sin\frac{\theta}{2}\,e^{-i\varphi/2} \\ \cos\frac{\theta}{2}\,e^{i\varphi/2} \end{pmatrix}$$

Problem 8.2.2. Show that, if $\hat{\vec{A}}(\hat{A}_x, \hat{A}_y, \hat{A}_z)$ and $\hat{\vec{B}}(\hat{B}_x, \hat{B}_y, \hat{B}_z)$ are vectorial operators in coordinate space, and $\hat{\vec{\sigma}} = (\hat{\sigma}_x, \hat{\sigma}_y, \hat{\sigma}_z)$ denotes the Pauli matrices, the following identity holds:

$$(\hat{\vec{\sigma}}\cdot\hat{\vec{A}})(\hat{\vec{\sigma}}\cdot\hat{\vec{B}}) = (\hat{\vec{A}}\cdot\hat{\vec{B}})\hat{I} + i\hat{\vec{\sigma}}\cdot(\hat{\vec{A}}\times\hat{\vec{B}})$$

where $\hat{I}$ is the 2×2 unit matrix.

(Solution): We use the properties (8.50), (8.51) and (8.52) of the Pauli matrices, which immediately yield:

$$\sigma_x\sigma_y = i\sigma_z, \quad \sigma_y\sigma_z = i\sigma_x, \quad \sigma_z\sigma_x = i\sigma_y$$

and the fact that $\hat{\vec{A}}$ and $\hat{\vec{B}}$ both commute with $\hat{\vec{\sigma}}$. We can then successively write:

$$(\hat{\vec{\sigma}}\cdot\hat{\vec{A}})(\hat{\vec{\sigma}}\cdot\hat{\vec{B}}) = (\hat{\sigma}_x\hat{A}_x + \hat{\sigma}_y\hat{A}_y + \hat{\sigma}_z\hat{A}_z)(\hat{\sigma}_x\hat{B}_x + \hat{\sigma}_y\hat{B}_y + \hat{\sigma}_z\hat{B}_z)$$

$$= (\hat{A}_x\hat{B}_x + \hat{A}_y\hat{B}_y + \hat{A}_z\hat{B}_z)\hat{I} + \hat{\sigma}_y\hat{\sigma}_z(\hat{A}_y\hat{B}_z - \hat{A}_z\hat{B}_y)$$

$$+ \hat{\sigma}_z\hat{\sigma}_x(\hat{A}_z\hat{B}_x - \hat{A}_x\hat{B}_z) + \hat{\sigma}_x\hat{\sigma}_y(\hat{A}_x\hat{B}_y - \hat{A}_y\hat{B}_x)$$

$$= (\hat{A}_x\hat{B}_x + \hat{A}_y\hat{B}_y + \hat{A}_z\hat{B}_z)\hat{I} + i\hat{\sigma}_x(\hat{A}_y\hat{B}_z - \hat{A}_z\hat{B}_y)$$

$$+ i\hat{\sigma}_y(\hat{A}_z\hat{B}_x - \hat{A}_x\hat{B}_z) + i\hat{\sigma}_z(\hat{A}_x\hat{B}_y - \hat{A}_y\hat{B}_x)$$

$$= (\hat{\vec{A}}\cdot\hat{\vec{B}})\hat{I} + i\hat{\vec{\sigma}}\cdot(\hat{\vec{A}}\times\hat{\vec{B}})$$

Problem 8.2.3. Show that the Pauli matrices have the property:

$$\hat{\sigma}_x\hat{\sigma}_y\hat{\sigma}_z = i$$

Problem 8.2.4. Prove that if $\vec{e}_r$ is the unit vector in a direction defined by the spherical polar coordinates θ and φ and $\hat{\vec{\sigma}}(\hat{\sigma}_x, \hat{\sigma}_y, \hat{\sigma}_z)$ denotes the Pauli matrices, we have:

$$(\vec{e}_r\cdot\hat{\vec{\sigma}})^2 = \hat{I}$$

where $\hat{I}$ is the 2×2 unit matrix.

8.3. Addition of Angular Momenta

A total angular momentum can be associated with operators which represent a simultaneous rotation of both space and spin coordinates, in other words with the change of state functions which reflect both spatial and spin properties. The angular momentum states can be written in the form:

$$\Phi_{nlmm_s} = \Psi_{nlm}(r,\theta,\varphi)\chi = \Psi_{nlm}a_+\chi_+ + \Psi_{nlm}a_-\chi_- = \begin{pmatrix} a_+\Psi_{nlm} \\ a_-\Psi_{nlm} \end{pmatrix} = \begin{pmatrix} \Phi_+ \\ \Phi_- \end{pmatrix} \quad (8.64)$$

where, according to Postulate 3, $|a_+\Psi_{nlm}|^2$ and $|a_-\Psi_{nlm}|^2$ represent the probability densities to find the particle in the volume element between r and $r + dr$, θ and $\theta + d\theta$ and φ and $\varphi + d\varphi$, with the spin component along the z-axis equal to $\frac{1}{2}\hbar$ and $-\frac{1}{2}\hbar$ respectively. It is obvious that the spatial distribution function, irrespective of the spin orientation, is:

$$P(r,\theta,\varphi) = |\Psi_{nlm}|^2 \left(|a_+|^2 + |a_-|^2\right) = |\Psi_{nlm}|^2$$

and the probabilities to have a given spin orientation, irrespective of localization, are:

$$P_{\frac{1}{2}} = \int_\infty |a_+\Psi_{nlm}|^2 \, dV = |a_+|^2, \qquad P_{-\frac{1}{2}} = \int_\infty |a_-\Psi_{nlm}|^2 \, dV = |a_-|^2$$

The operators on the eigenstates Φ_{nlmm_s} can be represented in the matrix form as:

$$\left(\hat{A}\right) = \hat{A}_0 \hat{I} + \hat{A}_x \hat{\sigma}_x + \hat{A}_y \hat{\sigma}_y + \hat{A}_z \hat{\sigma}_z = \begin{pmatrix} \hat{A}_0 + \hat{A}_z & \hat{A}_x - i\hat{A}_y \\ \hat{A}_x + i\hat{A}_y & \hat{A}_0 - \hat{A}_z \end{pmatrix} \quad (8.65)$$

which is a generalization of Eq.(8.59), where the elements are no longer numbers, but operators on the wave functions Ψ_{nlm} in coordinate space.

The change of the state function (8.64), under a rotation $\vec{\omega} = \vec{e}_z \varphi_0$ through a finite angle φ_0 about the z-axis, can be represented, in view of Eqs.(8.6) and (8.64) that define the finite rotation generators for spatial wave functions and spinors, in the form:

$$\Phi^{\varphi_0}_{nlmm_s} = \Psi_{nlm}(r,\theta,\varphi+\varphi_0)\chi^{\varphi_0} = e^{i(\hat{L}_z+\hat{S}_z)\varphi_0/\hbar}\Psi_{nlm}(r,\theta,\varphi)\chi = e^{i\hat{J}_z\varphi_0/\hbar}\Phi_{nlmm_s} \quad (8.66)$$

We introduce the *total angular momentum operator*, defined as:

$$\hat{\vec{J}} = \hat{\vec{L}} + \hat{\vec{S}} \quad (8.67)$$

whose components will satisfy the commutation relations (8.7), since $\hat{L}$ depends on spatial coordinates and $\hat{S}$ does not, so that they commute. Equation (8.67) shows that a rotation of both space and spin coordinates about the z-axis can be represented in terms of the $\hat{J}_z$ component of the total angular momentum. The possible angular momentum states of a system can be defined in terms of a set of common eigenfunctions:

$$\Phi_{nlmm_s} \equiv \Phi_{mm_s}$$

with eigenvalues $l(l+1)\hbar^2$, $m\hbar$, $s(s+1)\hbar^2$, $m_s\hbar$ of the four operators $\hat{L}^2, \hat{L}_z, \hat{S}^2, \hat{S}_z$ which represent compatible observables. There are $2l + 1$ allowed values of m and two values of m_s, for a given l and s, and therefore $2(2l + 1)$ possible eigenfunctions Φ_{mm_s}. Squaring Eq.(8.67) gives:

$$\hat{J}^2 = \hat{L}^2 + \hat{S}^2 + 2\hat{L} \cdot \hat{S} \tag{8.68}$$

which shows that the observables L^2, S^2, J^2 and J_z are also compatible. As a result, the states of the total angular momentum can also be represented in terms of the simultaneous eigenfunctions:

$$\Psi_{nljm_j} \equiv \Psi_{jm_j}$$

where j and m_j are quantum numbers defined by the eigenvalue equations (8.23) and (8.24). As there are $2j + 1$ possible values of m_j for each value of j, compatible with a given l and s, the total number of eigenfunctions is:

$$\sum_j (2j+1) = 2(2l+1) \tag{8.69}$$

since the number of angular momentum states must be the same in both representations. When the z-component of $\hat{L}$ and $\hat{S}$ is diagonal, so is that of $\hat{J}$ and we may write:

$$\hat{J}_z \Phi_{mm_s} = (m + m_s)\hbar \Phi_{mm_s} \tag{8.70}$$

which gives, by consideration of Eq.(8.24):

$$m_j = m + m_s \tag{8.71}$$

Since the maximum values of m and m_s are l and s respectively, that of m_j will be $l + s$, and in view of Eq.(8.20) have:

$$j_{max} = (m+m_s)_{max} = l+s, \qquad \Psi_{l+s,l+s} = \Phi_{ls} = R_n(r)Y_{ll}(\theta,\varphi)\chi \qquad (8.72)$$

For $m_j = l+s-1$, there are two mutually orthogonal Φ_{mm_s} states, corresponding to the possible choices:

$$j = l+s, \qquad m_j = l+s-1 = j-1: \quad \Psi_{l+s,l+s-1}$$

$$j = l+s-1, \quad m_j = l+s-1 = j \qquad : \quad \Psi_{l+s-1,l+s-1}$$

For $m_j = l+s-2$ there will be three mutually orthogonal Φ_{mm_s} states denoted by:

$$\Phi_{l,s-2}, \quad \Phi_{l-1,s-1}, \quad \Phi_{l-2,s}$$

from which three different linear combinations can be built, each with the given m_j value:

$$j = l+s, \qquad m_j = l+s-2 = j-2 \;:\; \Psi_{l+s,l+s-2}$$

$$j = l+s-1, \quad m_j = l+s-2 = j-1 \;:\; \Psi_{l+s-1,l+s-2}$$

$$j = l+s-2, \quad m_j = l+s-2 = j \quad\;:\; \Psi_{l+s-2,l+s-2}$$

This process may be continued until the minimum value of j, which is the minimum value of $|m_j|$ is reached, namely:

$$j_{min} = l-s \qquad (8.73)$$

so that the possible values of j are:

$$l-s \le j \le l+s \qquad (8.74)$$

The relation between the two representations of the total angular momentum eigenstates is usually given by linking the two sets of eigenfunctions Ψ_{jm_j} and Φ_{mm_s}, in the form:

$$\Psi_{jm_j} = \sum_m \sum_{m_s} C^{jm_j}_{lm s m_s} \Phi_{mm_s} \qquad (8.75)$$

where the coefficients are called the *Clebsch-Gordan coefficients*.

8. Angular Momentum 311

EXAMPLE 8.4. Representation of angular momenta for $l = 1$ and $s = \frac{1}{2}$

Consider a spinning electron with $s = \frac{1}{2}$ in a p-state of the one electron atom with $l = 1$. In a classical vector representation we may think of the electron as having both orbital and spin angular momenta, with each producing a magnetic field due to the electron charge. We recall that the torque exerted by a magnetic field on a magnetic dipole moment will cause the axis of the angular momentum to precess around the field direction with a constant polar orientation. Thus the quantum condition (8.67) can be represented by Figure 8.1, which shows the angular momenta $\vec{L}$ and $\vec{S}$ precessing about the common axis of their mechanical resultant $\vec{J}$. The opening angles of the precession cones are adjusted so that the resultant has the length corresponding to the possible values $j = l + \frac{1}{2} = \frac{3}{2}$ and $j = l - \frac{1}{2} = \frac{1}{2}$ given by Eq.(8.74).

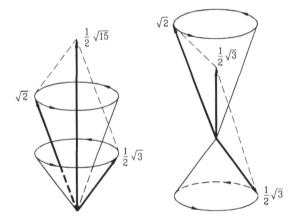

Figure 8.1. Precession of $\vec{L}$ and $\vec{S}$ around $\vec{J}$ for a spinning electron

There are six possible states of the angular momentum, as given by Eq.(8.69), which reads: $(2 \cdot \frac{3}{2} + 1) + (2 \cdot \frac{1}{2} + 1) = 2(2 \cdot 1 + 1) = 6$. The six possible Φ_{mm_s} states of angular momentum:

$$\Phi_{1,\frac{1}{2}}, \Phi_{1,-\frac{1}{2}}, \Phi_{0,\frac{1}{2}}, \Phi_{0,-\frac{1}{2}}, \Phi_{-1,\frac{1}{2}}, \Phi_{-1,-\frac{1}{2}}$$

are related to the spatial quantization given in Figure 8.2 (a), where the common axis of precession is taken as the z-axis for both m and m_s.

The quantum condition (8.71) allows us to give an alternative representation of the same states, in term of the six functions:

$$\Psi_{\frac{3}{2},\frac{3}{2}}, \Psi_{\frac{3}{2},-\frac{3}{2}}, \Psi_{\frac{3}{2},\frac{1}{2}}, \Psi_{\frac{3}{2},-\frac{1}{2}}, \Psi_{\frac{1}{2},\frac{1}{2}}, \Psi_{\frac{1}{2},-\frac{1}{2}}$$

which correspond to the spatial quantization of $\vec{J}$ in Figure 8.2 (b). Equation (8.72) shows that:

$$\Psi_{\frac{3}{2},\frac{3}{2}} = \Phi_{1,\frac{1}{2}} = R_n(r)Y_{11}(\theta,\varphi)\chi_+, \quad \Psi_{\frac{3}{2},-\frac{3}{2}} = \Phi_{-1,-\frac{1}{2}} = R_n(r)Y_{1,-1}(\theta,\varphi)\chi_-$$

whereas each of the remaining four states is given by a linear combination (8.75) of Φ_{mm_s}.

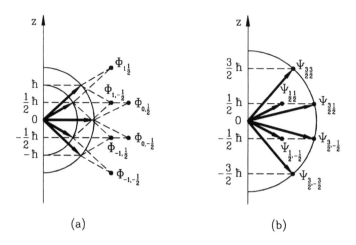

Figure 8.2. Spatial quantization of the total angular momentum

Using Eq.(8.31) we have:

$$\hat{J}_- \Psi_{\frac{3}{2},\frac{3}{2}} = \sqrt{\tfrac{3}{2}(\tfrac{3}{2}+1) - \tfrac{3}{2}(\tfrac{3}{2}-1)}\,\hbar\, \Psi_{\frac{3}{2},\frac{1}{2}} = \sqrt{3}\,\hbar\, \Psi_{\frac{3}{2},\frac{1}{2}}$$

so that:

$$\Psi_{\frac{3}{2},\frac{1}{2}} = \frac{1}{\sqrt{3}\,\hbar}\hat{J}_- \Psi_{\frac{3}{2},\frac{3}{2}} = \frac{1}{\sqrt{3}\,\hbar}(\hat{L}_- + \hat{S}_-)\Phi_{1,\frac{1}{2}}$$

Now from Eqs.(8.33) and (8.31) we obtain:

$$\hat{L}_- \Phi_{1,\frac{1}{2}} = \sqrt{1(1+1) - 1(1-1)}\,\hbar\,\Phi_{0,\frac{1}{2}} = \sqrt{2}\,\hbar\,\Phi_{0,\frac{1}{2}}$$

$$\hat{S}_- \Phi_{1,\frac{1}{2}} = \sqrt{\tfrac{1}{2}(\tfrac{1}{2}+1) - \tfrac{1}{2}(\tfrac{1}{2}-1)}\,\hbar\,\Phi_{1,-\frac{1}{2}} = \hbar\,\Phi_{1,-\frac{1}{2}}$$

and it follows that:

$$\Psi_{\frac{3}{2},\frac{1}{2}} = \frac{1}{\sqrt{3}}\left(\Phi_{1,-\frac{1}{2}} + \sqrt{2}\,\Phi_{0,\frac{1}{2}}\right) = \frac{R_n(r)}{\sqrt{3}}\left(Y_{11}\chi_- + \sqrt{2}\,Y_{10}\chi_+\right)$$

Similarly, since:

$$\hat{J}_- \Psi_{\frac{3}{2},\frac{1}{2}} = \sqrt{\tfrac{3}{2}(\tfrac{3}{2}+1) - \tfrac{1}{2}(\tfrac{1}{2}-1)}\,\hbar\,\Psi_{\frac{3}{2},-\frac{1}{2}} = 2\hbar\,\Psi_{\frac{3}{2},-\frac{1}{2}}$$

we obtain:

$$\Psi_{\frac{3}{2},-\frac{1}{2}} = \frac{1}{2\sqrt{3}\,\hbar}\hat{J}_-\left(\Phi_{1,-\frac{1}{2}} + \sqrt{2}\,\Phi_{0,\frac{1}{2}}\right) = \frac{1}{\sqrt{3}}\left(\sqrt{2}\,\Phi_{0,-\frac{1}{2}} + \Phi_{-1,\frac{1}{2}}\right) = \frac{R_n(r)}{\sqrt{3}}\left(\sqrt{2}\,Y_{10}\chi_- + Y_{1,-1}\chi_+\right)$$

8. Angular Momentum

Under the restriction $m+m_s = m_j$, Eq.(8.75) gives:

$$\Psi_{\frac{11}{22}} = C_1 \Phi_{1,-\frac{1}{2}} + C_2 \Phi_{0,\frac{1}{2}}, \quad \Psi_{\frac{1}{2},-\frac{1}{2}} = C_1' \Phi_{0,-\frac{1}{2}} + C_2' \Phi_{-1,\frac{1}{2}}$$

Since $\Psi_{\frac{11}{22}}$ is expressed in terms of the same Φ_{mm_s} states as $\Psi_{\frac{31}{22}}$, their orthogonality requires that: $C_1 + \sqrt{2}\, C_2 = 0$, and the normalization condition of $\Psi_{\frac{11}{22}}$ gives $C_1^2 + C_2^2 = 1$, so that we obtain:

$$\Psi_{\frac{11}{22}} = \frac{1}{\sqrt{3}}\left(\sqrt{2}\,\Phi_{1,-\frac{1}{2}} - \Phi_{0,\frac{1}{2}}\right) = \frac{R_n(r)}{\sqrt{3}}\left(\sqrt{2}\, Y_{11}\chi_- - Y_{10}\chi_+\right)$$

Similarly $\Psi_{\frac{1}{2},-\frac{1}{2}}$, which is orthogonal to $\Psi_{\frac{3}{2},-\frac{1}{2}}$, is obtained as:

$$\Psi_{\frac{1}{2},-\frac{1}{2}} = \frac{1}{\sqrt{3}}\left(\Phi_{0,-\frac{1}{2}} - \sqrt{2}\,\Phi_{-1,\frac{1}{2}}\right) = \frac{R_n(r)}{\sqrt{3}}\left(Y_{10}\chi_- - \sqrt{2}\, Y_{1,-1}\chi_+\right). \quad \clubsuit$$

It is convenient to solve the eigenvalue problem (8.23) for spin-$\frac{1}{2}$ particles in the matrix form, using the representation (8.65) which reads:

$$\hat{J}_z = \hat{L}_z + \hat{S}_z = \begin{pmatrix} \hat{L}_z + \hat{S}_z & 0 \\ 0 & \hat{L}_z - \hat{S}_z \end{pmatrix} = \begin{pmatrix} -i\hbar \dfrac{\partial}{\partial \varphi} + \frac{1}{2}\hbar & 0 \\ 0 & -i\hbar \dfrac{\partial}{\partial \varphi} - \frac{1}{2}\hbar \end{pmatrix}$$

$$\hat{J}^2 = \hat{L}^2 + \hat{S}^2 + 2(\hat{L}_x\hat{S}_x + \hat{L}_y\hat{S}_y + \hat{L}_z\hat{S}_z) = \begin{pmatrix} \hat{L}^2 + \frac{3}{4}\hbar^2 + \hbar\hat{L}_z & \hbar\hat{L}_- \\ \hbar\hat{L}_+ & \hat{L}^2 + \frac{3}{4}\hbar^2 + \hbar\hat{L}_z \end{pmatrix}$$

(8.76)

The eigenstates of $\hat{J}^2$ and $\hat{J}_z$ will be written in the general form (8.64):

$$\Psi_{jm_j} = \Psi_+ \chi_+ + \Psi_- \chi_- = \begin{pmatrix} \Psi_+ \\ \Psi_- \end{pmatrix} \tag{8.77}$$

so that the eigenvalue equation (8.23) for $\hat{J}_z$ reads:

$$\begin{pmatrix} -i\hbar \dfrac{\partial}{\partial \varphi} + \frac{1}{2}\hbar & 0 \\ 0 & -i\hbar \dfrac{\partial}{\partial \varphi} - \frac{1}{2}\hbar \end{pmatrix} \begin{pmatrix} \Psi_+ \\ \Psi_- \end{pmatrix} = m_j \hbar \begin{pmatrix} \Psi_+ \\ \Psi_- \end{pmatrix}$$

and is equivalent to:

$$-i\hbar\frac{\partial \Psi_+}{\partial \varphi} = \left(m_j - \tfrac{1}{2}\right)\hbar\Psi_+, \qquad -i\hbar\frac{\partial \Psi_-}{\partial \varphi} = \left(m_j + \tfrac{1}{2}\right)\hbar\Psi_-$$

Because m_j is half-integer, like j, for spin-$\tfrac{1}{2}$ particles, and Ψ_+ and Ψ_- should also be eigenfunctions of $\hat{L}^2$, their angular dependence will be given by the spherical harmonics $Y_{l,m_j \pm \tfrac{1}{2}}(\theta,\varphi)$, such that Eq.(8.77) becomes:

$$\Psi_{jm_j} = C_+ R(r) Y_{l,m_j-\tfrac{1}{2}}(\theta,\varphi)\chi_+ + C_- R(r) Y_{l,m_j+\tfrac{1}{2}}(\theta,\varphi)\chi_- \qquad (8.78)$$

The radial function $R(r)$ is a solution to Eq.(7.41) and C_+, C_- are the Clebsch-Gordan coefficients which are determined by the eigenvalue equation (8.23) for $\hat{J}^2$:

$$\begin{pmatrix} \hat{L}^2 + \tfrac{3}{4}\hbar^2 + \hbar\hat{L}_z & \hbar\hat{L}_- \\ \hbar\hat{L}_+ & \hat{L}^2 - \tfrac{3}{4}\hbar^2 - \hbar\hat{L}_z \end{pmatrix} \begin{pmatrix} C_+ R\,Y_{l,m_j-\tfrac{1}{2}} \\ C_- R\,Y_{l,m_j+\tfrac{1}{2}} \end{pmatrix} = j(j+1)\hbar^2 \begin{pmatrix} C_+ R\,Y_{l,m_j-\tfrac{1}{2}} \\ C_- R\,Y_{l,m_j+\tfrac{1}{2}} \end{pmatrix}$$

In view of Eqs.(8.33) for $\hat{L}_+$ and $\hat{L}_-$, we obtain the simultaneous equations:

$$C_+\left[l(l+1) - j(j+1) + m_j + \tfrac{1}{4}\right] + C_-\sqrt{l(l+1) - (m_j^2 - \tfrac{1}{4})} = 0$$

$$C_+\sqrt{l(l+1) - (m_j^2 - \tfrac{1}{4})} + C_-\left[l(l+1) - j(j+1) - m_j + \tfrac{1}{4}\right] = 0$$

The determinantal equation leads to the quantum numbers $j = l \pm \tfrac{1}{2}$. For $j = l + \tfrac{1}{2}$ we obtain:

$$C_+ = \sqrt{l + m_j + \tfrac{1}{2}}, \qquad C_- = \sqrt{l - m_j + \tfrac{1}{2}}$$

which substituted into Eq.(8.78) yield the normalization condition $|C_+|^2 + |C_-|^2 = 2l+1$ and therefore:

$$\Psi_{l+1/2, m_j} = \sqrt{\frac{l + m_j + \tfrac{1}{2}}{2l+1}}\, R(r) Y_{l,m_j-1/2}\,\chi_+ + \sqrt{\frac{l - m_j + \tfrac{1}{2}}{2l+1}}\, R(r) Y_{l,m_j+1/2}\,\chi_- \qquad (8.79)$$

In the same way we find:

$$\Psi_{l-\tfrac{1}{2}, m_j} = -\sqrt{\frac{l - m_j + \tfrac{1}{2}}{2l+1}}\, R(r) Y_{l,m_j-\tfrac{1}{2}}\,\chi_+ + \sqrt{\frac{l + m_j + \tfrac{1}{2}}{2l+1}}\, R(r) Y_{l,m_j+\tfrac{1}{2}}\,\chi_- \qquad (8.80)$$

Comparison with Eq.(8.75) shows that the Clebsch-Gordan coefficients are given by:

$$C^{l+\frac{1}{2},m_j}_{l,m_j-\frac{1}{2},\frac{1}{2},\frac{1}{2}} = C^{l-\frac{1}{2},m_j}_{l,m_j+\frac{1}{2},\frac{1}{2},-\frac{1}{2}} = \sqrt{\frac{l+m_j+\frac{1}{2}}{2l+1}}$$

$$C^{l+\frac{1}{2},m_j}_{l,m_j+\frac{1}{2},\frac{1}{2},-\frac{1}{2}} = -C^{l-\frac{1}{2},m_j}_{l,m_j-\frac{1}{2},\frac{1}{2},\frac{1}{2}} = \sqrt{\frac{l-m_j+\frac{1}{2}}{2l+1}}$$

(8.81)

EXAMPLE 8.5. The vector model of angular momentum

The representation of the total angular momentum eigenstates by two different sets of good quantum numbers can be understood in terms of two different schemes of coupling the orbital and spin angular momenta, in a *vector model* of atoms. If we place the one electron atom into a magnetic field which is parallel to the z-direction, the precession of $\vec{L}$ and $\vec{S}$, or $\vec{J}$, around the z-axis results in a constant value of their z component, which must be one of the possible eigenvalues $(m+m_s)\hbar$ or $m_j\hbar$.

The Ψ_{jm_j} representation of the total angular momentum implies that the operators $\hat{H}, \hat{L}^2, \hat{S}^2, \hat{J}^2$ and $\hat{J}_z$ commute, thus there is a set of common eigenfunctions with the following properties:

$$\hat{H}\Psi_{jm_j} = E_n \Psi_{jm_j}$$

$$\hat{L}^2 \Psi_{jm_j} = l(l+1)\hbar^2 \Psi_{jm_j}$$

$$\hat{S}^2 \Psi_{jm_j} = s(s+1)\hbar^2 \Psi_{jm_j} \qquad (8.82)$$

$$\hat{J}^2 \Psi_{jm_j} = j(j+1)\hbar^2 \Psi_{jm_j}$$

$$\hat{J}_z \Psi_{jm_j} = m_j \hbar \Psi_{jm_j}$$

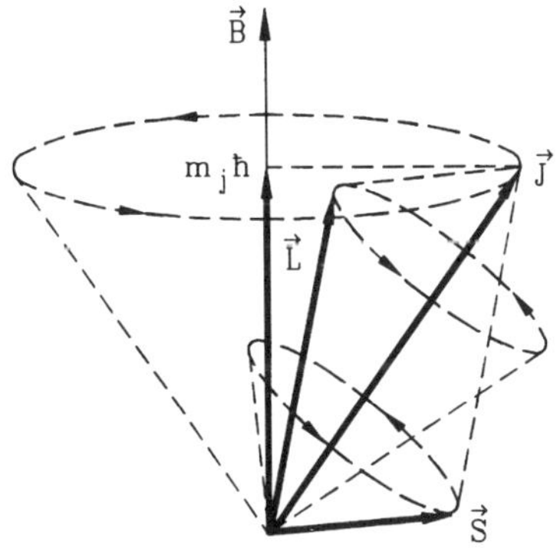

Figure 8.3. Precession of $\vec{J}$ around the field direction with $m_j\hbar$ as a constant of motion

The description (8.82) corresponds to an applied magnetic field which is weak compared with the extremely strong fields inside the atom, so that the total angular momentum will precess as a whole and adjust its polar orientation in such a way that its z component has one of the allowed values $m_j\hbar$. The good quantum numbers are l, s, j and m_j since $m\hbar$ and $m_s\hbar$ vary in time as a result of precession, as illustrated in Figure 8.3, so that Ψ_{jm_j} are the suitable eigenfunctions.

For an applied field stronger than the internal fields inside the atom, the coupling interaction between $\vec{L}$ and $\vec{S}$ is negligible and each of them adjusts so that their z component becomes a constant of motion.

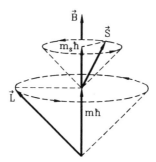

Figure 8.4. Independent precession of $\vec{L}$ and $\vec{S}$ around the field direction, with $m\hbar$ and $m_s\hbar$ as constants of motion

This is shown in Figure 8.4 where $\vec{L}$ and $\vec{S}$ precess independently about the z-axis. This implies that only $\hat{L}^2, \hat{S}^2, \hat{L}_z$ and $\hat{S}_z$ are quantified for strong fields, and therefore m and m_s are good quantum numbers.

The eigenstates of the angular momentum can be described by the set of common eigenfunctions Φ_{mm_s} of the operators $\hat{H}, \hat{L}^2, \hat{S}^2, \hat{L}_z$ and $\hat{S}_z$ which obey:

$$\hat{H}\Phi_{mm_s} = E_n\Phi_{mm_s}$$

$$\hat{L}^2\Phi_{mm_s} = l(l+1)\hbar^2\Phi_{mm_s}$$

$$\hat{S}^2\Phi_{mm_s} = s(s+1)\hbar^2\Phi_{mm_s} \qquad (8.83)$$

$$\hat{L}_z\Phi_{mm_s} = m\hbar\Phi_{mm_s}$$

$$\hat{S}_z\Phi_{mm_s} = m_s\hbar\Phi_{mm_s}$$

In both Eqs.(8.82) and (8.83), $\hat{H}$ denotes the central force Hamiltonian, defined by Eq.(7.1).

Problem 8.3.1. Find the expectation values of the operators $\hat{S}_i$, $\hat{L}_i$ and $\hat{J}_i$ ($i = x, y, z$) in the eigenstates of total angular momentum in one-electron atoms.

(Solution): Consider first the case $j = l + \frac{1}{2}$ for which the eigenstates (8.79) can be written as:

$$\Psi_{l+\frac{1}{2},m_j} = R_{nl}(r) \begin{pmatrix} \sqrt{\dfrac{l+m_j+\frac{1}{2}}{2l+1}}\, Y_{l,m_j-\frac{1}{2}} \\ \sqrt{\dfrac{l-m_j+\frac{1}{2}}{2l+1}}\, Y_{l,m_j+\frac{1}{2}} \end{pmatrix} = R_{nl}\begin{pmatrix} \psi_+ \\ \psi_- \end{pmatrix}$$

where $R_{nl}(r)$ are the normalized radial functions, Eq.(7.82), with no influence on the expectation values. We have:

$$\langle \hat{S}_z \rangle = \int (\psi_+^*, \psi_-^*) \frac{\hbar}{2}\begin{pmatrix} 1 & 0 \\ 0 & -1 \end{pmatrix}\begin{pmatrix} \psi_+ \\ \psi_- \end{pmatrix} d\Omega = \frac{\hbar}{2}\int (|\psi_+|^2 - |\psi_-|^2) d\Omega$$

$$= \frac{\hbar}{2}\frac{l+m_j+\frac{1}{2}}{2l+1}\int |Y_{l,m_j-\frac{1}{2}}|^2 d\Omega - \frac{\hbar}{2}\frac{l-m_j+\frac{1}{2}}{2l+1}\int |Y_{l,m_j+\frac{1}{2}}|^2 d\Omega$$

$$= \hbar\frac{m_j}{2l+1} = \frac{m_j \hbar}{2j}$$

In the same way, using Eq.(8.49), one obtains:

$$\langle \hat{S}_x \rangle = \frac{\hbar}{2}\int \psi_+^* \psi_- d\Omega - \frac{\hbar}{2}\int \psi_-^* \psi_+ d\Omega = 0$$

because the spherical harmonics corresponding to different values of the magnetic quantum number are orthogonal. Similarly we have $\langle \hat{S}_y \rangle = 0$.

For the operator $\hat{L}_z$ we can write:

$$\langle \hat{L}_z \rangle = \int (\psi_+^*, \psi_-^*)\begin{pmatrix} \hat{L}_z & 0 \\ 0 & \hat{L}_z \end{pmatrix}\begin{pmatrix} \psi_+ \\ \psi_- \end{pmatrix} d\Omega = \int (\psi_+^*, \psi_-^*)\begin{pmatrix} (m_j-\frac{1}{2})\hbar\,\psi_+ \\ (m_j+\frac{1}{2})\hbar\,\psi_- \end{pmatrix} d\Omega$$

$$= \hbar(m_j - \tfrac{1}{2})\frac{l+m_j+\frac{1}{2}}{2l+1}\int |Y_{l,m_j-\frac{1}{2}}|^2 d\Omega + \hbar(m_j + \tfrac{1}{2})\frac{l-m_j+\frac{1}{2}}{2l+1}\int |Y_{l,m_j+\frac{1}{2}}|^2 d\Omega$$

$$= \hbar m_j - \frac{\hbar}{2}\frac{2m_j}{2l+1} = m_j\hbar\left(1 - \frac{1}{2j}\right)$$

and from Eqs.(8.32) it follows that:

$$\langle \hat{L}_x \rangle = \langle \psi_{jm_j} | \hat{L}_x | \psi_{jm_j} \rangle = 0, \quad \langle \hat{L}_y \rangle = 0$$

We observe that:

$$\langle \hat{L}_z \rangle = m_j \hbar - \langle \hat{S}_z \rangle \quad \text{or} \quad \langle \hat{J}_z \rangle = m_j \hbar$$

as expected, and from Eq.(8.32), we have $\langle \hat{J}_x \rangle = \langle \hat{J}_y \rangle = 0$. For $j = l - \frac{1}{2}$ we similarly obtain:

$$\langle \hat{S}_z \rangle = -\frac{m_j \hbar}{2(j+1)}, \quad \langle \hat{L}_z \rangle = m_j \hbar \left[1 + \frac{1}{2(j+1)}\right], \quad \langle \hat{J}_z \rangle = m_j \hbar$$

Problem 8.3.2. Use the vector model of angular momentum to determine the allowed values for the angle between the $\vec{L}$ and $\vec{S}$ observables of an electron in one-electron atoms.

(Solution): The precession of $\vec{L}$ and $\vec{S}$ around $\vec{J}$ (see Figure 8.3) corresponds to Eq.(8.68), from which we have:

$$\hat{\vec{L}} \cdot \hat{\vec{S}} = \frac{1}{2}(\hat{J}^2 - \hat{L}^2 - \hat{S}^2)$$

or, in the eigenstates Ψ_{jm_j} of the total angular momentum:

$$\langle \hat{\vec{L}} \cdot \hat{\vec{S}} \rangle = \frac{\hbar^2}{2}[j(j+1) - l(l+1) - s(s+1)]$$

We may define the angle between the two vectors by:

$$\cos(\vec{L} \cdot \vec{S}) = \frac{\langle \hat{\vec{L}} \cdot \hat{\vec{S}} \rangle}{\sqrt{\langle \hat{L}^2 \rangle \langle \hat{S}^2 \rangle}} = \frac{j(j+1) - l(l+1) - s(s+1)}{2\sqrt{l(l+1)s(s+1)}}$$

where $j = l \pm \frac{1}{2}$ and $s = \frac{1}{2}$ for the atomic electron.

Problem 8.3.3. Find the probability to obtain the possible values of the orbital angular momentum components along the z-axis for a spinning atomic electron.

(Answer): $P(m_j - \frac{1}{2}) = \dfrac{l \pm m_j + \frac{1}{2}}{2l+1}, \quad P(m_j + \frac{1}{2}) = \dfrac{l \mp m_j + \frac{1}{2}}{2l+1} \quad$ for $j = l \pm \frac{1}{2}$.

8.4. Spin Magnetic Moment

The Hamiltonian operator of a spinning atomic electron can be written in the form:

$$\hat{H} = \frac{\hat{\vec{p}}^2}{2m_e} + U(r) = \frac{1}{2m_e}(\hat{\vec{\sigma}} \cdot \hat{\vec{p}})^2 + U(r) \tag{8.84}$$

in view of the Dirac identity (see Problem 8.2.2), which gives:

$$(\hat{\vec{\sigma}} \cdot \hat{\vec{p}})(\hat{\vec{\sigma}} \cdot \hat{\vec{p}}) = \hat{\vec{p}} \cdot \hat{\vec{p}} + i\hat{\vec{\sigma}} \cdot (\hat{\vec{p}} \times \hat{\vec{p}}) = \hat{\vec{p}}^2$$

We have seen (Example 7.5) that in the presence of an electromagnetic field, the appropriate Hamiltonian is obtained through the formal substitution:

$$\hat{\vec{p}} \to \hat{\vec{p}} + e\vec{A}$$

where $\vec{A}$ is the vector potential, so that Eq.(8.84) becomes:

$$\hat{H} = \frac{1}{2m_e}\left[\hat{\vec{\sigma}} \cdot (\hat{\vec{p}} + e\vec{A})\right]^2 + U(r)$$

$$= \frac{1}{2m_e}(\hat{\vec{p}} + e\vec{A})^2 + \frac{i}{2m_e}\hat{\vec{\sigma}} \cdot \left[(\hat{\vec{p}} + e\vec{A}) \times (\hat{\vec{p}} + e\vec{A})\right] + U(r) \tag{8.85}$$

We immediately obtain:

$$(\hat{\vec{p}} + e\vec{A}) \times (\hat{\vec{p}} + e\vec{A}) = e\hat{\vec{p}} \times \vec{A} + e\vec{A} \times \hat{\vec{p}} = -ie\hbar(\nabla \times \vec{A}) = -ie\hbar\vec{B}$$

which, on substitution into Eq.(8.85), gives:

$$\hat{H} = \frac{1}{2m_e}(\hat{\vec{p}} + e\vec{A})^2 + \frac{e\hbar}{2m_e}\hat{\vec{\sigma}} \cdot \vec{B} + U(r) = \frac{1}{2m_e}(\hat{\vec{p}} + e\vec{A})^2 + \frac{e}{m_e}\hat{\vec{S}} \cdot \vec{B} + U(r) \tag{8.86}$$

Comparison with Eq.(7.140) shows that there is an additional interaction energy of the form $-\hat{\vec{\mu}}_s \cdot \vec{B}$ between the magnetic field $\vec{B}$ and an intrinsic *spin magnetic moment*, represented by the operator:

$$\hat{\vec{\mu}}_s = -\frac{e}{m_e}\hat{\vec{S}} = -2\frac{e}{2m_e}\hat{\vec{S}} \tag{8.87}$$

where $g_s = 2$ is called the electron g-factor and, for spin, is twice as big as for orbital angular momentum, Eq.(7.132).

EXAMPLE 8.6. Spin-orbit interaction

Since the atomic electron is moving through the electric field of the nucleus, as illustrated in Figure 8.5, it always experiences the effect of an internally generated magnetic field $\vec{B}_i$ of the atom.

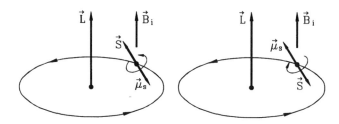

Figure 8.5. The spin and orbital motions of the electron

If $\vec{r}$ and $\vec{v}$ denote the position and velocity of the electron with respect to the nucleus +Ze, the nucleus appears to the electron to be moving around with a velocity $-\vec{v}$, so that the magnetic field experienced by the electron is classically described by:

$$\vec{B}_i = \frac{\mu_0}{4\pi}\frac{Ze(\vec{r}\times\vec{v})}{r^3} = \frac{Ze}{4\pi\varepsilon_0 m_e c^2}\frac{\vec{r}\times m_e \vec{v}}{r^3} = \frac{\vec{L}}{m_e ec^2}\frac{1}{r}\frac{d}{dr}\left(\frac{-Ze_0^2}{r}\right) = \frac{1}{m_e ec^2}\frac{1}{r}\frac{dU(r)}{dr}\vec{L} \quad (8.88)$$

where the potential energy $U(r)$, as given by Eq.(7.56), was inserted. The scalar product $-\vec{\mu}_s \cdot \vec{B}_i$ gives the increment in the potential energy of the spinning electron, in the frame of reference in which the electron is at rest. The interaction energy must, however, be multiplied by a factor of one half, called the Thomas factor, if measured in the frame of reference in which the nucleus is at rest, as a result of relativistic kinematical considerations. It follows that the interaction energy becomes:

$$-\frac{1}{2}\vec{\mu}_s \cdot \vec{B}_i = -\frac{1}{2}\left(-\frac{e}{m_e}\vec{S}\right)\cdot\left(\frac{1}{m_e ec^2}\frac{1}{r}\frac{dU(r)}{dr}\vec{L}\right) = \frac{1}{2m_e^2 c^2}\frac{1}{r}\frac{dU(r)}{dr}\vec{S}\cdot\vec{L} \quad (8.89)$$

This result gives the *spin-orbit coupling* term of the spinning electron Hamiltonian (8.86), which, by taking $\vec{A} = 0$ and considering the effect of the internal field $\vec{B}_i$ only, becomes:

$$\hat{H} = \hat{H}_0 + \frac{e}{m_e}\hat{\vec{S}}\cdot\vec{B}_i = \hat{H}_0 + \frac{1}{2m_e^2 c^2}\frac{1}{r}\frac{dU(r)}{dr}\hat{\vec{S}}\cdot\hat{\vec{L}} \quad (8.90)$$

where $\hat{H}_0$ denotes the central force Hamiltonian (7.1), in the absence of spin. For the eigenstates of the Hamiltonian (8.90), m and m_s are not good quantum numbers, because the components of

$\hat{\vec{L}}$ or $\hat{\vec{S}}$ will not commute with the spin-orbit term. Instead, using Eq.(8.68), it is immediately apparent that:

$$[\hat{\vec{S}} \cdot \hat{\vec{L}}, \hat{J}^2] = 0, \qquad [\hat{\vec{S}} \cdot \hat{\vec{L}}, \hat{J}_z] = 0$$

This implies that the appropriate eigenfunctions of the Hamiltonian (8.90) are Ψ_{jm_j} which satisfy Eqs.(8.82). The energy eigenvalues then result from:

$$\hat{H}\Psi_{jm_j} = \left[\hat{H}_0 + \frac{1}{2m_e^2 c^2} \frac{1}{r}\frac{dU(r)}{dr} \hat{\vec{S}} \cdot \hat{\vec{L}}\right]\Psi_{jm_j}$$

$$= \left[\hat{H}_0 + \frac{1}{4m_e^2 c^2} \frac{1}{r}\frac{dU(r)}{dr} (\hat{J}^2 - \hat{L}^2 - \hat{S}^2)\right]\Psi_{jm_j} \qquad (8.91)$$

which gives:

$$E_{nlj}\Psi_{jm_j} = E_n \Psi_{jm_j} + \frac{1}{4m_e^2 c^2}[j(j+1) - l(l+1) - s(s+1)]\hbar^2 \frac{1}{r}\frac{dU(r)}{dr}\Psi_{jm_j}$$

Using Ψ_{jm}^* as a multiplying factor in front of each term and integrating gives:

$$E_{nlj} = E_n + \frac{1}{2}[j(j+1) - l(l+1) - s(s+1)]\zeta_{nl} \qquad (8.92)$$

where ζ_{nl} denotes the expectation value:

$$\zeta_{nl} = \langle \Psi_{jm_j} | \frac{\hbar^2}{2m_e^2 c^2} \frac{1}{r}\frac{dU(r)}{dr} | \Psi_{jm_j}\rangle \qquad (8.93)$$

Equation (8.93) reduces to an average over the radial eigenfunctions $R_{nl}(r)$, given by Eq.(7.82), so that ζ depends on n and l only. If $l \neq 0$ there are two possible values of j for the spinning electron, namely $j = l \pm \frac{1}{2}$. It follows that the spin-orbit interaction splits a state with given quantum numbers n and l and energy E_{nl}, into two states with energies:

$$E_{nlj} = E_n + \frac{1}{2}l\zeta_{nl}, \qquad j = l + \frac{1}{2}$$

$$= E_n - \frac{1}{2}(l+1)\zeta_{nl}, \qquad j = l - \frac{1}{2} \qquad (8.94)$$

The s electron states with $l = 0$ are unaffected, whereas the splitting of all other states is proportional to $2l+1$. Thus the spin-orbit coupling completely removes the accidental degeneracy.

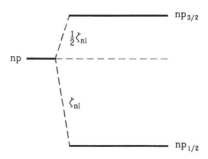

Figure 8.6. Fine structure of a *np*-state due to spin-orbit coupling

Eigenstates Ψ_{jm_j} of the spinning electron are labelled in the spectroscopic notation by *n*, *l* and *j* in the form:

$$n(l\ \text{symbol})_j$$

with the usual symbols for the allowed values of *l*. An example is given in Figure 8.6, which shows the fine structure of a *np*-state (with $l = 1$ for a spinning electron) which is split into a $np_{3/2}$-state for which $l = 1$ and $j = \frac{3}{2}$ and a $np_{1/2}$-state for which $l = 1$, $j = \frac{1}{2}$. When more electrons are present, the resultant orbital, spin and total angular momenta will be labelled by capitals and the notation of the atomic state is:

$$^{2S+1}(L\ \text{symbol})_J$$

The superscript $2S+1$ denotes the multiplicity of the level of a given *L*. ♛

In the presence of an uniform applied magnetic field $\vec{B}$, the first term of the Hamiltonian (8.86) can be approximated as given by Eq.(7.150). If we only consider the linear terms in $\vec{B}$ and take into account the influence of the internal field $\vec{B}_i$ on the spin magnetic moment, Eq.(8.90), the Hamiltonian operator (8.86) becomes:

$$\hat{H} = \hat{H}_0 + \frac{1}{2m_e^2 c^2}\frac{1}{r}\frac{dU(r)}{dr}\hat{\vec{S}}\cdot\hat{\vec{L}} + \frac{e}{2m_e}(\hat{\vec{L}} + 2\hat{\vec{S}})\cdot\vec{B} \tag{8.95}$$

Choosing $\vec{B}$ in the *z*-direction, Eq.(8.95) reduces to:

$$\hat{H} = \hat{H}_0 + \frac{1}{2m_e^2 c^2}\frac{1}{r}\frac{dU(r)}{dr}\hat{\vec{S}}\cdot\hat{\vec{L}} + \frac{eB}{2m_e}(\hat{L}_z + 2\hat{S}_z) \tag{8.96}$$

There are two cases to consider: the *weak-field case*, in which the energy due to the external magnetic field is small compared to the spin-orbit coupling, so that the eigenfunctions Ψ_{jm_j} are required to describe the eigenstates of the spinning electron and the *strong-field case*, in which the spin-orbit interaction becomes negligible and we can speak of a decoupling of $\vec{L}$ and $\vec{S}$ and of separate couplings to the external field, such that the electron eigenstates will be described in terms of Φ_{mm_s}.

In the *weak-field case*, from Eq.(8.96) we obtain:

$$\hat{H}\Psi_{jm_j} = \hat{H}_0\Psi_{jm_j} + \frac{1}{2m_e^2 c^2}\frac{1}{r}\frac{dU(r)}{dr}(\hat{\vec{S}}\cdot\hat{\vec{L}})\Psi_{jm_j} + \frac{eB}{2m_e}(\hat{J}_z + \hat{S}_z)\Psi_{jm_j}$$

or:

$$E_{nlj}\Psi_{jm_j} = E_n\Psi_{jm_j} + E_{SL}\Psi_{jm_j} + \frac{eB}{2m_e}m_j\hbar\Psi_{jm_j} + \frac{eB}{2m_e}\hat{S}_z\Psi_{jm_j}$$

where E_{SL} denotes the spin-orbit coupling energy given in Eq.(8.92). Putting $\Psi^*_{jm_j}$ as a multiplying factor in front of each term and integrating gives:

$$E_{nlj} = E_n + E_{SL} + \mu_B B m_j + \frac{eB}{2m_e}\langle\Psi_{jm_j}|\hat{S}_z|\Psi_{jm_j}\rangle \quad (8.97)$$

where μ_B is the Bohr magneton (7.134).

The calculation of the last matrix element requires the explicit knowledge of the eigenfunctions Ψ_{jm_j}. It reduces, however, to the previously solved eigenvalue problem for $\hat{J}_z$, if the average of $\hat{S}_z$ is evaluated by means of the vector model of angular momentum. Since the non-constant vector $\vec{S}$ precesses about the constant $\vec{J}$, as illustrated in Figure 8.3, only the component of $\vec{S}$ parallel to $\vec{J}$ is relevant, the other components averaging out. It follows that the average value of $\vec{S}$ can be represented in the vector model by:

$$\langle\vec{S}\rangle = \left(\vec{S}\cdot\frac{\vec{J}}{J}\right)\frac{\vec{J}}{J} = \frac{\vec{S}\cdot\vec{J}}{J^2}\vec{J}$$

which leads to:

$$\langle S_z\rangle = \frac{\vec{S}\cdot\vec{J}}{J^2}J_z = (g_L - 1)J_z \quad \text{or} \quad \hat{S}_z = \left\langle\frac{\hat{\vec{S}}\cdot\hat{\vec{J}}}{\hat{J}^2}\right\rangle\hat{J}_z = (g_L - 1)\hat{J}_z \quad (8.98)$$

where g_L is called the *Landé g-factor*. It can be calculated by squaring Eq.(8.67):

$$\hat{L}^2 = (\hat{\vec{J}} - \hat{\vec{S}})^2 \quad \text{or} \quad \hat{\vec{S}}\cdot\hat{\vec{J}} = \frac{1}{2}(\hat{J}^2 + \hat{S}^2 - \hat{L}^2)$$

In view of Eqs.(8.82), we then obtain, in terms of quantum numbers j, s and l, that:

$$g_L = 1 + \left\langle \frac{\hat{S}\cdot\hat{J}}{\hat{J}^2} \right\rangle = 1 + \frac{j(j+1) + s(s+1) - l(l+1)}{2j(j+1)} \qquad (8.99)$$

Substitution of this result into Eq.(8.98) yields:

$$\langle \Psi_{jm_j}|\hat{S}_z|\Psi_{jm_j}\rangle = (g_L - 1)\langle \Psi_{jm_j}|\hat{J}_z|\Psi_{jm_j}\rangle = (g_L - 1)m_j\hbar$$

so that Eq.(8.97) can be written as:

$$E_{nlj} = E_n + E_{SL} + \mu_B B g_L m_j \qquad (8.100)$$

Therefore, the effect of a weak external magnetic field is to split each energy level of the fine structure, with energy $E_{nl} + E_{SL}$, into $2j+1$ levels, labelled by the m_j values. This is known as the *anomalous Zeeman effect*. The Landé g-factor (8.99) gives directly the constant relative separation $\mu_B B g_L$ of Zeeman levels. For the one-electron atoms, where $j = l \pm \frac{1}{2}$, Eq.(8.99) reduces to:

$$g_L = 1 \pm \frac{1}{2l+1} \qquad (8.101)$$

so that the complete expression for the Zeeman levels is obtained by combining Eqs.(8.94) and (8.100) as:

$$E_{nlj=l+\frac{1}{2}} = E_n + \frac{1}{2}l\zeta_{nl} + \mu_B B \frac{2(l+1)}{2l+1} m_j$$

$$E_{nlj=l-\frac{1}{2}} = E_n - \frac{1}{2}(l+1)\zeta_{nl} + \mu_B B \frac{2l}{2l+1} m_j \qquad (8.102)$$

The energy level splitting is represented in Figure 8.7.

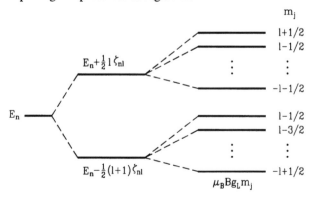

Figure 8.7. The anomalous Zeeman effect for one-electron atoms

In the *strong-field case*, in which the spin-orbit splitting is small compared to the magnetic splitting of the energy levels, and can be neglected to a first approximation, the Hamiltonian (8.96) reduces to:

$$\hat{H} = \hat{H}_0 + \frac{eB}{2m_e}(\hat{L}_z + 2\hat{S}_z) \qquad (8.103)$$

It is convenient in this case to obtain the energy eigenvalue equation using the eigenfunctions Φ_{mm_s} as follows:

$$\hat{H}\Phi_{mm_s} = \left[\hat{H}_0 + \frac{eB}{2m_e}(\hat{L}_z + 2\hat{S}_z)\right]\Phi_{mm_s} = \left[E_n + \frac{eB}{2m_e}(m+2m_s)\hbar\right]\Phi_{mm_s} \qquad (8.104)$$

so that:

$$E_{nlmm_s} = E_n + \mu_B B(m+2m_s) \qquad (8.105)$$

It is seen that the effect of a strong external magnetic field is to split a state with given quantum numbers n and l and energy E_n into $2l + 3$ states ($l \neq 0$), with equally spaced energy levels. The separation $\mu_B B$ of the energy levels is the same as that obtained in Eq.(7.151) in the case of the normal Zeeman effect.

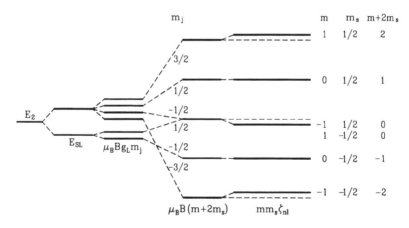

Figure 8.8. Transition from Zeeman to Paschen-Back levels of a 2p-state in the one-electron atom

The small correction term to the Zeeman levels (8.105), due to the spin-orbit interaction, must be specified in terms of the n, l, m and m_s quantum numbers, instead of n, l, j and m_j used in Eq.(8.92). This can be readily achieved for those states with $j = l + \frac{1}{2}$ of the one-electron atom, corresponding to the highest ($m = 1, m_s = \frac{1}{2}$) and the lowest ($m = -1, m_s = -\frac{1}{2}$) Zeeman level. The spin-orbit coupling (8.94) in these states can be rewritten as:

$$E_{SL} = \frac{1}{2}l\zeta_{nl} = m_s m \zeta_{nl} \qquad (8.106)$$

and it can be shown that Eq.(8.106) is valid for each Φ_{mm_s} state. Adding the correction (8.106) to Eq.(8.105), the strong-field energy levels become:

$$E_{nlmm_s} = E_n + \mu_B B(m + 2m_s) + mm_s \zeta_{nl} \qquad (8.107)$$

This result is called the *Paschen-Back effect*. The transition from the weak-field Zeeman levels (8.102) to the strong-field Paschen-Back levels (8.107) is illustrated in Figure 8.8 for a 2p-state of the spinning electron.

Problem 8.4.1. Show that the spin magnetic moment $\vec{\mu}_s$ can be expressed in terms of a probability current density:

$$\hat{\vec{j}}_s = -\frac{e}{m_e} \nabla \times (\Psi^* \hat{\vec{S}} \Psi)$$

in the same way as the orbital $\vec{\mu}_l$ and the induced $\vec{\mu}_d$ magnetic moments.

(Solution): Using the representative operator (8.87), the expectation value of the spin magnetic moment can be written as:

$$\vec{\mu}_s = -\frac{e}{m_e} \int_\infty \Psi^* \hat{\vec{S}} \Psi dV$$

The integral can be expressed, using the relation:

$$\int_\infty \vec{r} \times [\nabla \times (\Psi^* \hat{\vec{S}} \Psi)] dV = 2 \int_\infty \Psi^* \hat{\vec{S}} \Psi dV$$

which results from the properties of the vector product components. The x component, for instance, reads:

$$y\left[\frac{\partial}{\partial x}(\Psi^* \hat{S}_y \Psi) - \frac{\partial}{\partial y}(\Psi^* \hat{S}_x \Psi)\right] - z\left[\frac{\partial}{\partial z}(\Psi^* S_x \Psi) - \frac{\partial}{\partial x}(\Psi^* S_z \Psi)\right]$$

$$= \frac{\partial}{\partial x}\left[y(\Psi^* \hat{S}_y \Psi) + z(\Psi^* \hat{S}_z \Psi)\right] - \frac{\partial}{\partial y}\left[y(\Psi^* \hat{S}_x \Psi)\right] - \frac{\partial}{\partial z}\left[z(\Psi^* \hat{S}_x \Psi)\right] + 2(\Psi^* \hat{S}_x \Psi)$$

The Gauss theorem then can be used to reduce the volume integrals of the terms containing spatial derivatives to integrals over the surface bounding the volume and, when the volume integrals are taken over all space, Ψ and Ψ^* vanish on the surface at infinity. The only non-vanishing contribution comes from the last term on the right-hand side, as expected.

It follows that $\vec{\mu}_s$ can be written in a form similar to Eq.(7.128), namely:

$$\vec{\mu}_s = -\frac{1}{2}\frac{e}{m_e}\int_\infty \vec{r}\times\left[\nabla\times(\Psi^*\hat{\vec{S}}\Psi)\right]dV = \frac{1}{2}\int_\infty (\vec{r}\times\hat{\vec{j}}_s)dV$$

where the probability current density has the indicated form. We note that because of:

$$\nabla\cdot\hat{\vec{j}}_s = -\frac{e}{m_e}\nabla\cdot\left[\nabla\times(\Psi^*\hat{\vec{S}}\Psi)\right] = 0$$

the probability current density associated with the spin magnetic moment gives no contribution to the continuity equation for probability in external magnetic field, Eq.(7.143).

Problem 8.4.2. Find the average magnetic moment for the spinning electron in one-electron atoms.

(Solution): We have seen (Problem 7.4.2) that the total magnetic moment is given by:

$$\hat{\vec{\mu}} = -\frac{\partial \hat{H}}{\partial \vec{B}} = -\frac{e}{2m_e}(\hat{\vec{L}}+2\hat{\vec{S}})$$

where use has been made of Eq.(8.95) for the Hamiltonian in external magnetic field. The expectation values for the x and y components are equal to zero, in view of Problem 8.3.1, which shows that $\langle\hat{L}_x\rangle, \langle\hat{S}_x\rangle, \langle\hat{L}_y\rangle$ and $\langle\hat{S}_y\rangle$ are all zero. For the component along the z-axis we have:

$$\langle\hat{\mu}_z\rangle = -\frac{e}{2m_e}\left[\langle\hat{L}_z\rangle+2\langle\hat{S}_z\rangle\right] = -\frac{e}{2m_e}\left[\langle\hat{J}_z\rangle+\langle\hat{S}_z\rangle\right]$$

Using the results derived in Problem 8.3.1 we then obtain, for $j = l+\tfrac{1}{2}$:

$$\langle\hat{\mu}_z\rangle = -\frac{e\hbar}{2m_e}\left(m_j+\frac{m_j}{2j}\right) = -\mu_B m_j \frac{2j+1}{2j} = -\mu_B m_j g_L (j=l+\tfrac{1}{2})$$

where g_L is the Landé factor for the particular j value. For $j = l-\tfrac{1}{2}$ we similarly obtain:

$$\langle \hat{\mu}_z \rangle = -\frac{e\hbar}{2m_e}\left(m_j - \frac{m_j}{2(j+1)}\right) = -\mu_B m_j \frac{2j+1}{2(j+1)} = -\mu_B m_j g_L (j = l - \tfrac{1}{2})$$

Hence in general we can write:

$$\langle \hat{\mu}_z \rangle = -\mu_B m_j g_L(j)$$

Problem 8.4.3. Find the Landé factors for the s_j, p_j and d_j states of the one-electron atom.

(Answer): $\quad g_L = 2; \ g_L = \dfrac{2}{3}, \dfrac{4}{3}; \ g_L = \dfrac{4}{5}, \dfrac{6}{5}.$

9. APPROXIMATE METHODS

9.1. Stationary State Perturbation Theory

A large number of eigenvalue problems, such as the structure of multielectron atoms, lead to complex equations that have no solutions in closed form. If the conditions of the problem allow, by neglecting small terms, a simplification to a form that can be solved exactly, an approximate solution is then obtained by calculating the corrections due to terms that have been neglected. In many physical situations where the interaction energy represents a small disturbance or a *perturbation* of an idealized system, for which the eigenvalue problem is soluble, perturbation theory provides a systematic method of evaluating the small corrections and obtaining approximate solutions.

The Hamiltonian $\hat{H}$ of a given system is compared with the Hamiltonian $\hat{H}_0$ of an *unperturbed* system, of which the eigenvalues $E_k^{(0)}$ and the complete set of eigenfunctions $\Psi_k^{(0)}$ of the stationary states are already known:

$$\hat{H}_0 \Psi_k^{(0)} = E_k^{(0)} \Psi_k^{(0)} \qquad (9.1)$$

It is assumed that $\hat{H}$ includes small effects, neglected in $\hat{H}_0$ and represented by a perturbation operator $\hat{V}$:

$$\hat{H} = \hat{H}_0 + \hat{V} \qquad (9.2)$$

Approximate eigenvalues and eigenstates are derived for the Hamiltonian:

$$\hat{H}(\lambda) = \hat{H}_0 + \lambda \hat{V} \qquad (9.3)$$

where λ, called the *smallness parameter*, satisfies $0 \leq \lambda \leq 1$ such that $\lambda = 0$ corresponds to the unperturbed system $\hat{H}_0$ and $\lambda = 1$ corresponds to the actual system (9.2). The eigenvalues and eigenfunctions of $\hat{H}(\lambda)$, given by:

$$(\hat{H}_0 + \lambda \hat{V}) \Psi_k = E_k \Psi_k \qquad (9.4)$$

are assumed to be analytic functions of λ, expressible as convergent power series:

$$\Psi_k = \Psi_k^{(0)} + \lambda \Psi_k^{(1)} + \lambda^2 \Psi_k^{(2)} + \ldots$$

$$E_k = E_k^{(0)} + \lambda E_k^{(1)} + \lambda^2 E_k^{(2)} + \ldots$$
(9.5)

where $\Psi_k^{(1)}, \Psi_k^{(2)}$ and $E_k^{(1)}, E_k^{(2)}$ are the first and second order corrections to be determined. On substituting Eq.(9.5) into Eq.(9.4) and collecting the terms of like powers of λ one obtains:

$$(\hat{H}_0 - E_k^{(0)})\Psi_k^{(0)} + \left[(\hat{H}_0 - E_k^{(0)})\Psi_k^{(1)} + (\hat{V} - E_k^{(1)})\Psi_k^{(0)}\right]\lambda$$

$$+ \left[(\hat{H}_0 - E_k^{(0)})\Psi_k^{(2)} + (\hat{V} - E_k^{(1)})\Psi_k^{(1)} - E_k^{(2)}\Psi_k^{(0)}\right]\lambda^2 + \ldots = 0 \quad (9.6)$$

Since λ can take any value between $\lambda = 0$ and $\lambda = 1$, the coefficients of the various powers of λ must be equal to zero. The eigenvalues and eigenfunctions of the zeroth order equation are assumed known. The higher order equations must be solved in succession for $E_k^{(1)}, \Psi_k^{(1)}, E_k^{(2)}, \Psi_k^{(2)}, \ldots$ since the nth order corrections are obtained providing all the lower order terms have been previously calculated.

If the eigenstates $\Psi_k^{(0)}$ are *non-degenerate*, that is the eigenvalues $E_k^{(0)}$ are all distinct, they form a complete set, in terms of which we can make the expansion:

$$\Psi_k^{(1)} = \sum_{m=1}^{\infty} a_{km} \Psi_m^{(0)}$$
(9.7)

The *first-order corrections* result from the term in Eq.(9.6) containing λ to first order, which reads:

$$(\hat{H}_0 - E_k^{(0)})\Psi_k^{(1)} + (\hat{V} - E_k^{(1)})\Psi_k^{(0)} = 0$$
(9.8)

On substitution of Eq.(9.7) into Eq.(9.8), one obtains:

$$(\hat{H}_0 - E_k^{(0)})\sum_{m=1}^{\infty} a_{km}\Psi_m^{(0)} + (\hat{V} - E_k^{(1)})\Psi_k^{(0)} = 0$$

or, in view of Eq.(9.1):

$$E_k^{(1)}\Psi_k^{(0)} = \hat{V}\Psi_k^{(0)} + \sum_{m \neq k}(E_m^{(0)} - E_k^{(0)})a_{km}\Psi_m^{(0)}$$
(9.9)

Multiplying through $\Psi_i^{(0)*}$ and integrating gives:

$$E_k^{(1)}\delta_{ik} = \langle\Psi_i^{(0)}|\hat{V}|\Psi_k^{(0)}\rangle + (E_i^{(0)} - E_k^{(0)})a_{ki} \tag{9.10}$$

If we set $i = k$, the first-order eigenvalue correction to the unperturbed state $\Psi_k^{(0)}$ is obtained as the expectation value of the perturbation operator $\hat{V}$ in this state:

$$E_k^{(1)} = \langle\Psi_k^{(0)}|\hat{V}|\Psi_k^{(0)}\rangle = \int_\infty \Psi_k^{(0)*}(\vec{r})\,\hat{V}(\vec{r})\,\Psi_k^{(0)}(\vec{r})\,dV \tag{9.11}$$

In other words, the first-order corrections to the energy are given by the diagonal elements of the matrix representation of the perturbation energy $\hat{V}$, in the basis formed by the unperturbed states $\Psi_k^{(0)}$. If we set $i \neq k$, the coefficients a_{ki} can be obtained from Eq.(9.10) as:

$$a_{ki} = \frac{\langle\Psi_i^{(0)}|\hat{V}|\Psi_k^{(0)}\rangle}{E_k^{(0)} - E_i^{(0)}} = \frac{\int_\infty \Psi_i^{(0)*}(\vec{r})\,\hat{V}(\vec{r})\,\Psi_k^{(0)}(\vec{r})\,dV}{E_k^{(0)} - E_i^{(0)}} \tag{9.12}$$

and the coefficients a_{kk} have to be determined in a different manner. Since the eigenfunction (9.5) is determined up to a phase factor, we may choose the coefficients a_{kk} to be real, without loss of generality. The normalization of the eigenfunctions (9.5), to first order in λ, gives:

$$\langle\Psi_k|\Psi_k\rangle = \langle\Psi_k^{(0)} + \lambda\Psi_k^{(1)}|\Psi_k^{(0)} + \lambda\Psi_k^{(1)}\rangle \cong 1 + \lambda\langle\Psi_k^{(0)}|a_{kk}\Psi_k^{(0)}\rangle + \lambda\langle a_{kk}\Psi_k^{(0)}|\Psi_k^{(0)}\rangle$$

$$= 1 + 2\lambda\,Re\,a_{kk} = 1$$

so that the result is $a_{kk} = 0$. Substituting Eqs.(9.7), (9.11) and (9.12), Eqs.(9.5) give the first order results for the system under consideration, Eq.(9.2), that is for $\lambda = 1$, as:

$$\Psi_k = \Psi_k^{(0)} + \sum_{i \neq k} \frac{\langle\Psi_i^{(0)}|\hat{V}|\Psi_k^{(0)}\rangle}{E_k^{(0)} - E_i^{(0)}}\Psi_i^{(0)}$$

$$E_k = E_k^{(0)} + \langle\Psi_k^{(0)}|\hat{V}|\Psi_k^{(0)}\rangle \tag{9.13}$$

EXAMPLE 9.1. The anharmonic oscillator

Consider an oscillatory motion described by the Hamiltonian:

$$\hat{H} = \frac{\hat{p}^2}{2m} + \frac{1}{2}m\omega^2 x^2 + \alpha x^4 = \hat{H}_0 + \alpha x^4$$

which differs from the unperturbed Hamiltonian $\hat{H}_0$ of the harmonic oscillator by the anharmonic term αx^4. If α is assumed to be very small, the first-order correction to the energy levels (6.70) of the harmonic oscillator are given by Eq.(9.13) as:

$$E_n^{(1)} = \langle \Psi_n^{(0)} | \alpha x^4 | \Psi_n^{(0)} \rangle = \alpha \langle \psi_n | x^4 | \psi_n \rangle$$

and could be evaluated by inserting the form (6.71) of the unperturbed states ψ_n. However, the evaluation is facilitated by using the ladder operators, in terms of which we have, from Eq.(6.86):

$$\hat{x} = \frac{1}{\sqrt{2m\omega}}(\hat{a}+\hat{a}^+) \quad \text{or} \quad E_n^{(1)} = \frac{\alpha}{(2m\omega)^2} \langle \psi_n | (\hat{a}+\hat{a}^+)^4 | \psi_n \rangle$$

If we retain in the expansion of $(\hat{a}+\hat{a}^+)^4$ only the terms which contain two factors $\hat{a}$ and two factors $\hat{a}^+$ (see Problem 6.3.7), and use the recurrence relations (6.95), we obtain:

$$E_n^{(1)} = \frac{\alpha}{(2m\omega)^2} \langle \psi_n | \hat{a}^2 \hat{a}^{+2} + \hat{a}\hat{a}^+\hat{a}\hat{a}^+ + \hat{a}\hat{a}^{+2}\hat{a} + \hat{a}^+\hat{a}\hat{a}^+\hat{a} + \hat{a}^+\hat{a}^2\hat{a}^+ + \hat{a}^{+2}\hat{a}^2 | \psi_n \rangle$$

$$= \frac{3\alpha}{2}\left(\frac{\hbar}{m\omega}\right)^2 \left(n^2 + n + \frac{1}{2}\right)$$

and hence, the energy levels of the anharmonic oscillator are:

$$E_n = \left(n+\frac{1}{2}\right)\hbar\omega + \frac{3}{2}\left(n^2+n+\frac{1}{2}\right)\alpha\left(\frac{\hbar}{m\omega}\right)^2$$

We observe that, if instead of αx^4 we had an anharmonic term of the form αx^{2m+1}, with an odd power of x, such as αx (for the harmonic oscillator in a uniform electric field), or αx^3 (which is a perturbation appearing in diatomic molecules), there would be no energy change to the first order, because $\langle \psi_n | x^{2m+1} | \psi_n \rangle = 0$ for parity reasons. It is necessary, in such cases, to proceed to a second order calculation. ♪

The *second order corrections* can be obtained using a similar approach to the first order case. We start from the second order equation (9.6) which reads:

$$E_k^{(2)}\Psi_k^{(0)} = (\hat{H}_0 - E_k^{(0)})\Psi_k^{(2)} + (\hat{V} - E_k^{(1)})\Psi_k^{(1)} \tag{9.14}$$

where $\Psi_k^{(2)}$ is expanded in terms of the unperturbed functions $\Psi_k^{(0)}$ as:

$$\Psi_k^{(2)} = \sum_{n=1}^{\infty} b_{kn} \Psi_n^{(0)} \tag{9.15}$$

Inserting the first order corrections (9.7), multiplying Eq.(9.14) by $\Psi_i^{(0)*}$ and integrating, gives:

$$E_k^{(2)} \delta_{ik} = \sum_{n=1}^{\infty} (E_n^{(0)} - E_k^{(0)}) b_{kn} \delta_{in} + \sum_{m \neq k} a_{km} (\langle \Psi_i^{(0)} | \hat{V} | \Psi_m^{(0)} \rangle - E_k^{(1)} \delta_{im}) \qquad (9.16)$$

We then obtain for $i = k$ the second order eigenvalue corrections as:

$$E_k^{(2)} = \langle \Psi_k^{(0)} | \hat{V} | \Psi_k^{(1)} \rangle = \sum_{m \neq k} a_{km} \langle \Psi_k^{(0)} | \hat{V} | \Psi_m^{(0)} \rangle$$

$$= \sum_{m \neq k} \frac{\langle \Psi_k^{(0)} | \hat{V} | \Psi_m^{(0)} \rangle \langle \Psi_m^{(0)} | \hat{V} | \Psi_k^{(0)} \rangle}{E_k^{(0)} - E_m^{(0)}} = \sum_{m \neq k} \frac{|\langle \Psi_k^{(0)} | \hat{V} | \Psi_m^{(0)} \rangle|^2}{E_k^{(0)} - E_m^{(0)}} \qquad (9.17)$$

where we have used Eq.(9.12). If the unperturbed state of energy $E_k^{(0)}$ is taken to be the ground state of the system, such that $E_k^{(0)} - E_m^{(0)} < 0$, for all $m \neq k$, it follows that the second order correction to the ground state energy is always negative. Inserting Eq.(9.17) into Eq.(9.5) gives, for $\lambda = 1$:

$$E_k = E_k^{(0)} + \langle \Psi_k^{(0)} | \hat{V} | \Psi_k^{(0)} \rangle + \sum_{m \neq k} \frac{|\langle \Psi_k^{(0)} | \hat{V} | \Psi_m^{(0)} \rangle|^2}{E_k^{(0)} - E_m^{(0)}} \qquad (9.18)$$

If we make use of Eq.(9.16) for $i \neq k$, the coefficients of the expansion (9.15) and then the corresponding approximation to Ψ_k may be obtained. However, as the corrections to the eigenvalues and eigenfunctions become increasingly more complicated as one goes to higher-orders, it is common practice to avoid terms beyond the second-order correction in energy and beyond the first order to the wave function.

EXAMPLE 9.2. The Dalgarno perturbation method

The evaluation of the second order correction to the energy, Eq.(9.17), can be simplified provided we find a scalar function G of the variables that occur in $\hat{H}_0$, such that:

$$\hat{V} \Psi_k^{(0)} = [\hat{H}_0, G] \Psi_k^{(0)} \qquad (9.19)$$

where $\Psi_k^{(0)}$ is the stationary state defined by Eq.(9.1). If we multiply Eq.(9.19) by $\Psi_m^{(0)*}$ and integrate, one obtains:

$$\langle \Psi_m^{(0)} | \hat{V} | \Psi_k^{(0)} \rangle = (E_m^{(0)} - E_k^{(0)}) \langle \Psi_m^{(0)} | G | \Psi_k^{(0)} \rangle \qquad (9.20)$$

Substitution into Eq.(9.17) gives:

$$E_k^{(2)} = -\sum_{m \neq k} \langle \Psi_k^{(0)} | \hat{V} | \Psi_m^{(0)} \rangle \langle \Psi_m^{(0)} | G | \Psi_k^{(0)} \rangle = -\langle \Psi_k^{(0)} | \hat{V} G | \Psi_k^{(0)} \rangle \tag{9.21}$$

such that the second order correction reduces to the expectation value of $\hat{V}G$, taken with respect to the unperturbed wave function.

Consider, for instance, the perturbation introduced by an electric field $\vec{F}$, assumed in the z-direction, to the ground state level of the hydrogen atom, which acquires the energy:

$$\hat{V}(\vec{r}) = -eFz = -eFr\cos\theta \tag{9.22}$$

The unperturbed eigenfunctions are $\Psi_{100} = e^{-r/a_0}/\sqrt{\pi a^3}$ and there is no first order correction to the ground state level, because $\hat{V}(\vec{r})$ is of odd parity:

$$E_1^{(1)} = \int \Psi_{100}(\vec{r}) \hat{V}(\vec{r}) \Psi_{100}(\vec{r}) dV = 0$$

For the second order correction we determine the function G from Eq.(9.19), which reads:

$$-eFr\cos\theta \, \Psi_{100} = -\frac{\hbar^2}{2m} \nabla^2(G\Psi_{100}) - \left(\frac{e_0^2}{r} + E_1\right) G\Psi_{100} \tag{9.23}$$

where $E_1 = -e_0^2/2a_0$ is the ground state level, Eq.(7.80). It is apparent that:

$$\nabla^2(G\Psi_{100}) = G\nabla^2\Psi_{100} + 2\nabla\Psi_{100} \cdot \nabla G + \Psi_{100}\nabla^2 G$$

and substituting into Eq.(9.23) we are left with the following equation for G:

$$eFr\cos\theta \, \Psi_{100} = \frac{\hbar^2}{2m}(2\nabla\Psi_{100} \cdot \nabla G + \Psi_{100}\nabla^2 G)$$

Making use of the explicit form of Ψ_{100}, we have:

$$\nabla^2 G - \frac{2}{a_0}\frac{\partial G}{\partial r} = \frac{2m}{\hbar^2} eFr\cos\theta$$

because $\nabla e^{-r/a_0} \cdot \nabla G = -(\vec{e}_r \cdot \nabla G)/a_0 = -(\partial G/\partial r)/a_0$. A trial solution is $G = g(r)\cos\theta$, which leads to an equation for $g(r)$:

$$\frac{1}{r^2}\frac{d}{dr}\left(r^2 \frac{dg(r)}{dr}\right) - \frac{2}{r^2} g(r) - \frac{2}{a_0}\frac{dg(r)}{dr} = \frac{2m}{\hbar^2} eFr$$

from which it is a straightforward matter to obtain:

$$G = -\frac{m a_0 e}{\hbar^2} F\left(\frac{r^2}{2} + a_0 r\right)\cos\theta = -\frac{m a_0 e}{\hbar^2} F\left(\frac{r}{2} + a_0\right) z$$

Substituting this result into Eq.(9.21) gives:

$$E_1^{(2)} = -\frac{m a_0 e^2}{\hbar^2} F^2 \langle \Psi_{100} | \left(\frac{r}{2} + a_0\right) z^2 | \Psi_{100}\rangle$$

Since the ground state is spherically symmetric, there is no distinction between the z-, x- and y-axes, such that we can take $z^2 = (x^2 + y^2 + z^2)/3 = r^2/3$, which gives:

$$E_1^{(2)} = -\frac{m a_0 e^2}{\hbar^2} F^2 \frac{4\pi}{\pi a_0^3} \int_0^\infty e^{-2r/a_0}\left(\frac{r^3}{2} + a_0 r^2\right) r^2 dr = -\frac{9 m e^2}{4\hbar^2} F^2 a_0^4 = -9\pi\varepsilon_0 a_0^3 F^2 \quad (9.24)$$

where use has been made of the standard integral:

$$\int_0^\infty e^{-\alpha r} r^k dr = \frac{k!}{\alpha^{k+1}}$$

and of the definition (7.81) for the Bohr radius a_0. The energy shift of the ground level, Eq.(9.24), under an applied electric field F, is called the *Stark effect*, and can be interpreted in terms of a dipole moment $p = \alpha F$ acquired by the hydrogen atom in the direction of the field, related to the polarization energy by $p = -\partial E_1^{(2)}/\partial F$. The factor of proportionality:

$$\alpha = \frac{\partial p}{\partial F} = -\frac{\partial^2 E_1^{(2)}}{\partial F^2} = 18\pi\varepsilon_0 a_0^3 \quad (9.25)$$

is called the *electric polarizability* of hydrogen. ❧

If certain eigenstates $\Psi_k^{(0)}$ are *degenerate*, that is if there are n independent eigenfunctions $\Psi_{k\alpha}^{(0)}, \alpha = 1, 2, \ldots, n$ which all correspond to the same eigenvalue $E_k^{(0)}$, the formula (9.12) does not hold and one proceeds as follows. Assuming that the degenerate states $\Psi_{k\alpha}^{(0)}$ are orthonormal and allowing for degeneracy in our expansion (9.5) we obtain:

$$\Psi_k = \sum_{\alpha=1}^n A_{k\alpha} \Psi_{k\alpha}^{(0)} + \lambda \Psi_k^{(1)} + \ldots \quad (9.26)$$

If we substitute this expression into the eigenvalue equation (9.4), instead of the first-order equation (9.9), we obtain a set of n equations:

$$E_k^{(1)} \sum_{\alpha=1}^{n} A_{k\alpha} \Psi_{k\alpha}^{(0)} = \hat{V} \sum_{\alpha=1}^{n} A_{k\alpha} \Psi_{k\alpha}^{(0)} + \sum_{m \neq k} (E_m^{(0)} - E_k^{(0)}) a_{km} \Psi_m^{(0)}$$

where the expansion (9.7) must be written in terms of states $\Psi_m^{(0)}$ outside $\Psi_{k\alpha}^{(0)}$. In other words, $\Psi_m^{(0)}$ are states orthogonal to $\Psi_{k\alpha}^{(0)}$ for all α and $E_m^{(0)} - E_k^{(0)} \neq 0$. On multiplication by $\Psi_{k\gamma}^{(0)*}$ and integration one obtains, using orthonormality:

$$E_k^{(1)} A_{k\gamma} = \sum_{\alpha=1}^{n} A_{k\alpha} \langle \Psi_{k\gamma}^{(0)} | \hat{V} | \Psi_{k\alpha}^{(0)} \rangle \tag{9.27}$$

which is a set of homogenous equations for the coefficients $A_{k\alpha}$. Introducing the notation:

$$V_{\gamma\alpha} = \langle \Psi_{k\gamma}^{(0)} | \hat{V} | \Psi_{k\alpha}^{(0)} \rangle \tag{9.28}$$

we can rewrite Eq.(9.27) in the form:

$$\sum_{\alpha=1}^{n} V_{\gamma\alpha} A_{k\alpha} = E_k^{(1)} A_{k\gamma}$$

which is a matrix eigenvalue equation, analogous to Eq.(3.49):

$$\begin{pmatrix} V_{11} & V_{12} & \cdots & V_{1n} \\ V_{21} & V_{22} & \cdots & V_{2n} \\ \cdots & \cdots & \cdots & \cdots \\ V_{n1} & V_{n2} & \cdots & V_{nn} \end{pmatrix} \begin{pmatrix} A_{k1} \\ A_{k2} \\ \cdots \\ A_{kn} \end{pmatrix} = E_k^{(1)} \begin{pmatrix} A_{k1} \\ A_{k2} \\ \cdots \\ A_{kn} \end{pmatrix} \tag{9.29}$$

The condition for non-identically vanishing $A_{k\alpha}$ is the characteristic equation:

$$\det(\hat{V} - E_k^{(1)} \hat{I}) = \begin{vmatrix} V_{11} - E_k^{(1)} & V_{12} & \cdots & V_{1n} \\ V_{21} & V_{22} - E_k^{(1)} & \cdots & V_{2n} \\ \cdots & \cdots & \cdots & \cdots \\ V_{n1} & V_{n2} & \cdots & V_{nn} - E_k^{(1)} \end{vmatrix} = 0 \tag{9.30}$$

with n roots for $E_k^{(1)}$, which are denoted $E_{k\alpha}^{(1)}$, $\alpha = 1, 2, \ldots, n$. If the roots are all distinct, it is said that the degeneracy has been completely removed by the perturbation $\hat{V}$, otherwise it is said to be only partially removed. As the eigenvalues $E_{k\alpha}^{(1)}$ give the first order corrections to the energy, the formerly n-fold degenerate state splits up energetically into n close-lying states under the influence of perturbation, with energies (9.5) which can be written as:

$$E_{k\alpha} = E_k^{(0)} + E_{k\alpha}^{(1)} \qquad (9.31)$$

The insertion of each eigenvalue $E_{k\alpha}^{(1)}$ into the matrix equation (9.29) yields a set of n coefficients $A_{k\alpha}$ which are restricted by the normalization condition:

$$\sum_{\alpha=1}^{n} |A_{k\alpha}|^2 = 1$$

If *all* eigenstates are degenerate, we no longer have to consider the states $\Psi_m^{(0)}$ outside $\Psi_{k\alpha}^{(0)}$ in the expansion (9.7), so that Eq.(9.26) reduces to the first term, and therefore, the eigenfunction Ψ_k can be written as a linear combination of the degenerate states for each energy (9.31). Thus the first order results (9.13) of the perturbation theory for non-degenerate states are replaced by Eqs.(9.26) and (9.31) in the case of degenerate states, which are of special interest to the experimental physicist.

EXAMPLE 9.3. Splitting of a twofold degenerate state

Consider a twofold degenerate state of energy $E_k^{(0)}$ for which the matrix eigenvalue equation (9.29) reduces to:

$$\begin{pmatrix} V_{11} & V_{12} \\ V_{21} & V_{22} \end{pmatrix} \begin{pmatrix} A_1 \\ A_2 \end{pmatrix} = E^{(1)} \begin{pmatrix} A_1 \\ A_2 \end{pmatrix} \qquad (9.32)$$

where we have removed the index k, because it always appears in the same form, and $\alpha = 1, 2$. The characteristic equation (9.30) becomes:

$$\begin{vmatrix} V_{11} - E^{(1)} & V_{12} \\ V_{21} & V_{22} - E^{(1)} \end{vmatrix} = 0 \qquad (9.33)$$

As a further simplification, we assume that the perturbation operator is symmetric with respect to the degenerate eigenfunctions $\Psi_{k1}^{(0)}$ and $\Psi_{k2}^{(0)}$, such that:

$$K = V_{22} = V_{11} = \langle \Psi_{k1}^{(0)} | \hat{V} | \Psi_{k1}^{(0)} \rangle = \int \Psi_{k1}^{(0)*} \hat{V} \Psi_{k1}^{(0)} dV$$

$$J = V_{21} = V_{12} = \langle \Psi_{k1}^{(0)} | \hat{V} | \Psi_{k2}^{(0)} \rangle = \int \Psi_{k1}^{(0)*} \hat{V} \Psi_{k2}^{(0)} dV$$

It follows that $E_1^{(1)} = K + J$, $E_2^{(1)} = K - J$ and the degeneracy of the level $E_k^{(0)}$ is lifted according to Eq.(9.31), which reads:

$$E_{k1} = E_k^{(0)} + K + J, \quad E_{k2} = E_k^{(0)} + K - J$$

By insertion of $E_{k1}^{(1)} = K + J$ into Eq.(9.32), the normalized coefficients become $A_1 = A_2 = 1/\sqrt{2}$ so that the corresponding eigenfunctions (9.26) can be written as:

$$\Psi_{k1} = \frac{1}{\sqrt{2}}(\Psi_{k1}^{(0)} + \Psi_{k2}^{(0)})$$

Similarly $A'_1 = -A'_2 = 1/\sqrt{2}$ correspond to the eigenvalue $E_{k2}^{(1)} = K - J$ and it follows from Eq.(9.26) that:

$$\Psi_{k2} = \frac{1}{\sqrt{2}}(\Psi_{k1}^{(0)} - \Psi_{k2}^{(0)})$$

Problem 9.1.1. Calculate the perturbation of the first excited state ($n = 2$) of the hydrogen atom, introduced by an electrostatic field F.

(Solution): The perturbation energy, which is the potential energy of an electron in an electrostatic field, Eq.(9.22), is applied in this case to a four-fold degenerated state, described by the energy eigenfunctions derived from Eqs.(7.87) and (7.123):

$$\Psi_{200} = \Psi_{2s} = R_{2s}Y_{00} = \frac{1}{4\sqrt{2\pi a^3}}\left(2 - \frac{r}{a_0}\right)e^{-r/2a_0}$$

$$\Psi_{210} = \Psi_{2p_0} = R_{2p}Y_{10} = \frac{1}{4\sqrt{2\pi a_0^3}}\frac{r}{a_0}e^{-r/2a_0}\cos\theta$$

$$\Psi_{21,\pm 1} = \Psi_{2p_{\pm 1}} = R_{2p}Y_{1,\pm 1} = \frac{1}{8\sqrt{\pi a_0^3}}\frac{r}{a_0}e^{-r/2a_0}\sin\theta e^{\pm i\varphi}$$

all corresponding to the unperturbed energy $E_2 = -e_0^2/8a_0$. The elements (9.28) of the 4×4 perturbation matrix read:

$$\langle \Psi_{2lm} | -eFr\cos\theta | \Psi_{2lm'}\rangle$$

and it follows that the diagonal elements are zero, because $r\cos\theta$ is an odd function, while $|\Psi_{2lm}|^2$ is an even function. The matrix elements connecting different m values also vanish, and we are left with only two nonvanishing matrix elements:

$$\langle\Psi_{200}|-eFr\cos\theta|\Psi_{210}\rangle = \langle\Psi_{210}|-eFr\cos\theta|\Psi_{200}\rangle$$

$$= -\frac{eF}{32\pi a_0^4}\int_0^\infty\left(2-\frac{r}{a_0}\right)r^2 e^{-r/a_0}r^2 dr\cdot 2\pi\int_0^\pi\cos^2\theta\sin\theta\,d\theta$$

$$= 3eFa_0$$

Hence, the characteristic equation (9.30) becomes:

lm	00	10	11	1,−1
00	$-E_2^{(1)}$	$3eFa_0$	0	0
10	$3eFa_0$	$-E_2^{(1)}$	0	0
11	0	0	$-E_2^{(1)}$	0
1,−1	0	0	0	$-E_2^{(1)}$

$$= 0$$

and has the roots $\pm 3eFa_0$ and zero. The unperturbed level E_2 is thus split into three levels:

$$-e_0^2/8a_0 + 3eFa_0,\quad -e_0^2/8a_0,\quad -e_0^2/8a_0 - 3eFa_0$$

where the middle level is two-fold degenerate, with the corresponding eigenfunctions $\Psi_{2p_{\pm 1}}$. Inserting $E_2^{(1)} = \pm 3eFa_0$ into the matrix equation:

$$\begin{pmatrix} 0 & 3eFa_0 \\ 3eFa_0 & 0 \end{pmatrix}\begin{pmatrix} A_1 \\ A_2 \end{pmatrix} = E_2^{(1)}\begin{pmatrix} A_1 \\ A_2 \end{pmatrix}$$

we obtain $A_2 = \mp A_1$ respectively, so that the normalized eigenfunctions:

$$\Psi_{2s2p_0} = \frac{1}{\sqrt{2}}(\Psi_{2s} - \Psi_{2p_0}),\quad \Psi'_{2s2p_0} = \frac{1}{\sqrt{2}}(\Psi_{2s} + \Psi_{2p_0})$$

will correspond to the highest and lowest energy, respectively.

Problem 9.1.2. Calculate the energy shift of the energy levels of the harmonic oscillator under the anharmonic perturbation $\hat{V} = \alpha x^3$.

(Solution): As there is no energy change to first order, as discussed in Example 9.1, it is necessary to make the second order calculations, according to Eq.(9.17), and hence we need to evaluate the matrix elements $\langle\psi_n|x^3|\psi_m\rangle$. In view of Eqs.(6.95), we have:

$$\hat{x}\psi_n = \frac{1}{\sqrt{2m\omega}}(\hat{a}+\hat{a}^+)\psi_n = \left(\frac{\hbar}{2m\omega}\right)^{\frac{1}{2}}\left[\sqrt{n+1}\,\psi_{n+1}+\sqrt{n}\,\psi_{n-1}\right]$$

and successive application of this formula gives:

$$\hat{x}^3\psi_n = \left(\frac{\hbar}{2m\omega}\right)^{\frac{3}{2}}\left[\sqrt{(n+1)(n+2)(n+3)}\,\psi_{n+3}+3\sqrt{(n+1)^3}\,\psi_{n+1}\right.$$

$$\left.+3\sqrt{n^3}\,\psi_{n-1}+\sqrt{n(n-1)(n-2)}\,\psi_{n-3}\right]$$

In other words, only the matrix elements with $m = n\pm 3, n\pm 1$ are different from zero, and we obtain from Eq.(9.17) that:

$$E_n^{(2)} = -\alpha^2\left(\frac{\hbar}{2m\omega}\right)^3\left[\frac{1}{3\hbar\omega}(n+1)(n+2)(n+3)+\frac{9}{\hbar\omega}(n+1)^3-\frac{9}{\hbar\omega}n^3\right.$$

$$\left.-\frac{1}{3\hbar\omega}n(n-1)(n-2)\right] = -\frac{15\alpha^2}{4m}\left(\frac{\hbar}{m\omega^2}\right)^2\left(n^2+n+\frac{11}{30}\right)$$

Problem 9.1.3. Find the first order correction to the energy levels of a harmonic oscillator, due to the external perturbation energy $\hat{V} = \alpha x^2$.

(Answer): $\quad E_n^{(1)} = \dfrac{\alpha\hbar}{m\omega}\left(n+\dfrac{1}{2}\right)$

Problem 9.1.4. Calculate the second order Stark effect, determined by the perturbation $\hat{V} = qFx$, due to an external electric field F along the x-axis, on the energy levels of a harmonic oscillator, of natural frequency ω.

(Answer): $\quad E_n^{(2)} = -\dfrac{q^2F^2}{2m\omega^2}$

Problem 9.1.5. Considering α/r as a perturbation to the energy level E_n of the one-electron atom, show that $\langle 1/r\rangle_{nlm} = Z/n^2 a_0$, where a_0 is the Bohr radius.

Problem 9.1.6. Show that, by taking a perturbation α/r^2 to the centrifugal energy term of the radial equation of one-electron atoms, the first order calculation gives the expectation value $\langle 1/r^2\rangle_{nlm} = 2Z^2/(2l+1)n^3 a_0^2$.

9.2. The Variational Method

The perturbation theory is useful for a limited number of problems, as it provides small corrections to the already known energy eigenvalues and eigenfunctions of an unperturbed system. A more general approach to the estimation of energy levels, and particularly of the ground state energy, is based on the equivalence between the energy eigenvalue problem and that of finding the normalized wave functions Ψ:

$$\langle \Psi | \Psi \rangle = 1 \tag{9.34}$$

such that the average energy:

$$\langle \hat{H} \rangle = \langle \Psi | \hat{H} | \Psi \rangle = \int_\infty \Psi^*(\vec{r}) \hat{H} \Psi(\vec{r}) \, dV \tag{9.35}$$

is stationary with respect to arbitrary variations of Ψ. In other words the energy eigenvalue problem can be formulated as the variational equation:

$$\delta \langle \hat{H} \rangle = 0 \tag{9.36}$$

which must hold for a variation of Ψ by an arbitrary amount $\delta\Psi$. We observe that the normalization condition (9.34) restricts the problem to the discrete eigenvalue spectrum of $\hat{H}$, although the variational method is also valid for continuous eigenvalues. If the wave function Ψ is expanded in terms of a complete set of functions ψ_n:

$$\Psi = \sum_n a_n \psi_n \tag{9.37}$$

a variation $\delta\Psi$ of the function Ψ is obtained by replacing each coefficient a_n by $a_n + \delta a_n$. An arbitrary variation $\delta\Psi$ determines, to a first order approximation, a variation $\delta\langle \hat{H} \rangle$ of the form:

$$\delta \langle \hat{H} \rangle = \langle \Psi + \delta\Psi | \hat{H} | \Psi + \delta\Psi \rangle - \langle \Psi | \hat{H} | \Psi \rangle = \langle \delta\Psi | \hat{H} | \Psi \rangle + \langle \Psi | \hat{H} | \delta\Psi \rangle = 0 \tag{9.38}$$

but the possible variations $\delta\Psi$ are restricted to those for which the condition (9.34) is satisfied, which means that:

$$\delta \langle \Psi | \Psi \rangle = \langle \Psi + \delta\Psi | \Psi + \delta\Psi \rangle - \langle \Psi | \Psi \rangle = \langle \delta\Psi | \Psi \rangle + \langle \Psi | \delta\Psi \rangle = 0 \tag{9.39}$$

It is convenient to reduce the variational problem to an ordinary problem of extremum, by using the method of Lagrangeian multipliers, namely to ask that:

$$\delta\langle\hat{H}\rangle - \varepsilon\delta\langle\Psi|\Psi\rangle = 0 \qquad (9.40)$$

with respect to independent variations of Ψ, and find the value of the unknown multiplier ε. Substituting Eqs.(9.38) and (9.39) into Eq.(9.40) gives:

$$\langle\delta\Psi|\hat{H}\Psi\rangle - \varepsilon\langle\delta\Psi|\Psi\rangle + \langle\Psi|\hat{H}|\delta\Psi\rangle - \varepsilon\langle\Psi|\delta\Psi\rangle = 0$$

or, because $\hat{H}$ is a Hermitian operator:

$$\langle\delta\Psi|\hat{H}\Psi - \varepsilon\Psi\rangle + \langle\hat{H}\Psi - \varepsilon\Psi|\delta\Psi\rangle = 0 \qquad (9.41)$$

This equation can be satisfied for all possible variations $\delta\Psi$ if:

$$\hat{H}\Psi = \varepsilon\Psi, \quad \hat{H}\Psi^* = \varepsilon\Psi^* \qquad (9.42)$$

and this shows that ε must be real and must be an eigenvalue E_n of the Hamiltonian. The second equation is the complex conjugate of the first, and can be disregarded.

Hence, the variational problem has been reduced to the energy eigenvalue problem, as expected, and the corresponding wave function Ψ is an energy eigenvector:

$$\langle\Psi|\hat{H}|\Psi\rangle = \varepsilon\langle\Psi|\Psi\rangle = E \qquad (9.43)$$

If Ψ differs by an amount $\delta\Psi$ from the eigenfunction ψ_n of a stationary state E_n, we have to a first order approximation:

$$E = \langle\psi_n|\hat{H}|\psi_n\rangle + \langle\delta\Psi|\hat{H}|\psi_n\rangle + \langle\psi_n|\hat{H}|\delta\Psi\rangle = E_n\left(1 + \langle\delta\Psi|\psi_n\rangle + \langle\psi_n|\delta\Psi\rangle\right) = E_n \quad (9.44)$$

where use has been made of Eq.(9.39). In other words, if Ψ is a good approximation to the stationary state eigenfunction ψ_n, E provides a good approximation to E_n. In the special case of a Hamiltonian operator of the form (9.2), if the normalized eigenfunction ψ_n of the stationary state E_n is known, and satisfies Eq.(9.1), we can choose ψ_n as the trial function and Eq.(9.43) gives:

$$E = \langle\psi_n|\hat{H}|\psi_n\rangle = E_n + \langle\psi_n|\hat{V}|\psi_n\rangle \qquad (9.45)$$

This estimate of the energy eigenvalue in the presence of the perturbation energy $\hat{V}$ is the same as the first-order result of the perturbation theory, Eq.(9.13).

It is clear that the average energy E, Eq.(9.43), cannot be less than the ground state energy E_0 of the system. Using the expansion (9.37), we obtain from Eq.(9.43) that:

$$\langle\Psi|\hat{H}|\Psi\rangle = \sum_{n,m} a_n^* a_m \langle\psi_n|\hat{H}|\psi_m\rangle = \sum_n |a_n|^2 E_n \geq E_0 \sum_n |a_n|^2 = E_0 \qquad (9.46)$$

The equality holds good only if $\Psi = \psi_0$, where ψ_0 denotes the ground state eigenfunction, corresponding to the eigenvalue E_0.

Therefore, the variational method first consists of finding a suitable approximation Ψ, depending of a number of parameters $\lambda_1, \lambda_2, \ldots, \lambda_i$ to be determined for the ground state eigenfunction ψ_0, which is unknown. With this choice, we define:

$$E(\lambda_1, \lambda_2, \ldots, \lambda_i) = \int \Psi^*(\vec{r}, \lambda_1, \lambda_2, \ldots, \lambda_i) \hat{H} \Psi(\vec{r}, \lambda_1, \lambda_2, \ldots, \lambda_i) dV \quad (9.47)$$

As the parameters are varied, we obtain a variation in E, of the form:

$$\delta E = \sum_{k=1}^{i} \frac{\partial E(\lambda_1, \lambda_2, \ldots, \lambda_i)}{\partial \lambda_k} \delta \lambda_k$$

and we determine the set of values λ_i for which $E(\lambda_1, \lambda_2, \ldots, \lambda_i)$ has a minimum, by solving the equations:

$$\frac{\partial E(\lambda_1, \lambda_2, \ldots, \lambda_i)}{\partial \lambda_1} = \frac{\partial E(\lambda_1, \lambda_2, \ldots, \lambda_i)}{\partial \lambda_2} = \cdots = \frac{\partial E(\lambda_1, \lambda_2, \ldots, \lambda_i)}{\partial \lambda_i} = 0$$

If the lowest value of $E(\lambda_1, \lambda_2, \ldots, \lambda_i)$ is obtained for a set of solutions $\lambda_1^0, \lambda_2^0, \ldots, \lambda_i^0$, then $E_0 = E(\lambda_1^0, \lambda_2^0, \ldots, \lambda_i^0)$ and $\psi_0 = \Psi(\vec{r}, \lambda_1^0, \lambda_2^0, \ldots, \lambda_i^0)$ represent the best approximations to the ground state energy and eigenfunction that can be obtained from the trial function Ψ. The accuracy of the approximation is dependent on the appropriate choice of Ψ, which should take into account the eventual symmetry properties and other physical characteristics of each particular system.

EXAMPLE 9.4. Variational calculation of the ground state of hydrogen

As the ground state 1s of the hydrogen atom has spherical symmetry, we take our variational function as:

$$\Psi(\lambda) = \sqrt{\frac{\lambda^3}{\pi}} e^{-\lambda r}$$

where the normalization constant was obtained according to Eq.(9.34). Substituting into Eq.(9.47) gives:

$$E(\lambda) = \frac{\lambda}{\pi} \int e^{-\lambda r} \left(-\frac{\hbar^2}{2m_e} \nabla^2 - \frac{e_0^2}{r} \right) e^{-\lambda r} dV$$

$$= -\frac{\lambda^3}{\pi} \frac{\hbar^2}{2m_e} 4\pi \int_0^\infty e^{-\lambda r} \left[\frac{1}{r^2} \frac{d}{dr}\left(r^2 \frac{d}{dr} \right) \right] e^{-\lambda r} r^2 dr - \frac{\lambda^3}{\pi} 4\pi e_0^2 \int_0^\infty e^{-2\lambda r} r \, dr$$

where the two integrals reduce to:

$$\int_0^\infty e^{-\lambda r}\left[\frac{1}{r^2}\frac{d}{dr}\left(r^2\frac{d}{dr}\right)\right]e^{-\lambda r}r^2 dr = \int_0^\infty \left(\frac{d}{dr}e^{-\lambda r}\right)^2 r^2 dr = -\frac{1}{4\lambda}, \quad \int_0^\infty e^{-2\lambda r}r\,dr = \frac{1}{4\lambda^2}$$

It follows that:

$$E(\lambda) = \frac{\hbar^2}{2m_e}\left(\lambda^2 - 2\frac{m_e e_0^2}{\hbar^2}\lambda\right) = \frac{\hbar^2}{2m_e}\left(\lambda^2 - 2\frac{\lambda}{a_0}\right) = \frac{\hbar^2}{2m_e}\left(\lambda - \frac{1}{a_0}\right)^2 - \frac{\hbar^2}{2m_e a_0^2}$$

and so the minimum value of $E(\lambda)$ is obtained for $\lambda_0 = 1/a_0$, namely:

$$E_{min}(\lambda_0) = -\frac{e_0^2}{2a_0}$$

This is the exact value of the ground state energy of the hydrogen atom, Eq.(7.80). The corresponding approximation to the eigenfunction, $\Psi_0(\lambda_0) = e^{-r/a_0}/\sqrt{\pi a_0^3}$, also provides the exact answer, Ψ_{100}, in this simple case. ♦

Assuming that the E_0 and ψ_0 approximations to the ground state have been determined, we may successively apply the variational method to excited states, with the additional restriction of choosing for a given excited level E_k a trial wave function Ψ which is orthogonal to all the wave functions $\psi_0, \psi_1, \ldots, \psi_{k-1}$ previously found for lower levels. This can be accomplished by using the Gram-Schmidt orthogonalization process, described in Example 3.1. For the first excited state we can choose the trial function:

$$\Psi_1 = \Psi - \langle \psi_0 | \Psi \rangle \psi_0 \tag{9.48}$$

where Ψ is an arbitrary normalized wave function, such that Ψ_1 obviously has the properties:

$$\langle \Psi_1 | \Psi_1 \rangle = 1, \quad \langle \psi_0 | \Psi_1 \rangle = 0$$

Assuming that Ψ is given by the expansion (9.37), in the n-dimensional wave function space, Ψ_1 ranges through the space of $n-1$ dimensions of all the wave functions which are orthogonal to ψ_0:

$$\Psi_1 = \sum_{k=0}^{n-1} a_k \psi_k - \langle \psi_0 | \sum_{k=0}^{n-1} a_k \psi_k \rangle \psi_0 = \sum_{k=0}^{n-1} a_k \psi_k - a_0 \psi_0 = \sum_{k=1}^{n-1} a_k \psi_k$$

which implies that the expectation value of energy in this state is:

$$\langle\Psi_1|\hat{H}|\Psi_1\rangle = \sum_{k,m=1}^{n-1} a_k^* a_m \langle\psi_k|\hat{H}|\psi_m\rangle = \sum_{k=1}^{n-1} |a_k|^2 E_k \geq E_1 \sum_{k=1}^{n-1} |a_k|^2 = E_1 \qquad (9.49)$$

where E_1 is the energy eigenvalue of the first excited state of the system, and the equality holds if Ψ_1 coincides with the corresponding energy eigenfunction ψ_1. It follows that, using the trial wave function Ψ_1, we can obtain approximations to E_1 and its eigenfunction ψ_1 by minimization of the expectation value (9.49), as discussed before for the ground state variational problem.

EXAMPLE 9.5. Variational calculation of the first excited state of hydrogen

The first excited state $2s$ of the hydrogen atom can be calculated by using the trial function:

$$\Psi_1(\lambda,\mu) = N\left(1 - \mu\frac{r}{a_0}\right)\psi_0(\lambda) = \frac{N}{\sqrt{\pi a_0^3}}\left(1 - \mu\frac{r}{a_0}\right)e^{-\lambda r/a_0}$$

where ψ_0 has the form previously determined for the ground state. The constants N and μ can be expressed in terms of λ from the normalization and orthogonality conditions as follows:

$$N = \sqrt{\frac{3\lambda^5}{a_0^2 - \lambda a_0 + \lambda^2}}, \qquad \mu = \frac{\lambda+1}{3}$$

Hence, the function to be minimized reads:

$$E(\lambda) = 4\pi\frac{N^2}{\pi a_0^3}\int_0^\infty \left(1 - \frac{(\lambda+1)r}{a_0}\right)e^{-\lambda r/a_0}\left[-\frac{\hbar^2}{2m}\frac{1}{r^2}\frac{d}{dr}\left(r^2\frac{d}{dr}\right) - \frac{e_0^2}{r}\right]\left(1 - \frac{(\lambda+1)r}{a_0}\right)e^{-\lambda r/a_0}r^2 dr$$

$$= \frac{e_0^2}{a_0}\left[-\frac{\lambda}{2} + \frac{7\lambda^2}{6} - \frac{\lambda^2}{2(\lambda^2 - \lambda + 1)}\right]$$

The minimum condition on $E(\lambda)$ yields $\lambda_0 = 1/2$ such that:

$$E_{min}(\lambda_0) = -\frac{e_0^2}{8a_0}, \qquad \Psi_1(\lambda_0) = \frac{1}{\sqrt{8\pi a_0^3}}\left(1 - \frac{r}{2a_0}\right)e^{-r/2a_0}$$

and these are the exact values given by Eqs.(7.115) and (7.87) for E_{2s} and for the corresponding eigenfunction Ψ_{200} of the first excited state of hydrogen. ✤

Problem 9.2.1. Find the ground state energy of a harmonic oscillator using the variational method.

(Solution): A suitable trial function, normalized according to Eq.(9.34), is the Gaussian function:

$$\Psi(\lambda) = \sqrt{\frac{2\lambda}{\pi}} e^{-\lambda x^2}$$

such that:

$$\hat{H}\Psi(\lambda) = \sqrt{\frac{2\lambda}{\pi}} \left(-\frac{\hbar^2}{2m}\frac{d^2}{dx^2} + \frac{1}{2}m\omega^2 x^2 \right) e^{-\lambda x^2}$$

$$= \sqrt{\frac{2\lambda}{\pi}} \left[-\frac{\hbar^2}{2m}(4\lambda^2 x^2 - 2\lambda) + \frac{1}{2}m\omega^2 x^2 \right] e^{-\lambda x^2}$$

Substituting into Eq.(9.47) yields:

$$E(\lambda) = \sqrt{\frac{2\lambda}{\pi}} \int_{-\infty}^{\infty} \left[-\frac{\hbar^2}{2m}(4\lambda^2 x^2 - 2\lambda) + \frac{1}{2}m\omega^2 x^2 \right] e^{-2\lambda x^2} dx = \frac{\hbar^2}{2m}\lambda + \frac{1}{8}\frac{m\omega^2}{\lambda}$$

The condition that $E(\lambda)$ is a minimum reads:

$$\frac{dE(\lambda)}{d\lambda} = \frac{\hbar^2}{2m} - \frac{m\omega^2}{8\lambda^2} = 0 \quad \text{or} \quad \lambda_0 = \frac{m\omega}{2\hbar}$$

and it follows that $E(\lambda_0) = \hbar\omega/2$, as expected. The corresponding wave function reads:

$$\Psi(\lambda_0) = \sqrt{\frac{m\omega}{\pi\hbar}} e^{-m\omega x^2/2\hbar}$$

Problem 9.2.2. Find the ground state energy of hydrogen, using a Gaussian trial function of the form:

$$\psi(\lambda, r, \theta, \varphi) = N e^{-\lambda r^2}$$

(Solution): This trial function is different from the true wave function, considered in Example 9.4, but preserves the general features of spherical symmetry and the behaviour near the origin and in the asymptotic region. The normalization constant is obtained from Eq.(9.34), which reads:

$$4\pi N^2 \int_0^\infty e^{-2\lambda r^2} r^2 dr = 1 \quad \text{or} \quad N = \left(\frac{2\lambda}{\pi}\right)^{\frac{3}{4}}$$

We then obtain the expectation value:

$$E(\lambda) = \left(\frac{2\lambda}{\pi}\right)^{\frac{3}{2}} 4\pi \int_0^\infty e^{-\lambda r^2} \left[-\frac{\hbar^2}{2m_e} \frac{1}{r^2} \frac{d}{dr}\left(r^2 \frac{d}{dr}\right) - \frac{e_0^2}{r} \right] e^{-\lambda r^2} dr$$

$$= \frac{3\hbar^2}{m_e} \lambda - \left(\frac{8e_0^2}{\pi} \lambda\right)^{\frac{1}{2}}$$

The minimum value of $E(\lambda)$ is obtained as before:

$$\lambda_0 = \frac{8}{9\pi}\left(\frac{m_0 e_0^2}{\hbar^2}\right)^2 = \frac{8}{9\pi a_0^2} \quad \text{and} \quad E_{min}(\lambda_0) = -\frac{8}{3\pi} \frac{e_0^2}{2a_0} = 0.85 E_{1s}$$

This value is higher than but close to the exact energy eigenvalue for the ground state of hydrogen.

Problem 9.2.3. Calculate by the variational method the ground state energy for a particle in the infinite square potential well $0 \le x \le a$, for which $U(x) = 0$ inside the well and $U(x) = \infty$ outside, making use of a polynomial trial function.

(Answer): $E(\lambda_0) = \frac{10\hbar^2}{2m} = \frac{10}{\pi^2} E_0$

Problem 9.2.4. Find the ground state energy of a particle in a potential energy $U(x) = \alpha x^4$, by using a Gaussian trial function with normalized form:

$$\Psi(\lambda) - \sqrt{\frac{2\lambda}{\pi}} e^{-\lambda x^2}$$

(Answer): $E(\lambda_0) = \frac{3}{8}\left(\frac{6\alpha\hbar^4}{m^2}\right)^{\frac{1}{3}}$

Problem 9.2.5. Find an approximation to the ground state energy of the hydrogen atom by using the normalized trial function:

$$\Psi(\lambda) = \left(\frac{\lambda^5}{3\pi}\right)^{\frac{1}{2}} e^{-\lambda r}$$

(Answer): $E(\lambda_0) = \dfrac{3}{4} E_{1s}$

Problem 9.2.6. Calculate the first excited state energy of the harmonic oscillator by using the results of Problem 9.2.1 and the trial function:

$$\Psi_1(\lambda) = Nxe^{-\lambda x^2}$$

(Answer): $E_1(\lambda_0) = \dfrac{3}{2}\hbar\omega$

9.3. Time-Dependent Perturbation Theory

The energy is no longer a constant of motion for a system subjected to an external time-dependent force field. Assuming that such a perturbation is applied, at some particular time, to a system which is in a stationary state of definite energy, we will find this system, after a certain time, in a different state, that can always be expressed as a superposition of stationary states, and hence the possible energy values will be known with definite probabilities. In other words, the perturbation may result in the system undergoing transitions between stationary states, which are specified by the eigenvalues E_k and the complete set of eigenfunctions $\Psi_k(t)$ given by:

$$i\hbar \frac{\partial \Psi_k(t)}{\partial t} = \hat{H}_0 \Psi_k(t) = E_k \Psi_k(t) \tag{9.50}$$

A general solution of Eq.(9.50) has the form (5.62), which for the discrete eigenvalue spectrum reads:

$$\Psi^{(0)}(t) = \sum_k a_k^{(0)} \Psi_k(t) = \sum_k a_k e^{-iE_k t/\hbar} \psi_k \tag{9.51}$$

where $a_k^{(0)}$ are independent of time and restricted by:

$$\sum_k |a_k^{(0)}|^2 = 1 \tag{9.52}$$

During the presence of the time dependent perturbation $\hat{V}(t)$, the Hamiltonian of the system becomes:

$$\hat{H} = \hat{H}_0 + \lambda \hat{V}(t) \tag{9.53}$$

and, since the state of the system varies with time, we must adopt the time-dependent Schrödinger equation as a basis for the perturbation theory:

$$i\hbar \frac{\partial \Psi(t)}{\partial t} = \left[\hat{H}_0 + \lambda \hat{V}(t)\right] \Psi(t) \tag{9.54}$$

We assume that Eq.(9.54) has normalized solutions which can be expanded in terms of the $\Psi_k(t)$ as:

$$\Psi(t) = \sum_k a_k(t) \Psi_k(t) = \sum_k a_k(t) e^{-iE_k t/\hbar} \psi_k \tag{9.55}$$

where the normalization condition on $\Psi(t)$ gives:

$$\sum_k |a_k(t)|^2 = 1 \tag{9.56}$$

that is $|a_k(t)|^2$ gives the probability that the system is in the state $\Psi_k(t)$ at time t. Substituting Eq.(9.55) into Eq.(9.54) yields:

$$i\hbar \sum_k \left[\frac{da_k(t)}{dt} - \frac{iE_k}{\hbar} a_k(t)\right] e^{-iE_k t/\hbar} \psi_k = \sum_k \left[E_k + \lambda \hat{V}(t)\right] a_k(t) e^{-iE_k t/\hbar} \psi_k$$

which reduces to:

$$i\hbar \sum_k \frac{da_k(t)}{dt} e^{-iE_k t/\hbar} \psi_k = \lambda \sum_k \hat{V}(t) a_k(t) e^{-iE_k t/\hbar} \psi_k \tag{9.57}$$

Multiplication by ψ_n^* followed by integration gives, in view of the orthogonality of the stationary state eigenfunctions:

$$i\hbar \sum_k \frac{da_k(t)}{dt} e^{-iE_k t/\hbar} \delta_{nk} = \lambda \sum_k a_k(t) e^{-iE_k t/\hbar} \langle \psi_n | \hat{V}(t) | \psi_k \rangle \tag{9.58}$$

350 Quantum Physics

Denoting the matrix elements of the perturbation operator between the stationary states n and k by:

$$V_{nk}(t) = \langle \psi_n | \hat{V}(t) | \psi_k \rangle \tag{9.59}$$

we can rewrite Eqs.(9.58) as:

$$i\hbar \frac{da_n(t)}{dt} = \lambda \sum_k a_k(t) V_{nk}(t) e^{i(E_n - E_k)t/\hbar} \tag{9.60}$$

which is a set of simultaneous first order differential equations for the coefficients $a_n(t)$. If we assume that the system is in a stationary state ψ_i at some particular time $t = 0$ and then it is allowed to interact with the perturbation field, we obtain from Eq.(9.55) that:

$$\psi_i = \Psi(0) = \sum_k a_k(0) \psi_k$$

Multiplying by ψ_k^* and integrating gives:

$$a_k(0) = \delta_{ki} \tag{9.61}$$

that is $a_i(0) = 1$ and $a_k(0) = 0$ for $k \neq i$. This means that the term $k = i$ gives the greatest contribution to the sum and all other terms can be neglected, at $t = 0$, such that a formal solution of Eqs.(9.60) may be obtained by considering successive approximations of the form:

$$a_n(t) = \delta_{ni} + \lambda a_n^{(1)}(t) + \lambda^2 a_n^{(2)}(t) + \ldots \tag{9.62}$$

where $a_n^{(0)} = \delta_{ni}$. Substituting Eq.(9.62) for $a_n(t)$ and $a_k(t)$ into Eq.(9.60) we have:

$$\left[i\hbar \frac{da_n^{(1)}}{dt} - V_{ni}(t) e^{i(E_n - E_i)t/\hbar} \right] \lambda + \left[i\hbar \frac{da_n^{(2)}}{dt} - \sum_k a_k^{(1)} V_{nk}(t) e^{i(E_n - E_k)t/\hbar} \right] \lambda^2 + \ldots = 0 \tag{9.63}$$

This yields, to the first order in λ, the set of equations:

$$i\hbar \frac{da_n^{(1)}(t)}{dt} = V_{ni}(t) e^{i(E_n - E_i)t/\hbar} \tag{9.64}$$

and on integration we obtain:

$$a_n^{(1)}(t) = -\frac{i}{\hbar} \int_0^t V_{ni}(t') e^{i(E_n - E_i)t'/\hbar} dt' \tag{9.65}$$

Hence, to the first approximation we have:

$$a_n(t) = \delta_{ni} - \frac{i}{\hbar}\int_0^t V_{ni}(t')e^{i(E_n-E_i)t'/\hbar}\,dt' \qquad (9.66)$$

We may then consider the second order corrections, which from Eq.(9.63) result as:

$$a_n^{(2)}(t) = -\frac{i}{\hbar}\int_0^t \sum_k a_k^{(1)}(t')V_{nk}(t')e^{i(E_n-E_k)t'/\hbar}\,dt' \qquad (9.67)$$

where the $a_k^{(1)}(t')$ should be replaced by their expression (9.65). In this manner the coefficients $a_n(t)$, Eq.(9.66), can be calculated to any desired order.

The probability that the system is in the state ψ_n at time t, Eq.(3.33), in other words that $\Psi(t)$ coincides with ψ_n:

$$P_{in}(t) = |\langle \psi_n | \Psi(t) \rangle|^2 = |a_n(t)|^2 \qquad (9.68)$$

is stated as the *probability of direct transition* from state ψ_i to state ψ_n in time t, induced by the presence of the perturbation field.

It follows that time-dependent perturbation theory is concerned with the transitions which result from the perturbation between the stationary levels of the system, while time-independent perturbation theory calculates energy shifts in these levels. The most important applications where the perturbation is time-dependent involve the interaction of either the electric or magnetic field of electromagnetic radiation with the electric or magnetic moment of an atomic system of charged particles. The time-dependent perturbing effect of this interaction on the atomic stationary states will be discussed later on, in Chapter 11.

EXAMPLE 9.6. The Fermi golden rule

Consider a transition where a photon is emitted, that is $E_i > E_n$, we recall the Bohr condition (P.27) which reads:

$$E_i - E_n = \hbar \omega_{in} \qquad (9.69)$$

so that Eq.(9.66) can be rewritten as:

$$a_n(t) = -\frac{i}{\hbar}\int_0^t V_{ni}(t')e^{-i\omega_{in}t'}\,dt' \quad (n \neq i)$$

$$a_i(t) = 1 - \frac{i}{\hbar}\int_0^t V_{ii}(t')\,dt'$$

(9.70)

352   Quantum Physics

In the case of a *constant perturbation*, switched on at time $t = 0$, the integrals (9.70) become straightforward and we obtain:

$$a_n(t) = \frac{V_{ni}}{\hbar \omega_{in}}(e^{-i\omega_{in}t} - 1) \quad (n \neq i), \qquad a_i(t) = 1 - \frac{i}{\hbar}V_{ii}t \tag{9.71}$$

Substituting Eqs.(9.71) into Eq.(9.55) gives the wave function of the atomic system, to a first approximation, as:

$$\Psi(t) = e^{-iV_{ii}t/\hbar}\Psi_i(t) + \sum_{n \neq i}\frac{V_{ni}}{\hbar \omega_{in}}(e^{-i\omega_{in}t} - 1)\Psi_n(t) \tag{9.72}$$

The probability (9.68) of observing the system in the state ψ_n at time t, for the coefficients $a_n(t)$ given by Eq.(9.71), reduces to:

$$P_{in}(t) = \frac{|V_{ni}|^2}{\hbar^2 \omega_{in}^2} 4\sin^2(\omega_{in}t/2) \tag{9.73}$$

which means that transitions from state ψ_i to states ψ_n only take place when the matrix elements V_{ni} do not vanish. The plot of $P_{in}(t)$ as a function of ω_{in}, given in Figure 9.1 for a perturbation which is on for a time t, suggests that the probability behaves like a Dirac δ-function as t approaches infinity. We may represent the Dirac δ-function as:

$$\lim_{t \to \infty}\frac{\sin(\omega_{in}t/2)}{\pi\omega_{in}/2} = \lim_{t \to \infty}\frac{1}{2\pi}\int_{-t}^{t}e^{i\omega_{in}x/2}dx = \frac{1}{2\pi}\int_{-\infty}^{\infty}e^{i\omega_{in}x/2}dx = \delta(\omega_{in}/2) = 2\delta(\omega_{in}) \tag{9.74}$$

where use has been made of the property (1.46).

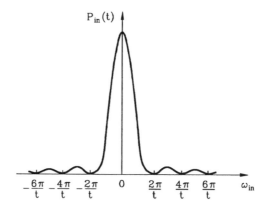

Figure 9.1. Dependence of the transition probability of frequency

It follows from Figure 9.1 that, by considering the large-t limit, we can take $\omega_{in} \cong 0$ and $[2\sin(\omega_{in}t/2)/\omega_{in}t] \to 1$, which gives:

$$\lim_{t\to\infty}\frac{4\sin^2(\omega_{in}t/2)}{\omega_{in}^2 t} = \lim_{t\to\infty}\left[\frac{\sin(\omega_{in}t/2)}{\omega_{in}t/2}\pi\frac{\sin(\omega_{in}t/2)}{\pi\omega_{in}/2}\right] = \pi\delta(\omega_{in}/2) = 2\pi\delta(\omega_{in}) \quad (9.75)$$

such that the transition probability becomes:

$$\lim_{t\to\infty} P_{in}(t) = \frac{2\pi t}{\hbar^2}|V_{ni}|^2\delta(\omega_{in}) = \frac{2\pi t}{\hbar}|V_{ni}|^2\delta(E_i - E_n) \quad (9.76)$$

Thus transitions takes place between states of equal unperturbed energies $E_n \to E_i$. The transition probability is proportional to the time t during which the perturbation is on, that is $P_{in}(t)$ approaches infinity as $t \to \infty$. Since no experiment lasts for an infinite time, it is the transition probability per unit time, called the *transition rate* $P_{i\to n}$, which is of particular interest. If E_n is part of a continuum, the transition takes place to any state lying in a finite range of N final states, rather than to a particular state. We may then define the transition rate as:

$$P_{i\to n} = \lim_{t\to\infty}\frac{1}{t}\int P_{in}(t)\,dN \quad (9.77)$$

The integral over the final states is usually transformed to one over energy if we define a *density of states* $\rho(E)$, representing the number of continuum states per energy interval, by:

$$dN = \rho(E_n)\,dE_n \quad (9.78)$$

so that, substituting Eqs.(9.78) and (9.76), Eq.(9.77) gives:

$$P_{i\to n} = \lim_{t\to\infty}\frac{1}{t}\int P_{in}(t)\rho(E_n)\,dE_n = \frac{2\pi}{\hbar}|V_{ni}|^2\rho(E_n)\delta(E_i - E_n) \quad (9.79)$$

This result is known as the *Fermi golden rule*, which states that transitions may occur only between states of equal energies, for which the matrix elements of the perturbation operator do not vanish, and that the transition rate is proportional to both the square moduli of these matrix elements and the density of final states. ✤

Problem 9.3.1. Assuming that a perturbation $V(t) = -\alpha x e^{-\gamma^2 t^2}$ is applied to a harmonic oscillator in the ground state, between $t = -\infty$ and $+\infty$, calculate to the first approximation the transition probability to a state $E_n = \hbar\omega(n + 1/2)$.

(Solution):   For $n \neq 0$ we obtain from Eq.(9.66):

$$a_n(\infty) = \frac{i\alpha}{\hbar} \int_{-\infty}^{\infty} \langle \psi_n | x | \psi_0 \rangle e^{-\gamma^2 t'^2} e^{in\omega t'} dt'$$

Using the energy representation of oscillatory motion, discussed in Example 5.6, we obtain that the only nonvanishing matrix elements of x are obtained for the transition $E_0 \to E_1$, namely:

$$\langle \psi_1 | x | \psi_0 \rangle = \langle \psi_0 | x | \psi_1 \rangle = \sqrt{\frac{\hbar}{2m\omega}}$$

Hence, we obtain:

$$a_1(\infty) = \frac{i\alpha}{\sqrt{2m\hbar\omega}} \int_{-\infty}^{\infty} e^{i\omega t' - \gamma^2 t'^2} dt' = \frac{i\alpha}{\sqrt{2m\hbar\omega}} \sqrt{\frac{\pi}{\gamma^2}} e^{-\omega^2/4\gamma^2}$$

which leads to the transition probability, Eq.(9.68), of the form:

$$P_{01}(\infty) = \frac{\pi \alpha^2}{2m\hbar\omega\gamma^2} e^{-\omega^2/2\gamma^2}$$

Problem 9.3.2. Calculate the probability of the transition $E_{1s} \to E_{2s}$ of the one electron atom, due to an abrupt change in the nuclear charge $Z \to Z \pm 1$, following $\beta^{\pm}$ decay.

(Solution): Because of the abrupt change at time $t = 0$, when a constant perturbation $\hat{V} = \pm e_0^2/r$ is switched on, we can integrate by parts in Eq.(9.66), and this gives for $n \neq i$:

$$a_n(t) = -V_{ni} \left[\frac{e^{i(E_n - E_i)t'/\hbar}}{E_n - E_i} \right]_0^t + \int_0^t \frac{\partial V_{ni}(t')}{\partial t'} \frac{e^{i(E_n - E_i)t'/\hbar}}{E_n - E_i} dt'$$

The time dependence considered for the matrix elements $V_{ni}(t')$ accounts for the effect of suddenly applying the constant perturbation at $t = 0$ and can be described by:

$$\frac{\partial V_{ni}(t')}{\partial t'} = V_{ni} \delta(t') \quad \text{or} \quad V_{ni}(t) = V_{ni} \int_{-\infty}^{t} \delta(t') dt' = V_{ni}, \quad t \geq 0$$

$$= 0, \quad t < 0$$

This leads to:

$$a_n(t) = \frac{V_{ni}}{E_i - E_n}\left[e^{i(E_n - E_i)t/\hbar} - 1\right] - \frac{V_{ni}}{E_n - E_i}$$

The first term corresponds to a transition probability of the form:

$$\frac{|V_{ni}|^2}{(E_i - E_n)^2} 2\left(1 - \cos\frac{E_n - E_i}{\hbar}t\right) = \frac{|V_{ni}|^2}{(E_i - E_n)^2} 4\sin^2\frac{E_n - E_i}{2\hbar}t$$

which, for large t values, behaves like the Dirac δ-function $\delta(E_n - E_i)$. In other words, we will obtain in this limit a transition probability different from zero only for energies $E_n \to E_i$. Neglecting this term, as we are interested in a first order transition to an excited state, and substituting $a_n(t)$ into Eq.(9.68) gives:

$$P_{in}(t) = \frac{|V_{ni}|^2}{(E_n - E_i)^2}$$

This result is valid when the perturbation energy $\hat{V} = \pm e_0^2/r$ is small, namely for large Z values only, and in this case we have $E_{2s} - E_{1s} = 3Z^2 e_0^2/8a_0$, which gives:

$$P_{1s \to 2s} \approx \frac{16a_0^2}{9Z^4 e_0^4}\left|\langle\Psi_{2s}|\pm\frac{e_0^2}{r}|\Psi_{1s}\rangle\right|^2 = \frac{2}{3}\frac{2^{11}}{9^4 Z^2}$$

We notice that this problem can also be solved by taking into account that $\Psi_{2s}(Z\pm 1)$ is a stationary state of the one electron atom with atomic number $Z\pm 1$, and thus it belongs to a complete set of eigenfunctions. We can then expand the initial eigenfunction:

$$\Psi_{1s}(Z) = \sqrt{\frac{Z^3}{\pi a_0^3}} e^{-Zr/a_0}$$

in terms of the complete set:

$$\Psi_{1s}(Z) = \sum_{nlm} a_n \Psi_{nlm}(Z\pm 1)$$

Multiplying this equation by Ψ_{200}^* and integrating gives:

$$a_2 = \int_\infty \Psi_{2s}^*(Z\pm 1)\Psi_{1s}(Z)\,dV$$

We thus obtain, using Eq.(7.87):

$$\int \Psi_{2s}^*(Z\pm 1)\Psi_{1s}(Z)dV = 4\pi\sqrt{\frac{Z^3(Z\pm 1)^3}{16\pi a_0^6}}\int_0^\infty \left[2-\frac{(Z\pm 1)r}{a_0}\right]e^{-(Z\pm 1)r/2a_0}e^{-Zr/a_0}r^2 dr$$

$$=\pm\frac{\sqrt{2^{11}Z^3(Z\pm 1)^3}}{(3Z\pm 1)^4}$$

which leads to a transition probability which is close to the approximate result, in the large Z limit:

$$P_{1s\to 2s} = \frac{2^{11}Z^3(Z\pm 1)^3}{(3Z\pm 1)^8} \approx \frac{2^{11}}{9^4 Z^2}$$

Problem 9.3.3. For the harmonic oscillator in the ground state, find the probability of the transition to the nth eigenstate $E_n = \hbar\omega(n+1/2)$, if a constant perturbation $V = -\alpha x$ is abruptly turned on.

(Answer): $$P_{0n} = \left(\frac{\alpha^2}{m\hbar\omega^3}\right)^n \frac{1}{2^n n!} e^{-\alpha^2/2m\hbar\omega^3}$$

Problem 9.3.4. A harmonic oscillator is in the ground state at $t = -\infty$. Find the transition probability to the first excited state, if a perturbation of the form:

$$V(t) = -\frac{\alpha x}{1+(\gamma t)^2}$$

is applied between $t = -\infty$ and $+\infty$.

(Answer): $$P_{01} = \frac{\pi^2\alpha^2}{2m\hbar\omega\gamma^2}e^{-2\omega/\gamma}$$

10. MANY-PARTICLE SYSTEMS

10.1. The Pauli Exclusion Principle

The postulates of quantum mechanics and their consequences, which have been developed to describe a single particle, are still valid when applied to a system of N particles. The system can be described by a wave function Ψ which depends on the time and on the $3N$ coordinates of all the particles:

$$\Psi = \Psi(\vec{r}_1,\ldots,\vec{r}_i,\ldots \vec{r}_N,t) = \Psi(x_1,y_1,z_1,\ldots,x_i,y_i,z_i,\ldots x_N,y_N,z_N,t) \tag{10.1}$$

If we interpret $|\Psi(\vec{r}_i,t)|^2$ as the probability density for finding, at time t, particle 1 at $\vec{r}_1,\ldots$, particle i at $\vec{r}_i,\ldots$ and particle N at $\vec{r}_N$, a normalization condition analogous to Eq.(1.37) holds:

$$\int_1 \cdots \int_i \cdots \int_N \Psi^*\Psi d\mathrm{V} = 1 \tag{10.2}$$

The integral is defined in a space with $3N$ dimensions where a point is given by $3N$ position coordinates $x_1,y_1,z_1,\ldots,x_i,y_i,z_i,\ldots,x_N,y_N,z_N$ and the volume element is:

$$d\mathrm{V} = d\mathrm{V}_1 \cdots d\mathrm{V}_i \cdots d\mathrm{V}_N = dx_1 dy_1 dz_1 \cdots dx_i dy_i dz_i \cdots dx_N dy_N dz_N$$

Physical observables of the system will be represented by Hermitian operators, obtained from the classical representation in terms of position and momentum variables as indicated by Postulate 2. Thus the many-particle Hamiltonian has the form:

$$\hat{H} = \sum_{i=1}^{N}\left[-\frac{\hbar^2}{2m_i}\nabla_i^2 + U_i(\vec{r}_i,t) + \sum_{j\neq i} U_{ij}(\vec{r}_i,\vec{r}_j)\right] \tag{10.3}$$

where ∇_i only acts on the coordinates of the ith particle, of mass m_i. The one-particle potential energy $U_i(\vec{r}_i,t)$ is due to the external fields in which this particle moves, and $U_{ij}(\vec{r}_i,\vec{r}_j)$ gives the mutual interaction between particles i and j. It is assumed that

358 Quantum Physics

position and momentum operators of different particles commute, and so do all operators describing single particle observables, when they refer to different particles.

The evolution of a many-particle system will then be formulated as a generalization of Eq.(5.24) in terms of Ψ and $\hat{H}$ given by Eqs.(10.1) and (10.3):

$$i\hbar \frac{\partial \Psi(t)}{\partial t} = \hat{H}\Psi(t) \qquad (10.4)$$

If the Hamiltonian (10.3) does not depend explicity on the time, it is usual to express the time dependence of the many-particle state function by means of the evolution operator $\hat{T}(t)$ defined as:

$$\Psi(t) = \Psi(\vec{r}_1, \ldots, \vec{r}_i, \ldots, \vec{r}_N, t) = \hat{T}(t)\Psi(\vec{r}_1, \ldots, \vec{r}_i, \ldots \vec{r}_N) \qquad (10.5)$$

which is formally derived from Eq.(10.4) as:

$$i\hbar \frac{\partial}{\partial t}\left[\hat{T}(t)\Psi\right] = i\hbar \frac{\partial \hat{T}(t)}{\partial t}\Psi = \hat{H}\hat{T}(t)\Psi \quad \text{or} \quad \hat{T}(t) = e^{-i\hat{H}t/\hbar} \qquad (10.6)$$

which is the analogue of Eq.(5.28). The evolution operator is *unitary*, Eq.(2.111), since:

$$\hat{T}(t)\hat{T}^{+}(t) = \hat{I} \qquad (10.7)$$

Substituting Eq.(10.5), it follows from Eq.(10.4) that the stationary states of N particle systems are given by the energy eigenvalue equation:

$$\hat{H}\Psi(\vec{r}_1, \ldots, \vec{r}_i, \ldots, \vec{r}_N) = E\Psi(\vec{r}_1, \ldots, \vec{r}_i, \ldots, \vec{r}_N) \qquad (10.8)$$

The solution to the N-particle problem is facilitated if we take into account the symmetries of many systems of physical interest which allow us to define constants of motion and criteria for simplifying the form (10.3) of the Hamiltonian operator. An alternative to complex explicit calculations is to associate with specific symmetry properties unitary operators $\hat{U}$, that do not involve time explicitly ($\partial \hat{U}/\partial t = 0$), which transform the state functions Ψ into $\hat{U}\Psi$, so that Eq.(10.4) gives:

$$i\hbar \frac{\partial}{\partial t}(\hat{U}\Psi) = i\hbar \hat{U}\frac{\partial \Psi}{\partial t} = \hat{U}\hat{H}\Psi = (\hat{U}\hat{H}\hat{U}^{+})\hat{U}\Psi$$

This is again a Schrödinger equation for the state $\hat{U}\Psi$ with the new Hamiltonian $\hat{H}' = \hat{U}\hat{H}\hat{U}^{+}$. If the Hamiltonian is remains unaltered by the transformation, we have:

$$\hat{H} = \hat{U}\hat{H}\hat{U}^{+} \quad \text{or} \quad [\hat{U}, \hat{H}] = 0 \qquad (10.9)$$

According to the evolution equation (5.3), this implies $d\hat{U}/dt = 0$ so that U is a constant of motion. If the unitary transformation $\hat{U}$, which leaves invariant the Hamiltonian, is a function of an observable, its constancy implies that the given observable is a constant in any state of the system.

As a first example, similar to the case of a single particle (see Problem 4.2.1), the *translation* of the N-particle system through a constant vector $\vec{r}_0$ will be described by the associated operator:

$$\hat{S}(\vec{r}_0) = e^{i\hat{\vec{P}}\cdot\vec{r}_0/\hbar} = \prod_{i=1}^{N} e^{i\hat{\vec{p}}_i\cdot\vec{r}_0/\hbar} = \prod_{i=1}^{N} e^{i(\hat{p}_{xi}x_0 + \hat{p}_{yi}y_0 + \hat{p}_{zi}z_0)/\hbar} \tag{10.10}$$

where $\vec{P} = \sum_{i=1}^{N} \vec{p}_i$ is the total momentum of the system. In view of Eqs.(10.3) and (10.9) we have:

$$\hat{S}(x_0)x_i \hat{S}^+(x_0) = \{x_i\hat{S}(x_0) + [\hat{S}(x_0), x_i]\}\hat{S}^+(x_0) = x_i - i\hbar \frac{\partial \hat{S}(x_0)}{\partial p_{x_i}} \hat{S}^+(x_0) = x_i + x_0$$

where use has been made of Eq.(4.14) with $\hat{S}(x_0) = f(p_{x_i})$. It follows that:

$$\hat{S}(\vec{r}_0)\vec{r}_i \hat{S}(\vec{r}_0) = \vec{r}_i + \vec{r}_0$$

which is a translation of the system through $\vec{r}_0$. Consequently, any function of the relative coordinates only will be invariant under translation. Hence, if we assume in Eq.(10.3) that the potential energy has the form:

$$U(\vec{r}_1,\ldots,\vec{r}_i,\ldots,\vec{r}_N) = \sum_{i=1}^{N}\sum_{j>i} U_{ij}(\vec{r}_i,\vec{r}_j) = \sum_{i=1}^{N}\sum_{j>i} U_{ij}(\vec{r}_i - \vec{r}_j) \tag{10.11}$$

it is left unchanged by translation. On the other hand, we always have:

$$\hat{S}(\vec{r}_0)f(\vec{p}_i)\hat{S}(\vec{r}_0) = f(\vec{p}_i)$$

if the unitary operator has the form (10.10). Hence, we obtain for the Hamiltonian (10.3) that:

$$\hat{S}(\vec{r}_0)\hat{H}\hat{S}^+(\vec{r}_0) = \hat{H}$$

which means that the translation operator (10.10), and therefore the total momentum $\vec{P}$, is a constant of motion, provided the condition (10.11) is satisfied. We observe, however, that this is not the case in the presence of external fields, when the one-particle potential energies do not depend on relative coordinates only.

Similarly, a *rotation* through a constant angle φ_0, about the z-axis, will be described by the associated operator of the form (8.6), which reads:

$$\hat{R}(\varphi_0) = e^{i\hat{J}_z \varphi_0/\hbar} = \prod_{i=1}^{N} e^{i\hat{L}_{zi}\varphi_0/\hbar} \qquad (10.12)$$

Since a rotation through φ_0 linearly transforms the coordinates and momentum components of all particles, the operator (10.12), and hence J_z, is a constant of motion, if the Hamiltonian (10.3) can be written in terms of scalar products of coordinate and momentum vectors only. This is valid for any component of $\vec{J}$, since the direction of the z-axis is arbitrary. In the presence of external fields with an axis of symmetry, only the component of $\vec{J}$ along this axis will be a constant of motion. If the total angular momentum includes the intrinsic spin of the particle, Eq.(8.67), the rotation operator (10.12) takes the form:

$$\hat{R}(\varphi_0) = e^{i\hat{J}_z \varphi_0/\hbar} = e^{i(\hat{L}_z + \hat{S}_z)\varphi_0/\hbar} = \prod_{i=1}^{N} e^{i(\hat{L}_{zi} + \hat{S}_{zi})\varphi_0/\hbar} \qquad (10.13)$$

where $\hat{S}_z$ denotes the total spin component along the z-axis.

The wave function of a system of N particles, each with spin, must be modified in the manner showed by Eq (8.64), and will be a function of $3N$ position coordinates and N spin coordinates:

$$\Phi = \Phi(\vec{r}_1 m_{s1}, \ldots, \vec{r}_i m_{si}, \ldots, \vec{r}_N m_{sN}, t)$$

$$= \Psi(\vec{r}_1, \ldots, \vec{r}_i, \ldots, \vec{r}_N, t) \chi(m_{s1}, \ldots, m_{si}, \ldots, m_{sN}) \qquad (10.14)$$

where the interactions between the spin and translational motion of the particles have been neglected. The spin wave function χ has the usual meaning, such that $|\chi|^2$ gives the probability density for particle 1 to have spin orientation $m_{s1}, \ldots$, and for particle N to have spin orientation m_{sN}. This probability is independent of the state of translational motion described by the wave function (10.1).

A special constant of motion must be introduced if the N particles of the system are *identical*, such that any observable associated with them depends in the same way on the variables describing the particles. In other words, any observable is *symmetric* in the variables of N identical particles. Since a classical particle has a definite trajectory, it is possible to distinguish identical particles by following their paths. In quantum mechanics the uncertainty principle makes it impossible to define precisely the particle trajectory, so that a given particle, in a system of N identical particles, cannot be labelled. We will be able to determine the state of a system, but we will not be able to relate a state function

Ψ_i to a particular particle i. This means that no change of the physical state of a system can be observed if two particles i and j are exchanged. The operation of exchanging the particle indices i and j is represented by a unitary operator $\hat{P}_{ij}$:

$$\hat{P}_{ij}\Phi(\vec{r}_1 m_{s1},\ldots,\vec{r}_i m_{si},\ldots,\vec{r}_j m_{sj},\ldots,\vec{r}_N m_{sN}) = \Phi(\vec{r}_1 m_{s1},\ldots,\vec{r}_j m_{sj},\ldots,\vec{r}_i m_{si},\ldots,\vec{r}_N m_{sN})$$

$$= e^{i\delta}\Phi(\vec{r}_1 m_{s1},\ldots,\vec{r}_i m_{si},\ldots,\vec{r}_j m_{sj},\ldots,\vec{r}_N m_{sN}) \quad (10.15)$$

where the normalized function Φ must only change by a phase factor if particles are relabelled. A second exchange of the two particles results in obtaining the original wave function, that is:

$$\hat{P}_{ij}^2 \Phi = \hat{P}_{ij}(\hat{P}_{ij}\Phi) = \hat{P}_{ij}(e^{i\delta}\Phi) = e^{2i\delta}\Phi = \Phi$$

which gives $\delta = n\pi$, and therefore $e^{i\delta} = e^{in\pi} = \pm 1$. This means that the wave functions must have a *definite symmetry*, since they either remain unchanged or change sign upon the exchange of any two particles:

$$\hat{P}_{ij}\Phi^s(\vec{r}_1 m_{s1},\ldots,\vec{r}_i m_{si},\ldots,\vec{r}_j m_{sj},\ldots,\vec{r}_N m_{sN}) = \Phi^s(\vec{r}_1 m_{s1},\ldots,\vec{r}_j m_{sj},\ldots,\vec{r}_i m_{si},\ldots,\vec{r}_N m_{sN})$$

$$\hat{P}_{ij}\Phi^a(\vec{r}_1 m_{s1},\ldots,\vec{r}_i m_{sa},\ldots,\vec{r}_j m_{sj},\ldots,\vec{r}_N m_{sN}) = -\Phi^a(\vec{r}_1 m_{s1},\ldots,\vec{r}_j m_{sj},\ldots,\vec{r}_i m_{si},\ldots,\vec{r}_N m_{sN}) \quad (10.16)$$

where Φ^s is said to be *symmetric* and Φ^a is *antisymmetric* with respect to the exchange of any two particles.

The definite symmetry of the wave functions under the exchange of identical particles appears as a requirement of the *principle of indistinguishability*. To make this clear, we consider a two-particle system.

EXAMPLE 10.1. The principle of indistinguishability

Consider a system of two identical particles, described by the wave function (10.14), which reads:

$$\Phi = \Psi(\vec{r}_1,\vec{r}_2,t)\chi(m_{s1},m_{s2})$$

If we exclude the spin, the Hamiltonian of the system has the form (10.3), which reduces to:

$$\hat{H}(\vec{r}_1,\vec{r}_2) = \left[-\frac{\hbar^2}{2m_e}\nabla_1^2 + U(\vec{r}_1,t)\right] + \left[-\frac{\hbar^2}{2m_e}\nabla_2^2 + U(\vec{r}_2,t)\right] + U_{12}(\vec{r}_{12})$$

Assuming that the interaction between particles gives a potential energy depending on their separation only, the Hamiltonian is symmetric with respect to the exchange of the two particles:

$$\hat{H}(\vec{r}_1,\vec{r}_2) = \hat{H}(\vec{r}_2,\vec{r}_1)$$

The time dependent Schrödinger equation (10.4) can be written either in terms of $\Psi(\vec{r}_1,\vec{r}_2,t)$:

$$i\hbar\frac{\partial\Psi(\vec{r}_1,\vec{r}_2,t)}{\partial t} = \hat{H}(\vec{r}_1,\vec{r}_2)\Psi(\vec{r}_1,\vec{r}_2,t)$$

or in terms of $\Psi(\vec{r}_2,\vec{r}_1,t)$:

$$i\hbar\frac{\partial\Psi(\vec{r}_2,\vec{r}_1,t)}{\partial t} = \hat{H}(\vec{r}_2,\vec{r}_1)\Psi(\vec{r}_2,\vec{r}_1,t) = \hat{H}(\vec{r}_1,\vec{r}_2)\Psi(\vec{r}_2,\vec{r}_1,t)$$

Since both wave functions satisfy the same evolution equation, we must take their linear combination as a general solution:

$$\Psi_{12}(\vec{r}_1,\vec{r}_2,t) = A\Psi(\vec{r}_1,\vec{r}_2,t) + B\Psi(\vec{r}_2,\vec{r}_1,t), \quad \Psi_{12}(\vec{r}_2,\vec{r}_1,t) = B\Psi(\vec{r}_1,\vec{r}_2,t) + A\Psi(\vec{r}_2,\vec{r}_1,t) \quad (10.17)$$

The principle of indistinguishability requires that the wave function of the two-particle system predicts no observable change of properties if the particles are interchanged. In other words, no change of the probability density is envisaged after interchange of the particles:

$$|\Psi_{12}(\vec{r}_1,\vec{r}_2,t)|^2 = |\Psi_{12}(\vec{r}_2,\vec{r}_1,t)|^2 \quad (10.18)$$

Substituting Eqs.(10.17), Eq.(10.18) immediately gives the following conditions on the complex coefficients:

$$A^*A = B^*B, \quad A^*B = B^*A$$

The first condition indicates that the only difference between A and B must be a phase factor:

$$A = Ce^{i\alpha}, \quad B = Ce^{i\beta}$$

whereas the second condition gives $e^{2i(\alpha-\beta)} = 1$ or $\alpha - \beta = n\pi$, where n is an arbitrary integer, so that:

$$B = Ce^{i(\alpha-n\pi)} = Ae^{-in\pi} = \pm A$$

Thus there only exists one symmetric $(B = A)$ and one antisymmetric $(B = -A)$ linear combination (10.17), which read:

$$\Psi^s(\vec{r}_1,\vec{r}_2,t) = A[\Psi(\vec{r}_1,\vec{r}_2,t) + \Psi(\vec{r}_2,\vec{r}_1,t)] = \Psi^s(\vec{r}_2,\vec{r}_1,t)$$

$$\Psi^a(\vec{r}_1,\vec{r}_2,t) = A[\Psi(\vec{r}_1,\vec{r}_2,t) - \Psi(\vec{r}_2,\vec{r}_1,t)] = -\Psi^a(\vec{r}_2,\vec{r}_1,t)$$

(10.19)

This result follows directly from the indistinguishability statement (10.18) for a two-particle system. ♪

Since the interaction between identical particles is always symmetric under their exchange, the Hamiltonian is left unchanged by the unitary operator $\hat{P}_{ij}$, that is:

$$\hat{P}_{ij}\hat{H}\hat{P}_{ij}^+ = \hat{H} \quad \text{or} \quad [\hat{H},\hat{P}_{ij}]=0$$

Hence, the exchange operator is associated to a constant of motion. In other words, *the symmetry or antisymmetry of a state function is permanent.* Experiment has shown that the particular state of symmetry of a system is directly related to the intrinsic spin of the identical particles. Empirically it has been demonstrated that:

(i) systems of identical particles with integer spins, called *bosons*, are described by symmetric wave functions;

(ii) systems of identical particles with half-integer spins, called *fermions*, are described by antisymmetric wave functions.

The symmetry requirements (10.16) imply that the antisymmetric wave function vanishes when two particles have the same set of coordinates $\vec{r}_i m_{si} = \vec{r}_j m_{sj} = \vec{r} m_s$:

$$\Phi^a(\vec{r}_1 m_{s1},\ldots,\vec{r} m_s,\ldots,\vec{r} m_s,\ldots,\vec{r}_N m_{sN}) = -\Phi^a(\vec{r}_1 m_{s1},\ldots,\vec{r} m_s,\ldots,\vec{r} m_s,\ldots,\vec{r}_N m_{sN})$$

or:

$$\Phi^a(\vec{r}_1 m_{s1},\ldots,\vec{r} m_s,\ldots,\vec{r} m_s,\ldots,\vec{r}_N m_{sN}) = 0$$

This means that a state where two fermions occupy the same position in space and have the same spin cannot exist, as formulated by the **Pauli exclusion principle:** *two identical fermions can never simultaneously occupy the same state $\vec{r} m_s$.* This statement has a direct effect on the spatial distribution of fermions, which unlike that of bosons, is restricted by the condition of statistical repulsion: if one fermion is situated at a given position $\vec{r}$, it excludes all the others with the same spin state m_s from that position.

A special formulation of the Pauli exclusion principle, suitable for discussing the structure of multielectron atoms, can be derived for systems of N identical noninteracting particles. The particles are assumed to move independently in the field, which gives rise to the single particle potential energies $U(\vec{r}_i)$, so that Eq.(10.3) reduces to:

$$\hat{H} = \sum_{i=1}^{N}\hat{H}_i = \sum_{i=1}^{N}\left[-\frac{\hbar^2}{2m}\nabla_i^2 + U(\vec{r}_i)\right] \tag{10.20}$$

The solution of the Schrödinger equation $i\hbar\partial\Phi(t)/\partial t = \hat{H}\Phi(t)$ can be written in terms of the stationary states Φ in a form analogous to Eq.(10.5):

$$\Phi(t) = e^{-i\hat{H}t/\hbar}\Phi = e^{-iEt/\hbar}\Phi \tag{10.21}$$

where:

$$(\hat{H}_1 + \ldots + \hat{H}_N)\Phi(\vec{r}_1 m_{s1},\ldots,\vec{r}_N m_{sN}) = E\Phi(\vec{r}_1 m_{s1},\ldots,\vec{r}_N m_{sN}) \quad (10.22)$$

We consider that each single-particle eigenvalue problem has known solutions, which are assumed to form an orthonormal set of stationary states:

$$\hat{H}_i \varphi_{E_i}(\vec{r}_i, m_{si}) = E_i \varphi_{E_i}(\vec{r}_i, m_{si}) \quad (10.23)$$

The Schrödinger equation (10.21) is then solved by the product of one-particle wave functions:

$$\Phi(\vec{r}_1 m_{s1},\ldots,\vec{r}_N m_{sN}) = \varphi_{E_1}(\vec{r}_1 m_{s1})\cdots\varphi_{E_N}(\vec{r}_N m_{sN}) \quad (10.24)$$

and the total energy is obtained as:

$$E = \sum_i E_i \quad (10.25)$$

The same energy eigenvalue E will, however, be obtained for $N!$ eigenfunctions (10.24), which can be generated by permutation of N particles. None of these eigenfunctions is, in general, symmetric or antisymmetric with respect to particle interchange, but their sum is clearly symmetric, so that the symmetric wave function, for *bosons*, results as:

$$\Phi^s = \frac{1}{\sqrt{N!}} \sum_{n=1}^{N!} \hat{P}_{ij}^{n-1} \varphi_{E_1}(\vec{r}_1 m_{s1})\cdots\varphi_{E_N}(\vec{r}_N m_{sN}) \quad (10.26)$$

The antisymmetrical state is formed by inserting a $\pm$ sign in Eq.(10.26), namely + for every permutation involving the exchange of an even number of pairs of identical particles and $-$ for every permutation involving an odd number of exchanges, that is:

$$\Phi^a = \frac{1}{\sqrt{N!}} \sum_{n=1}^{N!} (-\hat{P}_{ij})^{n-1} \varphi_{E_1}(\vec{r}_1 m_{s1})\cdots\varphi_{E_N}(\vec{r}_N m_{sN}) \quad (10.27)$$

An equivalent way of representing the state (10.27) of N identical *fermions* is the *Slater determinant*:

$$\Phi^a = \frac{1}{\sqrt{N!}} \begin{vmatrix} \varphi_{E_1}(\vec{r}_1 m_{s1}) & \varphi_{E_1}(\vec{r}_2 m_{s2}) & \cdots & \varphi_{E_1}(\vec{r}_N m_{sN}) \\ \varphi_{E_2}(\vec{r}_1 m_{s1}) & \varphi_{E_2}(\vec{r}_2 m_{s2}) & \cdots & \varphi_{E_2}(\vec{r}_N m_{sN}) \\ \cdots & \cdots & & \cdots \\ \varphi_{E_N}(\vec{r}_1 m_{s1}) & \varphi_{E_N}(\vec{r}_2 m_{s2}) & \cdots & \varphi_{E_N}(\vec{r}_N m_{sN}) \end{vmatrix} \quad (10.28)$$

as it can be verified by expanding the determinant. If two fermions are at the same position $\vec{r}$ in space, and have the same spin m_s, two columns in Eq.(10.28) will be identical and the eigenfunction Φ^a will vanish, in agreement with the previous formulation of the Pauli exclusion principle. Furthermore, if two of the occupied one-particle states φ_E are the same, the determinant (10.28) has two equal rows, and again vanishes. This leads to the special formulation of the **Pauli exclusion principle:** *in any configuration, the completely defined one-particle states cannot be occupied by more than one fermion.* In other words, no two fermions can be assigned the same set of quantum numbers.

EXAMPLE 10.2. Wave functions for a two-particle system

For the two-particle system considered in the previous example, the wave function takes the form:

$$\Psi(\vec{r}_1,\vec{r}_2,t) = e^{-iEt/\hbar}\Psi(\vec{r}_1,\vec{r}_2) = e^{-i(E_1+E_2)t/\hbar}\psi_{E_1}(\vec{r}_1)\psi_{E_2}(\vec{r}_2)$$

Exchange of the particles gives:

$$\Psi(\vec{r}_2,\vec{r}_1,t) = e^{-i(E_1+E_2)t/\hbar}\psi_{E_1}(\vec{r}_2)\psi_{E_2}(\vec{r}_1)$$

so that the symmetric and antisymmetric linear combinations become:

$$\Psi^s(\vec{r}_1,\vec{r}_2,) = A[\psi_{E_1}(\vec{r}_1)\psi_{E_2}(\vec{r}_2) + \psi_{E_1}(\vec{r}_2)\psi_{E_2}(\vec{r}_1)]$$

$$\Psi^a(\vec{r}_1,\vec{r}_2,) = A[\psi_{E_1}(\vec{r}_1)\psi_{E_2}(\vec{r}_2) - \psi_{E_1}(\vec{r}_2)\psi_{E_2}(\vec{r}_1)]$$

We recall that spin was excluded and further assume that the one-particle stationary states form an orthonormal set:

$$\int \psi_{E_1}^*(\vec{r}_1)\psi_{E_2}(\vec{r}_1)dV_1 = \int \psi_{E_1}^*(\vec{r}_2)\psi_{E_2}(\vec{r}_2)dV_2 = \delta_{E_1 E_2}$$

For $E_1 \neq E_2$ the normalization condition (10.2) on either Ψ^s or Ψ^a gives:

$$\iint \Psi^{s*}\Psi^s\, dV_1\, dV_2 = \iint \Psi^{a*}\Psi^a\, dV_1\, dV_2 = 2A^2 = 1 \tag{10.29}$$

so that the stationary state functions become:

$$\Psi^s = \frac{1}{\sqrt{2}}\left[\psi_{E_1}(\vec{r}_1)\psi_{E_2}(\vec{r}_2) + \psi_{E_1}(\vec{r}_2)\psi_{E_2}(\vec{r}_1)\right] \tag{10.30}$$

$$\Psi^a = \frac{1}{\sqrt{2}}\left[\psi_{E_1}(\vec{r}_1)\psi_{E_2}(\vec{r}_2) - \psi_{E_1}(\vec{r}_2)\psi_{E_2}(\vec{r}_1)\right] = \frac{1}{\sqrt{2}}\begin{vmatrix}\psi_{E_1}(\vec{r}_1) & \psi_{E_1}(\vec{r}_2)\\ \psi_{E_2}(\vec{r}_1) & \psi_{E_2}(\vec{r}_2)\end{vmatrix} \tag{10.31}$$

366 Quantum Physics

respectively. For $E_1 = E_2 = E$, the normalization condition (10.29) reads:

$$\iint \Psi^{s*} \Psi^s \, dV_1 \, dV_2 = 4A^2 = 1$$

so that the wave function of a two-boson system is:

$$\Psi^s = \frac{1}{2}\left[\psi_E(\vec{r}_1)\psi_E(\vec{r}_2) + \psi_E(\vec{r}_2)\psi_E(\vec{r}_1)\right] = \psi_E(\vec{r}_1)\psi_E(\vec{r}_2)$$

The antisymmetric wave function (10.31) vanishes for $E_1 = E_2 = E$, as required by the Pauli exclusion principle. ♣

Problem 10.1.1. Find the dependence of the wave functions on the spin variables for a system of two noninteracting electrons.

(Solution): Using single particle wave functions, which reflect both spatial and spin properties in the form:

$$\Phi_E(\vec{r}, m_s) = \psi_E(\vec{r})\chi_{m_s}$$

where $m_s = \pm\frac{1}{2}$ and the corresponding χ_{m_s} have the form (8.56), the antisymmetric state of the two-electron system is given by Eq.(10.28) as:

$$\Phi^a = \frac{1}{\sqrt{2}}\begin{vmatrix} \psi_{E_1}(\vec{r}_1)\chi_{m_{s1}}^{(1)} & \psi_{E_1}(\vec{r}_2)\chi_{m_{s1}}^{(2)} \\ \psi_{E_2}(\vec{r}_1)\chi_{m_{s2}}^{(1)} & \psi_{E_2}(\vec{r}_2)\chi_{m_{s2}}^{(2)} \end{vmatrix}$$

It is immediate to write Φ^a in terms of the symmetric and antisymmetric spatial wave functions, given by Eq.(10.30) and (10.31), namely:

$$\Phi^a = \frac{1}{\sqrt{2}}\left[\psi_{E_1}(\vec{r}_1)\psi_{E_2}(\vec{r}_2)\chi_{m_{s1}}^{(1)}\chi_{m_{s2}}^{(2)} - \psi_{E_1}(\vec{r}_2)\psi_{E_2}(\vec{r}_1)\chi_{m_{s1}}^{(2)}\chi_{m_{s2}}^{(1)}\right]$$

$$= \frac{1}{2\sqrt{2}}\left[\psi_{E_1}(\vec{r}_1)\psi_{E_2}(\vec{r}_2) + \psi_{E_1}(\vec{r}_2)\psi_{E_2}(\vec{r}_1)\right]\left[\chi_{m_{s1}}^{(1)}\chi_{m_{s2}}^{(2)} - \chi_{m_{s1}}^{(2)}\chi_{m_{s2}}^{(1)}\right]$$

$$+ \frac{1}{2\sqrt{2}}\left[\psi_{E_1}(\vec{r}_1)\psi_{E_2}(\vec{r}_2) - \psi_{E_1}(\vec{r}_2)\psi_{E_2}(\vec{r}_1)\right]\left[\chi_{m_{s1}}^{(1)}\chi_{m_{s2}}^{(2)} + \chi_{m_{s1}}^{(2)}\chi_{m_{s2}}^{(1)}\right]$$

$$= \frac{1}{\sqrt{2}}(\Psi^s\chi^a + \Psi^a\chi^s)$$

10. Many-Particle Systems

The antisymmetric spin function is nonvanishing for $m_{s1} \neq m_{s2}$ only, so that we can take $m_{s1} = +\frac{1}{2}$, $m_{s2} = -\frac{1}{2}$, which gives:

$$\chi^a = \frac{1}{\sqrt{2}}(\chi_+^{(1)}\chi_-^{(2)} - \chi_+^{(2)}\chi_-^{(1)})$$

This is an eigenfunction of the operators $\hat{S}^2$ and $\hat{S}_z$ of the two-electron system, defined by:

$$\hat{S}^2 = (\hat{\vec{S}}_1 + \hat{\vec{S}}_2)^2 = \hat{S}_1^2 + \hat{S}_2^2 + 2\hat{\vec{S}}_1 \cdot \hat{\vec{S}}_2 = \tfrac{3}{2}\hbar^2 + 2\hat{\vec{S}}_1 \cdot \hat{\vec{S}}_2, \quad \hat{S}_z = \hat{S}_{1z} + \hat{S}_{2z}$$

It is first obvious that:

$$\hat{S}_z \chi^a = \frac{1}{\sqrt{2}}(\hat{S}_{1z} + \hat{S}_{2z})\chi_+^{(1)}\chi_-^{(2)} - \frac{1}{\sqrt{2}}(\hat{S}_{1z} + \hat{S}_{2z})\chi_+^{(2)}\chi_-^{(1)}$$

$$= \frac{1}{\sqrt{2}}\left[(\hat{S}_{1z}\chi_+^{(1)})\chi_-^{(2)} + \chi_+^{(1)}\hat{S}_{2z}\chi_-^{(2)} - \chi_+^{(2)}\hat{S}_{1z}\chi_-^{(1)} - (\hat{S}_{2z}\chi_+^{(2)})\chi_-^{(1)}\right]$$

$$= \frac{\hbar}{2\sqrt{2}}(\chi_+^{(1)}\chi_-^{(2)} - \chi_+^{(1)}\chi_-^{(2)} + \chi_+^{(2)}\chi_-^{(1)} - \chi_+^{(2)}\chi_-^{(1)}) = 0$$

It is then convenient to use the property:

$$2\hat{\vec{S}}_1 \cdot \hat{\vec{S}}_2 = (S_{1x} + iS_{1y})(S_{2x} - iS_{2y}) + (S_{1x} - iS_{1y})(S_{2x} + iS_{2y}) + 2S_{1z}S_{2z}$$

such that:

$$\hat{S}^2 = \tfrac{3}{2}\hbar^2 + \hat{S}_{1+}\hat{S}_{2-} + \hat{S}_{1-}\hat{S}_{2+} + 2S_{1z}S_{2z}$$

where (for $i = 1, 2$) we have:

$$\hat{S}_{i+}\chi_+^{(i)} = \hbar\begin{pmatrix} 0 & 1 \\ 0 & 0 \end{pmatrix}\begin{pmatrix} 1 \\ 0 \end{pmatrix} = 0, \quad \hat{S}_{i+}\chi_-^{(i)} = \hbar\begin{pmatrix} 0 & 1 \\ 0 & 0 \end{pmatrix}\begin{pmatrix} 0 \\ 1 \end{pmatrix} = \hbar\begin{pmatrix} 1 \\ 0 \end{pmatrix} = \hbar\chi_+^{(i)}$$

$$\hat{S}_{i-}\chi_-^{(i)} = 0 \quad , \quad \hat{S}_{i-}\chi_+^{(i)} = \hbar\chi_-^{(i)}$$

It is now immediate to show that $\hat{S}^2\chi^a = 0$, and therefore the antisymmetric spin function describes a spin state with two antiparallel spins and total spin equal to zero. The symmetric spin function:

$$\chi^s = \frac{1}{\sqrt{2}}(\chi_{m_{s1}}^{(1)}\chi_{m_{s2}}^{(2)} + \chi_{m_{s1}}^{(2)}\chi_{m_{s2}}^{(1)})$$

368   Quantum Physics

describes three possible spin states of the two electron system, corresponding to the two spins either parallel ($m_{s1} = m_{s2} = \frac{1}{2}$ and $m_{s1} = m_{s2} = -\frac{1}{2}$) or antiparallel ($m_{s1} = \frac{1}{2}, m_{s2} = -\frac{1}{2}$) to one another. The normalized symmetric spin functions read:

$$\chi_1^s = \chi_+^{(1)}\chi_+^{(2)}, \quad \chi_0^s = \frac{1}{\sqrt{2}}(\chi_+^{(1)}\chi_-^{(2)} + \chi_+^{(2)}\chi_-^{(1)}), \quad \chi_{-1}^s = \chi_-^{(1)}\chi_-^{(2)}$$

and are all eigenfunctions of $\hat{S}^2$ and $\hat{S}_z$. For instance:

$$\hat{S}^2\chi_1^s = \tfrac{3}{2}\hbar^2 \chi_+^{(1)}\chi_+^{(2)} + (\hat{S}_{1+}\chi_+^{(1)})(\hat{S}_{2-}\chi_+^{(2)})$$

$$+ (\hat{S}_{1-}\chi_+^{(1)})(\hat{S}_{2+}\chi_+^{(2)}) + 2(\hat{S}_{1z}\chi_+^{(1)})(\hat{S}_{2z}\chi_+^{(2)})$$

$$= \left(\tfrac{3}{2}\hbar^2 + \tfrac{1}{2}\hbar^2\right)\chi_+^{(1)}\chi_+^{(2)} = 2\hbar^2 \chi^{(s)}$$

$$\hat{S}_z\chi_1^s = (\hat{S}_{1z}\chi_+^{(1)})\chi_+^{(2)} + \chi_+^{(1)}(\hat{S}_{2z}\chi_+^{(2)}) = \tfrac{1}{2}\hbar\chi_+^{(1)}\chi_+^{(2)} + \tfrac{1}{2}\hbar\chi_+^{(1)}\chi_+^{(2)} = \hbar\chi_1^s$$

In other words $\chi_+^{(1)}$ describes a state of total spin $S = 1$ which points in the direction of the z-axis. Similarly, χ_0^s and χ_{-1}^s describe states of total spin $S = 1$ with orientations with respect to the z-axis given by the eigenvalues 0 and -1 of $\hat{S}_z$ respectively.

Problem 10.1.2. Find the symmetric and antisymmetric eigenfunctions of the total spin operators $\hat{S}^2$ and $\hat{S}_z$ for a system of two noninteracting bosons with spin $s_1 = s_2 = 1$.

(Solution):   The symmetric state of the two-boson system, including both spatial and spin properties, has the form:

$$\Phi^s = \frac{1}{\sqrt{2}}\left[\psi_{E_1}(\vec{r}_1)\psi_{E_2}(\vec{r}_2)\chi_{m_{s1}}^{(1)}\chi_{m_{s2}}^{(2)} + \psi_{E_1}(\vec{r}_2)\psi_{E_2}(\vec{r}_1)\chi_{m_{s1}}^{(2)}\chi_{m_{s2}}^{(1)}\right]$$

$$= \frac{1}{2\sqrt{2}}\left[\psi_{E_1}(\vec{r}_1)\psi_{E_2}(\vec{r}_2) + \psi_{E_1}(\vec{r}_2)\psi_{E_2}(\vec{r}_1)\right]\left[\chi_{m_{s1}}^{(1)}\chi_{m_{s2}}^{(2)} + \chi_{m_{s1}}^{(2)}\chi_{m_{s2}}^{(1)}\right]$$

$$+ \frac{1}{2\sqrt{2}}\left[\psi_{E_1}(\vec{r}_1)\psi_{E_2}(\vec{r}_2) - \psi_{E_1}(\vec{r}_2)\psi_{E_2}(\vec{r}_1)\right]\left[\chi_{m_{s1}}^{(1)}\chi_{m_{s2}}^{(2)} - \chi_{m_{s1}}^{(2)}\chi_{m_{s2}}^{(1)}\right]$$

$$= \frac{1}{\sqrt{2}}(\Psi^s\chi^s + \Psi^a\chi^a)$$

10. Many-Particle Systems

The normalized symmetric spin functions have the form:

$$\chi^s_{m_{s1}m_{s2}} = \frac{1}{\sqrt{2}}(\chi^{(1)}_{m_{s1}}\chi^{(2)}_{m_{s2}} + \chi^{(2)}_{m_{s1}}\chi^{(1)}_{m_{s2}}), \quad m_{s1} \neq m_{s2}$$

$$\chi^s_{m_{si}m_{si}} = \chi^{(1)}_{m_{si}}\chi^{(2)}_{m_{si}}$$

where $m_{si} = -1, 0, 1$ and:

$$\chi^{(i)}_1 = \begin{pmatrix} 1 \\ 0 \\ 0 \end{pmatrix}, \quad \chi^{(i)}_0 = \begin{pmatrix} 0 \\ 1 \\ 0 \end{pmatrix}, \quad \chi^{(i)}_{-1} = \begin{pmatrix} 0 \\ 0 \\ 1 \end{pmatrix}$$

Since $\hat{S}_z = \hat{S}_{1z} + \hat{S}_{2z}$ we obtain $\hat{S}_z \chi^s = (m_{s1} + m_{s2})\hbar\chi^s$ which means that $M_s = m_{s1} + m_{s2} = 0, \pm 1, \pm 2$. We also have:

$$\hat{S}^2 = \hat{S}_1^2 + \hat{S}_2^2 + \hat{S}_{1+}\hat{S}_{2-} + \hat{S}_{1-}\hat{S}_{2+} + 2\hat{S}_{1z}\hat{S}_{2z} = 4\hbar^2 + \hat{S}_{1+}\hat{S}_{2-} + \hat{S}_{1-}\hat{S}_{2+} + 2\hat{S}_{1z}\hat{S}_{2z}$$

where, for each particle $i = 1, 2$, we have:

$$\hat{S}_{i+}\chi^{(i)}_1 = \frac{\hbar}{\sqrt{2}}\begin{pmatrix} 0 & 2 & 0 \\ 0 & 0 & 2 \\ 0 & 0 & 0 \end{pmatrix}\begin{pmatrix} 1 \\ 0 \\ 0 \end{pmatrix} = 0, \quad \hat{S}_{i+}\chi^{(i)}_0 = \sqrt{2}\,\hbar\chi^{(i)}_1, \quad \hat{S}_{i+}\chi^{(i)}_{-1} = \sqrt{2}\,\hbar\chi^{(i)}_0$$

$$\hat{S}_{i-}\chi^{(i)}_1 = \sqrt{2}\,\hbar\chi^{(i)}_0, \quad \hat{S}_{i-}\chi^{(i)}_0 = \sqrt{2}\,\hbar\chi^{(i)}_{-1}, \quad \hat{S}_{i-}\chi^{(i)}_{-1} = 0$$

It immediately follows that:

$$\hat{S}^2 \chi^s_{11} = 6\hbar^2 \chi^s_{11}, \quad \hat{S}^2 \chi^s_{10} = 6\hbar^2 \chi^s_{10}$$

$$\hat{S}^2 \chi^s_{-1,-1} = 6\hbar^2 \chi^s_{-1,-1}, \quad \hat{S}^2 \chi^s_{-1,0} = 6\hbar^2 \chi^s_{-1,0}$$

such that we have the symmetric spin states:

$$S = 2, \ M_S = 2 \ : \ \chi^s_{11} = \chi^{(1)}_1 \chi^{(2)}_1$$

$$S = 2, \ M_S = 1 \ : \ \chi^s_{10} = \frac{1}{\sqrt{2}}(\chi^{(1)}_1 \chi^{(2)}_0 + \chi^{(2)}_1 \chi^{(1)}_0)$$

$$S = 2, \ M_S = -1: \ \chi^s_{-1,0} = \frac{1}{\sqrt{2}}(\chi^{(1)}_{-1} \chi^{(2)}_0 + \chi^{(2)}_{-1} \chi^{(1)}_0)$$

$$S = 2, \ M_S = -2: \ \chi^s_{-1,-1} = \chi^{(1)}_{-1} \chi^{(2)}_{-1}$$

On the other hand, we obtain:

$$\hat{S}^2 \chi^s_{1,-1} = \hat{S}^2 \chi^s_{00} = 4\hbar^2 \chi^s_{00} + 2\hbar^2 \chi^s_{1,-1}$$

which means that $\chi^s_{1,-1}$ and χ^s_{00} are not eigenfunctions of $\hat{S}^2$. However we may consider the linear combination $\chi^s = 2\chi^s_{00} + \chi^s_{1,-1}$ which is eigenfunction of $\hat{S}^2$ with eigenvalue $6\hbar^2$, and of $\hat{S}_z$ with eigenvalue $M_S = 0$. This leads to the normalized symmetric spin function:

$$S = 2, \quad M_S = 0 : \quad \chi^s = \frac{1}{\sqrt{6}} (2\chi^{(1)}_0 \chi^{(2)}_0 + \chi^{(1)}_1 \chi^{(2)}_{-1} + \chi^{(2)}_1 \chi^{(1)}_{-1})$$

Another possible linear combination is the difference $\chi^s_{00} - \chi^s_{1,-1}$, for which:

$$\hat{S}^2 (\chi^s_{00} - \chi^s_{1,-1}) = 0, \quad \hat{S}_z (\chi^s_{00} - \chi^s_{1,-1}) = 0$$

and defines an eigenstate with $S = 0$, $M_S = 0$. It follows that spin functions of the two-boson system are symmetric eigenfunctions of $\hat{S}^2$ and $\hat{S}_z$ for $S = 0, 2$.

The normalized antisymmetric functions read:

$$\chi^a_{m_{s1} m_{s2}} = \frac{1}{\sqrt{2}} (\chi^{(1)}_{m_{s1}} \chi^{(2)}_{m_{s2}} - \chi^{(2)}_{m_{s1}} \chi^{(1)}_{m_{s2}})$$

and clearly vanish for $m_{s1} = m_{s2}$, which means that the possible values of $M_S = m_{s1} + m_{s2}$ ($m_{s1} \ne m_{s2}$) are now restricted to 0 ($m_{s1} = 1, m_{s2} = -1$) and ± 1 ($m_{s1} = 0, m_{s2} = \pm 1$). It is immediate to show that:

$$\hat{S}^2 \chi^a_{1,-1} = 2\hbar^2 \chi^a_{1,-1}, \quad \hat{S}^2 \chi^a_{01} = 2\hbar^2 \chi^a_{01}, \quad \hat{S}^2 \chi^a_{0,-1} = 2\hbar^2 \chi^a_{0,-1}$$

such that the antisymmetric spin states of the system will only correspond to $S = 1$, namely:

$$S = 1, \quad M_S = 1 : \quad \chi^a_{01} = \frac{1}{\sqrt{2}} (\chi^{(1)}_0 \chi^{(2)}_1 - \chi^{(2)}_0 \chi^{(1)}_1)$$

$$S = 1, \quad M_S = 0 : \quad \chi^a_{1,-1} = \frac{1}{\sqrt{2}} (\chi^{(1)}_1 \chi^{(2)}_{-1} - \chi^{(2)}_1 \chi^{(1)}_{-1})$$

$$S = 1, \quad M_S = -1 : \quad \chi^a_{0,-1} = \frac{1}{\sqrt{2}} (\chi^{(1)}_0 \chi^{(2)}_{-1} - \chi^{(2)}_0 \chi^{(1)}_{-1})$$

Problem 10.1.3. Find the ratio of the number of symmetrical spin states to that of antisymmetrical spin states, for a system of two identical particles of spin s.

(Answer): $(s+1)/s$

Problem 10.1.4. Show that, for a two-electron system moving in a central potential, the singlet spin state corresponds to even values L of its total angular momentum and the triplet states have odd L values.

10.2. The Helium Atom

The helium atom, consisting of a nucleus of charge $Ze = 2e$ and two orbiting electrons, is the simplest multielectron atom. If the nucleus is placed at the origin and the spin-interaction terms are ignored, the Hamiltonian operator may be written as:

$$\hat{H} = -\frac{\hbar^2}{2m_e}\left(\nabla_1^2 + \nabla_2^2\right) - Ze_0^2\left(\frac{1}{r_1} + \frac{1}{r_2}\right) + \frac{e_0^2}{|\vec{r}_1 - \vec{r}_2|} = \hat{H}_1 + \hat{H}_2 + \hat{V} \quad (10.32)$$

where $\hat{V}$ denotes the mutual electrostatic interaction of the two electrons, which will be treated as a perturbation. The Hamiltonians $\hat{H}_1$ and $\hat{H}_2$ are single particle operators for the two individual electrons which satisfy the eigenvalue equations:

$$\hat{H}_i \varphi_{n_i l_i m_i m_{si}} = \left(-\frac{\hbar^2}{2m_e}\nabla_i^2 - \frac{Ze_0^2}{r_i}\right)\varphi_{n_i l_i m_i m_{si}} = E_{n_i}\varphi_{n_i l_i m_i m_{si}} \quad (10.33)$$

where $i = 1, 2$. The eigenstates $\varphi_{n_i l_i m_i m_{si}}$ have the form (8.64), previously derived for the angular momentum states:

$$\varphi_{n_i l_i m_i m_{si}}(\vec{r}_i) = R_{n_i l_i}(r) Y_{l_i m_i}(\theta, \varphi) \chi_{m_{si}} \quad (10.34)$$

and correspond to the degenerate energy levels (7.80):

$$E_{n_i} = -13.6\frac{Z^2}{n_i^2}(\text{eV}) = -\frac{54.4}{n_i^2}(\text{eV}) \quad (10.35)$$

where use has been made of the a_0 value, given by Eq.(7.81). Thus, if the mutual electrostatic interaction in the Hamiltonian, $\hat{V}$, is ignored, the idealized helium atom can be regarded as a system of two identical fermions, of energy:

$$E = E_{n_1} + E_{n_2} = -54.4\left(\frac{1}{n_1^2} + \frac{1}{n_2^2}\right)(\text{eV}) \quad (10.36)$$

with the corresponding eigenfunction represented by the Slater determinant (10.28) which reads:

$$\Phi = \frac{1}{\sqrt{2}}\begin{vmatrix} \varphi_{n_1 l_1 m_1 m_{s1}}(\vec{r}_1) & \varphi_{n_1 l_1 m_1 m_{s1}}(\vec{r}_2) \\ \varphi_{n_2 l_2 m_2 m_{s2}}(\vec{r}_1) & \varphi_{n_2 l_2 m_2 m_{s2}}(\vec{r}_2) \end{vmatrix} \quad (10.37)$$

The ground state configuration of this system contains both electrons in the lowest energy level, that is $n_1 = n_2 = 1$, $l_1 = l_2 = 0$, $m_1 = m_2 = 0$, $m_{s1} = \frac{1}{2}$, $m_{s2} = -\frac{1}{2}$. The ground state energy (10.36) is $E_g = -108.8$ eV and:

$$\Phi_g = \frac{1}{\sqrt{2}}\begin{vmatrix} \psi_{100}(\vec{r}_1)\chi_+^{(1)} & \psi_{100}(\vec{r}_2)\chi_+^{(2)} \\ \psi_{100}(\vec{r}_1)\chi_-^{(1)} & \psi_{100}(\vec{r}_2)\chi_-^{(2)} \end{vmatrix}$$

$$= \frac{1}{\sqrt{2}}\psi_{100}(\vec{r}_1)\psi_{100}(\vec{r}_2)\begin{vmatrix} \chi_+^{(1)} & \chi_+^{(2)} \\ \chi_-^{(1)} & \chi_-^{(2)} \end{vmatrix} = \psi_{100}(\vec{r}_1)\psi_{100}(\vec{r}_2)\chi^S \quad (10.38)$$

The spatial part is symmetrical whereas the spin function χ^S, which is antisymmetrical under spin label interchange:

$$\chi^S = \frac{1}{\sqrt{2}}(\chi_+^{(1)}\chi_-^{(2)} - \chi_-^{(1)}\chi_+^{(2)}) \quad (10.39)$$

defines a *singlet* spin state, of total spin $S = 0$, such that $2S + 1 = 1$. As both the spatial and spin functions are unique, the ground state is non-degenerate.

The first excited state contains one electron in the ground state level $n = 1$, $l = 0$, $m = 0$, having the spatial wave function ψ_{100}, and another in the energy level defined by $n = 2$, with $l = 0$, $m = 0$ or $l = 1$, $m = 0, \pm 1$ and corresponding spatial wave functions ψ_{2lm}. The configuration is fourfold degenerate, because of exchange and spin degeneracy, with total energy $E_e = -68$ eV, as given by Eq.(10.36). Two of the corresponding Slater determinants (10.37) are products of a spatial function and a spin function, both normalized and of definite symmetry:

$$\Phi_{lm} = \frac{1}{\sqrt{2}}\begin{vmatrix} \psi_{100}(\vec{r}_1)\chi_+^{(1)} & \psi_{100}(\vec{r}_2)\chi_+^{(2)} \\ \psi_{2lm}(\vec{r}_1)\chi_+^{(1)} & \psi_{2lm}(\vec{r}_2)\chi_+^{(2)} \end{vmatrix}$$

$$= \frac{1}{\sqrt{2}}[\psi_{100}(\vec{r}_1)\psi_{2lm}(\vec{r}_2) - \psi_{2lm}(\vec{r}_1)\psi_{100}(\vec{r}_2)]\chi_+^{(1)}\chi_+^{(2)} = \Psi^a\chi_+^{(1)}\chi_+^{(2)} \quad (10.40)$$

and:

$$\Phi^I_{lm} = \frac{1}{\sqrt{2}} \begin{vmatrix} \psi_{100}(\vec{r}_1)\chi^{(1)}_- & \psi_{100}(\vec{r}_2)\chi^{(2)}_- \\ \psi_{2lm}(\vec{r}_1)\chi^{(1)}_- & \psi_{2lm}(\vec{r}_2)\chi^{(2)}_- \end{vmatrix}$$

$$= \frac{1}{\sqrt{2}}[\psi_{100}(\vec{r}_1)\psi_{2lm}(\vec{r}_2) - \psi_{2lm}(\vec{r}_1)\psi_{100}(\vec{r}_2)]\chi^{(1)}_-\chi^{(2)}_- = \Psi^a \chi^{(1)}_-\chi^{(2)}_- \quad (10.41)$$

whereas the other two are not of this form, namely:

$$\Phi^{II}_{lm} = \frac{1}{\sqrt{2}} \begin{vmatrix} \psi_{100}(\vec{r}_1)\chi^{(1)}_+ & \psi_{100}(\vec{r}_2)\chi^{(2)}_- \\ \psi_{2lm}(\vec{r}_1)\chi^{(1)}_+ & \psi_{2lm}(\vec{r}_2)\chi^{(2)}_- \end{vmatrix}$$

$$\Phi^{III}_{lm} = \frac{1}{\sqrt{2}} \begin{vmatrix} \psi_{100}(\vec{r}_1)\chi^{(1)}_- & \psi_{100}(\vec{r}_2)\chi^{(2)}_+ \\ \psi_{2lm}(\vec{r}_1)\chi^{(1)}_- & \psi_{2lm}(\vec{r}_2)\chi^{(2)}_+ \end{vmatrix} \quad (10.42)$$

The influence of the mutual electrostatic interaction $\hat{V}$ is taken into account by using the stationary state perturbation theory. For the non-degenerate ground state, because of the spin-independence of the perturbation operator $\hat{V}$, the first-order energy correction (9.11) is obtained (see Problem 10.2.1) as:

$$K_{10} = \langle \Phi_g | \hat{V} | \Phi_g \rangle = \iint \psi^*_{100}(\vec{r}_1)\psi^*_{100}(\vec{r}_2) \frac{e_0^2}{|\vec{r}_1 - \vec{r}_2|} \psi_{100}(\vec{r}_1)\psi_{100}(\vec{r}_2) dV_1 dV_2$$

$$= e_0^2 \iint \frac{|\psi_{100}(\vec{r}_1)|^2 |\psi_{100}(\vec{r}_2)|^2}{|\vec{r}_1 - \vec{r}_2|} dV_1 dV_2 = \frac{5}{8}\frac{Ze_0^2}{a_0} \quad (10.43)$$

The positive energy shift K_{10} represents the repulsive electrostatic interaction between the two densities $\rho_1 = e|\psi_{100}(\vec{r}_1)|^2$ and $\rho_2 = e|\psi_{100}(\vec{r}_2)|^2$. The magnitude of K_{10} given by Eq.(10.43) leads to:

$$E_{10} = E_g + K_{10} = -\frac{Ze_0^2}{a_0}\left(Z - \frac{5}{8}\right) \quad (10.44)$$

For $Z = 2$ we obtain a correction of 34 eV, such that $E_{10} = -108.8 + 34 = -74.8$ (eV). This value is approximately equal to the experimental ground state energy of $E_{10} = -79$ eV. The difference is due to the correction magnitude, which represents about 30% of E_g.

EXAMPLE 10.3. Variational calculation of the ground state of helium

A better approximation to the ground state energy can be obtained by applying the variational method, with the trial spatial function:

$$\Psi(\vec{r}_1,\vec{r}_2,Z^*) = \psi_{100}(\vec{r}_1,Z^*)\psi_{100}(\vec{r}_2,Z^*) = \frac{1}{\pi}\left(\frac{Z^*}{a_0}\right)^3 e^{-Z^*(r_1+r_2)/a_0}$$

where Z^* is taken to be a variable parameter. In this way we might expect to obtain, by minimizing the expectation value of energy, a magnitude Z^* smaller than the real nuclear charge Z. Such a value would account for the physical situation found in the helium atom, where each one of the two electrons moves in a potential energy which equals the nuclear attraction $-Ze_0^2/r$ reduced by the screening action of the second electron. The Hamiltonian operator, Eq.(10.32), can be rewritten as:

$$\hat{H} = -\frac{\hbar^2}{2m_e}(\nabla_1^2+\nabla_2^2) - Ze_0^2\left(\frac{1}{r_1}+\frac{1}{r_2}\right) + \frac{e_0^2}{|\vec{r}_1-\vec{r}_2|} = \hat{H}(Z^*) + (Z^*-Z)e_0^2\left(\frac{1}{r_1}+\frac{1}{r_2}\right)$$

Substituting this expression into the basic equation (9.47) of the variational method, we obtain:

$$E(Z^*) = \iint \Psi^*(\vec{r}_1,\vec{r}_2,Z^*)\left[\hat{H}(Z^*) + (Z^*-Z)e_0^2\left(\frac{1}{r_1}+\frac{1}{r_2}\right)\right]\Psi(\vec{r}_1,\vec{r}_2,Z^*)dV_1dV_2$$

$$= -\frac{Z^*e_0^2}{a_0}\left(Z^*-\frac{5}{8}\right) + \frac{(Z^*-Z)e_0^2}{\pi^2}\left(\frac{Z^*}{a_0}\right)^6 (4\pi)^2 \int_0^\infty\int_0^\infty e^{-Z^*(r_1+r_2)/a_0}\left(\frac{1}{r_1}+\frac{1}{r_2}\right)r_1^2 r_2^2 dr_1 dr_2$$

$$= -\frac{Z^*e_0^2}{a_0}\left(Z^*-\frac{5}{8}\right) + (Z^*-Z)Z^*\frac{2e_0^2}{a_0}$$

$$= -\frac{e_0^2}{a_0}\left[Z^{*2} - \frac{5}{8}Z^* - 2(Z^*-Z)Z^*\right] \tag{10.45}$$

where use has been made of the energy expectation value (10.44), as given by the first order perturbation theory. The minimum value of $E(Z^*)$ results from:

$$\frac{dE(Z^*)}{dZ^*} = -\frac{e_0^2}{a_0}\left(-2Z^*-\frac{5}{8}+2Z\right) = 0 \quad \text{or} \quad Z_0^* = Z-\frac{5}{16}$$

where Z_0^* so determined can be regarded as an effective nuclear charge for the electron motion in helium. It follows from Eq.(10.45) that the ground state energy is approximated by:

$$E_{10} = E_{min}(Z_0^*) = -\left(Z-\frac{5}{16}\right)^2 \frac{e_0^2}{a_0} \tag{10.46}$$

For $Z=2$ we obtain $E_{10} = -77.38$ eV, a value which is closer to the experimental result than that given by the perturbation method, Eq.(10.44). ♪

For the degenerate excited state of helium, the first order corrections to the energy level E_e are the solutions of the characteristic equation (9.30) where, in view of Eq.(10.40), the matrix elements of the perturbation operator read:

$$V_{11} = \langle \Phi_{lm} | \hat{V} | \Phi_{lm} \rangle = \int\int \Psi^{a*} \chi_+^{(1)+} \chi_+^{(2)+} \hat{V} \Psi^a \chi_+^{(1)} \chi_+^{(2)} dV_1 dV_2$$

$$= e_0^2 \int\int \frac{|\psi_{100}(\vec{r}_1)|^2 |\psi_{2lm}(\vec{r}_2)|^2}{|\vec{r}_1 - \vec{r}_2|} dV_1 dV_2$$

$$- e_0^2 \int\int \frac{\psi_{100}^*(\vec{r}_1) \psi_{2lm}^*(\vec{r}_2) \psi_{2lm}(\vec{r}_1) \psi_{100}(\vec{r}_2)}{|\vec{r}_1 - \vec{r}_2|} dV_1 dV_2 = K_{2l} - J_{2l} \qquad (10.47)$$

and similarly, from Eq.(10.41):

$$V_{22} = \langle \Phi_{lm}' | \hat{V} | \Phi_{lm}' \rangle = \int \Psi^{a*} \chi_-^{(1)+} \chi_-^{(2)+} \hat{V} \Psi^a \chi_-^{(1)} \chi_-^{(2)} dV_1 dV_2 = K_{2l} - J_{2l} \quad (10.48)$$

since the total spin functions $\chi_+^{(1)} \chi_+^{(2)} (S=1, M_S=1)$ and $\chi_-^{(1)} \chi_-^{(2)} (S=1, M_S=-1)$ are assumed orthogonal and normalized. The term K_{2l} corresponds to a repulsive electrostatic interaction between the charge distribution of the two electrons, analogous to that denoted by K_{10} in Eq.(10.43), while the second term J_{2l}, called the *exchange integral*, has no classical interpretation. The magnitude of the exchange integral depends on the product $\psi_{100} \psi_{2lm}$, that is on the overlapping of the two wave functions. It can be shown, using Eqs.(10.42), that:

$$\langle \Phi_{lm}^{II} | \hat{V} | \Phi_{lm}^{II} \rangle = \langle \Phi_{lm}^{III} | \hat{V} | \Phi_{lm}^{III} \rangle = K_{2l}$$

$$\langle \Phi_{lm}^{II} | \hat{V} | \Phi_{lm}^{III} \rangle = \langle \Phi_{lm}^{III} | \hat{V} | \Phi_{lm}^{II} \rangle = -J_{2l}$$
(10.49)

whereas all the other elements of the characteristic determinant vanish, because of the orthogonality of the spin functions, arising from the spin-independent nature of the perturbation operator $\hat{V}$. Thus the characteristic equation (9.30) reduces to:

$$\det(\hat{V} - E^{(1)} \hat{I}) = \begin{vmatrix} K_{2l} - J_{2l} - E^{(1)} & 0 & 0 & 0 \\ 0 & K_{2l} - J_{2l} - E^{(1)} & 0 & 0 \\ 0 & 0 & K_{2l} - E^{(1)} & -J_{2l} \\ 0 & 0 & -J_{2l} & K_{2l} - E^{(1)} \end{vmatrix}$$

$$= (K_{2l} - J_{2l} - E^{(1)})^2 \begin{vmatrix} K_{2l} - E^{(1)} & -J_{2l} \\ -J_{2l} & K_{2l} - E^{(1)} \end{vmatrix}$$

$$= (K_{2l} - J_{2l} - E^{(1)})^3 (K_{2l} + J_{2l} - E^{(1)}) = 0 \qquad (10.50)$$

which has a triple root $E_T^{(1)}$ and a single root $E_S^{(1)}$:

$$E_T^{(1)} = K_{2l} - J_{2l}, \qquad E_S^{(1)} = K_{2l} + J_{2l} \qquad (10.51)$$

Substituting $E_T^{(1)}$ into the matrix eigenvalue equation:

$$\begin{pmatrix} K_{2l} & -J_{2l} \\ -J_{2l} & K_{2l} \end{pmatrix} \begin{pmatrix} A_1 \\ A_2 \end{pmatrix} = E^{(1)} \begin{pmatrix} A_1 \\ A_2 \end{pmatrix} \qquad (10.52)$$

we obtain $A_1 = A_2 = 1/\sqrt{2}$, and the corresponding eigenfunction, which defines a state with $S = 1$, $M_S = 0$, can then be written as:

$$\Phi_{lm}^{IV} = \frac{1}{\sqrt{2}} (\Phi_{lm}^{II} + \Phi_{lm}^{III})$$

$$= \frac{1}{\sqrt{2}} [\psi_{100}(\vec{r}_1) \psi_{2lm}(\vec{r}_2) - \psi_{2lm}(\vec{r}_1) \psi_{100}(\vec{r}_2)] \frac{1}{\sqrt{2}} (\chi_+^{(1)} \chi_-^{(2)} + \chi_-^{(1)} \chi_+^{(2)})$$

$$= \Psi^a \frac{1}{\sqrt{2}} (\chi_+^{(1)} \chi_-^{(2)} + \chi_-^{(1)} \chi_+^{(2)}) \qquad (10.53)$$

By combining the space-antisymmetric, spin-symmetric wave functions (10.40), (10.41) and (10.53) we can represent the *triplet* state $S = 1$, $M_S = 0, \pm 1$ of helium, for a given l, as:

$$\Phi_{lm}^T = \frac{1}{\sqrt{2}} [\psi_{100}(\vec{r}_1) \psi_{2lm}(\vec{r}_2) - \psi_{2lm}(\vec{r}_1) \psi_{100}(\vec{r}_2)] \chi^T = \Psi^a \chi_{M_S}^T \qquad (10.54)$$

where χ^T is the triplet spin function:

$$\chi_{M_S}^T = \chi_+^{(1)} \chi_+^{(2)} \qquad (M_S = 1)$$

$$= \frac{1}{\sqrt{2}} (\chi_+^{(1)} \chi_-^{(2)} + \chi_-^{(1)} \chi_+^{(2)}) \quad (M_S = 0)$$

$$= \chi_-^{(1)} \chi_-^{(2)} \qquad (M_S = -1) \qquad (10.55)$$

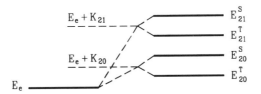

Figure 10.1. Ground level shift and first excited level splitting in helium

Similarly, by substituting $E_S^{(1)}$, Eq.(10.52) gives $A_{S1} = -A_{S2} = 1/\sqrt{2}$ so that the corresponding *singlet* state is:

$$\Phi_{lm}^S = \frac{1}{\sqrt{2}}(\Phi_{lm}^{II} - \Phi_{lm}^{III})$$

$$= \frac{1}{\sqrt{2}}[\psi_{100}(\vec{r}_1)\psi_{2lm}(\vec{r}_2) + \psi_{2lm}(\vec{r}_1)\psi_{100}(\vec{r}_2)]\frac{1}{\sqrt{2}}(\chi_+^{(1)}\chi_-^{(2)} - \chi_-^{(1)}\chi_+^{(2)})$$

$$= \frac{1}{\sqrt{2}}[\psi_{100}(\vec{r}_1)\psi_{2lm}(\vec{r}_2) + \psi_{2lm}(\vec{r}_1)\psi_{100}(\vec{r}_2)]\chi^S = \Psi^s \chi^S \qquad (10.56)$$

This is a space-symmetric, spin-antisymmetric wave function, like the ground-state eigenfunction (10.38).

Therefore, when the interaction between the two electrons is taken into account, there is a shift K_{10} of the ground state level and the first excited states split into a triplet E_{21}^T and a singlet E_{21}^S (Figure 10.1):

$$E_{20}^T = E_e + K_{20} - J_{20}, \qquad E_{20}^S = E_e + K_{20} + J_{20}$$

$$E_{21}^T = E_e + K_{21} - J_{21}, \qquad E_{21}^S = E_e + K_{21} + J_{21}$$

where $E_e = -68$ eV. Since both K_{21} and J_{21} are positive, the energy shift of the triplet state level E_{21}^T is lower than that of the singlet state level E_{21}^S by an amount $2J_{21}$. In other words, for a given excited configuration of helium, the states of highest spin have the lowest energy. It also follows that, as a result of the interaction between the two electrons, the accidental degeneracy of E_e is removed.

378   Quantum Physics

Problem 10.2.1. Show that the first order energy correction to the ground state energy of helium, due to the repulsive electrostatic interaction between the two electrons, has the magnitude:

$$K_{10} = e_0^2 \iint \frac{|\psi_{100}(\vec{r}_1)|^2 \, |\psi_{100}(\vec{r}_2)|^2}{|\vec{r}_1 - \vec{r}_2|} dV_1 \, dV_2 = \frac{5}{8} \frac{Z e_0^2}{a_0}$$

(Solution):   Using $\Psi_{100}(\vec{r})$ as given by Eqs.(7.35) and (7.87) we obtain:

$$K_{10} = \frac{e_0^2}{\pi^2} \left(\frac{Z}{a_0}\right)^6 \iiiint \frac{1}{|\vec{r}_1 - \vec{r}_2|} e^{-2Z(r_1+r_2)/a_0} r_1^2 \, dr_1 \, d\Omega_1 \, r_2^2 \, dr_2 \, d\Omega_2$$

$$= \frac{1}{32\pi^2} \frac{Z e_0^2}{a_0} \iiiint \frac{e^{-(\rho_1+\rho_2)}}{|\vec{\rho}_1 - \vec{\rho}_2|} \rho_1^2 \, d\rho_1 \, d\Omega_1 \, \rho_2^2 \, d\rho_2 \, d\Omega_2$$

where $\vec{\rho}_{1,2} = 2Z \vec{r}_{1,2}/a_0$ are dimensionless variables. If we choose the z-axis along the direction of $\vec{\rho}_1$, such that θ denotes the angle between $\vec{\rho}_1$ and $\vec{\rho}_2$, we can write:

$$K_{10} = \frac{1}{32\pi^2} \frac{Z e_0^2}{a_0} \iint e^{-\rho_1} I(\rho_1) \rho_1^2 \, d\rho_1 \, d\Omega_1 = \frac{1}{8\pi} \frac{Z e_0^2}{a_0} \int_0^\infty e^{-\rho_1} I(\rho_1) \rho_1^2 \, d\rho_1$$

where:

$$I(\rho_1) = 2\pi \int_0^\infty \int_0^\pi \frac{e^{-\rho_2} \rho_2^2 \, d\rho_2 \, \sin\theta \, d\theta}{(\rho_1^2 - 2\rho_1\rho_2 \cos\theta + \rho_2^2)^{\frac{1}{2}}}$$

The integration over θ is straightforward:

$$\int_0^\pi \frac{\sin\theta \, d\theta}{(\rho_1^2 - 2\rho_1\rho_2 + \rho_2^2)^{\frac{1}{2}}} = \frac{1}{\rho_1\rho_2} \int_0^\pi \frac{d}{d\theta} (\rho_1^2 + \rho_2^2 - 2\rho_1\rho_2 \cos\theta)^{\frac{1}{2}} \, d\theta$$

$$= \frac{1}{\rho_1\rho_2} (\rho_1 + \rho_2 - |\rho_1 - \rho_2|) = \frac{2}{\rho_1} \quad \text{if} \quad \rho_2 < \rho_1$$

$$= \frac{2}{\rho_2} \quad \text{if} \quad \rho_2 > \rho_1$$

It follows that $I(\rho_1)$ can be written in the form:

$$I(\rho_1) = 4\pi\left[\frac{1}{\rho_1}\int_0^{\rho_1} e^{-\rho_2}\rho_2^2\, d\rho_2 + \int_{\rho_1}^{\infty} e^{-\rho_2}\rho_2\, d\rho_2\right] = \frac{4\pi}{\rho_1}\left[2 - e^{-\rho_1}(2+\rho_1)\right]$$

and hence, we have:

$$K_{10} = \frac{1}{2}\frac{Ze_0^2}{a_0}\left[2\int_0^{\infty} e^{-\rho_1}\rho_1\, d\rho_1 - 2\int_0^{\infty} e^{-2\rho_1}\rho_1\, d\rho_1 - \int_0^{\infty} e^{-2\rho_1}\rho_1^2\, d\rho_1\right] = \frac{5}{8}\frac{Ze_0^2}{a_0}$$

Problem 10.2.2. Show that the probability density $P(\vec{r})$ of finding one electron at the position $\vec{r}$, irrespective of the position of the other electron, in the first excited state of helium, is given by:

$$P(\vec{r}) = |\psi_{100}(\vec{r})|^2 + |\psi_{2lm}(\vec{r})|^2$$

(Solution): Disregarding the spin coordinates, the wave function of the first excited state of helium can be written, in view of Eqs.(10.54) and (10.56), in the form:

$$\Psi(\vec{r}_1, \vec{r}_2) = \frac{1}{\sqrt{2}}\left[\psi_{100}(\vec{r}_1)\psi_{2lm}(\vec{r}_2) + (-1)^S \psi_{2lm}(\vec{r}_1)\psi_{100}(\vec{r}_2)\right]$$

where $S = 0$ or 1, and $|\psi(\vec{r}_1, \vec{r}_2)|^2$ represents the probability density for the position distribution of the two electrons:

$$|\Psi(\vec{r}_1, \vec{r}_2)|^2 = \frac{1}{2}\Big[|\psi_{100}(\vec{r}_1)|^2 |\psi_{2lm}(\vec{r}_2)|^2 + |\psi_{2lm}(\vec{r}_1)|^2 |\psi_{100}(\vec{r}_2)|^2$$

$$+ (-1)^{(S)} \psi_{100}(\vec{r}_1)\psi_{2lm}^*(\vec{r}_1)\psi_{2lm}(\vec{r}_2)\psi_{100}^*(\vec{r}_2)$$

$$+ (-1)^S \psi_{100}^*(\vec{r}_1)\psi_{2lm}(\vec{r}_1)\psi_{2lm}^*(\vec{r}_2)\psi_{100}(\vec{r}_2)\Big]$$

The probability density $P(\vec{r})$ of finding one electron at a definite position $\vec{r}$ is defined in terms of $\Psi(\vec{r}_1, \vec{r}_2)$ as:

$$P(\vec{r}) = \int \left|\Psi(\vec{r}_1, \vec{r}_2)\right|^2_{\vec{r}_1 = \vec{r}} dV_2 + \int \left|\Psi(\vec{r}_1, \vec{r}_2)\right|^2_{\vec{r}_2 = \vec{r}} dV_1$$

where integration is performed over the position coordinates of the other electron. Substituting $|\Psi(\vec{r}_1, \vec{r}_2)|^2$ gives:

$$P(\vec{r}) = |\psi_{100}(\vec{r})|^2 \int |\psi_{2lm}(\vec{r}_2)|^2 dV_2 + |\psi_{2lm}(\vec{r})|^2 \int |\psi_{100}(\vec{r}_1)|^2 dV_1$$

$$+\frac{(-1)^S}{2}\psi_{100}(\vec{r})\psi^*_{2lm}(\vec{r})\left[\int \psi_{2lm}(\vec{r}_2)\psi^*_{100}(\vec{r}_2)dV_2 + \int \psi_{2lm}(\vec{r}_1)\psi^*_{100}(\vec{r}_1)dV_1\right]$$

$$+\frac{(-1)^S}{2}\psi^*_{100}(\vec{r})\psi_{2lm}(\vec{r})\left[\int \psi^*_{2lm}(\vec{r}_2)\psi_{100}(\vec{r}_2)dV_2 + \int \psi^*_{2lm}(\vec{r}_1)\psi_{100}(\vec{r}_1)dV_1\right]$$

$$=|\psi_{100}(\vec{r})|^2 + |\psi_{2lm}(\vec{r})|^2$$

where the first two integrals are equal to unity, while all the others vanish, due to the orthonormalization properties of the spatial wave functions $\psi_{100}(\vec{r})$ and $\psi_{2lm}(\vec{r})$ for a single electron.

Problem 10.2.3. Show that the energy separation between the two energy levels of the $1s2s$ electron configuration of helium is:

$$E^S_{20} - E^T_{20} = 2J_{20} = \left(\frac{2}{3}\right)^6 \frac{e_0^2}{a_0}$$

by calculating the exchange integral J_{20} introduced in Eq.(10.47).

10.3. Multielectron Atoms

To calculate the energy levels of many-electron systems we consider the Hamiltonian for the motion of Z electrons about a nucleus of mass M and charge Ze:

$$\hat{H} = \sum_{i=1}^{Z}\left(-\frac{\hbar^2}{2m_e}\nabla_i^2 - \frac{Ze_0^2}{r_i}\right) + \sum_{i=1}^{Z}\sum_{j>i}\frac{e_0^2}{|\vec{r}_i - \vec{r}_j|} = \sum_{i=1}^{Z}\hat{H}_i + \hat{V} \qquad (10.57)$$

which is an obvious generalization of Eq.(10.32), including the kinetic energy of the electrons, the electrostatic interaction between the nucleus and each of the Z electrons, and the mutual interaction between all pairs of electrons $\hat{V}$. The single particle operators $\hat{H}_i$ for the individual electrons are assumed to satisfy the eigenvalue equations (10.33). It is no longer a good approximation, in atoms with Z much greater than 2, to treat the term $\hat{V}$ as a perturbation, because we have already seen in the case of helium, Eq.(10.44), that the first order energy correction is not very small. A better approximation scheme is provided by the variational method, which accounts for the motion of each electron in a central potential energy, consisting of the nuclear attraction $-Ze_0^2/r_i$ reduced by the screening action of the other electrons (see Example 10.3).

It is convenient to choose first, for simplicity, a trial wave function which is the product of single electron wave functions:

$$\Phi = \varphi_{E_1}(\vec{r}_1 m_{s1}) \ldots \varphi_{E_i}(\vec{r}_i m_{si}) \ldots \varphi_{E_Z}(\vec{r}_Z m_{sZ}) \quad (10.58)$$

where it is assumed that all the $\varphi_{E_i}(\vec{r}_i m_{si})$ are normalized and orthogonal, either in spatial or spin coordinates, such that there is no more than one electron in the same state $\vec{r}_i m_{si}$. Because of spin-independence of the Hamiltonian (10.58), we obtain the expectation value of energy in the form:

$$E = \int_1 \cdots \int_i \cdots \int_Z \Phi^* \hat{H} \Phi \, dV_1 \cdots dV_i \cdots dV_Z$$

$$= \int_1 \cdots \int_i \cdots \int_Z \varphi_{E_1}^*(\vec{r}_1) \cdots \varphi_{E_i}^*(\vec{r}_i) \cdots \varphi_{E_Z}^*(\vec{r}_Z) \left(\sum_{i=1}^Z \hat{H}_i + \hat{V} \right) \varphi_{E_1}(\vec{r}_1) \cdots \varphi_{E_i}(\vec{r}_i) \cdots \varphi_{E_Z}(\vec{r}_Z) dV_1 \cdots dV_i \cdots dV_Z$$

$$= \sum_{i=1}^Z \left[\int \varphi_{E_i}^*(\vec{r}_i) \hat{H}_i \varphi_{E_i}(\vec{r}_i) dV_i + \sum_{j>i} \int \varphi_{E_i}^*(\vec{r}_i) \left(e_0^2 \int \frac{|\varphi_{E_j}(\vec{r}_j)|^2}{|\vec{r}_i - \vec{r}_j|} dV_j \right) \varphi_{E_i}(\vec{r}_i) dV_i \right]$$

$$= \sum_{i=1}^Z \int \varphi_{E_i}^*(\vec{r}_i) \left(\hat{H}_i + \sum_{j>i} \hat{K}_{ij} \right) \varphi_{E_i}(\vec{r}_i) dV_i \quad (10.59)$$

The operator $\hat{K}_{ij}$ represents the effect of the electron density $\rho_j = e|\varphi_{E_j}(\vec{r}_j)|^2$ on the electron at $\vec{r}_i$, such that:

$$\hat{K}_{ij} \varphi_{E_i}(\vec{r}_i) = e_0^2 \int \frac{|\varphi_{E_j}(\vec{r}_j)|^2}{|\vec{r}_i - \vec{r}_j|} \varphi_{E_i}(\vec{r}_i) dV_j \quad (10.60)$$

and was introduced in order to apply the variational method to the single electron problem. Assuming that an arbitrary variation $\delta\Phi$ is exclusively due to the variation of $\varphi_{E_i}(\vec{r}_i)$, namely :

$$\delta\Phi = \varphi_{E_1}(\vec{r}_1) \cdots \delta\varphi_{E_i}(\vec{r}_i) \cdots \varphi_{E_Z}(\vec{r}_Z)$$

we must find the minimum of the energy E_i, given by:

$$E_i = \int \varphi_{E_i}^*(\vec{r}_i) \left(\hat{H}_i + \sum_{j \neq i} \hat{K}_{ij} \right) \varphi_{E_i}(\vec{r}_i) dV_i \quad (10.61)$$

which includes all the terms of Eq.(10.59) involving the single electron function $\varphi_{E_i}(\vec{r}_i)$. As discussed in Section 9.2, the variational equation (9.38), which reads:

$$\delta E_i = \langle \delta\varphi_{E_i} | \hat{H}_i + \sum_{j \neq i} \hat{K}_{ij} | \varphi_{E_i} \rangle + \langle \varphi_{E_i} | \hat{H}_i + \sum_{j \neq i} \hat{K}_{ij} | \delta\varphi_{E_i} \rangle = 0 \quad (10.62)$$

is restricted to the variations $\delta\varphi_{E_i}$ compatible with the normalization condition on φ_{E_i}, Eq.(9.39), namely:

$$\delta\langle \varphi_{E_i} | \varphi_{E_i} \rangle = \langle \delta\varphi_{E_i} | \varphi_{E_i} \rangle + \langle \varphi_{E_i} | \delta\varphi_{E_i} \rangle = 0 \quad (10.63)$$

By using the Lagrangian multiplier ε, such that:

$$\delta E_i - \varepsilon \delta \langle \varphi_{E_i} | \varphi_{E_i} \rangle = 0$$

we obtain:

$$\langle \delta\varphi_{E_i} | (\hat{H}_i + \sum_{j \neq i} \hat{K}_{ij}) \varphi_{E_i} - \varepsilon \varphi_{E_i} \rangle + \langle (\hat{H} + \sum_{j \neq i} \hat{K}_{ij}) \varphi_{E_i} - \varepsilon \varphi_{E_i} | \delta\varphi_{E_i} \rangle = 0$$

which means that φ_{E_i} satisfies the eigenvalue problem:

$$\left(\hat{H}_i + \sum_{j \neq i} \hat{K}_{ij} \right) \varphi_{E_i}(\vec{r}_i) = \varepsilon \varphi_{E_i}(\vec{r}_i)$$

Multiplying by $\varphi_{E_i}^*$ and integrating over $\vec{r}_i$, we obtain from Eq.(10.61) that $\varepsilon = E_i$, so that the energy eigenfunction and eigenvalue of the ith electron are determined by solving the single particle equation:

$$\left(\hat{H}_i + \sum_{j \neq i} \hat{K}_{ij} \right) \varphi_{E_i}(\vec{r}_i) = E_i \, \varphi_{E_i}(\vec{r}_i) \quad (10.64)$$

The set of Z simultaneous equations of the form (10.64), written for each electron of the atom, are called the *Hartree equations*. It is clear that the terms $\hat{K}_{ij} \, \varphi_{E_i}(\vec{r}_i)$, having the explicit form (10.60), can only be calculated if $\varphi_{E_j}(\vec{r}_j)$ are already known, which implies that the Hartree equations cannot be simultaneously solved in closed form. However, an iterative procedure can be applied, assuming an initial set of functions $\varphi_{E_i}^{(0)}(r_i)$ to obtain the potential energies $K_{ij}^{(0)} = \langle \varphi_{E_i}^{(0)} | \hat{K}_{ij} | \varphi_{E_i}^{(0)} \rangle$ and then solving numerically Eqs.(10.64) for a new set of single electron functions $\varphi_{E_i}^{(1)}(r_i)$. This procedure is repeated until the wave functions $\varphi_{E_i}^{(k)}(\vec{r}_i)$ and the eigenvalues $E_i^{(k)}$ have values which are no longer different from those used in the equations solved for them.

The Hartree equation for the ith electron, Eq.(10.64), can be rewritten in terms of the average potential arising from all the other electrons and the nucleus:

$$V_{eff}(\vec{r}_i) = \sum_{j \neq i} \hat{K}_{ij} - \frac{Ze_0^2}{r_i} = \sum_{j \neq i} e_0^2 \int \frac{|\varphi_{E_j}(\vec{r}_j)|^2}{|\vec{r}_i - \vec{r}_j|} dV_j - \frac{Ze_0^2}{r_i} \quad (10.65)$$

A convenient approximation consists of replacing $V_{eff}(\vec{r}_i)$ by its average over the angle, which is a central potential $V_{eff}(r_i) = \int V_{eff}(\vec{r}_i) d\Omega_i / 4\pi$. Substituting into Eq.(10.64), gives the so-called *central field approximation* to the Hartree equations:

$$\left(-\frac{\hbar^2}{2m_e}\nabla_i^2 + V_{eff}(r_i)\right)\varphi_{E_i}(\vec{r}_i) = E_i \, \varphi_{E_i}(\vec{r}_i) \quad (10.66)$$

Because the Hamiltonian contains only central forces and does not involve the spin, it commutes with both the orbital and spin angular momentum operators, such that the solutions $\varphi_{E_i}(\vec{r}_i)$ are eigenstates of the angular momentum, of the form (10.34), called *orbitals*:

$$\varphi_{E_i}(\vec{r}_i m_{s_i}) = R_{n_i l_i}(r_i) Y_{l_i m_i}(\theta_i, \varphi_i) \chi_{m_{s_i}} \quad (10.67)$$

The expectation value of energy, in a state of the multielectron atom which is written as a simple product of orbitals, Eq.(10.58), is obtained from Eq.(10.59) as:

$$E = \sum_{i=1}^{Z} E_{n_i} + \sum_{i=1}^{Z}\sum_{j>i} \int \varphi_{E_i}^*(\vec{r}_i) \hat{K}_{ij} \varphi_{E_i}(\vec{r}_i) dV_i = \sum_{i=1}^{Z} E_{n_i} + \frac{1}{2}\sum_{i=1}^{Z}\sum_{j \neq i} \int \varphi_{E_i}^*(\vec{r}_i) \hat{K}_{ij} \varphi_{E_i}(\vec{r}_i) dV_i$$

$$= \sum_{i=1}^{Z} E_{n_i} + \frac{1}{2}\sum_{i=1}^{Z}\sum_{j \neq i} e_0^2 \iint_{i\;j} \frac{|\varphi_{E_j}(\vec{r}_j)|^2 \, |\varphi_{E_i}(\vec{r}_i)|^2}{|\vec{r}_i - \vec{r}_j|} dV_i \, dV_j \quad (10.68)$$

where each electron pair is counted only once, although the summation is unrestricted except for the requirement that $i \neq j$. If E' denotes the expectation energy of the system where one particular orbital, say $\varphi_{E_k}(\vec{r}_k)$, is left out:

$$E' = \sum_{\substack{i=1 \\ i \neq k}}^{Z} E_{n_i} + \frac{1}{2}\sum_{\substack{i=1 \\ i \neq k}}^{Z}\sum_{j \neq i} e_0^2 \iint_{i\;j} \frac{|\varphi_{E_j}(\vec{r}_j)|^2 \, |\varphi_{E_i}(\vec{r}_i)|^2}{|\vec{r}_i - \vec{r}_j|} dV_i \, dV_j$$

$$= E - E_{n_k} - \sum_{j \neq k} e_0^2 \iint_{k\ j} \frac{|\varphi_{E_j}(\vec{r}_j)|^2 |\varphi_{E_k}(\vec{r}_k)|^2}{|\vec{r}_k - \vec{r}_j|} dV_k\, dV_j = E - E_k \qquad (10.69)$$

it is clear that each eigenvalue E_i of the Hartree equations (10.64) provides the magnitude of the *ionization* energy $\Delta E_i = E_\infty - E_i$, which is required to remove one electron from the corresponding orbital of the multielectron atom.

EXAMPLE 10.4. The Hartree-Fock approximation

If the variational method is applied to the multielectron atom by using as a trial function a Slater determinant, Eq.(10.28), we obtain the so-called *Hartree-Fock approximation*, that fully accounts for the Pauli exclusion principle. Although more calculation is required, one obtains a similar form of the expectation value of energy, namely:

$$E = \int \Phi^{a*} \left(\sum_{i=1}^{Z} \hat{H}_i + \hat{V} \right) \Phi^a dV_1 \cdots dV_i \cdots dV_Z$$

$$= \sum_{i=1}^{Z} \int \varphi_{E_i}^*(\vec{r}_i) \left(\hat{H}_i + \sum_{j>i} \hat{K}_{ij} - \sum_{j>i} \hat{J}_{ij} \right) \varphi_{E_i}(\vec{r}_i)\, dV_i \qquad (10.70)$$

where the $\hat{K}_{ij}$ are the same as defined by Eq.(10.60). This expression only differs from Eq.(10.59) by the addition of the *exchange* term, involving the operator:

$$\hat{J}_{ij}\, \varphi_{E_i}(\vec{r}_i) = e_0^2 \int \frac{\varphi_{E_i}^*(\vec{r}_j)\, \varphi_{E_j}(\vec{r}_i)}{|\vec{r}_i - \vec{r}_j|}\, \varphi_{E_i}(\vec{r}_i)\, dV_j \qquad (10.71)$$

It then follows, using the same procedure applied before to derive the Hartree equations (10.64), that we now obtain:

$$\left(\hat{H}_i + \sum_{j \neq i} \hat{K}_{ij} - \sum_{j \neq i} \hat{J}_{ij} \right) \varphi_{E_i}(\vec{r}_i) = E_i\, \varphi_{E_i}(\vec{r}_i) \qquad (10.72)$$

which represent a set of Z simultaneous equations called the *Hartree-Fock equations*. They can be similarly solved by iteration for the orbitals $\varphi_{E_i}(\vec{r}_i)$ and yield eigenvalues E_i which differ by 10-20% from the ionization energies obtained from the Hartree equations. ☙

The central field approximation allows one to consider electrons as independent particles, each moving in the spherically symmetric field which is due partly to the nucleus and partly to all the other electrons. Thus, in order to characterize the state of a multielectron atom it is necessary to specify the quantum numbers n, l, m and m_s of each atomic orbital (10.67). The corresponding energy eigenvalue E_i increases as n increases.

However, since the potential energy (10.65) no longer has the $1/r_i$ form, the accidental degeneracy of all the orbitals with a given n is removed, so that E_i are functions of both n and l. In other words, there will be a splitting for different l values for a given n, with the lowest level corresponding to $l = 0$ and an increasing energy as l increases. This is expected in view of the results obtained for the radial motion in the one-electron atom, Eq.(7.119), where a significant probability for the electron to approach the nucleus and to feel the full nuclear attraction was found for small l values, while the electron is kept away from the nucleus by the centrifugal barrier as l increases. Although the l-dependent splitting is smaller than the splitting between different n values for low Z, it increases with increasing n, such that for larger Z values we see from Figure 10.2 that there are ns energy levels lower than $(n-1)d$ or $(n-2)f$ levels.

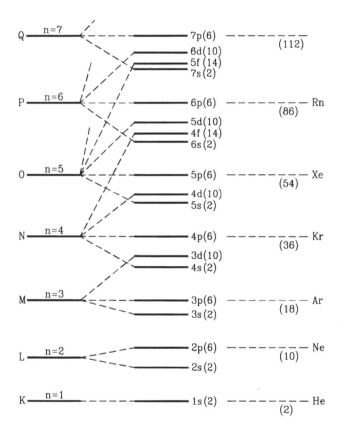

Figure 10.2. Energy levels for electrons in multielectron atoms

By listing the n and l values, in the order of increasing electron energy, we obtain the *electron configuration* of the multielectron atom. The orbitals which have the same value n are said to form a *shell* which is denoted by K, L, M, N, O, P, Q as $n = 1, 2, 3, 4, 5, 6, 7$ respectively. The orbitals with the same value for both n and l form a *subshell* and are called *equivalent*, as they all correspond to the same energy eigenvalue E_{nl}. We have

386 Quantum Physics

shown before, Eq.(8.69), that the maximum number of equivalent electrons, obtained by summing over either m and m_s or j and m_j angular momentum quantum numbers, is $2(2l+1)$. A subshell which contains the maximum number of electrons, namely two in a given s-subshell, six in a p-subshell, ten in a d-subshell, fourteen in an f-subshell, etc., is called a *closed* subshell. The maximum number of electrons in a shell, which is then also said to be closed, is $\sum_{l=0}^{n-1} 2(2l+1) = 2n^2$. Consequently, regularities in the shell model of multielectron atoms might be expected for $Z = 2, 10, 28, 60$ and 110. However, the experimentally observed regularities, illustrated in Figure 10.2, which are expressed by the so-called *atomic magic numbers* $Z = 2, 10, 18, 36, 54$ and 86, are determined by the maximum number of equivalent electrons $2(2l+1)$ rather than by the closed shells.

EXAMPLE 10.5. Electron configuration of the low Z atoms

Since the repulsive electrostatic interaction in Eq.(10.65) is always positive, the average potential energy of the electron at any distance r_i from the nucleus, Eq.(10.68), is smaller than that of the corresponding electron in the one-electron atom. Consequently, the energy eigenvalues E_i of the form (10.61) are smaller than E_{n_i} given by Eq.(10.35). For low Z values, however, we may express each eigenvalue E_i in a form similar to Eq.(10.35), namely:

$$E_i = -\frac{e_0^2}{2a_0}\frac{Z_{eff}^2}{n_i^2} = -13.6\frac{Z_{eff}^2}{n_i^2} \quad (\text{eV}) \tag{10.73}$$

where the difference $Z - Z_{eff}$ is called the *shielding constant* and gives the effect of the other $Z - 1$ electrons in reducing the electrostatic interaction between the nucleus and the electron in the orbital $\varphi_{E_i}(\vec{r}_i)$. As the energy E_i depends on n and l only, we can determine the electron configuration of any new atom, formed by adding one electron to an already known ground state configuration, by assigning to the additional electron the quantum numbers n and l which make the ground state energy of the new atom a minimum.

Hydrogen $(Z = 1)$ has a single electron in the ground state configuration $1s^1$ $(n = 1, l = 0)$. The energy required to remove the electron represents the *ionization* energy of the one-electron atom, defined in view of Eq.(10.35) by:

$$\Delta E_n = E_\infty - E_n = \frac{e_0^2}{2a_0}\frac{Z^2}{n^2} \tag{10.74}$$

For hydrogen we obtain $\Delta E_1 = e_0^2/2a_0 = 13.6$ eV.

Helium $(Z = 2)$ has the ground state configuration $1s^2$. The ionization energy in a multielectron atom, defined as the energy required to remove the *most weakly bound electron*, is derived from Eq.(10.73) as:

$$\Delta E_i = E_\infty - E_i = \frac{e_0^2}{2a_0}\frac{Z_{eff}^2}{n_i^2} \tag{10.75}$$

The experimental ionization energies for the lightest atoms are represented in Figure 10.3. Since the total energy of the ground state of helium is $E_{10} = -79$ eV and the ionization energy of a single electron in a $1s$ orbit about the $Z = 2$ nucleus is given by Eq.(10.74) as $\Delta E_1 = 4e_0^2/2a_0 = 54.4$ eV, the ionization energy (10.75) must be the difference between the two, namely $\Delta E_{10} = |E_{10}| - \Delta E_1 = 24.6$ eV. This gives $Z_{eff} = \sqrt{24.6/13.6} = 1.35$ which means that each electron moves in the field of a central charge $Z_{eff} e = 1.35 e$. The shielding constant is $Z - Z_{eff} = 0.65$. The configuration of helium $1s^2$ represents the closed shell K. All the atoms whose electrons form closed shells are chemically inert and are called *noble gases*.

Lithium $(Z = 3)$ is obtained by adding one electron to the $1s^2$ configuration. Using the ionization energy $\Delta E = 5.4$ eV (Figure 10.3), Eq.(10.75) gives $Z_{eff} = \sqrt{5.4/13.6} = 0.63$ for a third electron in the K shell and $Z_{eff} = 2\sqrt{5.4/13.6} = 1.26$ if it belongs to the L shell, with $n = 2$. The first value implies a shielding factor $Z - Z_{eff} = 2.37$, which exceeds the charge of the two inner electrons in the configuration $1s^2$. Thus we must assume that the electron configuration of helium is $1s^2 2s^1$. The outer electron moves in the field of a central charge $Z_{eff} e = 1.26 e$. The shielding constant of the inner electrons $Z - Z_{eff} = 1.74$. Atoms having electron configuration similar to lithium, namely $Na(\cdots 3s^1)$, $K(\cdots 4s^1)$, $Rb(\cdots 5s^1)$ and $Cs(\cdots 6s^1)$ are called *alkali* atoms. All of them have one electron more than a noble gas and find it energetically favourable to contribute this electron in chemical reactions. This property is typical of a metallic behaviour.

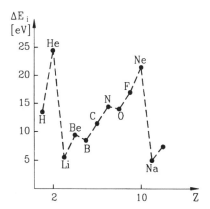

Figure 10.3. Ionization energies for low Z value atoms

Beryllium $(Z = 4)$ has an additional electron which leads to the configuration $1s^2 2s^2$ and an experimental value of 9.3 eV for the ionization energy. From Eq.(10.75) it follows that $Z_{eff} = 2\sqrt{9.3/13.6} = 1.65$, which gives a shielding constant $Z - Z_{eff} = 2.35$. As 1.74 is the contribution of the inner configuration $1s^2$ to the shielding, it follows that the contribution of the second $2s$ electron is 0.61. In other words, each of the two equivalent electrons in the subshell $2s^2$ screens 0.61 of a proton charge, a value which is close to the shielding constant 0.65 of an electron in the subshell $1s^2$ of helium.

Boron ($Z = 5$) has an ionization energy of 8.3 eV. From Eq.(10.73) it follows that $Z_{eff} = 2\sqrt{8.3/13.6} = 1.56$, if the fifth electron belongs to the L shell, and $Z_{eff} = 3\sqrt{8.3/13.6} = 2.34$, if it belongs to the M shell, with $n = 3$. The latter value implies that an electron which is expected to be further away from the nucleus than an electron in an L shell, moves in the field of a larger central charge $Z_{eff}\,e = 2.34\,e$ than that determined for electrons in the $2s$ subshell, so that the electron must have $n = 2$. Since the exclusion principle forbids a $2s^3$ configuration, the suggestion is that the fifth electron gives the boron configuration $1s^2 2s^2 2p^1$.

The $2p$ subshell fills up starting with boron and ending with neon. Figure 10.3 shows an almost monotonic increase in the ionization energy of these atoms, as the increased electrostatic repulsion decreases the radius of the subshell. The atoms also become gradually smaller as Z increases in this region. The electron configurations are $1s^2 2s^2 2p^2$ for *carbon* ($Z = 6$), $1s^2 2s^2 2p^3$ for *nitrogen* ($Z = 7$), $1s^2 2s^2 2p^4$ for *oxygen* ($Z = 8$) and $1s^2 2s^2 2p^5$ for *fluorine* ($Z = 9$). Atoms with configurations similar to fluorine, that is $\text{Cl}(\cdots 3p^5)$, $\text{Br}(\cdots 4p^5)$ and $\text{I}(\cdots 5p^5)$ are called *halogens*. All have one electron less than a complete np^6 subshell and chemically behave as non-metals, finding it energetically favourable to acquire an electron in chemical reactions.

Neon ($Z = 10$) has closed shells K and L with the configuration $1s^2 2s^2 2p^6$. The $2p^6$ subshell is a spherically symmetric configuration with a total angular momentum of zero, since the six occupied $2p$ states include all possible orientations of the spin and orbital angular momenta. For this reason neon, like helium, is chemically inert. The other noble gases, all having high ionization energies, are $\text{Ar}(\cdots 3p^6)$, $\text{Kr}(\cdots 4p^6)$, $\text{Xe}(\cdots 5p^6)$ and $\text{Rn}(\cdots 6p^6)$, with electronic configurations which account for their atomic magic numbers $Z = 18, 36, 54$ and 86 respectively.

The information provided by the electron configuration is insufficient for a complete description of the state of the atom, because it does not specify how the orbital and spin angular momenta of the individual electrons are combined to give the total angular momentum of the atom. Experiment shows that different orbitals corresponding to the same electron configuration are split, such that we have to consider a classification of the energy eigenstates according to the eigenvalues of angular momenta which are constants of motion. Neglecting the weak interaction, it can be assumed that the total orbital angular momentum $\vec{L}$ and the total spin angular momentum $\vec{S}$, defined by:

$$\vec{L} = \sum_{i=1}^{Z} \vec{L}_i, \quad \vec{S} = \sum_{i=1}^{Z} \vec{S}_i \qquad (10.76)$$

are separately conserved, which means that the operators $\hat{L}^2, \hat{L}_z, \hat{S}^2$ and $\hat{S}_z$ commute with the Hamiltonian. In this approximation, $\vec{L}$ and $\vec{S}$ precess independently about the z-axis, such that $M = \sum_{i=1}^{Z} m_i$ and $M_S = \sum_{i=1}^{Z} m_{si}$ are conserved, as shown before in Figure 8.4 for one-electron atoms. Since the magnitudes $L(L+1)\hbar^2$ and $S(S+1)\hbar^2$ are well defined for a given energy level, this will be completely specified by both the electron configuration and the *LS* values. The spectroscopic description of an energy level of definite *L* and *S* is $^{2S+1}(L$ symbol), corresponding to a multiplet of $(2L + 1)(2S + 1)$ orbitals which are degenerate. The *L* symbols are $S(L = 0)$, $P(L = 1)$, $D(L = 2)$, $F(L = 3)$ and the superscript indicates the spin multiplicity which defines *singlet* ($S = 0$), *doublet* ($S = \frac{1}{2}$), *triplet* ($S = 1$) or *quadruplet* ($S = \frac{3}{2}$) spin states. Experiment shows that orbitals with different *L* and *S* which belong to the same configuration are split by fractions of an electron volt, and this is one order of magnitude less than the separation of a few electron volts between different electron configurations.

The spectral terms *LS*, corresponding to equivalent electrons of a configuration, are simply determined from the total magnetic quantum numbers M and M_S. In the case of a two electron configuration, for example, we have $M = m_1 + m_2$ and $M_S = m_{s1} + m_{s2}$. For an s^2 configuration (such as the ground state of helium $1s^2$) it follows that $M = 0$, and hence $L = 0$, and $M_S = 0$ because $m_{s1} \neq m_{s2}$ due to the exclusion principle, which implies $S = 0$. This means that the s^2 configuration is always a singlet spin state ^1S, as described by Eq.(10.39).

For an *ss* configuration (like the first excited state of helium $1s^1 2s^1$) we have again $m_1 = m_2 = 0$ and $m_{s1}, m_{s2} = \pm \frac{1}{2}$, but the Pauli exclusion principle no longer restricts the possible combinations between the quantum numbers, which leads to $M = 0$ and $M_S = 1, 0, 0, -1$. Hence, we have $L = 0$ and $S = 0$ ($M_S = 0$), $S = 1$ ($M_S = 1, 0, -1$), which gives a splitting of the energy level into a singlet ^1S term and a triplet spin state ^3S described by Eq.(10.55).

For an *sp* configuration, where $m_1 = 0, m_2 = 1, 0, -1$ and $m_{s1}, m_{s2} = \pm \frac{1}{2}$ there are twelve possible orbitals and no restrictions due to the exclusion principle. We similarly obtain $M = 1, 0, -1$ and $M_S = 1, 0, 0, -1$ and therefore $L = 1$ and $S = 0, 1$. It follows that the energy level is split into a singlet term ^1P (3-fold degenerate) and a triplet term ^3P (9-fold degenerate).

The calculation is more complex for a p^2 configuration, where $m_1, m_2 = 1, 0, -1$ and $m_{s1}, m_{s2} = \pm \frac{1}{2}$. Since the equivalent orbitals have to be specified by different m or/and m_s, the 36 possible combinations of the quantum numbers are restricted to a number of 15 different states of the configuration. We find the cases: $M_L = \pm 2$, $M_S = 0$ which indicate a 5-fold degenerate singlet term with $L = 2$, $S = 0$ denoted by ^1D; then $M_L = \pm 1$, $M_S = \pm 1$ which lead to $L = 1$ and $S = 1$ corresponding the triplet state ^3P (9-fold degenerate); and finally $M_L = 0, M_S = 0$ which define the singlet state ^1S. The separation of the *LS* terms is indicated in Figure 10.4 (a). The order of terms of different multiplicity is obtained by solving the Hartree-Fock equations (10.72), but it was established earlier by the two empirical *Hund rules*:

(i) The lowest energy in a given configuration corresponds to the *LS* term with the largest *S* value.

(ii) For a given *S*, the *LS* term with the largest *L* value has the lowest energy.

Both rules account for increasing the average electronic distance, so reducing the average repulsive interaction of the electron, and hence the energy eigenvalue, either by increasing S or by increasing L for a given S. As seen in Figure 10.4 (a), the order of the terms in increasing energy is ^3P, ^1D and ^1S in a p^2 configuration.

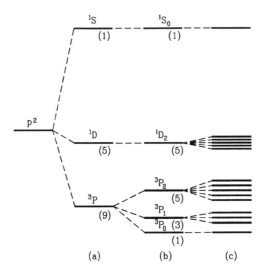

Figure 10.4. Splitting of the p^2 configuration in LS coupling by (a) electrostatic interaction, (b) spin-orbit interaction and (c) Zeeman effect

If we consider the presence of spin-orbit interactions, introduced in Example 8.5 for one-electron atoms, the Hamiltonian of the multielectron atom (10.57) becomes:

$$\hat{H} = \sum_{i=1}^{Z}\left(-\frac{\hbar^2}{2m_e}\nabla_i^2 - \frac{Ze_0^2}{r_i} + \sum_{j>i}\frac{e_0^2}{|\vec{r}_i - \vec{r}_j|}\right) + \sum_{i=1}^{Z}\alpha(r_i)\hat{\vec{L}}_i \cdot \hat{\vec{S}}_i \qquad (10.77)$$

If the last term can be treated as a small perturbation, which is the situation found for most atoms, the interaction is called *normal* or *LS coupling*.

In a vector model representation, given in Figure 10.5 for the two-electron system, although the $\vec{L}_i \cdot \vec{S}_i$ interaction results in $\vec{L}_i$ and $\vec{S}_i$ precessing about their resultant, this precession is much slower than that of the $\vec{L}_i$ about $\vec{L}$ and that of the $\vec{S}_i$ about $\vec{S}$, determined by the stronger electrostatic interactions. Consequently, only the components of the $\vec{L}_i$ parallel to $\vec{L}$ and those of the $\vec{S}_i$ parallel to $\vec{S}$ are relevant, the other components averaging out. It follows that we may replace the independent $\vec{L}_i$ and $\vec{S}_i$ angular momenta by their average values along the $\vec{L}$ and $\vec{S}$ directions:

$$\vec{L}_i \to \left(\frac{\vec{L}_i \cdot \vec{L}}{L^2}\right)\vec{L}, \qquad \vec{S}_i \to \left(\frac{\vec{S}_i \cdot \vec{S}}{S^2}\right)\vec{S}$$

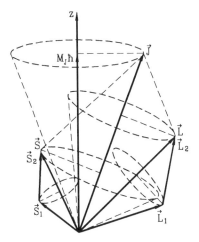

Figure 10.5. Vector representation of *LS* coupling

Hence, the spin-orbit energy in Eq.(10.77) becomes:

$$E_{SL} = \langle \sum_{i=1}^{z} \alpha(r_i) \hat{\vec{L}}_i \cdot \hat{\vec{S}}_i \rangle = \left\langle \left[\sum_{i=1}^{z} \alpha(r_i) \left(\frac{\hat{\vec{L}}_i \cdot \hat{\vec{L}}}{\hat{L}^2} \right) \left(\frac{\hat{\vec{S}}_i \cdot \hat{\vec{S}}}{\hat{S}^2} \right) \right] \hat{\vec{L}} \cdot \hat{\vec{S}} \right\rangle = \alpha(S,L) \langle \hat{\vec{L}} \cdot \hat{\vec{S}} \rangle$$

$$= \frac{1}{2}\alpha(L,S) \langle \hat{J}^2 - \hat{L}^2 - \hat{S}^2 \rangle = \frac{1}{2}\zeta(L,S)\left[J(J+1) - L(L+1) - S(S+1) \right] \quad (10.78)$$

It is clear that $\hat{\vec{L}} \cdot \hat{\vec{S}}$ commutes with $\hat{L}^2$ and $\hat{S}^2$, such that L and S still characterize the energy levels, but M and M_S vary in time as a result of precession, as seen in Figure 10.5. The constant vector is now the total angular momentum $\hat{\vec{J}} = \hat{\vec{L}} + \hat{\vec{S}}$, of magnitude $J(J+1)\hbar^2$, where the possible values of J are as given by Eq.(8.74):

$$J = L+S, L+S-1, \ldots, |L-S| \quad (10.79)$$

Hence, in the presence of the spin-orbit interaction, the energy levels are completely specified by the *LSJ* values, in the form $^{2S+1}(L \text{ symbol})_J$. The $(2L+1)(2S+1)$-fold degenerate *LS* multiplets are split into levels with defined *J*, given by Eq.(10.79), which are known as the *fine structure* of the energy spectrum. The separation between the fine structure levels is of the order of hundredths of an electron volt. Figure 10.4 (b) illustrates the fine structure of a p^2 configuration, where the spin-orbit interaction breaks the 9-fold degenerate multiplet ^3P with $L = 1$, $S = 1$ and produces the $^3P_0, ^3P_1$ and $^3P_2$ levels corresponding to the possible values $J = 0, 1, 2$ inserted into Eq.(10.78). The order of these levels is established by a third empirical rule, derived from spin-orbit calculations:

(iii) If the configuration is no more than half filled, then the lowest energy corresponds to the smallest J value, $J = |L - S|$. If the configuration is more than half filled then the lowest energy corresponds to the maximum value $J = L + S$.

In a further approximation we may consider the atomic magnetic moment of the atom, associated to the total angular momentum as discussed in Section 8.5 for one-electron atoms:

$$\hat{\vec{\mu}} = -\frac{\mu_B}{\hbar} \sum_{i=1}^{Z} (\hat{\vec{L}}_i + 2\hat{\vec{S}}_i) = -\frac{\mu_B}{\hbar} (\hat{\vec{L}} + 2\hat{\vec{S}}) = -\frac{\mu_B}{\hbar} (\hat{\vec{J}} + \hat{\vec{S}}) \qquad (10.80)$$

Since $\vec{S}$ precesses about the constant vector $\vec{J}$, as shown in Figure 10.5, one may replace $\vec{S}$ by its average value $\vec{S} \to [(\vec{S} \cdot \vec{J})/J^2]\vec{J}$ which, on substitution into Eq.(10.80), gives:

$$\hat{\vec{\mu}} = -\frac{\mu_B}{\hbar}\left(1 + \left\langle \frac{\hat{\vec{S}} \cdot \hat{\vec{J}}}{\hat{J}^2} \right\rangle\right)\hat{\vec{J}} = -\frac{\mu_B}{\hbar} g \hat{\vec{J}} = -\gamma \hat{\vec{J}} \qquad (10.81)$$

where γ is called the gyromagnetic ratio and the Landé g-factor for LS coupling is defined in the form:

$$g = 1 + \left\langle \frac{\hat{\vec{S}} \cdot \hat{\vec{J}}}{\hat{J}^2} \right\rangle = 1 + \frac{J(J+1) + S(S+1) - L(L+1)}{2J(J+1)} \qquad (10.82)$$

This implies, as discussed for one-electron atoms, Eq.(8.100), that in the presence of a weak magnetic field B, each LSJ multiplet of the fine structure splits into $2J + 1$ magnetic sublevels, with a constant relative separation $\mu_B B g$. The Zeeman magnetic splitting completely removes the degeneracy, as illustrated for the p^2 configuration in Figure 10.4 (c).

EXAMPLE 10.6. The jj coupling interaction

As shown in the case of one-electron atoms, the internal magnetic field experienced by the electron, Eq.(8.88), and hence the magnitude of the spin-orbit interaction, Eq.(8.93), linearly depends on the atomic number Z. This implies, for atoms with large Z values, that the spin-orbit interaction of each electron might dominate over its electrostatic interaction with the other electrons. In this case, the Hamiltonian (10.77) should be written as:

$$\hat{H} = \sum_{i=1}^{Z}\left(-\frac{\hbar^2}{2m_e}\nabla_i^2 - \frac{Ze_0^2}{r_i} + \alpha(r_i)\hat{\vec{L}}_i \cdot \hat{\vec{S}}_i\right) + \sum_{i=1}^{Z}\sum_{j>i} \frac{e_0^2}{|\vec{r}_i - \vec{r}_j|} \qquad (10.83)$$

If the last term can be treated as a perturbation, the interaction is called *jj coupling* and has the vector model representation plotted in Figure 10.6 (a). As discussed before in Example 8.4, the good quantum numbers which describe the electron orbitals are n_i, l_i, j_i and m_{ji} and the constants of motion for the atoms are:

$$\hat{J} = \sum \hat{J}_i, \qquad \hat{J}_z = \sum \hat{J}_{iz} \qquad (10.84)$$

The spin-orbit energy is obtained from Eq.(10.83) by summing up the single electron contributions of the form (8.92), which gives:

$$E_{SL} = \sum_{i=1}^{Z} \frac{1}{2} \zeta_i \left[j_i(j_i+1) - l_i(l_i+1) - s_i(s_i+1) \right] \qquad (10.85)$$

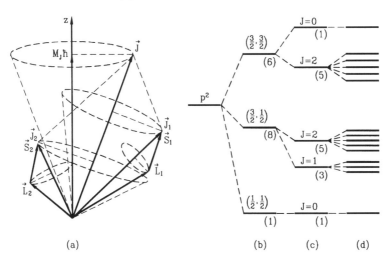

Figure 10.6. Vector representation of *jj* coupling (a) and the corresponding splitting of the p^2 configuration by (b) spin-orbit interaction, (c) electrostatic interaction and (d) Zeeman effect

For the two-electron configuration p^2, where $j_1, j_2 = \frac{3}{2}, \frac{1}{2}$, we obtain from Eq.(10.85), by taking $\zeta_1 = \zeta_2 = \zeta$ for equivalent electrons, that:

$$E_{SL} = \zeta \qquad \text{for} \qquad j_1 = \tfrac{3}{2}, \ j_2 = \tfrac{3}{2}$$

$$= -\tfrac{1}{2}\zeta \qquad j_1 = \tfrac{3}{2}, \ j_2 = \tfrac{1}{2}$$

$$= -2\zeta \qquad j_1 = \tfrac{1}{2}, \ j_2 = \tfrac{1}{2}$$

which leads to the energy levels $(j_1 j_2)$ illustrated in Figure 10.5 (b). According to the exclusion principle, the configuration $(n 1\tfrac{3}{2} m_{j1})(n 1\tfrac{3}{2} m_{j2})$ is only allowed with different m_j values,

394 Quantum Physics

such that $M_j = m_{j1} + m_{j2}$ can only assume the values $2, 1, 0, 0, -1, -2$, which means that $J = 0, 2$. Similarly we obtain the possible values $J = 1, 2$ for the configuration $(n1\frac{3}{2}m_{j1})(n1\frac{1}{2}m_{j2})$ and $J = 0$ for $(n1\frac{1}{2}m_{j1})(n1\frac{1}{2}m_{j2})$. As a result, the electrostatic perturbation breaks the spin-orbit multiplets into the fine structure levels, with defined J values, which are plotted in Figure 10.6 (c).

The Zeeman splitting of the fine structure levels is determined by the atomic magnetic moment, which for jj coupling reads:

$$\hat{\vec{\mu}} = -\frac{\mu_B}{\hbar}\sum_{i=1}^{Z}(\vec{L}_i + 2\vec{S}_i) = -\frac{\mu_B}{\hbar}\sum_{i=1}^{Z}(\hat{\vec{J}}_i + \hat{\vec{S}}_i) \tag{10.86}$$

In view of the precession of $\vec{S}_i$ about $\vec{J}_i$, we may use the results obtained for one-electron atoms:

$$\hat{\vec{\mu}} = -\frac{\mu_B}{\hbar}\sum_{i=1}^{Z}g_i\hat{\vec{J}}_i = -\frac{\mu_B}{\hbar}\sum_{i=1}^{Z}g_i\left\langle\frac{\hat{\vec{J}}_i\cdot\hat{\vec{J}}}{\hat{J}^2}\right\rangle\hat{\vec{J}} = -\frac{\mu_B}{\hbar}g\hat{\vec{J}} = -\gamma\hat{\vec{J}} \tag{10.87}$$

where, using Eq.(8.101), we have:

$$g = \sum_{i=1}^{Z}\left(1\pm\frac{1}{2l_i+1}\right)\left\langle\frac{\hat{\vec{J}}_i\cdot\hat{\vec{J}}}{\hat{J}^2}\right\rangle \tag{10.88}$$

Each level of the fine structure is again split by the Zeeman interaction into $2J+1$ magnetic levels (Figure 10.6 (d)), although the Landé g-factor for jj coupling, defined by Eq.(10.88), is different from that given by Eq.(10.82) for LS coupling. ♦

Problem 10.3.1. Find the fine structure energy levels for a d^2 electron configuration.

(Solution): The two d equivalent electrons have $n_1 = n_2 = n$, $l_1 = l_2 = 2$ such that in LS coupling we obtain:

$$L = l_1 + l_2,\ldots,|l_1 - l_2| = 4, 3, 2, 1, 0$$

$$S = s_1 + s_2,\ldots,|s_1 - s_2| = 1, 0$$

The triplet state $S = 1$, where $M_S = 1, 0, -1$, involves orbitals which have equal quantum numbers $m_{s1} = m_{s2} = \frac{1}{2}$, giving $M_S = 1$, and $m_{s1} = m_{s2} = -\frac{1}{2}$ corresponding to $M_S = -1$. Hence, the Pauli exclusion principle requires that $m_1 \neq m_2$, and this excludes a triplet state for $L = 4, 2$ and 0 values, because for

$L = 0$ we have $M = 0$ and then $m_1 = m_1 = 0$, $L = 2$ means that $M = 2, 1, 0, -1, -2$, where the values 2, 0 and -2 imply equal magnetic quantum numbers $m_1 = m_2 = 1$, $m_1 = m_2 = 0$ and $m_1 = m_2 = -1$ respectively, and similar considerations apply to $L = 4$. It follows that the only possible triplet spin states are:

$$^3P(S = 1, L = 1), \quad ^3F(S = 1, L = 3)$$

As discussed in the case of helium, the triplet states are symmetric in the spin variables, Eq.(10.55), and hence their spatial part is antisymmetric, such that the wave functions of the ^3P and ^3F states have the form:

$$\Phi_{LS} = \frac{1}{\sqrt{2}} \begin{vmatrix} \psi_{nl_1}(\vec{r}_1) & \psi_{nl_1}(\vec{r}_2) \\ \psi_{nl_2}(\vec{r}_1) & \psi_{nl_2}(\vec{r}_2) \end{vmatrix} \chi_{M_S}^T$$

The spin-orbit interaction breaks the triplet LS levels, according to the possible J values given by Eq.(10.79) into the fine structure energy levels $^3P_0, ^3P_1, ^3P_2$ and $^3F_2, ^3F_3$ and $^3F_4$ respectively. There are no restrictions on L for the singlet spin state $S = 0$, so that we obtain the LS terms:

$$^1S(S = 0, L = 0), \quad ^1D(S = 0, L = 2), \quad ^1G(S = 0, L = 4)$$

These levels are described by wave functions which are the product of a symmetric spatial part and an antisymmetric spin function, Eq.(10.39), of the form:

$$\Phi'_{LS} = \frac{1}{2}\left[\psi_{nl_1}(\vec{r}_1)\psi_{nl_2}(\vec{r}_2) + \psi_{nl_1}(\vec{r}_2)\psi_{nl_2}(\vec{r}_1)\right] \begin{vmatrix} \chi_+^{(1)} & \chi_+^{(2)} \\ \chi_-^{(1)} & \chi_-^{(2)} \end{vmatrix}$$

In the presence of the spin-orbit interaction, the singlet LS levels are relabelled as the $^1S_0, ^1D_2$ and $^1G_2$ fine structure terms.

Problem 10.3.2. For an atom with p^3 electron configuration find how many Zeeman levels it has, and obtain expressions for their separation, if placed in a weak magnetic field B.

(Solution): The LS terms in a p^3 configuration might have $L = 3, 2, 1, 0$ and $S = \frac{3}{2}, \frac{1}{2}$. The exclusion principle eliminates the $L = 3$ state, which implies a symmetric spatial part of the wave function and hence a completely antisymmetric spin function, which is impossible to form with more than two spins $\frac{1}{2}$. The allowed combinations are $L = 2, S = \frac{1}{2}$ or ^2D, which involve $(2L + 1)(2S + 1) = 10$ states; $L = 1, S = \frac{1}{2}$ or ^2P (6 states); and $L = 0, S = \frac{3}{2}$ or ^4S (4 states).

This gives a total of 20 orbitals. The spin-orbit interaction splits these terms into fine structure multiplets, according to the possible J values (10.79), each having $2J+1$ components, namely: $^2D_{\frac{5}{2}}$ (6 components), $^2D_{\frac{3}{2}}(4)$, $^2P_{\frac{3}{2}}(4)$, $^2P_{\frac{1}{2}}(2)$ and $^4S_{\frac{3}{2}}(4)$. The Zeeman effect completely removes the degeneracy of the fine structure terms. The magnetic separation $\mu_B B g$ is a function of the Landé g-factor, and then depends on L, S and J, as given by Eq.(10.82). We thus have:

Fine structure levels:	$^2D_{\frac{5}{2}}$	$^2D_{\frac{3}{2}}$	$^2P_{\frac{3}{2}}$	$^2P_{\frac{1}{2}}$	$^4S_{\frac{3}{2}}$
Number of Zeeman levels:	6	4	4	2	4
Energy spacing:	$\frac{6}{5}\mu_B B$	$\frac{4}{5}\mu_B B$	$\frac{4}{3}\mu_B B$	$\frac{2}{3}\mu_B B$	$2\mu_B B$

Problem 10.3.3. Use the vector model representation of jj coupling to determine the maximum and minimum values of the angle between two observables $\vec{j}_1$ and $\vec{j}_2$, in terms of their quantum numbers j_1 and j_2.

(Answer): $\cos(\vec{j}_1 \cdot \vec{j}_2)_{max} = \sqrt{\dfrac{j_1 j_2}{(j_1+1)(j_2+1)}}$, $\cos(\vec{j}_1 \cdot \vec{j}_2)_{min} = -\sqrt{\dfrac{(j_1+1)j_2}{j_1(j_2+1)}}$

Problem 10.3.4. Find the components along the direction of a weak magnetic field of the atomic magnetic moment in the LSJ state $^3F_3$.

(Answer): $\mu_z = 0, \pm\dfrac{13}{12}\mu_B, \pm\dfrac{13}{6}\mu_B, \pm\dfrac{13}{4}\mu_B.$

Problem 10.3.5. Find the LS terms for a f^2 electron configuration.

(Answer): $^3P, \ ^3F, \ ^3H, \ ^1S, \ ^1D, \ ^1I.$

10.4. The Shell Model of the Nucleus

Using an extension of our picture of the atom, the atomic nucleus can be treated as a many-particle system which contains almost all the mass of the atom and a charge of $+Ze$. Nuclear masses are measured in terms of the *atomic mass unit*, u, which is close to the rest mass m_H of a hydrogen atom in its ground state. The *mass number* A of a nucleus is the integer nearest to the ratio between the nuclear mass and the atomic mass unit. Except for hydrogen, A is always greater than the *atomic number* Z. The nucleus is a system of Z *protons*, of mass m_p and charge $+e$, and $N = A - Z$ *neutrons*, of mass m_n almost equal to that of a proton and zero charge. Hence, the mass of the nucleus is given, to a first approximation, by the sum of masses of the *nucleons* (protons and neutrons) and its charge is equal to the total charge of the protons. A nuclear species, or *nuclide*, is therefore specified by the symbol ${}^A_Z X$, where X is the chemical symbol, or simply by ${}^A X$ since, given X, the atomic number Z is known and we can always find N from $A - Z$. Nuclides of the same Z but different N are called *isotopes*, those of the same N and different Z are known as *isotones* and those of the same mass number A, but different Z and N, are said to be *isobars*.

Since the nucleons are held together by strong attractive forces, it is convenient to introduce the binding energy $B(A,Z)$ of a nucleus, which represents the energy that is released in bringing the nucleons together or, conversely, the work necessary to dissociate the nucleus into separate nucleons. It is common practice to define the nuclear binding energy according to the energy-inertia relation of special relativity, in terms of atomic rather than nuclear masses as:

$$B(A,Z) = [Zm_H + (A - Z)m_n - M(A,Z)]c^2 = \Delta M c^2 \quad (10.89)$$

where $M(A,Z)$ is the atomic mass of the isotope and ΔM is the so-called mass defect. If the state of infinite separation of the nucleons is taken to be the zero level of energy, the total energy of the ground state of a nucleus is $-B(A,Z)$. Nuclear energies are conveniently measured in MeV (1 MeV = 10^6 eV), where the electron-volt is the common atomic unit of energy. Equation (10.89) gives the energy equivalent of an atomic mass unit as $1u = 931.44$ MeV, so that one can work either with masses or energies as convenient. The binding energy per nucleon:

$$f = \frac{B(A,Z)}{A} \quad (10.90)$$

is often called the *binding fraction* and its plot as a function of the mass number, given in Figure 10.7 (a), is usually used for a systematic study of nuclear binding energy. It is seen that the binding fraction of most nuclei is about 8 MeV and is approximately independent of A, except for the lightest nuclei. This implies that a nucleon in a large nucleus is not bound to more nucleons than in a small one, in other words there is a saturation of the interaction between nucleons. Hence, the nuclear forces have a range which is of the order of the diameter of one nucleon. Saturation indicates effects which keep nucleons apart from each other, which can be understood if, in addition to mass and charge, we consider the *spin angular momentum* of the nucleons. It is found experimentally that protons and neutrons have an intrinsic angular momentum $\frac{1}{2}\hbar$, like electrons, so that the Pauli exclusion principle forbids two nucleons of the same kind to occupy states with identical quantum numbers, and this leads to the saturation of nuclear forces. The nuclei are most tightly bound near $A \approx 60$ (Fe, Ni), where f is a maximum. In light nuclei, a single nucleon is attracted by a few other nucleons, thus the nucleon separation is large and the stability is reduced. In heavy nuclei the decrease of stability is due to the Coulomb repulsion between protons, which becomes important for large Z, and hence large A.

Under the assumption that nuclear forces between neutrons are identical to nuclear forces between protons, the energy states of neutrons and protons must be identical. However, the Coulomb interaction between the protons results in raising the bottom of the proton well by $eV(r)$, where $V(r)$ is the electrostatic potential inside the nucleus, and this gives rise to a potential energy barrier for protons at the nuclear boundary. There is no Coulomb barrier for neutrons, as illustrated in Figure 10.7 (b), which shows a simplified energy level diagram for protons and neutrons, consistent with the Pauli exclusion principle and modified to account for the electrostatic interaction.

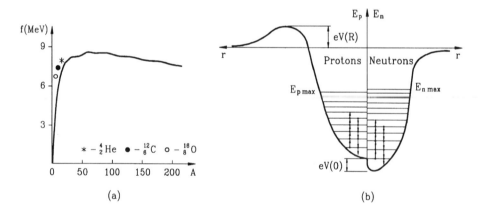

Figure 10.7. (a) Binding fraction as a function of nuclear mass number (b) Schematic energy level diagram for nucleons

All the evidence indicates that there are periodicities in the nuclear binding energies which might be due to a shell structure similar to the atomic shell structure. For neutrons, closed shells have been associated with the *magic numbers* $N = 2, 8, 20, 28, 50, 82$ and 126 and, for protons, with $Z = 2, 8, 20, 50$ and 82, because the nuclei with one or both of the magic numbers show a particular stability, when compared with other nuclei. The magic numbers can be derived using the *single-particle shell model* where the many-nucleon problem is reduced to a single-particle problem, under the assumption that, despite the strong overall attraction between nucleons, which provides the nuclear binding energy, each nucleon moves in a nuclear potential due to all the other nucleons. In other words, the short-range nuclear forces should average out to produce the nuclear potential, so that all couplings between nucleons can be neglected. If the nuclear potential energy is represented, in the simplest case, by a spherically symmetric function $U(r)$, which is only dependent on r, the situation becomes similar to the one-electron atom problem, where the electron moving in the effective potential energy (7.44) results in a series of electron shells, with stable closed configurations.

The nuclear states for single nucleons are thus obtained by solving the Schrödinger energy equation which, as described in Chapter 7, can be separated for any shape of the central force potential $U(r)$, into an angular equation, having the spherical harmonics as solutions, and a radial equation. If an orbital angular momentum $\hat{L}$, specified by a quantum number l according to the quantum condition (7.19), is assigned to the nucleon, the radial equation takes the form (7.43):

$$\frac{d^2\xi(r)}{dr^2} + \frac{2m}{\hbar^2}\left[E - U(r) - \frac{l(l+1)\hbar^2}{2mr^2}\right]\xi(r) = 0 \qquad (10.91)$$

where $\xi(r) = rR(r)$ and m stands for m_p or m_n. Since the nuclear forces are short-range, the nucleons experience an average potential which is nearly constant inside the nucleus and falls sharply to zero near the nuclear surface. The simplest forms of $U(r)$ which simulate this behaviour (see Figure 10.8) are the harmonic oscillator potential:

$$U(r) = -V_0 + \frac{1}{2}m\omega^2 r^2 \qquad (10.92)$$

and the square well potential:

$$U(r) = -V_0, \qquad r \leq R$$
$$= 0, \qquad r > R \qquad (10.93)$$

Substituting Eq.(10.92), the radial equation takes the form:

$$\frac{d^2\xi(r)}{dr^2} + \left\{\frac{2m}{\hbar}\left[E + V_0 - \frac{m\omega^2}{2}r^2\right] - \frac{l(l+1)}{r^2}\right\}\xi(r) = 0 \qquad (10.94)$$

For $r \to 0$, Eq.(10.94) can be approximated by:

$$\frac{d^2\xi(r)}{dr^2} - \frac{l(l+1)}{r^2}\xi(r) = 0 \quad \text{or} \quad \xi(r) = Nr^{l+1} \qquad (10.95)$$

and for $r \to \infty$, it reduces to:

$$\frac{d^2\xi(r)}{dr^2} + \frac{2m}{\hbar^2}\left(E + V_0 - \frac{m\omega^2}{2}r^2\right)\xi(r) = 0$$

which is similar to Eq.(6.62) for a quantum oscillator. Hence, the asymptotically valid solutions must take the form (6.65), which reads:

$$\xi(r) \sim e^{-sr^2/2} \qquad (10.96)$$

where $s = m\omega/\hbar$. A trial solution, valid for all r, should include the restrictions near the origin and in the asymptotic region, Eqs.(10.95) and (10.96), such that it can be chosen in the form:

$$\xi(r) = r^{l+1} e^{-sr^2/2} \eta(r) \qquad (10.97)$$

On substitution into Eq.(10.94), and multiplication by $\eta(r)/\xi(r)$, one obtains:

$$\frac{d^2\eta(r)}{dr^2} + 2\frac{d\eta(r)}{dr}\left(\frac{l+1}{r} - sr\right) - s\eta(r)\left[2l + 3 - \frac{2}{\hbar\omega}(E+V_0)\right] = 0$$

or:

$$\frac{d^2\eta(r)}{dr^2} + \frac{d\eta(r)}{dr}\left(\frac{2q-1}{r} - 2sr\right) - 4ps\eta(r) = 0 \qquad (10.98)$$

where $2l + 3 = 2q$ and $2l + 3 - \frac{2}{\hbar\omega}(E+V_0) = 4p$. We look for $\eta(r)$ in the form of a power series:

$$\eta(r) = \sum_k a_k r^k$$

which, on insertion into Eq.(10.98), yields the recurrence relation:

$$(k+2)(k+1)a_{k+2} + (2q-1)(k+2)a_{k+2} - 2ska_k - 4psa_k = 0$$

and this reads:

$$a_{k+2} = \frac{2s(k+2p)}{(k+2q)(k+2)} a_k \qquad (10.99)$$

It follows that the power series assumes the form of the confluent hypergeometric function (7.73), namely:

$$\eta(r) = F(p,q;sr^2) = 1 + \frac{p}{q}\frac{sr^2}{1!} + \frac{p(p+1)}{q(q+1)}\frac{(sr^2)^2}{2!} + \frac{p(p+1)(p+2)}{q(q+1)(q+2)}\frac{(sr^2)^3}{3!} + \ldots$$

which must terminate after a finite number of terms, such that the asymptotic behaviour is satisfied. This means that p must be equated to a negative integer, say $-i$:

$$2l + 3 - \frac{2}{\hbar \omega}(E + V_0) = -4i$$

We obtain the energy eigenvalues:

$$E_n = -V_0 + \hbar \omega \left(2i + l + \frac{3}{2}\right) = -V_0 + \hbar \omega \left(n + \frac{3}{2}\right) \tag{10.100}$$

where $n = 2i + l$ is a *quantum number* and hence, if n is even, $l = n, n-2, \cdots, 0$ and if n is odd, $l = n, n-2, \cdots, 1$. The energy levels (10.100) are thus equally spaced. The ground state $(n = 0)$:

$$E_0 = -V_0 + \frac{3}{2} \hbar \omega \tag{10.101}$$

is not degenerate, since it can only be formed with $i = 0$, $l = 0$. Because there are $2l + 1$ values of the magnetic quantum number in a given l state, the higher energy levels are degenerate. If we include spin, the total number of protons or neutrons in a given l state is $2(2l + 1)$, so that the degeneracy of the level n is $(n + 1)(n + 2)$.

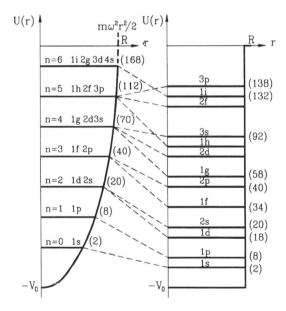

Figure 10.8. Energy levels for protons or neutrons in an oscillator potential and in an infinite square well potential

A nucleon state is normally specified using the appropriate spectroscopic symbol $s, p, d, f, g, \ldots$ for its l value and a number in front of the letter, which is 1 if the symbol following appears for the first time, 2 if it appears for the second time, and so on. Hence, the sequence of energy levels from the bottom of the parabolic well, given in Figure 10.8, is: $1s$ ($n = 0$; $l = 0$); $1p$ ($n = 1$; $l = 1$); $1d$, $2s$ ($n = 2$; $l = 2, 0$); $1f$, $2p$ ($n = 3$; $l = 3,1$); $1g$, $2d$, $3s$ ($n = 4$; $l = 4, 2, 0$); $1h$, $2f$, $3p$ ($n = 5$; $l = 5, 3, 1$); $1i$, $2g$, $3d$, $4s$ ($n = 6$; $l = 6, 4, 2, 0$). Note that the number in front of the spectroscopic symbol is not the quantum number n, as in the designation of atomic spectral lines. It is seen from Figure 10.8 that the oscillator potential function (10.92) leads to the magic numbers 2, 8, 20, 40, 70, 112, 168 which are not identical with the nuclear magic numbers mentioned earlier. Use of the square well potential (10.93) changes the level spacing and slightly changes the order of levels, as indicated in Figure 10.8, predicting the shell closure at 2, 8, 18, 20, 34, 40, 58, 92, 132, 138 which still does not correspond to the experimental evidence.

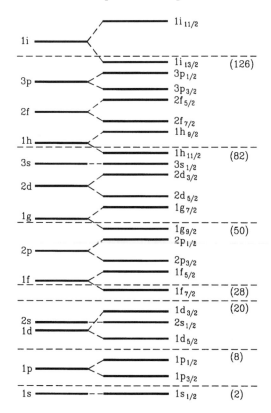

Figure 10.9. Nucleon levels in the single particle shell model

A more realistic scheme takes into account the possible splitting of each energy level into two, according to whether the orbital and spin angular momenta $\hat{\vec{L}}$ and $\hat{\vec{S}}$ of the nucleon in the level are in the same or opposite directions. Such an additional energy could arise from a spin-orbit coupling of the unpaired nucleon, which is postulated to be analogous to that introduced for electrons in an atom, Eq.(8.90):

$$-\frac{\lambda}{2m^2c^2}\frac{1}{r}\frac{dU(r)}{dr}\hat{S}\cdot\hat{L} \tag{10.102}$$

where m stands for m_p or m_n, and $U(r)$ is the nuclear potential energy. We notice that, since an interaction energy of the form used in Eq.(8.90) is too small and also of the wrong sign to give the expected splitting, the sign was changed and a dimensionless constant λ was introduced in Eq.(10.102), to fit the data. This means that the origin of the spin-orbit coupling of nucleons in the nucleus must be different from the electromagnetic interaction discussed in Section 8.4. With the spin-orbit coupling we obtain a splitting of the energy levels E_n, similar to that derived in Eq.(8.94), namely:

$$E_{nl} = E_n - \frac{1}{2}l\zeta \quad (j = l+\tfrac{1}{2}), \qquad E_{nl} = E_n + \frac{1}{2}(l+1)\zeta \quad (j = l-\tfrac{1}{2}) \tag{10.103}$$

where E_n are the energy eigenvalues in the absence of spin-orbit interaction and ζ is a parameter having the dimensions of energy:

$$\zeta = \frac{\lambda\hbar^2}{2m^2c^2}\frac{1}{r}\frac{dU(r)}{dr} \tag{10.104}$$

The level splitting (10.103) is such that the level with total angular momentum $j = l+\tfrac{1}{2}$ lies below the level with $j = l-\tfrac{1}{2}$, which is the opposite of what happens for electrons, but agrees with experimental evidence. The resulting states are specified by adding the half-integral values j as subscripts of the spectroscopic symbols. The correct magic numbers for large l can be obtained if we count the levels with largest l and $j = l+\tfrac{1}{2}$ together with the group of lower energy. The procedure is illustrated in Figure 10.9, starting from the energy levels in a square well potential, shown in Figure 10.8.

EXAMPLE 10.7. Magnetic hyperfine interactions

If each nucleon is assigned a total angular momentum, which represents the sum of the orbital and intrinsic angular momenta, and is subject to the same rules of quantization as the total angular momentum of the single electron states, Eqs.(8.23) and (8.24), the nuclear shell model implies that, because of the Pauli exclusion principle, like nucleons combine in pairs in such a way that their angular momenta cancel. This is consistent with the experimental data on the total angular momentum of nuclei, called the *nuclear spin* $\vec{I}$, which can be described as made up of the spins of the nucleons and of their orbital angular momenta, following the *LS* coupling model, as discussed in Section 10.3. We therefore postulate that:

$$|\vec{I}| = \sqrt{I(I+1)}\,\hbar$$

$$I_z = m_I\hbar, \quad m_I = -I, -I+1, \ldots, I-1, I \tag{10.105}$$

where, as established by experiment:

(i) nuclei with even A and even Z have $I = 0$;

(ii) nuclei with even A and odd Z have integral non-zero spin varying from $\hbar$ to $7\hbar$;

(iii) nuclei with odd A have half-integral spin between $\frac{1}{2}\hbar$ and $\frac{9}{2}\hbar$.

The first result is a natural consequence of the assumption that in a particular level the protons pair with antiparallel spins, as do the neutrons. The nuclear spin of even A odd Z nuclei would be due to the last proton and neutron, and because their angular momenta may combine in different ways, the most stable spin configuration can only be found by calculation. The nuclear spin of an odd A nucleus is due entirely to the angular momentum of the last unpaired nucleon $j = l \pm \frac{1}{2}$.

The nuclear spin vector $\hat{I}$ combines with the angular momentum vector $\hat{J}$ of the electrons in the outer atom to get a complete atomic angular momentum vector:

$$\hat{F} = \hat{L} + \hat{S} + \hat{I} = \hat{J} + \hat{I} \tag{10.106}$$

such that the possible values of F are given by a rule similar to Eq.(10.79), namely:

$$F = J + I, J + I - 1, \ldots, |J - I| \tag{10.107}$$

As there is always a magnetic dipole moment associated with the angular momentum of any charged particle, the nucleus has a magnetic moment which is introduced, in formal analogy with Eq.(10.81), as:

$$\hat{\mu}_I = \frac{\mu_N}{\hbar} g_N \hat{I} = \gamma_N \hat{I} \tag{10.108}$$

where μ_N is the *nuclear magneton*:

$$\mu_N = \frac{e\hbar}{2m_p} \tag{10.109}$$

The nuclear g-factor (and hence the nuclear gyromagnetic ratio γ_N) cannot yet be computed from theory, unlike the Landé g-factor, Eq.(10.82). Because the nuclear magneton is about two thousand times smaller than the Bohr magneton μ_B, Eq.(7.134), the magnetic interaction between the magnetic moments of the nucleus and the electrons is weaker than the spin-orbit coupling, which causes the fine structure of the atomic energy levels. Each of these levels will be split further by JI coupling into $2I + 1$ or $2J + 1$ closely spaced sublevels, according to wether $J \geq I$ or $J \leq I$. This is known as the *hyperfine structure* of the electron energy levels.

The magnetic hyperfine interaction can be represented, in a similar manner to the classical dipolar interaction, by the scalar product of the angular momenta $\hat{\mu}_J \cdot \hat{\mu}_I$, that is:

$$\hat{H} = a \hat{J} \cdot \hat{I} = \frac{A}{\hbar^2} \hat{J} \cdot \hat{I} \tag{10.110}$$

where A is called the *hyperfine coupling constant*. It is immediately apparent from Eq.(10.106) that:

$$\hat{J}\cdot\hat{I} = \frac{1}{2}(\hat{F}^2 - \hat{J}^2 - \hat{I}^2)$$

so that the expectation energy values of the hyperfine structure are given by:

$$E_F = \langle \hat{H} \rangle = \frac{A}{2}[F(F+1) - J(J+1) - I(I+1)] \qquad (10.111)$$

The energy spacings between successive hyperfine levels, with the same J and I but different F values, are then obtained in a form called the *interval rule*:

$$E_{F+1} - E_F = A(F+1) \qquad (10.112)$$

which allows F to be determined, so making it possible to find I. ♉

Problem 10.4.1. Determine the nuclear spin I predicted by the nuclear shell model for the following odd A nuclides: $^{7}_{3}\text{Li}$, $^{15}_{7}\text{N}$, $^{39}_{19}\text{K}$, $^{59}_{27}\text{Co}$ and $^{23}_{11}\text{Na}$.

(Solution): The level diagram in Figure 10.9, where each level $n(l \text{ symbol})_j$ may contain $2j+1$ nucleons, is valid for nucleons of one kind, and hence it is convenient to separate the total number of neutrons and protons accommodated in the scheme. All the nuclides involved have an even number of neutrons, which combine in pairs with antiparallel spins, giving a resultant zero angular momentum. In the case of $^{7}_{3}\text{Li}$, for example, the closed shell $1s_{\frac{1}{2}}$ contains two neutrons and the other two are accommodated in the $1p_{\frac{3}{2}}$ state. Hence we are concerned with the occupation numbers for protons only. The proton configuration is obtained from Figure 10.9 in the form:

$$^{7}_{3}\text{Li}: \quad (1s_{\frac{1}{2}})^2 (1p_{\frac{3}{2}})^1$$

$$^{15}_{7}\text{N}: \quad (1s_{\frac{1}{2}})^2 (1p_{\frac{3}{2}})^4 (1p_{\frac{1}{2}})^1$$

$$^{39}_{19}\text{K}: \quad (1s_{\frac{1}{2}})^2 (1p_{\frac{3}{2}})^4 (1p_{\frac{1}{2}})^2 (1d_{\frac{5}{2}})^6 (2s_{\frac{1}{2}})^2 (1d_{\frac{3}{2}})^3$$

$$^{59}_{27}\text{Co}: \quad (1s_{\frac{1}{2}})^2 (1p_{\frac{3}{2}})^4 (1p_{\frac{1}{2}})^2 (1d_{\frac{5}{2}})^6 (2s_{\frac{1}{2}})^2 (1d_{\frac{3}{2}})^4 (1f_{\frac{7}{2}})^7$$

The nuclear spin is determined by the angular momentum of the last unpaired proton, such that $I = \frac{1}{2}$ for $^{15}_{7}\text{N}$, $I = \frac{3}{2}$ for $^{7}_{3}\text{Li}$ and $^{39}_{19}\text{K}$, and $I = \frac{7}{2}$ for $^{59}_{27}\text{Co}$, as experimentally measured.

It is worth noting that the prediction based on the nuclear shell model breaks down in a few cases, for example $^{23}_{11}\text{Na}$, which has the proton configuration of:

$$^{23}_{11}\text{Na}: \; (1s_{\frac{1}{2}})^2 (1p_{\frac{3}{2}})^4 (1p_{\frac{1}{2}})^2 (1d_{\frac{5}{2}})^3$$

It would be expected to have a spin $I = \frac{5}{2}$, but the measured value is $I = \frac{3}{2}$. In this case the pairing is broken and the three protons in the $1d_{\frac{5}{2}}$ level combine their angular momenta to produce a nuclear spin of $\frac{3}{2}$.

Problem 10.4.2. Calculate the magnetic moment for odd A nuclides, assuming that it is due entirely to the magnetic moment of the last unpaired nucleon.

(Solution): By analogy with the magnetic moment for the one-electron atom (see Section 8.5) we have:

$$\hat{\vec{\mu}} = \hat{\vec{\mu}}_l + \hat{\vec{\mu}}_s = \frac{\mu_N}{\hbar}(g_l \hat{\vec{l}} + 2g_s \hat{\vec{s}})$$

where $g_l = 1$ for a proton and 0 for a neutron (which makes no contribution to the magnetic moment by virtue of its orbital motion, as it has zero electric charge). The spin g-factors were found experimentally to be $g_s = 2.79$ for the proton and $g_s = -1.91$ for the neutron. Since only the components of $\vec{l}$ and $\vec{s}$ parallel to $\vec{j}$ are relevant, the other components averaging out, we have:

$$\hat{\vec{\mu}} = \frac{\mu_N}{\hbar}\left(g_l \left\langle \frac{\hat{\vec{l}} \cdot \hat{\vec{j}}}{\hat{j}^2} \right\rangle + 2g_s \left\langle \frac{\hat{\vec{s}} \cdot \hat{\vec{j}}}{\hat{j}^2} \right\rangle\right) \hat{\vec{j}}$$

where:

$$\left\langle \frac{\hat{\vec{l}} \cdot \hat{\vec{j}}}{\hat{j}^2} \right\rangle = \frac{j(j+1) + l(l+1) - s(s+1)}{2j(j+1)}$$

$$\left\langle \frac{\hat{\vec{s}} \cdot \hat{\vec{j}}}{\hat{j}^2} \right\rangle = \frac{j(j+1) + s(s+1) - l(l+1)}{2j(j+1)}$$

Because $j = l \pm \frac{1}{2}$, we immediately obtain:

$$\hat{\mu} = \frac{\mu_N}{\hbar}\left[g_l + \frac{2g_s - g_l}{2j}\right]\hat{j}, \quad j = l + \tfrac{1}{2}$$

$$\hat{\mu} = \frac{\mu_N}{\hbar}\left[g_l - \frac{2g_s - g_l}{2(j+1)}\right]\hat{j}, \quad j = l - \tfrac{1}{2}$$

If the magnetic moment is determined by the last unpaired proton (odd A and odd Z), this gives:

$$\hat{\mu} = \frac{\mu_N}{\hbar}\left(1 + \frac{2.29}{j}\right)\hat{j} \quad (j = l+\tfrac{1}{2}), \quad \hat{\mu} = \frac{\mu_N}{\hbar}\left(1 - \frac{2.29}{j+1}\right)\hat{j} \quad (j = l-\tfrac{1}{2})$$

and for an odd number of neutrons (odd A and even Z) we obtain:

$$\hat{\mu} = \frac{\mu_N}{\hbar}\left(-\frac{1.91}{j}\right)\hat{j} \quad (j = l+\tfrac{1}{2}), \quad \hat{\mu} = \frac{\mu_N}{\hbar}\left(\frac{1.91}{j+1}\right)\hat{j} \quad (j = l-\tfrac{1}{2})$$

Problem 10.4.3. Solve the energy eigenvalue problem for the ground state of the deuteron, which consists of a proton and a neutron, assuming that the interaction is described by the square well potential energy defined in Eq.(10.93).

(Answer): $\quad E = V_0 + \dfrac{\pi^2 \hbar^2}{4 m_p R^2}$

Problem 10.4.4. Calculate the nuclear spin of the following odd A nuclides: $^{17}_{8}\text{O}$, $^{67}_{30}\text{Zn}$ and $^{207}_{82}\text{Pb}$, according to the shell model predictions.

(Answer): $\quad I = \tfrac{5}{2}, \tfrac{5}{2}$ and $\tfrac{1}{2}$ respectively.

11. ATOMIC RADIATION

11.1. Radiative Transitions

The states of the atom, determined by the motion of electrons, are stationary in the presence of all internal atomic forces and of interactions with constant external fields. When electrons interact with the field associated with electromagnetic radiation, the atomic states are no longer stationary, and the time-dependent perturbation treatment has to be applied to calculate the effect produced by the radiation, which is a function of time. Consequently, as shown in Section 9.3, the time evolution of the atom will be described in terms of transitions between the stationary states of the unperturbed atomic system.

We may use as a model, for simplicity, the one-electron atom which is described, under the influence of electromagnetic field, by a Hamiltonian of the form given in Eq.(7.140), where the electromagnetic radiation in a vacuum is completely determined by a transverse vector potential $\vec{A}$ ($\nabla \cdot \vec{A} = 0$). Using Eq.(7.141), and disregarding the term proportional to $\vec{A}^2$, because the perturbation due to electromagnetic waves is very small, the Hamiltonian of the atom in the presence of radiation reduces to:

$$\hat{H} = \frac{1}{2m}(\hat{\vec{p}} + e\vec{A})^2 + U(\vec{r}) = -\frac{\hbar^2}{2m}\nabla^2 + U(\vec{r}) - \frac{ie\hbar}{2m}\vec{A}\cdot\nabla = \hat{H}_0 - \frac{ie\hbar}{2m}\vec{A}\cdot\nabla \quad (11.1)$$

where $\hat{H}_0$ is the Hamiltonian of the system in the absence of an electromagnetic field. The perturbation operator depends on the vector potential $\vec{A}$, and hence on the electric and magnetic fields of the radiation, given by:

$$\vec{E} = -\frac{\partial \vec{A}}{\partial t}, \quad \vec{B} = \nabla \times \vec{A} \quad (11.2)$$

which implies that the classical Maxwell equations are satisfied if the vector potential is a solution of the homogenous wave equation:

$$\left(\nabla^2 - \frac{1}{c^2}\frac{\partial^2}{\partial t^2}\right)\vec{A} = 0, \quad \nabla \cdot \vec{A} = 0 \quad (11.3)$$

For monochromatic radiation, Eq.(11.3) has a simple plane wave solution:

$$\vec{A} = \vec{A}_0 e^{i(\vec{k}\cdot\vec{r}-\omega t)} + \vec{A}_0^* e^{-i(\vec{k}\cdot\vec{r}-\omega t)}, \quad \vec{A}_0 \cdot \vec{k} = 0 \tag{11.4}$$

where $\vec{A}_0$ is a constant vector and $\omega = ck$. It follows that the perturbation operator in Eq.(11.1) can simply be written in the *harmonic* form:

$$\hat{V}(\vec{r},t) = -\frac{ie\hbar}{m}\vec{A}\cdot\nabla = \hat{V}(\vec{r})e^{-i\omega t} + \hat{V}^*(\vec{r})e^{i\omega t} \tag{11.5}$$

where:

$$\hat{V}(\vec{r}) = -\frac{ie\hbar}{m} e^{i\vec{k}\cdot\vec{r}} \vec{A}_0 \cdot \nabla \tag{11.6}$$

When applied to atomic transitions, Eq.(11.6) takes a simpler form, because the electromagnetic energy of the light waves involved is of the order of the electron energies, namely $\hbar\omega = \hbar ck \approx e_0^2/r$ or $kr \approx e_0^2/\hbar c \approx 10^{-3}$. Hence, if we consider the expansion:

$$e^{i\vec{k}\cdot\vec{r}} = 1 + i\vec{k}\cdot\vec{r} - \frac{1}{2}(\vec{k}\cdot\vec{r})^2 + \ldots \tag{11.7}$$

it is a good approximation to replace the exponential in Eq.(11.6) by unity, for almost all the atomic transitions observed, which gives:

$$\hat{V}(\vec{r}) = -\frac{ie\hbar}{m}\vec{A}_0 \cdot \nabla \tag{11.8}$$

This is called the *electric dipole approximation* because the perturbation (11.8) can be reduced to the coupling energy between the atomic electric dipole moment $\vec{D} = -e\vec{r}$ and the electric field of the radiation:

$$\vec{E} = -\frac{\partial \vec{A}}{\partial t} = -\frac{\partial}{\partial t}(\vec{A}_0 e^{-i\omega t}) = i\omega \vec{A}_0 e^{-i\omega t} \tag{11.9}$$

which has a spatially constant amplitude over the atomic region of size r. This becomes immediately apparent if we recall that, for any Hamiltonian $\hat{H}_0$, the evolution equation (5.3) yields:

$$[\vec{r}, \hat{H}_0] = i\hbar \frac{d\vec{r}}{dt} = \frac{i\hbar}{m}\hat{\vec{p}} = \frac{\hbar^2}{m}\nabla$$

Hence, the matrix elements of $\hat{V}(\vec{r})$ between two stationary states of the unperturbed atomic system take the form:

$$V_{ni} = \langle \psi_n | \hat{V}(\vec{r}) | \psi_i \rangle = -\frac{ie}{\hbar} \vec{A}_0 \cdot \langle \psi_n | \frac{\hbar^2}{m} \nabla | \psi_i \rangle = -\frac{ie}{\hbar} \vec{A}_0 \cdot \langle \psi_n | \vec{r} \hat{H}_0 - \hat{H}_0 \vec{r} | \psi_i \rangle$$

$$= \frac{i}{\hbar} \vec{A}_0 \cdot (E_i - E_n) \langle \psi_n | -e\vec{r} | \psi_i \rangle = i\omega_{in} \vec{A}_0 \cdot \langle \psi_n | -e\vec{r} | \psi_i \rangle$$

$$= \langle \psi_n | \vec{D} \cdot \vec{E}_0(\omega_{in}) | \psi_i \rangle \tag{11.10}$$

where use has been made of Eq.(11.9). This result suggests that the operator (11.8) can also be written in terms of the dipole moment operator:

$$\hat{V}(\vec{r}) = \vec{D} \cdot \vec{E}(\omega) = (\vec{\varepsilon} \cdot \vec{D}) E_0(\omega) = \hat{D}(\vec{r}) E_0(\omega) \tag{11.11}$$

where $\vec{\varepsilon} = \vec{E}_0 / |\vec{E}_0|$ is a unit complex vector which defines the orientation of the radiation field. The dipole approximation accounts for all major features of the atomic transitions, which have been confirmed by experiment.

Since the V_{ni} defined by Eq.(11.10) are real, the matrix elements of the perturbation operator (11.5) involved in the time-dependent perturbation treatment, Eq.(9.59), can be written as:

$$V_{ni}(t) = \langle \psi_n | \hat{V}(\vec{r}) | \psi_i \rangle e^{-i\omega t} + \langle \psi_n | \hat{V}^*(\vec{r}) | \psi_i \rangle e^{i\omega t} = V_{ni}(e^{i\omega t} + e^{-i\omega t}) \tag{11.12}$$

It then follows, to a first approximation, as described by Eq.(9.66), that:

$$a_n(t) = -\frac{i}{\hbar} V_{ni} \int_0^t \left[e^{-i(\omega_{in}-\omega)t'} + e^{-i(\omega_{in}+\omega)t'} \right] dt' = \frac{V_{ni}}{\hbar} \left[\frac{e^{-i(\omega_{in}-\omega)t'}}{\omega_{in}-\omega} + \frac{e^{-i(\omega_{in}+\omega)t'}}{\omega_{in}+\omega} \right]_0^t$$

$$= \frac{V_{ni}}{\hbar} \left[\frac{e^{-i(\omega_{in}-\omega)t} - 1}{\omega_{in}-\omega} + \frac{e^{-i(\omega_{in}+\omega)t} - 1}{\omega_{in}+\omega} \right] \tag{11.13}$$

If the atomic system is bathed in a radiation field of frequency $\omega = \omega_{ni} = -\omega_{in}$, which defines $E_n > E_i$ from the Bohr condition (9.69), the transition from state ψ_i to state ψ_n describes a process of *induced absorption*, where the energy of the system is increased by energy transfer from the radiation field to the atom. When $\omega = \omega_{ni} = -\omega_{in}$ the second term on the right-hand side of Eq.(11.13) grows without limit, so that the first term can be neglected.

If, however, $\omega_{in} > 0$, so that $E_i > E_n$, the transition from state ψ_i to state ψ_n is accompanied by a decrease in the energy of the system. Such a process is termed *induced emission* where energy is transferred to a radiation field of frequency $\omega = \omega_{in}$. Induced emission is described by the first term in Eq.(11.13), which has the dominant contribution to $a_n(t)$ for this value of ω.

If we consider the effect of each term in Eq.(11.13) on the transition probability between states, the only change which occurs in Eq.(9.71) is a shift of $\pm \omega$ in the resonance frequency, which reads:

$$a_n^\pm(t) = \frac{V_{ni}}{\hbar(\omega_{in} \pm \omega)}[e^{-i(\omega_{in} \pm \omega)t} - 1] \tag{11.14}$$

If we proceed as shown in Example 9.6, we obtain a transition probability which is similar to that given by Eq.(9.76), namely:

$$\lim_{t \to \infty} P_{in}^\pm(t) = \frac{2\pi t}{\hbar}|V_{ni}|^2 \, \delta(E_i - E_n \pm \hbar\omega) \tag{11.15}$$

We observe that $P_{in}^+(t)$, which vanishes unless the final energy E_n is greater than the initial energy E_i (in other words $E_n = E_i + \hbar\omega$), describes the *absorption* of energy by the system, whereas $P_{in}^-(t)$ represents the probability of energy *emission*, restricted by the condition $E_i = E_n + \hbar\omega$. The transition rate is then derived from Eq.(11.15) if we divide by t and integrate over either the initial states, the final states or the radiation frequencies, which may be elements of a continuum. If we separate the contribution to the transition rate from the electric dipole moment, as suggested by Eq.(11.11), which leads to:

$$|V_{ni}|^2 = |D_{ni}|^2 \, |E_0(\omega_{in})|^2 = |D_{ni}|^2 \, \frac{I(\omega_{in})}{\varepsilon_0 c} \tag{11.16}$$

it becomes apparent that the transition rate is proportional to the classical intensity I of the radiation.

Because the electromagnetic radiation from ordinary light sources contains a range of frequency components rather than being strictly monochromatic, it is suitable to describe it in terms of a continuous superposition of harmonic waves, of finite extension defined by a field amplitude $E(\omega)$, which depends on frequency over a specified range of values, as discussed in Example 1.1. In other words, the usual physical situation is that of a transition between two discrete bound states induced by a perturbation which is continuous in ω, so corresponding to the wave packet representation of radiation, Eq.(1.7), where $k\xi = (vk)(\xi/v) = \omega t$. Hence, we have to consider the absorption and emission processes that are induced by a perturbation:

$$\hat{V}(t) = \hat{D}(\vec{r}) \int_{-\infty}^{\infty} E_0(\omega) e^{\pm i\omega t} d\omega \tag{11.17}$$

Consequently, the transition rate will be defined as:

$$P_{i \to n} = \lim_{t \to \infty} \frac{1}{t} \int_{-\infty}^{\infty} P_{in}^{\pm}(t) d\omega \qquad (11.18)$$

and, on substitution of Eqs.(11.15) and (11.16) into Eq.(11.18), we obtain:

$$P_{i \to n} = \frac{2\pi}{\hbar^2} |D_{ni}|^2 \int_{-\infty}^{\infty} |E_0(\omega)|^2 \hbar \delta(E_i - E_n \pm \hbar\omega) d\omega$$

$$= \frac{2\pi}{\hbar^2} |D_{ni}|^2 \left| E_0\left(\frac{E_i - E_n}{\hbar}\right) \right|^2 = \frac{2\pi}{\hbar^2} |D_{ni}|^2 |E_0(\omega_{in})|^2 \qquad (11.19)$$

This shows that the transition rate is the same for both the induced absorption and induced emission. The radiative transitions can only be induced between stationary states for which the matrix elements D_{ni} do not vanish, and by radiation fields which contain the corresponding resonant frequency component $\omega = \omega_{in}$.

EXAMPLE 11.1. Selection rules for electric dipole transitions

If we consider the stationary states ψ_{jm_j} of an isolated atomic system, which are determined by the symmetry properties under rotation, we may assume that the angular momentum is conserved during the interaction with the radiation field, such that each transition will be associated with a definite change in the quantum numbers j and m_j. The transition rate (11.19) is determined by the matrix elements D_{in} of the dipole moment operator $\hat{\vec{D}}$, which is a vector operator whose transformation under rotation is defined by the commutation relations (8.5). It is convenient to express the properties of $\hat{\vec{D}}$ by means of its components in a Cartesian coordinate system attached to the atom, namely:

$$\hat{D}_+ = \hat{D}_x + i\hat{D}_y, \qquad \hat{D}_- = \hat{D}_x - i\hat{D}_y, \qquad \hat{D}_z \qquad (11.20)$$

which, on substitution into Eq.(8.5), lead to:

$$[\hat{J}_+, \hat{D}_+] = 0, \qquad [\hat{J}_-, \hat{D}_+] = -2\hbar\hat{D}_z, \qquad [\hat{J}_z, \hat{D}_+] = \hbar\hat{D}_+$$

$$[\hat{J}_+, \hat{D}_-] = 2\hbar\hat{D}_z, \qquad [\hat{J}_-, \hat{D}_-] = 0, \qquad [\hat{J}_z, \hat{D}_-] = -\hbar\hat{D}_- \qquad (11.21)$$

$$[\hat{J}_+, \hat{D}_z] = -\hbar\hat{D}_+, \qquad [\hat{J}_-, \hat{D}_z] = \hbar\hat{D}_-, \qquad [\hat{J}_z, \hat{D}_z] = 0$$

From the commutation relations containing $\hat{J}_z$, the selection rules for m_j on the matrix elements of $\vec{D}$ in the jm_j representation are immediately apparent. For instance we have:

$$\hat{J}_z \hat{D}_+ \psi_{jm_j} = \hat{D}_+ \hat{J}_z \psi_{jm_j} + \hbar \hat{D}_+ \psi_{jm_j} = (m_j + 1)\hbar \hat{D}_+ \psi_{jm_j}$$

where use has been made of Eq.(8.24). This shows that $\hat{D}_+ \psi_{jm_j}$ is an eigenfunction of $\hat{J}_z$ with eigenvalue $(m_j + 1)\hbar$, such that the matrix element:

$$\langle \psi_{j'm'_j} | \hat{D}_+ | \psi_{jm_j} \rangle = \langle \psi_{j'm'_j} | \hat{D}_+ \psi_{jm_j} \rangle = \delta_{m'_j, m_j+1} \qquad (11.22)$$

is zero unless $m'_j = m+1$ or $\Delta m = m'_j - m_j = 1$. In a similar way, the second and third commutation relations from the last column of Eqs.(11.21) give:

$$\langle \psi_{j'm'_j} | \hat{D}_- | \psi_{jm_j} \rangle = \delta_{m'_j, m_j-1}, \qquad \langle \psi_{j'm'_j} | \hat{D}_z | \psi_{jm_j} \rangle = \delta_{m'_j m_j} \qquad (11.23)$$

In other words, electric dipole radiative transitions can only be induced between stationary states which obey the *selection rule* for m_j:

$$\Delta m_j = m'_j - m_j = 0, \pm 1 \qquad (11.24)$$

Assuming Δm values which satisfy Eq.(11.24), the selection rule on j can be determined from the first commutation rule for $\hat{D}_+$ in Eqs.(11.21), which gives:

$$\langle \psi_{j', m_j+1} | \hat{J}_+ \hat{D}_+ - \hat{D}_+ \hat{J}_+ | \psi_{j, m_j-1} \rangle = \sqrt{j'(j'+1) - m_j(m_j+1)} \; \hbar \langle \psi_{j'm_j} | \hat{D}_+ | \psi_{j, m_j-1} \rangle$$

$$- \sqrt{j(j+1) - m_j(m_j-1)} \; \hbar \langle \psi_{j', m_j+1} | \hat{D}_+ | \psi_{j m_j} \rangle = 0$$

where use has been made of Eqs.(8.31) and (8.30). This provides a recurrence relation of the form:

$$\langle \psi_{j'm_j} | \hat{D}_+ | \psi_{j, m_j-1} \rangle \sqrt{(j' - m_j)(j' + m_j + 1)} \; \hbar = \langle \psi_{j', m_j+1} | \hat{D}_+ | \psi_{jm_j} \rangle \sqrt{(j + m_j)(j - m_j + 1)} \; \hbar$$

which is convenient to be rewritten in terms of factorials, as:

$$\langle \psi_{j'm_j} | \hat{D}_+ | \psi_{j, m_j-1} \rangle \sqrt{\frac{(j' - m_j)!(j' + m_j + 1)!}{(j' - m_j - 1)!(j' + m_j)!}} \; \hbar = \langle \psi_{j', m_j+1} | \hat{D}_+ | \psi_{jm_j} \rangle \sqrt{\frac{(j + m_j)!(j - m_j + 1)!}{(j + m_j - 1)!(j - m_j)!}} \; \hbar$$

and also:

$$\frac{\langle \psi_{j'm_j} | \hat{D}_+ | \psi_{j, m_j-1} \rangle}{\sqrt{\frac{(j' + m_j)!(j - m_j + 1)!}{(j' - m_j)!(j + m_j - 1)!}} \; \hbar} = \frac{\langle \psi_{j', m_j+1} | \hat{D}_+ | \psi_{jm_j} \rangle}{\sqrt{\frac{(j' + m_j + 1)!(j - m_j)!}{(j' - m_j - 1)!(j + m_j)!}} \; \hbar} = \langle j' \| \hat{D} \| j \rangle \qquad (11.25)$$

We observe that the second ratio is obtained from the first by replacing $m_j \to m_j + 1$, and therefore their common value, denoted by $\langle j' \| \hat{D} \| j \rangle$ and known as the *reduced matrix element*, is independent of m_j. Consequently:

$$\langle \psi_{j', m_j+1} | \hat{D}_+ | \psi_{jm_j} \rangle = \sqrt{\frac{(j'+m_j+1)!(j-m_j)!}{(j'-m_j-1)!(j+m_j)!}} \hbar \langle j' \| \hat{D} \| j \rangle \qquad (11.26)$$

This matrix element is non-vanishing provided the arguments of the factorials in the denominator are both non-negative, which gives:

$$-j \le m_j \le j'-1 \quad \text{or} \quad j'+j \ge 1 \qquad (11.27)$$

and it is finite provided the arguments of the factorials in the numerator are also non-negative, for all the allowed m_j values, which leads to:

$$j'-j+1 \ge 0, \quad j-j'+1 \ge 0 \quad \text{or} \quad -1 \le \Delta j \le 1 \qquad (11.28)$$

It follows that the selection rules on j are:

$$\Delta j = 0, \pm 1 \qquad (11.29)$$

observing, however, that no $j = 0 \to j = 0$ transition is allowed, because of the condition (11.27). From Eqs.(11.24), (11.26) and (11.29) it is immediately apparent that there are three nonzero matrix elements of $\hat{D}_1$ only, namely:

$$\langle \psi_{j+1, m_j+1} | \hat{D}_+ | jm_j \rangle = \sqrt{(j+m_j+1)(j+m_j+2)} \, \hbar \langle j+1 \| \hat{D} \| j \rangle$$

$$\langle \psi_{j, m_j+1} | \hat{D}_+ | jm_j \rangle = \sqrt{(j-m_j)(j+m_j+1)} \, \hbar \langle j \| \hat{D} \| j \rangle \qquad (11.30)$$

$$\langle \psi_{j-1, m_j+1} | \hat{D}_+ | jm_j \rangle = \sqrt{(j-m_j)(j-m_j-1)} \, \hbar \langle j-1 \| \hat{D} \| j \rangle$$

The nonvanishing matrix elements of $\hat{D}_z$ can be then found from Eqs.(11.30) and the commutation relation $\hat{D}_z = [\hat{D}_+, \hat{J}_-]/2\hbar$, Eq.(11.21), in the form:

$$\langle \psi_{j+1, m_j} | \hat{D}_z | jm_j \rangle = -\sqrt{(j+m_j+1)(j-m_j+1)} \, \hbar \langle j+1 \| \hat{D} \| j \rangle$$

$$\langle \psi_{jm_j} | \hat{D}_z | jm_j \rangle = m_j \, \hbar \langle j \| \hat{D} \| j \rangle \qquad (11.31)$$

$$\langle \psi_{j-1, m_j} | \hat{D}_z | jm_j \rangle = \sqrt{(j+m_j)(j-m_j)} \, \hbar \langle j-1 \| \hat{D} \| j \rangle$$

Similarly, the matrix elements of $\hat{D}_-$ result from Eqs.(11.31) and the commutator $\hat{D}_- = [\hat{J}_-, \hat{D}_z]/\hbar$ as:

$$\langle \psi_{j+1,m_j-1} | \hat{D}_- | \psi_{jm_j} \rangle = -\sqrt{(j-m_j+1)(j-m_j+2)}\, \hbar \langle j+1 \| \hat{D} \| j \rangle$$

$$\langle \psi_{j,m_j-1} | \hat{D}_- | \psi_{jm_j} \rangle = \sqrt{(j+m_j)(j-m_j+1)}\, \hbar \langle j \| \hat{D} \| j \rangle \qquad (11.32)$$

$$\langle \psi_{j-1,m_j-1} | \hat{D}_- | \psi_{jm_j} \rangle = \sqrt{(j+m_j)(j+m_j-1)}\, \hbar \langle j-1 \| \hat{D} \| j \rangle$$

The selection rules (11.24) and (11.29), which are determined by the symmetry properties of the quantum states, are complemented by another rule concerning the definite parity $\pi_i, \pi_n (= \pm 1)$ of these states:

$$\hat{\Pi} \psi_i = \pi_i \psi_i, \qquad \hat{\Pi} \psi_n = \pi_n \psi_n$$

Since the electric dipole operator is a polar vector, which changes sign under the parity operation:

$$\hat{\Pi} \hat{\vec{D}} \hat{\Pi}^{-1} = -\hat{\vec{D}}$$

we have:

$$D_{ni} = \langle \psi_n | \vec{\varepsilon} \cdot \hat{\vec{D}} | \psi_i \rangle = -\langle \psi_n | \vec{\varepsilon} \cdot (\hat{\Pi} \hat{\vec{D}} \hat{\Pi}^{-1}) | \psi_i \rangle = -\langle \hat{\Pi} \psi_n | \vec{\varepsilon} \cdot \hat{\vec{D}} | \hat{\Pi} \psi_i \rangle = -\pi_n \pi_i D_{ni}$$

It follows that the matrix element vanishes if ψ_n and ψ_i are states of the same parity, such that $\pi_n \pi_i = 1$. In other words, the electric dipole transitions are only allowed between states of opposite parity. ✥

The electric dipole radiation is emitted or absorbed with a definite state of *polarization*, when one of the selection rules is satisfied. This is immediately apparent in the case of one electron atoms, where the total angular momentum reduces to the orbital angular momentum, if the spin is disregarded. It follows that the parity of the stationary states is $(-1)^l$. Although the selection rule (11.29) allows radiative transitions corresponding to $\Delta l = 0$, or $l' = l$, such transitions are forbidden for the electric dipole radiation by the selection rule on the parity, and hence we have for all the allowed transitions:

$$l' = l \pm 1 \qquad (11.33)$$

In terms of the conservation of angular momentum, Eq.(11.33) shows that there are only two independent states for the radiated photon, which has an intrinsic angular momentum of $\hbar$. Polarized photons possess a component of angular momentum along the z-axis of either $+\hbar$ or $-\hbar$, corresponding to only two states of polarization. These properties of the photon are consistent with the classical approach.

EXAMPLE 11.2. Polarization and intensity of atomic radiation

The polarization state of electric dipole radiation can be described in terms of the unit vector $\vec{\varepsilon}$ which defines the orientation of the electric field. We may assume that the emitted radiation is observed along a direction $\vec{e}_r$, defined by the spherical polar coordinates θ and φ with respect to the Cartesian coordinate system attached to the atom, such that the propagation vector $\vec{k}$ is taken parallel to $\vec{e}_r$. The polarization vector $\vec{\varepsilon}$ of the transverse electric field of the radiation is then defined in terms of the unit vectors $\vec{e}_\theta$ and $\vec{e}_\varphi$, introduced before in Figure 2.1, which determine the plane normal to $\vec{e}_r$:

$$\vec{\varepsilon} = \varepsilon_\theta \vec{e}_\theta + \varepsilon_\varphi \vec{e}_\varphi \tag{11.34}$$

It is convenient to obtain a relation between the components of the electric dipole operator $\hat{\vec{D}}$ in the two coordinate systems:

$$\hat{\vec{D}} = \hat{D}_x \vec{e}_x + \hat{D}_y \vec{e}_y + \hat{D}_z \vec{e}_z = \hat{D}_\theta \vec{e}_\theta + \hat{D}_\varphi \vec{e}_\varphi \tag{11.35}$$

and this will be immediately apparent if the spherical polar unit vectors $\vec{e}_r, \vec{e}_\theta$ and $\vec{e}_\varphi$ are expressed in terms of the Cartesian unit vectors $\vec{e}_x, \vec{e}_y, \vec{e}_z$ (see Example 2.1) in the form:

$$\vec{e}_r = \sin\theta \cos\varphi \, \vec{e}_x + \sin\theta \sin\varphi \, \vec{e}_y + \cos\theta \, \vec{e}_z$$

$$\vec{e}_\theta = \cos\theta \cos\varphi \, \vec{e}_x + \cos\theta \sin\varphi \, \vec{e}_y - \sin\theta \, \vec{e}_z \tag{11.36}$$

$$\vec{e}_\varphi = -\sin\varphi \, \vec{e}_x + \cos\varphi \, \vec{e}_y$$

Combining Eqs.(11.35) and (11.36) we obtain:

$$\hat{D}_\theta = \hat{D}_x \vec{e}_x \cdot \vec{e}_\theta + \hat{D}_y \vec{e}_y \cdot \vec{e}_\theta + \hat{D}_z \vec{e}_z \cdot \vec{e}_\theta = \frac{1}{2}\cos\theta(\hat{D}_+ e^{-i\varphi} + \hat{D}_- e^{i\varphi}) - \hat{D}_z \sin\theta$$

$$\tag{11.37}$$

$$\hat{D}_\varphi = \hat{D}_x \vec{e}_x \cdot \vec{e}_\varphi + \hat{D}_y \vec{e}_y \cdot \vec{e}_\varphi + \hat{D}_z \vec{e}_z \cdot \vec{e}_\varphi = \frac{i}{2}(\hat{D}_- e^{i\varphi} - \hat{D}_+ e^{-i\varphi})$$

such that we obtain the polarization of the $\hat{D}_+, \hat{D}_-$ and $\hat{D}_z$ operators, defined by Eq.(11.20) and involved in the allowed transitions, in the form:

$$\hat{\vec{D}} = \frac{1}{2}(\vec{e}_\theta \cos\theta - i\vec{e}_\varphi) e^{-i\varphi} \hat{D}_+ + \frac{1}{2}(\vec{e}_\theta \cos\theta + i\vec{e}_\varphi) e^{i\varphi} \hat{D}_- - \vec{e}_\theta \sin\theta \, \hat{D}_z \tag{11.38}$$

If we consider the special case where the atomic radiation is observed in the direction $\theta = 0$, along the z-axis of the atomic coordinate system ($\vec{e}_r \equiv \vec{e}_z$), and take for simplicity $\varphi = 0$, such that $\vec{e}_\theta \equiv \vec{e}_x$ and $\vec{e}_\varphi \equiv \vec{e}_y$, one obtains:

$$\hat{D} = \frac{1}{2}(\vec{e}_x - i\vec{e}_y)\hat{D}_+ + \frac{1}{2}(\vec{e}_x + i\vec{e}_y)\hat{D}_- = \frac{1}{\sqrt{2}}\vec{e}_-\hat{D}_+ + \frac{1}{\sqrt{2}}\vec{e}_+\hat{D}_- \quad (11.39)$$

The unit vectors $\vec{e}_\pm$ correspond to the independent states of polarization allowed for the radiated photon, illustrated in Figure 11.1, which are respectively called the left and right circular polarization. For instance, using Eqs.(11.4) and (11.9) we obtain the electric vector:

$$\vec{E}_+ = i\omega|\vec{A}_0|(\vec{e}_+ e^{-i\omega t} - \vec{e}_+^* e^{i\omega t}) = \sqrt{2}\,\omega|\vec{A}_0|(\vec{e}_x \sin\omega t - \vec{e}_y \cos\omega t)$$

which rotates counterclockwise with respect to the observer looking toward the source. This radiation state is referred to as left-circular polarization. The right-circular polarization state is similarly defined in terms of the orthogonal unit vector $\vec{e}_-$.

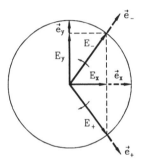

Figure 11.1. Configuration of the independent polarization states for the radiated photon

In the case of transverse observation ($\theta = \pi/2$) of the atomic radiation, where $\vec{e}_\theta \equiv -\vec{e}_z$, Eq.(11.38) reduces to:

$$\hat{D} = \vec{e}_z \hat{D}_z \quad (11.40)$$

and this corresponds to a state of linear polarization, in the direction $\vec{e}_z$. This state can always be resolved into two orthogonal circularly polarized components, in the plane normal to the direction of observation.

Combining Eq.(11.38) with the selection rules discussed in Example 11.1, the transition rates (11.19) corresponding to the allowed electric dipole transitions can be written as:

$$P(\Delta m_j = +1) = \frac{2\pi}{\hbar^2 \varepsilon_0 c} I(\omega) \frac{1}{4}(1+\cos^2\theta)|\langle \psi_{j',m_j+1}|\hat{D}_+|\psi_{jm_j}\rangle|^2$$

$$P(\Delta m_j = 0) = \frac{2\pi}{\hbar^2 \varepsilon_0 c} I(\omega) \sin^2\theta |\langle \psi_{j'm_j}|\hat{D}_z|\psi_{jm_j}\rangle|^2 \quad (11.41)$$

$$P(\Delta m_j = -1) = \frac{2\pi}{\hbar^2 \varepsilon_0 c} I(\omega) \frac{1}{4}(1+\cos^2\theta)|\langle \psi_{j',m_j-1}|\hat{D}_-|\psi_{jm_j}\rangle|^2$$

418 Quantum Physics

The matrix elements of the electric dipole operator components are given by Eqs.(11.30) to (11.32) in terms of the reduced matrix elements, such that the *relative* intensities can be determined for any given transition.

Problem 11.1.1. Find the selection rules for the orbital and magnetic quantum numbers, l and m, in an electric dipole transition between the stationary states of one-electron atoms.

(Solution): The perturbation operator (11.11) can be expressed either in Cartesian or in spherical polar coordinates, having the origin at the centre of the atom, in the form:

$$(-e\vec{r}) \cdot \vec{E} = exE_x + eyE_y + ezE_z = -er\sin\theta(E_x\cos\varphi + E_y\sin\varphi) - er\cos\theta E_z$$

$$= -\frac{e}{2}r\sin\theta e^{i\varphi}(E_x - iE_y) - \frac{e}{2}r\sin\theta e^{-i\varphi}(E_x + iE_y) - er\cos\theta E_z$$

$$= \sum_k (-er_k)E_k = \sum_k (-er_k)E_{0k}(\omega)e^{-i\omega t}$$

were k stands for the three components of the interaction operator during induced absorption. On substitution into Eq.(11.19) one obtains:

$$P_{i \to n} = \frac{2\pi e^2}{\hbar^2}|\langle\psi_n|r_k|\psi_i\rangle|^2 |E_{0k}(\omega_{in})|^2$$

and we may describe the stationary states by the central field wave functions (7.84). Hence, in spherical coordinates, the matrix elements take the form:

$$\langle\Psi_{n'l'm'}|\tfrac{1}{2}r\sin\theta e^{i\varphi}|\Psi_{nlm}\rangle = \frac{1}{2}\int_0^\infty R^*_{n'l'}R_{nl}r^3 dr$$

$$\times \int_0^\pi P_{l'}^{m'}(\cos\theta)P_l^m(\cos\theta)\sin\theta\, d(\cos\theta)\int_0^{2\pi} e^{-i(m'-m-1)\varphi}d\varphi$$

$$\langle\Psi_{n'l'm'}|\tfrac{1}{2}r\sin\theta e^{-i\varphi}|\Psi_{nlm}\rangle = \frac{1}{2}\int_0^\infty R^*_{n'l'}R_{nl}r^3 dr$$

$$\times \int_0^\pi P_{l'}^{m'}(\cos\theta)P_l^m(\cos\theta)\sin\theta\, d(\cos\theta)\int_0^{2\pi} e^{-i(m'-m+1)\varphi}d\varphi$$

$$\langle \Psi_{n'l'm'} | r\cos\theta | \Psi_{nlm} \rangle = \int_0^\infty R_{n'l'}^* R_{nl} r^3 dr$$

$$\times \int_0^\pi P_{l'}^{m'}(\cos\theta) P_l^m(\cos\theta) \cos\theta \, d(\cos\theta) \int_0^{2\pi} e^{-i(m'-m)\varphi} d\varphi$$

The selection rule on the magnetic quantum number results from integration over φ, where the result vanishes unless $m' - m = -1$, $m' - m = 1$ or $m' = m$ respectively. In other words the allowed transitions for electric dipole radiation are restricted to a change in the magnetic quantum number:

$$\Delta m = 0, \pm 1$$

The selection rule on l can be derived from the integrals in θ, substituting the recurrence relations involving the associated Legendre functions, which are apparent from their definition (7.32), namely:

$$(2l+1)\sin\theta \, P_l^m(\cos\theta) = P_{l+1}^{m+1}(\cos\theta) - P_{l-1}^{m+1}(\cos\theta)$$

$$(2l+1)\cos\theta \, P_l^m(\cos\theta) = (l-m+1) P_{l+1}^m(\cos\theta) + (l+m) P_{l-1}^m(\cos\theta)$$

The first and second integrals in θ vanish unless $l' = l \pm 1$, and hence, the allowed transitions for electric dipole radiation are also restricted by:

$$\Delta l = \pm 1$$

which is the same result found before by considering the parity of the stationary states.

Problem 11.1.2. Find the relative intensities of the weak-field Zeeman patterns for the $2^2 P_{\frac{3}{2}} \to 1^2 S_{\frac{1}{2}}$ transitions that occur in the spectra of one-electron atoms.

(Solution): There are four initial states $2^2 P_{\frac{3}{2}}$ of odd parity ($l = 1$), assumed to be equally populated, which in the jm_j representation read:

$$\psi_{\frac{3}{2},\frac{3}{2}} = |\tfrac{3}{2},\tfrac{3}{2}\rangle, \quad \psi_{\frac{3}{2},\frac{1}{2}} = |\tfrac{3}{2},\tfrac{1}{2}\rangle, \quad \psi_{\frac{3}{2},-\frac{1}{2}} = |\tfrac{3}{2},-\tfrac{1}{2}\rangle, \quad \psi_{\frac{3}{2},-\frac{3}{2}} = |\tfrac{3}{2},-\tfrac{3}{2}\rangle$$

and two final states $1^2 S_{\frac{1}{2}}$, of even parity ($l = 0$), namely:

$$\psi_{\frac{1}{2},\frac{1}{2}} = |\tfrac{1}{2},\tfrac{1}{2}\rangle, \quad \psi_{\frac{1}{2},-\frac{1}{2}} = |\tfrac{1}{2},-\tfrac{1}{2}\rangle$$

plus one radiated photon with $J = 1$ and $M = \Delta m_j = 1, 0, -1$. Six allowed transitions are observed, rather than eight, as it might be expected, due to the selection rule on m_j. The relative intensities of the allowed transitions are given by Eqs.(11.41), where the factors which are common to all of them may be disregarded. Substituting the squares of the matrix elements, as given by Eqs.(11.30) to (11.32) for the case $j' = j+1$, we obtain:

$$I(M = +1) \sim \frac{1}{4}(1+\cos^2\theta)(j+m_j+1)(j+m_j+2)$$

$$I(M = 0) \sim \sin^2\theta(j-m_j+1)(j+m_j+1)$$

$$I(M = -1) \sim \frac{1}{4}(1+\cos^2\theta)(j-m_j+1)(j-m_j+2)$$

For the six allowed transitions, this becomes:

$$I(|\tfrac{3}{2},\tfrac{3}{2}\rangle \to |\tfrac{1}{2},\tfrac{1}{2}\rangle) = I(|\tfrac{3}{2},-\tfrac{3}{2}\rangle \to |\tfrac{1}{2},-\tfrac{1}{2}\rangle) = \tfrac{3}{2}(1+\cos^2\theta)$$

$$I(|\tfrac{3}{2},\tfrac{1}{2}\rangle \to |\tfrac{1}{2},\tfrac{1}{2}\rangle) = I(|\tfrac{3}{2},-\tfrac{1}{2}\rangle \to |\tfrac{1}{2},-\tfrac{1}{2}\rangle) = 2\sin^2\theta$$

$$I(|\tfrac{3}{2},\tfrac{1}{2}\rangle \to |\tfrac{1}{2},-\tfrac{1}{2}\rangle) = I(|\tfrac{3}{2},-\tfrac{1}{2}\rangle \to |\tfrac{1}{2},\tfrac{1}{2}\rangle) = \tfrac{1}{2}(1+\cos^2\theta)$$

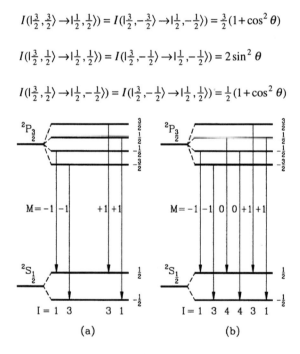

Figure 11.2. Zeeman splitting of the $2^2P_{\tfrac{3}{2}} \to 1^2S_{\tfrac{1}{2}}$ spectral line:
(a) longitudinal observation; (b) transverse observation

If we take the axis of quantization $\vec{e}_z$ in the direction of the magnetic field $\vec{B}$, the intensities for longitudinal observation of the Zeeman pattern ($\theta = 0$) are found in the ratios 1:3:0:0:3:1, as illustrated in Figure 11.2 (a). In other words, the intensity of the linearly polarized radiation, emitted in transitions for which

$\Delta m_j = 0$ is predicted to be zero in the direction $\theta = 0$, which agrees with the observed Zeeman patterns of spectral lines. For transverse observation ($\theta = \pi/2$) we obtain a six-line pattern, with relative intensities 1:3:4:4:3:1, as illustrated in Figure 11.2 (b), and this is also confirmed by experiment.

Problem 11.1.3. Show that the Zeeman pattern for the $\Delta j = 0$ transition $2^2P_{\frac{1}{2}} \to 1^2S_{\frac{1}{2}}$ in one-electron atoms consists of equally intense lines, and find the number of lines observed for $\theta = 0$ and $\theta = \pi/2$.

(Answer): $2(\theta = 0);\ 4(\theta = \pi/2)$

Problem 11.1.4. Find the intensity ratio of the two transitions $2^2P_{\frac{3}{2}} \to 1^2S_{\frac{1}{2}}$ and $2^2P_{\frac{1}{2}} \to 1^2S_{\frac{1}{2}}$, corresponding to the splitting of the $2p \to 1s$ spectral line by spin-orbit coupling.

(Answer): $I(2^2P_{\frac{3}{2}} \to 1^2S_{\frac{1}{2}}) / I(2^2P_{\frac{1}{2}} \to 1^2S_{\frac{1}{2}}) = 2$

Problem 11.1.5. An electron confined to the region $0 \le x \le \alpha$ by an infinitely deep square well potential, has the normalized eigenfunctions $\psi_n(x)$ of the form found in Problem 3.1.3. Calculate the matrix elements of the electric dipole moment operator $\hat{D} = e(x - \alpha/2)$, taken between the states with quantum numbers n and m.

(Answer): $D_{nm} = \dfrac{e\alpha}{\pi^2}\left[\dfrac{\cos(n-m)\pi - 1}{(n-m)^2} - \dfrac{\cos(n+m)\pi - 1}{(n+m)^2}\right]$

11.2. Spontaneous Emission

Consider a system consisting of a large number of identical atoms, in a radiation field of spectral energy density $u(\omega, T)$ in the frequency range $\omega + d\omega$, given by:

$$u_\omega = u(\omega, T) = \frac{1}{2}\left[\varepsilon_0 \langle E^2 \rangle + \frac{1}{\mu_0}\langle B^2 \rangle\right] = \varepsilon_0 \langle E^2 \rangle = \frac{1}{2}\varepsilon_0 |E_0(\omega)|^2 \quad (11.42)$$

where the thermodynamic equilibrium at temperature T is obtained by absorption and emission of radiation. In the electric dipole approximation (11.19), if the radiation is incident upon the atom from all directions and with random polarization, the transition rate must include independent contributions from the matrix element of each component of the electric dipole moment, that is:

$$P_{i \to n} = \frac{2\pi e^2}{\hbar^2} |E_{0k}(\omega_{in})|^2 \sum_k |\langle \psi_n | r_k | \psi_i \rangle|^2 \tag{11.43}$$

If r_k are taken to be the Cartesian components of $\vec{r}$, we also have:

$$|E_{0x}(\omega_{in})|^2 = |E_{0y}(\omega_{in})|^2 = |E_{0z}(\omega_{in})|^2 = \frac{1}{3}|E_0(\omega_{in})|^2$$

This gives:

$$P_{i \to n} = \frac{2\pi e^2}{3\hbar^2} |E_0(\omega_{in})|^2 \sum_k |\langle \psi_n | r_k | \psi_i \rangle|^2$$

$$= \frac{4\pi e^2}{3\varepsilon_0 \hbar^2} u(\omega_{in}, T) \sum_k |\langle \psi_n | r_k | \psi_i \rangle|^2 = B_{i \to n} u(\omega_{in}, T)$$

where $B_{i \to n}$ is defined as:

$$B_{i \to n} = \frac{4\pi e^2}{3\varepsilon_0 \hbar^2} \sum_k |\langle \psi_n | r_k | \psi_i \rangle|^2 \tag{11.44}$$

for the electric dipole transition. If $E_n > E_i$, $B_{i \to n}$ is associated with an absorption process and is called the *Einstein coefficient of absorption*.

If the system in the stationary state ψ_n makes a transition to a lower state ψ_i, the transition rate may similarly be written as:

$$P_{n \to i} = B_{n \to i} u(\omega_{ni}, T) \tag{11.45}$$

where $B_{n \to i}$ is called the *Einstein coefficient of induced emission*. We have shown earlier that there is the same transition rate for induced absorption and induced emission, hence we can write:

$$B_{i \to n} = B_{n \to i} \tag{11.46}$$

Let there be N_i atoms in the energy level E_i and N_n atoms in the energy level E_n. Since no net energy is absorbed or emitted by the atoms, the spectral energy density $u(\omega_{in}, T) = u(\omega_{ni}, T)$ remains constant and Eq.(11.46) leads to:

11. Atomic Radiation

$$N_i B_{i \to n} u(\omega_{in}, T) = N_n B_{n \to i} u(\omega_{ni}, T) = N_n B_{i \to n} u(\omega_{in}, T)$$

This result is inconsistent, however, with the occupation number of the levels given by the classical Boltzman distribution law for equilibrium to exist between the atoms and the radiation field:

$$\frac{N_n}{N_i} = \frac{e^{-E_n/k_B T}}{e^{-E_i/k_B T}} = e^{-\hbar \omega_{ni}/k_B T} \tag{11.47}$$

Hence, we must assume that there exists a process, independent of the presence of the radiation field $u(\omega_{ni}, T)$, which makes the lower energy levels more densely populated. The concept of *spontaneous emission* was formally introduced in terms of an intrinsic transition rate:

$$P_{n \to i} = A_{n \to i} \tag{11.48}$$

called the *Einstein coefficient of spontaneous emission*. The equation which describes the balance between absorption and emission at thermal equilibrium:

$$N_i B_{i \to n} u(\omega_{in}, T) = N_i B_{n \to i} u(\omega_{ni}, T) = N_n \{ B_{n \to i} u(\omega_{ni}, T) + A_{n \to i} \} \tag{11.49}$$

may be solved for $u(\omega_{ni}, T)$:

$$u(\omega_{ni}, T) = \frac{A_{n \to i}}{B_{n \to i}} \frac{1}{N_i/N_n - 1} = \frac{A_{n \to i}}{B_{n \to i}} \frac{1}{e^{\hbar \omega_{ni}/k_B T} - 1} \tag{11.50}$$

which is expected to obey the Planck radiation formula (P.18), and this yields:

$$A_{n \to i} = \frac{\hbar \omega_{ni}^3}{\pi^2 c^3} B_{n \to i} \tag{11.51}$$

Hence, the rate of spontaneous emission for electric dipole radiation is obtained from Eqs.(11.44) and (11.51) as:

$$A_{n \to i} = \frac{8 e^2 \omega_{ni}^3}{3 \varepsilon_0 h c^3} \sum_k |\langle \psi_n^{(0)} | r_k | \psi_i^{(0)} \rangle|^2 \tag{11.52}$$

The spontaneous emission of photons, with a constant rate $A_{n \to i}$, results in a finite lifetime of the excited state ψ_n for the ensemble of atoms. If we assume that N_n atoms were certainly in the excited state at a particular time t, the probability per unit time that a decay will occur after a short time is proportional to N_n and the rate of spontaneous emission:

$$\frac{dN_n}{dt} = -A_{n\to i} N_n \quad \text{or} \quad N_n(t) = N_n(0)\, e^{-A_{n\to i} t} = N_n(0)\, e^{-t/\langle t \rangle} \quad (11.53)$$

This shows that the Einstein coefficient defined by Eq.(11.52) represents the *decay constant* of the excited state, which is a measure of the *mean lifetime* $\langle t \rangle = 1/A_{n\to i}$. It follows from Eq.(11.53) that there is a time dependence of the radiation intensity:

$$|E|^2 = |E_0(\omega_{ni})|^2\, e^{-A_{n\to i} t}$$

which makes any field component vary about the central frequency ω_{ni} of the transition as:

$$E = E_0(\omega_{ni})\, e^{-i\omega_{ni} t - A_{n\to i} t/2} = \int_{-\infty}^{\infty} E_0(\omega)\, e^{-i\omega t}\, d\omega \quad (11.54)$$

where use has been made of Eq.(11.17). The Fourier transform of Eq.(11.54) is:

$$E_0(\omega) = \frac{E_0(\omega_{ni})}{2\pi} \int_0^{\infty} e^{i(\omega - \omega_{ni}) - A_{n\to i} t/2}\, d\omega = \frac{E_0(\omega_{ni})}{2\pi} \frac{1}{-i(\omega - \omega_{ni}) + \frac{1}{2} A_{n\to i}}$$

which gives the spectral intensity in the so-called Lorentzian form:

$$|E_0(\omega)|^2 = \frac{|E_0(\omega_{ni})|^2}{4\pi^2} \frac{1}{(\omega - \omega_{ni})^2 + \frac{1}{4} A_{n\to i}^2} \quad (11.55)$$

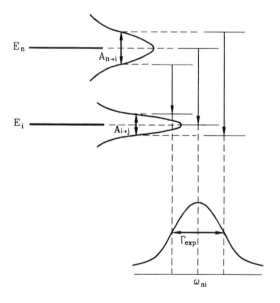

Figure 11.3. Natural and experimental linewidth in the electric dipole transitions

Equation (11.55) shows that there is a *natural linewidth* of the spontaneous emission, determined by the Einstein coefficient $A_{n \to i}$, as illustrated in Figure 11.3. Since in general both the upper and lower states of electric dipole transitions have finite lifetimes, the linewidth Γ of the frequency distribution of radiation, experimentally observed, is obtained as a sum of the two natural linewidths corresponding to the decay of these states.

EXAMPLE 11.3. Amplification of radiation

When a radiation beam propagates through an optical medium, each process of absorption by a single atom substracts a photon of energy $\hbar\omega_{ni}$ from the beam, while each process of induced emission results in the addition of the same energy, so that the net time rate of change of the spectral energy density is:

$$\frac{du_\omega}{dt} = (N_n - N_i) P_{i \to n} \hbar\omega_{ni} = (N_n - N_i) B_{i \to n} u_\omega \hbar\omega_{ni}$$

If we make use of Eqs.(11.42) and (11.16), this equation can be rewritten in terms of the average energy flow, given by the spectral intensity $I(\omega)$, as:

$$\frac{dI(\omega)}{dx} = (N_n - N_i) B_{i \to n} \frac{\hbar\omega_{ni}}{c} I(\omega) = \alpha I(\omega) \tag{11.56}$$

Integrating over a distance x, travelled by the radiation beam, gives:

$$I(\omega) = I_0(\omega) e^{\alpha x} \tag{11.57}$$

The Boltzman distribution law, illustrated in Figure 11.4, shows that $N_n = N_i e^{-\hbar\omega_{ni}/k_B T} \ll N_i$, that is $\alpha < 0$ and the radiation will be attenuated as it passes through the medium, because the loss due to absorption exceeds the gain due to the induced emission.

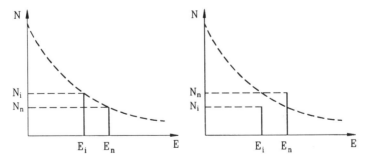

Figure 11.4. Change in the Boltzman distribution by population inversion

Experiment shows, however, that a *population inversion* can be produced, when $N_n > N_i$ for $E_n > E_i$, as in Figure 11.4. Substituting Eq.(11.51), Eq.(11.56) gives α, called the *gain constant*, as:

$$\alpha = (N_n - N_i)\frac{\lambda^2}{4} A_{n \to i} \qquad (11.58)$$

If the population inversion exists, the radiation will be amplified as it passes through the medium. Methods for producing the population inversion in optical media have been developed in *laser* systems. ∮

Problem 11.2.1. A charged particle executes harmonic oscillations along the x-axis with angular frequency ω. Calculate the rate of spontaneous emission and the allowed frequencies of the emission spectrum.

(Solution): For a simple harmonic oscillator, the Einstein coefficient for spontaneous emission, Eq.(11.52), simplifies to:

$$A_{n \to i} = \frac{8e^2 \omega_{ni}^3}{3\varepsilon_0 hc^3} |\langle \psi_n | \hat{x} | \psi_i \rangle|^2$$

where the stationary states satisfy Eq.(6.94). If we write the position operator, in view of Eqs.(6.86), in the form $\hat{x} = (\hat{a} + \hat{a}^+)/\sqrt{2m\omega}$, it follows that the matrix element vanishes unless $i = n \pm 1$:

$$\langle \psi_n | \hat{x} | \psi_i \rangle = \left(\frac{\hbar}{2m\omega}\right)^{\frac{1}{2}} \sqrt{n+1} \qquad \text{if} \qquad i = n+1$$

$$= \left(\frac{\hbar}{2m\omega}\right)^{\frac{1}{2}} \sqrt{n} \qquad \text{if} \qquad i = n-1$$

Hence, the only spontaneous transitions which occur are those with $\omega_{ni} = \omega$, in which the quantum number changes by one unit because of jumps from any given level to the next lower level in the series. In other words, thus the only frequency radiated is the frequency of classical oscillation. We obtain spontaneous emission with the constant rate:

$$A_{n \to n-1} = \frac{2}{3} \frac{ne^2}{\pi \varepsilon_0 mc^3} \omega^2$$

Problem 11.2.2. Calculate the Einstein coefficient for the spontaneous decay of hydrogen from the 2p excited state.

(Solution): The decay corresponds, from parity considerations, to the radiative transition $2p \to 1s$ only, with the Einstein coefficient (11.52) which can be written as:

$$A_{2p \to 1s} = \frac{8e^2 \omega_{ni}^2}{\varepsilon_0 hc^3} |\langle \Psi_{100} | \vec{r} | \Psi_{21m} \rangle|^2$$

where $\vec{r} = r\vec{e}_r$, and $\vec{e}_r$ is defined by Eq.(11.36). Using the wave functions (7.123), it is convenient to separate the evaluation of the radial and angular parts of the matrix element:

$$\langle \Psi_{100} | \vec{r} | \Psi_{21m} \rangle = \int_0^\infty R_{10}^*(r) R_{21}(r) r^3 dr \int_0^\pi \int_0^{2\pi} Y_{00}^* \vec{e}_r Y_{1m} \sin\theta \, d\theta \, d\varphi$$

Integrating the radial part gives:

$$\int_0^\infty R_{10}^*(r) R_{21}(r) r^3 dr = \frac{1}{\sqrt{6}} \int_0^\infty \frac{1}{a_0^{3/2}} e^{-r/a_0} \frac{1}{a_0^{5/2}} e^{-r/2a_0} r^4 dr$$

$$= \frac{1}{\sqrt{6} a_0^4} \int_0^\infty e^{-3r/2a_0} r^4 dr = \left(\frac{3}{2}\right)^{\frac{1}{2}} \frac{2^8}{3^5} a_0$$

Combining Eqs.(11.36) and (7.36), we obtain:

$$\vec{e}_r = \sqrt{\frac{4\pi}{3}} \left[\frac{1}{\sqrt{2}} (\vec{e}_x + i\vec{e}_y) Y_{1,-1} - \frac{1}{\sqrt{2}} (\vec{e}_x - i\vec{e}_y) Y_{11} + \vec{e}_z Y_{10} \right]$$

$$= \sqrt{\frac{4\pi}{3}} (\vec{e}_+ Y_{1,-1} - \vec{e}_- Y_{11} + \vec{e}_z Y_{10})$$

such that the angular integral simplifies to:

$$\int_0^\pi \int_0^{2\pi} Y_{00}^* \vec{e}_r Y_{1m} \sin\theta d\theta d\varphi = \frac{1}{\sqrt{3}} \int_0^\pi \int_0^{2\pi} (\vec{e}_+ Y_{1,-1} - \vec{e}_- Y_{11} + \vec{e}_z Y_{10}) Y_{1m} \sin\theta d\theta d\varphi$$

$$= \frac{1}{\sqrt{3}} (\vec{e}_+ \delta_{m1} - \vec{e}_- \delta_{m,-1} + \vec{e}_z \delta_{m0})$$

This leads to:

$$|\langle \Psi_{100} | \vec{r} | \Psi_{21m} \rangle| = \frac{3}{2} \frac{2^{16}}{3^{10}} \frac{a_0^2}{3} (|\vec{e}_+|^2 \delta_{m1} + |\vec{e}_-|^2 \delta_{m,-1} + |\vec{e}_z|^2 \delta_{m0})$$

$$= \frac{3}{2} \frac{2^{16}}{3^{10}} \frac{a_0^2}{3} \frac{1}{3} = \frac{2^{15}}{3^{11}} a_0^2$$

where it was assumed, for simplicity, that the ensemble of hydrogen atoms were randomly distributed, with equal probability 1/3 with respect to m.

The transition rate $A_{2p \to 1s}$ also depends on the radiation frequency, given by:

$$\omega_{ni} = \frac{E_{21m} - E_{100}}{\hbar} = \frac{e_0^2}{2\hbar a_0}(1 - \frac{1}{4}) = \frac{3}{8}\frac{e_0^2}{\hbar a_0}$$

where use has been made of Eq.(7.80) for the energy levels of hydrogen. This finally yields, on substitution:

$$A_{2p \to 1s} = \frac{8e^2}{\varepsilon_0 hc^3}\left(\frac{3}{8}\frac{e_0^2}{\hbar a_0}\right)^3 \frac{2^{15}}{3^{11}} a_0^2 = \frac{2^{10}}{3^8} \frac{e_0^8}{\hbar^4 c^3} \frac{1}{a_0}$$

which is a transition rate of about $10^9\, s^{-1}$, corresponding to a mean lifetime of $10^{-9}\, s$ for the excited state of hydrogen.

Problem 11.2.3. Find the ratio of the probabilities for the spontaneous transitions $P_{nl \to n',l+1}$ and $P_{nl \to n',l-1}$ which may occur in hydrogen.

(Answer): $\quad \dfrac{P_{nl \to n', l+1}}{P_{nl \to n', l-1}} = \dfrac{l+1}{l}$

Problem 11.2.4. Show that spontaneous emission from free electrons is not compatible with the conservation laws for energy and momentum.

11.3. Magnetic Resonance

Magnetic resonance consists of the absorption of electromagnetic energy, through stimulated transitions between the electron or nuclear spin levels, in the presence of a weak applied magnetic field B. This phenomenon is used to investigate the ground-state Zeeman splitting of the multiplets of the fine structure produced by a static magnetic field $\vec{B} = B\vec{e}_z$. We have seen in Section 10.3 that the electron magnetic moment $\hat{\vec{\mu}}$, Eq.(10.81), interacts with the magnetic field, acquiring an energy:

$$\hat{H} = -\hat{\vec{\mu}} \cdot \vec{B} = \gamma B \hat{J}_z \qquad (11.59)$$

The arrangement for an electron spin resonance (ESR) experiment is based on the fact that this magnetic interaction not only causes a level splitting into $2j + 1$ magnetic sublevels, but also induces a time dependence of the angular momentum eigenstates, of the form (5.60), which reads:

$$\Psi_{jm_j}(t) = e^{-i\hat{H}t/\hbar}\psi_{jm_j} = e^{-i(\gamma B \hat{J}_z)t/\hbar}\psi_{jm_j} = e^{i(-\gamma Bt)\hat{J}_z/\hbar}\psi_{jm_j} = \hat{R}^+(\gamma Bt)\psi_{jm_j} \quad (11.60)$$

Comparison with Eq.(8.6) shows that $\hat{R}(\gamma Bt)$ is a rotation operator about the z-axis, and it is convenient to set:

$$\omega_L = \gamma B = \frac{\mu_B}{\hbar} gB \quad (11.61)$$

In other words, the angular momentum $\hat{\vec{J}}$ in each of its eigenstates $\Psi_{jm_j}(t)$ precesses about the static field $\vec{B}$ with the classical Larmor frequency ω_L (see Example P.4), and so does the expectation value of the magnetic moment:

$$\langle \hat{\vec{\mu}}(t)\rangle = \langle \Psi_{jm_j}(t)|\hat{\vec{\mu}}|\Psi_{jm_j}(t)\rangle = R(\gamma Bt)\langle\hat{\vec{\mu}}\rangle R^+(\gamma Bt)$$

For an infinitesimal time interval dt we may approximate:

$$\langle \hat{\vec{\mu}}(t)\rangle = (1+\frac{i}{\hbar}\gamma Bdt \hat{J}_z)\langle\hat{\vec{\mu}}\rangle(1-\frac{i}{\hbar}\gamma Bdt \hat{J}_z) \approx \langle\hat{\vec{\mu}}\rangle + \frac{i}{\hbar}\gamma Bdt \langle[\hat{J}_z,\vec{\mu}]\rangle$$

which gives the following equation for the time derivative of the expectation value:

$$\frac{d}{dt}\langle\hat{\vec{\mu}}\rangle = -\frac{i}{\hbar}\gamma^2 B\langle[\hat{J}_z,\hat{\vec{J}}]\rangle = \gamma^2 B\langle \hat{J}_y\vec{e}_x - \hat{J}_x\vec{e}_y\rangle$$

The last result follows from the commutation relations (8.2), and represents the expansion of the cross product $\hat{\vec{J}} \times \vec{B}$, such that we obtain the equation of motion for the expectation value of the magnetic moment in the form:

$$\frac{d}{dt}\langle\hat{\vec{\mu}}\rangle = \gamma \vec{B} \times \langle\hat{\vec{\mu}}\rangle \quad (11.62)$$

which is the same as the classical equation of motion, derived in Example P4. The physical situation is described by the vector model of magnetic resonance, illustrated for the two-level system ($J = S = \frac{1}{2}$) in Figure 11.5 (a). The electron resonance transition corresponds to the $\langle\hat{\vec{\mu}}\rangle$ vector passing from one orientation to the other. This requires a torque, which is obtained by applying a magnetic field $\vec{B}_1$ in the xy plane. It is obvious that no effective deflection of $\langle \hat{\vec{\mu}} \rangle$ will be obtained if $\vec{B}_1$ is stationary in a given direction, because when $\langle \hat{\vec{\mu}} \rangle$ has a component along $\vec{B}_1$ it is tipped away from its cone of rotation, but it is then restored to the same cone when its component becomes antiparallel to $\vec{B}_1$. A transition only occurs if $\vec{B}_1$ rotates in the xy plane in phase with the precession of $\langle \hat{\vec{\mu}} \rangle$, that is with the Larmor frequency ω_L. This can be achieved by applying a time-dependent electromagnetic field $\vec{B}_1(t)$ perpendicular to the static field $\vec{B}$:

$$\vec{B}_1(t) = 2\vec{B}_1 \cos \omega t \qquad (11.63)$$

such that the total field will be composed of the static field in the z-direction and a rotating field in the xy plane:

$$B_x = B_1 \cos \omega t, \quad B_y = B_1 \sin \omega t, \quad B_z = B \qquad (11.64)$$

Resonant absorption is caused by an electromagnetic frequency $\omega = \omega_L$, and this condition can be formulated, using Eq.(11.61), as:

$$\hbar \omega = \mu_B g B \qquad (11.65)$$

The resonance condition is reached either by varying ω or ω_L (i.e. the static field B) in a spectrometer shown in Figure 11.5 (b).

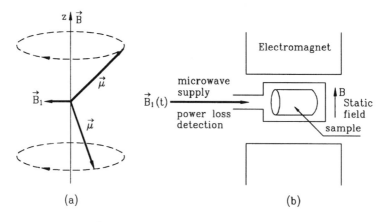

Figure 11.5. (a) Precession of the magnetic moments in the two-spin states and a rotating field B_1 in the xy plane; (b) schematic ESR spectrometer

If typical values for a completely free electron are put into Eq.(11.65), it is seen that, in a static field of 1 Tesla, the required resonance frequency is 28 GHz, which is a frequency in the microwave region. The macroscopic sample should therefore be surrounded by a microwave cavity and the extremely high tuning accuracy of high frequency equipment, and its sensitivity to power loss, enable the ESR technique to probe the electron energy levels in the finest detail.

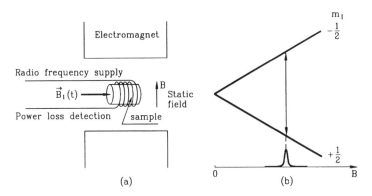

Figure 11.6. (a) Schematic NMR spectrometer, (b) proton spin levels and single line NMR spectrum

A resonance equation similar to Eq.(11.65) is obtained in the case of nuclear magnetic resonance (NMR), which involves the nuclear Zeeman levels, in the form:

$$\hbar\omega = \mu_N g_N B \qquad (11.66)$$

where g_N is the dimensionless nuclear g-factor and μ_N is the nuclear magneton, Eq.(10.109). Substitution of typical figures for the proton case in Eq.(11.66), gives a resonance frequency of 42.6 MHz, for the same field of 1 Tesla, which is in the radio frequency region. Consequently, in a NMR spectrometer, as shown in Figure 11.6 (a), the sample is simply surrounded by a radio frequency coil. In practice one usually works with a fixed frequency ω, sweeping the magnetic field through the resonance value and obtaining a spectrum in which the absorption of energy is plotted as a function of magnetic field strength (Figure 11.6 (b)).

Assuming that an oscillating magnetic field (11.63) acts on the atom, across the static field $\vec{B}$, in the x direction, the resulting time-dependent perturbation is:

$$\hat{V}(t) = 2\mu_B g B_1 \hat{J}_x \cos\omega t = \mu_B g B_1 \hat{J}_x (e^{i\omega t} + e^{-i\omega t}) \qquad (11.67)$$

If we take the matrix elements between two Zeeman angular momentum eigenstates ψ_{jm_j} and $\psi_{jm'_j}$, we obtain:

$$\hat{V}_{m_j m'_j}(t) = \mu_B g B_1 \langle \psi_{jm'_j} | \hat{J}_x | \psi_{jm_j} \rangle (e^{i\omega t} + e^{-i\omega t}) \qquad (11.68)$$

which has a form similar to Eq.(11.12). It follows that the transition rate from state ψ_{jm_j} to state $\psi_{jm'_j}$ can be derived, using Eq.(11.15), in the form:

$$P_{m_j \to m'_j} = \frac{2\pi}{\hbar^2} \mu_B^2 g^2 B_1^2 |\langle \psi_{jm'_j}|\hat{J}_x|\psi_{jm_j}\rangle|^2 \delta(\omega_{m_j m'_j} - \omega) \tag{11.69}$$

where we may further use Eq.(8.44) for the nonvanishing matrix elements of $\hat{J}_x$ and $\gamma = \mu_B g/\hbar$ to obtain:

$$P_{m_j \to m_j \pm 1} = \frac{\pi}{2} \gamma^2 B_1^2 [j(j+1) - m_j(m_j \pm 1)] \delta(\omega_{m_j, m_j \pm 1} - \omega) \tag{11.70}$$

These radiative transitions involve the perturbation produced by the magnetic dipole moment, and hence are called *magnetic dipole transitions*. The selection rule on m_j, implied by Eq.(11.70), is $\Delta m_j = \pm 1$. We notice that the magnetic dipole $\hat{\mu}$ and the angular momentum $\hat{J}$ operators are axial vectors, which are not changed under the parity operation:

$$\hat{\Pi} \hat{\mu} \hat{\Pi} = \hat{\mu}, \quad \hat{\Pi} \hat{J} \hat{\Pi} = \hat{J}$$

It follows that:

$$\langle \psi_{jm'_j}|\hat{J}_x|\psi_{jm_j}\rangle = \langle \psi_{jm'_j}|\hat{\Pi}\hat{J}_x\hat{\Pi}|\psi_{jm_j}\rangle = \pi_{m'_j}\pi_{m_j}\langle \psi_{jm'_j}|\hat{J}_x|\psi_{jm_j}\rangle$$

such that the transition rate vanishes if $\psi_{jm'_j}$ and ψ_{jm_j} are states of opposite parity, for which $\pi_{m'_j}\pi_{m_j} = -1$. This means that the magnetic dipole transitions are allowed between states of the same parity only, as it is always the case with those involved in magnetic resonance experiments.

The transition rate for the NMR spectral lines is calculated in precisely the same way as for electron resonance, and this gives:

$$P_{m_I \to m'_I} = \frac{\pi}{2} \gamma_N^2 B_1^2 [I(I+1) - m_I(m_I \pm 1)] \delta(\omega_{m_I, m_I \pm 1} - \omega) \tag{11.71}$$

Since μ_N is much smaller than μ_B, this transition rate is a factor of about 10^{-5} smaller for NMR transitions as compared to ESR, Eq.(11.70).

EXAMPLE 11.4. Electron resonance transitions in hydrogen

A realistic description of the hydrogen atom must take into account the influence on the electron energy levels of the magnetic hyperfine interaction between the magnetic moments of the electron and proton, discussed in Example 10.7. For the ground state of hydrogen, where $\hat{J} = \hat{S}$, the interaction operator (10.110) reduces to the form $a\hat{S}\cdot\hat{I}$. In the presence of an external magnetic field $\vec{B} = B\vec{e}_z$, the electron and the nucleus both interact with the static field, such that the complete perturbation is described by the Hamiltonian:

$$\hat{H} = \gamma B \hat{S}_z - \gamma_N B \hat{I}_z + a(\hat{S}_x \hat{I}_x + \hat{S}_y \hat{I}_y + \hat{S}_z \hat{I}_z)$$

$$= \frac{\mu_B}{\hbar} g B \hat{S}_z - \frac{\mu_N}{\hbar} g_N B \hat{I}_z + \frac{A}{\hbar^2} \hat{S}_z \hat{I}_z + \frac{A}{2\hbar^2}(\hat{S}_+ \hat{I}_- + \hat{S}_- \hat{I}_+) \quad (11.72)$$

This is known as the *spin Hamiltonian*, because it operates on the electron and nuclear spin variables only. Since the eigenvalues of the spin Hamiltonian correspond to the various allowed values $m_s = \pm\frac{1}{2}$ ($S = \frac{1}{2}$) and $m_I = \pm\frac{1}{2}$ ($I = \frac{1}{2}$), it is convenient to use as a basis of representation the simple product function $\Phi_{m_s m_I} = \chi^{(e)}_{m_s} \chi^{(N)}_{m_I}$, where it is immediately apparent that the nonvanishing matrix elements are:

$$\langle \Phi_{\frac{1}{2}\frac{1}{2}} | \hat{H} | \Phi_{\frac{1}{2}\frac{1}{2}} \rangle = \tfrac{1}{2} \mu_B g B - \tfrac{1}{2} \mu_N g_N B + \tfrac{1}{4} A = \tfrac{1}{2}(\varepsilon_e - \varepsilon_N) + \tfrac{1}{4} A$$

$$\langle \Phi_{\frac{1}{2},-\frac{1}{2}} | \hat{H} | \Phi_{\frac{1}{2},-\frac{1}{2}} \rangle = \tfrac{1}{2} \mu_B g B + \tfrac{1}{2} \mu_N g_N B - \tfrac{1}{4} A = \tfrac{1}{2}(\varepsilon_e + \varepsilon_N) - \tfrac{1}{4} A$$

$$\langle \Phi_{-\frac{1}{2},\frac{1}{2}} | \hat{H} | \Phi_{-\frac{1}{2},\frac{1}{2}} \rangle = -\tfrac{1}{2} \mu_B g B - \tfrac{1}{2} \mu_N g_N B - \tfrac{1}{4} A = -\tfrac{1}{2}(\varepsilon_e + \varepsilon_N) - \tfrac{1}{4} A$$

$$\langle \Phi_{-\frac{1}{2},-\frac{1}{2}} | \hat{H} | \Phi_{-\frac{1}{2},-\frac{1}{2}} \rangle = -\tfrac{1}{2} \mu_B g B + \tfrac{1}{2} \mu_N g_N B + \tfrac{1}{4} A = -\tfrac{1}{2}(\varepsilon_e - \varepsilon_N) + \tfrac{1}{4} A$$

$$\langle \Phi_{-\frac{1}{2},\frac{1}{2}} | \hat{H} | \Phi_{\frac{1}{2},-\frac{1}{2}} \rangle = \tfrac{1}{2} A$$

$$\langle \Phi_{\frac{1}{2},-\frac{1}{2}} | \hat{H} | \Phi_{-\frac{1}{2},\frac{1}{2}} \rangle = \tfrac{1}{2} A$$

It follows that the characteristic equation for energy, Eq.(3.8), reads:

$$\begin{vmatrix} \tfrac{1}{2}(\varepsilon_e - \varepsilon_N) + \tfrac{1}{4} A - E & 0 & 0 & 0 \\ 0 & \tfrac{1}{2}(\varepsilon_e + \varepsilon_N) - \tfrac{1}{4} A - E & \tfrac{1}{2} A & 0 \\ 0 & \tfrac{1}{2} A & -\tfrac{1}{2}(\varepsilon_e + \varepsilon_N) - \tfrac{1}{4} A - E & 0 \\ 0 & 0 & 0 & -\tfrac{1}{2}(\varepsilon_e - \varepsilon_N) + \tfrac{1}{4} A - E \end{vmatrix} = 0$$

This provides the following energy eigenvalues, where the terms in A^2 can be neglected to a first approximation, because the hyperfine coupling energy is very much less than the electron Zeeman energy:

$$E_1 = \tfrac{1}{2} \mu_B g B - \tfrac{1}{2} \mu_N g_N B + \tfrac{1}{4} A$$

$$E_2 \approx \tfrac{1}{2} \mu_B g B + \tfrac{1}{2} \mu_N g_N B - \tfrac{1}{4} A + \frac{A^2}{4(\mu_B g B + \mu_N g_N B)} \approx \tfrac{1}{2} \mu_B g B + \tfrac{1}{2} \mu_N g_N B - \tfrac{1}{4} A$$

$$E_3 \approx -\tfrac{1}{2}\mu_B g B - \tfrac{1}{2}\mu_N g_N B - \tfrac{1}{4}A - \frac{A^2}{4(\mu_B g B + \mu_N g_N B)} \approx -\tfrac{1}{2}\mu_B g B - \tfrac{1}{2}\mu_N g_N B - \tfrac{1}{4}A$$

$$E_4 = -\tfrac{1}{2}\mu_B g B + \tfrac{1}{2}\mu_N g_N B + \tfrac{1}{4}A \qquad (11.73)$$

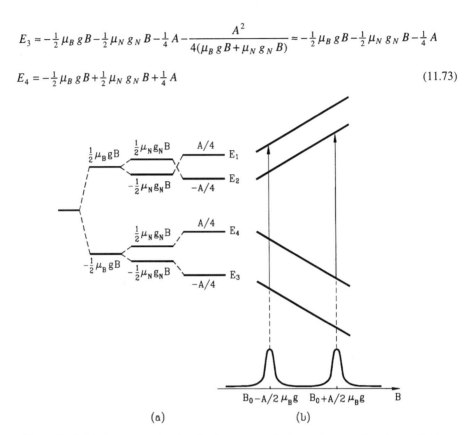

Figure 11.7. (a) Spin energy levels of hydrogen (b) ESR transitions as a function of field strength at fixed frequency

The magnetic energy levels are illustrated in Figure 11.7 (a). There are two electron resonance transitions allowed by Eq.(11.70), namely $\Phi_{\frac{1}{2},\frac{1}{2}} \to \Phi_{-\frac{1}{2},\frac{1}{2}}$ and $\Phi_{\frac{1}{2},-\frac{1}{2}} \to \Phi_{-\frac{1}{2},-\frac{1}{2}}$ with $\Delta m_s = \pm 1$ and $\Delta m_I = 0$. Because the $\hat{J}_x \equiv \hat{S}_x$ operator only acts on the electron spin, the nuclear spin part can be factored out, and it follows from Eq.(11.70) that the two transitions have equal rates:

$$P = \frac{\pi}{2}\gamma^2 B_1^2 \delta(\omega_{m_s m_s'} - \omega)$$

Hence, as illustrated in Figure 11.7 (b), the ESR spectrum consists of two lines of equal intensities. The resonance condition is obtained from Eqs.(11.73) as:

$$\hbar\omega = E_1 - E_3 = \mu_B g B_1 + \tfrac{1}{2}A \quad \text{or} \quad B_1 = \frac{\hbar\omega}{\mu_B g} - \frac{A}{2\mu_B g} = B_0 - \frac{A}{2\mu_B g}$$

$$\hbar\omega = E_2 - E_4 = \mu_B g B_2 - \tfrac{1}{2}A \quad \text{or} \quad B_2 = \frac{\hbar\omega}{\mu_B g} + \frac{A}{2\mu_B g} = B_0 + \frac{A}{2\mu_B g}$$

such that the line position provides information on the magnitude A of hyperfine coupling.

Problem 11.3.1. Find the equations that describe the precession about a static magnetic field B of the expectation value of the electron spin, for the two level system with $J = S = \frac{1}{2}$.

(Solution): Let us assume that $\vec{B} = B\vec{e}_z$ and consider the special case where the electron spin has the initial polarization along the x-axis, described by the spin-up eigenfunction (8.60):

$$\chi_+^x = \frac{1}{\sqrt{2}}(\chi_+ + \chi_-) = \frac{1}{\sqrt{2}}\begin{pmatrix}1\\1\end{pmatrix}$$

It is then immediately apparent, using Eq.(8.49), that:

$$\langle \hat{S}_x \rangle = \langle \chi_+^x | \hat{S}_x | \chi_+^x \rangle = \frac{1}{\sqrt{2}}(1,1)\frac{\hbar}{2}\begin{pmatrix}0 & 1\\1 & 0\end{pmatrix}\frac{1}{\sqrt{2}}\begin{pmatrix}1\\1\end{pmatrix} = \frac{\hbar}{2}$$

$$\langle \hat{S}_y \rangle = \frac{1}{\sqrt{2}}(1,1)\frac{\hbar}{2}\begin{pmatrix}0 & -i\\i & 0\end{pmatrix}\frac{1}{\sqrt{2}}\begin{pmatrix}1\\1\end{pmatrix} = 0$$

$$\langle \hat{S}_z \rangle = \frac{1}{\sqrt{2}}(1,1)\frac{\hbar}{2}\begin{pmatrix}1 & 0\\0 & -1\end{pmatrix}\frac{1}{\sqrt{2}}\begin{pmatrix}1\\1\end{pmatrix} = 0$$

At a later time t, we obtain from Eqs.(11.60) and (11.61) the spin state function:

$$\chi_+^x(t) = e^{-i\gamma B \hat{S}_z t/\hbar}\chi_+^x = \frac{1}{\sqrt{2}}\begin{pmatrix}e^{-i\omega_L t/2}\\e^{i\omega_L t/2}\end{pmatrix}$$

which yields the precession equations:

$$\langle \hat{S}_x \rangle_t = \langle \chi_+^x(t) | \hat{S}_x | \chi_+^x(t) \rangle = \frac{1}{\sqrt{2}}(e^{i\omega_L t/2}, e^{-i\omega_L t/2})\frac{\hbar}{2}\begin{pmatrix}0 & 1\\1 & 0\end{pmatrix}\frac{1}{\sqrt{2}}\begin{pmatrix}e^{-i\omega_L t/2}\\e^{i\omega_L t/2}\end{pmatrix}$$

$$= \frac{\hbar}{4}(e^{i\omega_L t} + e^{-i\omega_L t}) = \frac{\hbar}{2}\cos\omega_L t$$

$$\langle \hat{S}_y \rangle_t = \frac{1}{\sqrt{2}}(e^{i\omega_L t/2}, e^{-i\omega_L t/2})\frac{\hbar}{2}\begin{pmatrix}0 & -i\\i & 0\end{pmatrix}\frac{1}{\sqrt{2}}\begin{pmatrix}e^{-i\omega_L t/2}\\e^{i\omega_L t/2}\end{pmatrix}$$

$$= \frac{\hbar}{4i}(e^{i\omega_L t} - e^{-i\omega_L t}) = \frac{\hbar}{2}\sin\omega_L t$$

436 Quantum Physics

$$\langle \hat{S}_z \rangle_t = \frac{1}{\sqrt{2}}(e^{i\omega_L t/2}, e^{-i\omega_L t/2}) \frac{\hbar}{2}\begin{pmatrix} 1 & 0 \\ 0 & -1 \end{pmatrix} \frac{1}{\sqrt{2}}\begin{pmatrix} e^{-i\omega_L t/2} \\ e^{i\omega_L t/2} \end{pmatrix} = 0$$

It follows that the spin precesses in the xy plane, with the Larmor frequency.

Problem 11.3.2. Show that the electron resonance spectrum of hydrogen, in the absence of an external field, consists of a single line.

(Solution): In the absence of the applied field B, the spin Hamiltonian of the hydrogen atoms, Eq.(11.72), reduces to the isotropic hyperfine interaction:

$$\hat{H} = \frac{A}{\hbar^2}\hat{S}_z\hat{I}_z + \frac{A}{2\hbar^2}(\hat{S}_+\hat{I}_- + \hat{S}_-\hat{I}_+)$$

such that the characteristic equation of the energy eigenvalue problem simplifies to:

$$\begin{vmatrix} \frac{1}{4}A - E & 0 & 0 & 0 \\ 0 & -\frac{1}{4}A - E & \frac{1}{2}A & 0 \\ 0 & \frac{1}{2}A & -\frac{1}{4}A - E & 0 \\ 0 & 0 & 0 & \frac{1}{4}A - E \end{vmatrix} = (\tfrac{1}{4}A - E)^3(-\tfrac{3}{4}A - E) = 0$$

The energy level $E = -\tfrac{3}{4}A$ corresponds to an antisymmetric singlet spin state:

$$\Phi^a = \frac{1}{\sqrt{2}}(\Phi_{\frac{1}{2},-\frac{1}{2}} - \Phi_{-\frac{1}{2},\frac{1}{2}}) = \frac{1}{\sqrt{2}}(\chi_+^{(e)}\chi_-^{(N)} - \chi_-^{(e)}\chi_+^{(N)})$$

while the "triplet" energy state with $E = \tfrac{1}{4}A$ is described by:

$$\Phi_1^s = \Phi_{\frac{1}{2}\frac{1}{2}} = \chi_+^{(e)}\chi_+^{(N)}$$

$$\Phi_0^s = \frac{1}{\sqrt{2}}(\Phi_{\frac{1}{2},-\frac{1}{2}} + \Phi_{-\frac{1}{2},\frac{1}{2}}) = \frac{1}{\sqrt{2}}(\chi_+^{(e)}\chi_-^{(N)} + \chi_-^{(e)}\chi_+^{(N)})$$

$$\Phi_{-1}^s = \Phi_{-\frac{1}{2},-\frac{1}{2}} = \chi_-^{(e)}\chi_-^{(N)}$$

The physical situation is similar to that found for a system of two interacting electrons in helium, Eqs.(10.39) and (10.55), such that it is convenient to introduce the total angular momentum, Eq.(10.106), which reads:

$$\hat{F} = \hat{S} + \hat{I}$$

The singlet spin state is non-degenerate, as $F = 0$ and $2F + 1 = 1$, while the triplet spin state corresponds to $F = 1$ and $2F + 1 = 3$, and hence the three symmetric spin functions have been labelled by the quantum numbers $m_F = m_s + m_I = 0, \pm 1$. It follows that there is one singlet-triplet allowed transition, which results in a single line zero-field ESR spectrum, with frequency $\omega = A/\hbar$.

Problem 11.3.3. A particle of spin $J = \frac{1}{2}$ is in a magnetic field described by Eqs.(11.64). If at time $t = 0$ the spin points along the positive z-axis, find the time elapsed until the spin will point along the negative z direction.

(Answer): $t = \pi/\gamma B_1$

Problem 11.3.4. Find the separation of the hyperfine levels in zero-field for the one-electron atom with $S = \frac{1}{2}$ and $I > \frac{1}{2}$.

(Answer): $E = -\frac{1}{4} A \pm \frac{1}{2}(I + \frac{1}{2})A$

12. NUCLEAR RADIATION

12.1. Radioactive Decay Law

Dynamic nuclear properties are associated with transitions from some initial to some final system of nucleons. The transitions will occur spontaneously through *radioactive decay*, if the total energy of the final system is less than that of the initial system. In the approximation where the interaction energy involved in the transition is small, and can be treated as a constant perturbation, which is applied for a time t, the transition rate from an initial state Ψ_i into a final state Ψ_f can be expressed by the Fermi golden rule (9.79), which reads:

$$P_{i \to f} = \frac{2\pi}{\hbar} |\langle \Psi_f | \hat{V} | \Psi_i \rangle|^2 \frac{dn}{dE} \tag{12.1}$$

where $\hat{V}$ is the operator for the interaction and dn/dE is the number of final states in the interval E to $E + dE$ of the total energy, called the density of final states.

The three radioactive decay modes, known as α-rays (or α-particles, which are nuclei of $^4_2$He, consisting of two protons and two neutrons), β-rays (or β-particles, which are electrons) and γ-rays (electromagnetic radiation), have a common time dependence, which can be derived if it is assumed that the decay is statistical in nature. In other words the probability that a *parent* (initial) nuclear state decays is of the same magnitude at any time for any nucleus. If N nuclei are in a given nuclear state at time t and λdt is the probability of decay in the time interval dt, the decrease in the number N in the short time dt is given by:

$$-dN = N\lambda dt \tag{12.2}$$

which, on integration, yields the radioactive decay law:

$$N(t) = N_0 e^{\lambda t} = N_0 e^{-t/\langle t \rangle} \tag{12.3}$$

where $N(t)$ is the number of nuclei present at time t in the given state, $N_0 = N(0)$, λ is the *decay constant* and $\langle t \rangle$ is the *mean lifetime* of the state, given by:

$$\langle t \rangle = \frac{1}{N_0} \int_0^{N_0} t \, dN = \frac{1}{N_0} \int_\infty^0 t(-N_0 e^{-\lambda t} \lambda) dt = \int_0^\infty \lambda t e^{-\lambda t} dt = \frac{1}{\lambda} \quad (12.4)$$

Instead of counting the number of nuclei $N(t)$, it is easier to count the rate at which decays occur in the radioactive sample, which is called the *activity*:

$$\left| \frac{dN}{dt} \right| = \lambda N_0 e^{-\lambda t} = \lambda N(t) \quad (12.5)$$

and obviously follows the exponential decay law (12.3). It is also convenient to introduce the *half-life* $t_{1/2}$ as the time interval in which the activity is reduced to one-half:

$$t_{1/2} = \frac{\ln 2}{\lambda} = \langle t \rangle \ln 2 \quad (12.6)$$

The probability of radioactive decay through the direct transition of the nucleus from an initial state Ψ_i to a final state Ψ_f, in time t, is given by:

$$P_{if}(t) = |a_f(t)|^2 \quad (12.7)$$

where the wave function amplitudes $a_f(t)$ are obtained from Eq.(9.60), which can be approximated for $\lambda = 1$, as discussed in Section 9.3, by:

$$i\hbar \frac{da_f(t)}{dt} = a_i(t) V_{fi}(t) e^{i(E_f - E_i)t/\hbar} \quad (12.8)$$

Since the probability of finding our decaying system in the initial state Ψ_i must decrease with time according to the radioactive decay law (12.3), namely:

$$|\Psi_i(t)|^2 = |\Psi_i|^2 e^{-\lambda t} \quad (12.9)$$

we choose:

$$a_i(t) = e^{-\lambda t/2} \quad (12.10)$$

rather than $a_i = 1$, as considered in Eq.(9.64). Substituting this result into Eq.(12.8), and assuming a constant perturbation switched on at time $t = 0$, gives:

$$i\hbar \frac{da_f(t)}{dt} = V_{fi} e^{-\lambda t/2 + i(E_f - E_i)t/\hbar} \qquad (12.11)$$

and on integration we get an oscillatory behaviour of the wave function amplitude:

$$a_f(t) = \frac{\hbar V_{fi}\left[e^{i(E_f - E_i + i\hbar\lambda/2)t/\hbar} - 1\right]}{(E_f - E_i) + i\hbar\lambda/2} \qquad (12.12)$$

At the end of the nuclear transition we should consider the amplitude corresponding to $t \to \infty$:

$$a_f(\infty) = \frac{\hbar V_{fi}}{(E_f - E_i) + i\hbar\lambda/2} \qquad (12.13)$$

which yields the transition probability (12.7) in the form:

$$P_{if}(\infty) = \frac{\hbar^2 |V_{fi}|^2}{(E_f - E_i)^2 + \frac{1}{4}\Gamma^2} \qquad (12.14)$$

Hence, the probability to observe the nucleus with energy E_f in the vicinity of E_i follows a Lorentzian distribution, and this implies that the intensity $I(\omega)$ of nuclear radiation can be described by Eq.(11.55), derived for unstable electron states. By analogy, $\Gamma = \hbar\lambda = \hbar/\langle t \rangle$ gives the natural linewidth of the emitted radiation, which is the same as the *width* of the decaying nuclear state. Nuclear states which are populated in ordinary decays, typically have lifetimes (12.4) greater than 10^{-12} sec, corresponding to $\Gamma(eV) = 0.66 \times 10^{-15}/\langle t \rangle$, and this is of the order of 10^{-3} eV. Since their width is small compared with the energy spacing of nuclear levels, which is of the order of 10^{-3} MeV, the nuclear decaying states may be regarded as discrete quasi-stationary states.

Problem 12.1.1. Derive the observed line profile of nuclear radiation in the case when detection is restricted to a time t_0 which is short compared to $\langle t \rangle$.

(Solution): If observations are made up to a given time t_0, Eq.(12.12) gives:

$$a_f(t_0) = \frac{\hbar V_{fi}}{(E_f - E_i) + i\Gamma/2}\left[e^{-\Gamma t_0/2\hbar} e^{i(E_f - E_i)t_0/\hbar} - 1\right]$$

and therefore, we obtain the intensity:

$$I(E) = \frac{\hbar^2 |V_{fi}|^2}{(E-E_i)^2 + \frac{1}{4}\Gamma^2} \left[1 + e^{-\Gamma t_0/\hbar} - 2e^{-\Gamma t_0/2\hbar} \cos\left(\frac{E-E_i}{\hbar} t_0\right) \right]$$

For $t_0 \ll \langle t \rangle = \hbar/\Gamma$ or $\Gamma t_0/\hbar \ll 1$, the observed line profile has the form:

$$I(E) \approx 2I_0 \frac{1-\cos\left(\frac{E-E_i}{\hbar} t_0\right)}{(E-E_i)^2} = I_0 \frac{\sin^2\left(\frac{E-E_i}{2\hbar} t_0\right)}{(E-E_i)^2}$$

The linewidth is approximately given by $\hbar/t_0$, which is much larger than Γ, and hence the spread of the observed energy distribution is much increased.

Problem 12.1.2. Find the line profile and the linewidth Δ for the gamma radiation emitted by an ensemble of radioactive nuclei of mass M, at equilibrium at temperature T, assuming that the emitters behave like an ideal gas.

(Solution): For isolated atoms it is sufficient to consider the translational motion of the nucleus along the direction in which the γ-ray is emitted, say Ox, because the decay frequency $\omega_{if} = (E_i - E_f)/\hbar$ remains unaffected by the components of motion perpendicular to this axis. The observed frequency ω is different from ω_{if} due to the Doppler effect (see Problem P.2.1):

$$\frac{\omega - \omega_{if}}{\omega_{if}} = \pm \frac{v_x}{c}$$

Hence, the velocity distribution, which is described in one dimension by the Boltzmann distribution law:

$$P(v_x) = \sqrt{\frac{M}{2\pi k_B T}} e^{-Mv_x^2/2k_B T}$$

results in a distribution of frequencies, such that the intensity distribution has a Gaussian profile:

$$I(\omega) = I_0 e^{-Mc^2(\omega-\omega_{if})^2/2k_B T \omega_{if}^2}$$

with linewidth Δ given by the condition $I_0/2 = I_0 e^{-Mc^2\Delta^2/2k_B T \omega_{if}^2}$. Therefore the Lorentzian distribution in energy of the emitted γ-ray, with natural linewidth Γ, is broadened into a Gaussian distribution, of width:

$$\Delta = \frac{\omega_{if}}{c}\sqrt{(2k_B \ln 2)\frac{T}{M}}$$

It is apparent that the Doppler line broadening is reduced for γ-rays emitted by heavy nuclides at low temperature.

Problem 12.1.3. If at time $t = 0$ we have N_0 atoms of a radioactive nuclide X_1, which decays through the series:

$$X_1 \xrightarrow{\lambda_1} X_2 \xrightarrow{\lambda_2} X_3 \xrightarrow{\lambda_3} \cdots$$

show that the number of atoms $N_i(t)$, present at any time t $(i = 1, 2, 3, \ldots)$, can be expressed in the form:

$$N_i(t) = \sum_{j=1}^{i} A_{ij} e^{-\lambda_j t}$$

and find the recurrence relation for the A_{ij}.

(Answer): $\quad A_{ij} = \dfrac{\lambda_{i-1}}{\lambda_i - \lambda_j} A_{i-1, j}$

12.2. Alpha Decay

Alpha decay is the process of spontaneous emission of an α-particle (^4_2He), represented by the equation:

$$^A_Z X \to {^{A-4}_{Z-2}} X' + \alpha \tag{12.15}$$

where the atomic number, and hence, the chemical nature of the residual nucleus $^{A-4}_{Z-2} X'$ is different from its parent $^A_Z X$. The *separation energy* S_α needed to remove an α-particle from the parent nucleus is defined in terms of the nuclear binding energies (10.89) as:

$$S_\alpha = B(A, Z) - B(A-4, Z-2) - B(\alpha) = Af(A) + (A-4)f(A-4) - B(\alpha)$$

$$= (A-4)\big[f(A) - f(A-4)\big] + 4f(A) - B(\alpha)$$

For $A > 60$ the binding fraction $f(A)$, Eq.(10.90), varies sufficiently smoothly with A for its derivative with respect to A to exist, so that:

$$S_\alpha \cong 4(A-4)\frac{df}{dA} + 4f(A) - B(\alpha) \tag{12.16}$$

The binding energy of the α-particle is $B(\alpha) = 28$ MeV and for heavy nuclei $f(A) \cong 7.5$ MeV, so $4f(A) - B(\alpha) \cong 2$ MeV. Since for large A we have $df/dA < 0$ (see Figure 10.7 (a)) we find that S_α becomes negative for $A > 150$ and consequently many heavy nuclei are α radioactive. Experiment shows that α emitters with large decay energies $Q_\alpha = -S_\alpha$ have short half-lives $t_{1/2}$. The reciprocal is also true and this result is known as the *Geiger-Nutall law*:

$$\ln \lambda = a - bQ_\alpha^{-1/2} = a - \frac{b'}{v_\alpha} \tag{12.17}$$

where v_α is the velocity with which the α-particle leaves the nucleus, expressed for the nonrelativistic case, where $Q_\alpha = mv_\alpha^2/2$.

The mechanism of alpha decay has been explained by Gamov using the one-particle model, which assumes that the α-particle exists within the nucleus and moves in the field of the residual nucleus. The corresponding potential energy is given by the superposition of the nuclear and electrostatic potentials, resulting in the Coulomb barrier shown in Figure 12.1, which can be represented, to a good approximation, by:

$$U(r) = -V_0, \quad r < R$$
$$= \frac{zZ'e_0^2}{r}, \quad r \geq R \tag{12.18}$$

where $z = 2$ for an α-particle, $Z' = Z - 2$ is the atomic number of the residual nucleus and $r = R$ defines the boundary of the parent nucleus.

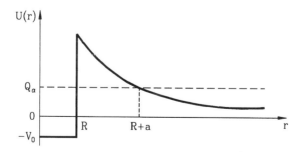

Figure 12.1. Coulomb barrier for alpha decay

444   Quantum Physics

We have seen before, in Example 6.4, that a particle can penetrate a potential energy barrier of finite width and height, with a probability given by the transmission coefficient τ, Eq.(6.50). A good approximation for α-particle tunnelling through the barrier, shown in Figure 12.1, will then be expressed in the form:

$$\tau = \tau_0 e^{-2G} \tag{12.19}$$

where, in view of Eq.(6.50), the so called *Gamov factor* G is given by:

$$G = \int_R^{R+a} \sqrt{\frac{2m_\alpha}{\hbar^2}[U(r) - Q_\alpha]}\, dr = \int_R^{R+a} \sqrt{\frac{2m_\alpha}{\hbar^2}\left(\frac{2Z'e_0^2}{r} - Q_\alpha\right)}\, dr \tag{12.20}$$

It is clear from Figure 12.1 that the distance a is specified by the condition:

$$\frac{2Z'e_0^2}{(R+a)} = Q_\alpha \tag{12.21}$$

so that it is convenient to introduce the energy ratio:

$$\frac{1}{x} = \frac{2Z'e_0^2}{rQ_\alpha} = \frac{R+a}{r} \tag{12.22}$$

It follows that Eq.(12.20) may successively be rewritten as:

$$G = \sqrt{\frac{2m_\alpha Q_\alpha}{\hbar^2}} \frac{2Z'e_0^2}{Q_\alpha} \int_{R/(R+a)}^{1} \sqrt{\frac{1}{x} - 1}\, dx$$

$$= \sqrt{\frac{2m_\alpha}{\hbar^2 Q_\alpha}}\, 2Z'e_0^2 \left[\sqrt{x(1-x)} - \arccos\sqrt{x}\right]_{R/(R+a)}^{1}$$

$$= \sqrt{\frac{2m_\alpha}{\hbar^2 Q_\alpha}}\, 2Z'e_0^2 \left[\arccos\sqrt{\frac{R}{R+a}} - \sqrt{\frac{R}{R+a}\left(1 - \frac{R}{R+a}\right)}\right] \tag{12.23}$$

For small energies Q_α, we have $R \ll a$ (see Figure 12.1), and we can approximate the last term in Eq.(12.23) by a series expansion, which gives:

$$G = \sqrt{\frac{2m_\alpha}{\hbar^2 Q_\alpha}}\, RU(R)\left(\frac{\pi}{2} - 2\sqrt{\frac{R}{R+a}}\right) = \sqrt{\frac{2m_\alpha}{\hbar^2 Q_\alpha}}\, RU(R)\left[\frac{\pi}{2} - 2\sqrt{\frac{Q_\alpha}{U(R)}}\right]$$

$$= \frac{\pi RU(R)}{\hbar v_\alpha} - \frac{2R}{\hbar}\sqrt{2m_\alpha U(R)} \tag{12.24}$$

Substituting Eq.(12.24), the probability (12.19) for tunnelling the barrier is thus obtained in terms of v_α, R and Z'. The decay constant λ of an α-emitter is given by the product of the tunnelling probability τ and the frequency v_i/R with which the α-particle hits the barrier:

$$\lambda \approx \frac{v_i}{R} \tau \qquad (12.25)$$

where v_i denotes the velocity of the α-particle *inside* the nucleus. By combining Eqs.(12.25), (12.24) and (12.19) we then obtain:

$$\ln \lambda \approx \ln \frac{v_i}{R} - \frac{2\pi R U(R)}{\hbar v_\alpha} + \frac{4R}{\hbar} \sqrt{2m_\alpha U(R)} \qquad (12.26)$$

and this correctly reproduces the Geiger-Nutall law (12.17) to within a factor of 50. Due to the strong dependence of λ on R, Eq.(12.26) can be used to evaluate the nuclear radius.

The Gamov theory was developed for heavy nuclei in their ground state $l = 0$. For $l > 0$ a centrifugal potential is added to the potential energy $U(r)$, as shown before in Eq.(10.91). As a result, the height of the potential barrier, Eq.(12.18), is increased, and so is the effective thickness a of the barrier. Instead of Eq.(12.20), the Gamov factor must be expressed as:

$$G = \int_R^{R+a} \sqrt{\frac{2m_\alpha}{\hbar^2} \left(\frac{2Z'e_0^2}{r} + \frac{l(l+1)\hbar^2}{2m_\alpha r^2} - Q_\alpha \right)} \, dr \qquad (12.27)$$

It is, however, clear that, for large Z, the influence of the centrifugal term is small, because the ratio of the centrifugal to the Coulomb barrier height is:

$$\frac{l(l+1)\hbar^2}{4m_\alpha Z' e_0^2 R} \cong 0.002 \, l(l+1)$$

and, consequently, λ does not depend strongly on the quantum number l. The simplifying assumptions of the one-particle model seem therefore justified by the small influence of the neglected factors on the decay constant λ.

Problem 12.2.1. The variation of nuclear radius as a function of A is well described by $R = R_0 A^{1/3}$. Calculate the nuclear unit radius R_0 in terms of the Z, A, Q_α and $t_{1/2}$ values for a given α-emitting isotope.

Quantum Physics

(Solution): It is convenient to use the Gamov factor, Eq.(12.24) written in the form:

$$G = \sqrt{\frac{2m_\alpha}{\hbar^2 Q_\alpha}} \, 2Z'e_0^2 \left(\frac{\pi}{2} - 2\sqrt{\frac{R}{R+a}} \right) = \frac{e^2 \sqrt{2m_\alpha}}{4\varepsilon_0 \hbar} \frac{Z'}{\sqrt{Q_\alpha}} - \sqrt{\frac{4m_\alpha e^2}{\pi \varepsilon_0 \hbar^2}} \sqrt{Z'R}$$

As the energies of α-particles range from 4 to 9 MeV, and the radius is of the order of femtometers (fm) or 10^{-15} m, this formula simplifies to:

$$G = 1.97 \frac{Z'}{\sqrt{Q_\alpha}} - 1.49 \sqrt{Z'R}$$

with Q_α measured in MeV and R in fm. The corresponding velocities of the emitted α-particles are about 10^7 m/s, and assuming velocities v_i of the same order inside the nucleus, we obtain:

$$\log t_{1/2} = \log \frac{\ln 2}{\lambda} \approx \log \frac{v_i}{R} - \frac{1}{2.3}\left(3.94 \frac{Z'}{\sqrt{Q_\alpha}} - 2.98 \sqrt{Z'R} \right)$$

or:

$$R_0 = \left(\frac{17 + 0.77 \log t_{1/2}}{\sqrt{Z-2}} + 1.32 \sqrt{\frac{Z-2}{Q_\alpha}} \right)^2 A^{-\frac{1}{3}}$$

if $t_{1/2}$ is measured in seconds. The available data yield R_0 in the range 1.4 to 1.5 fm, which represents, however, an overestimate, due to the fact that the size of the α-particles has not been considered. It is obvious that variations of several orders of magnitude in the half-life of α-emitting isotopes, which ranges from 10^{-7} s to 10^9 years, result in a small change in the R_0 value.

Problem 12.2.2. The experimental energies of the α-particles emitted in the decays of ^{226}Ra and ^{226}Th are 4.9 and 6.5 MeV respectively. Find the ratio of their half-lives.

(Answer): $t_{1/2}(^{226}\text{Ra}) / t_{1/2}(^{226}\text{Th}) = 2.5 \times 10^7$

Problem 12.2.3. Derive an approximate expression for the Gamov factor, in terms of the wave number $k = m_\alpha v_\alpha / \hbar$, which can be introduced for the α-rays at large distances from the nucleus.

(Answer): $G \approx e^{-\pi k(R+a)}$

12.3. Beta Decay

The term β-decay describes processes in which a nucleus makes an isobaric transition, where the mass number A remains constant and the atomic number Z increases or decreases by one. These processes are associated with the emission of electrons e^- (negative beta decay, β^-) or positrons e^+ (positive beta decay, β^+). The second of these is always simultaneous with the capture by the nucleus of an orbital atomic electron (electron capture). Although positive or negative electrons are emitted in β-decay, there is strong evidence that they cannot form a constituent part of nuclei. We are then led to the assumption that electrons are created at the moment the nucleus decays. Since the only change between the initial and final nucleus is that a neutron has been changed into a proton, in a β^--decay, or conversely, in a β^+-decay, the basic decay processes can be assigned to nucleons:

$$_{Z}^{A}X \rightarrow {}_{Z+1}^{A}X' + e^- \quad \text{or} \quad {}_{0}^{1}n \rightarrow {}_{1}^{1}p + e^-$$

$$_{Z}^{A}X \rightarrow {}_{Z-1}^{A}X' + e^+ \quad \text{or} \quad {}_{1}^{1}p \rightarrow {}_{0}^{1}n + e^+$$

$$_{Z}^{A}X + e^- \rightarrow {}_{Z-1}^{A}X' \quad \text{or} \quad {}_{1}^{1}p + e^- \rightarrow {}_{0}^{1}n$$

Such two-body processes would predict that each β-particle should be emitted with a well-defined energy. However, it is observed experimentally that electrons in β-decay have a continuous energy distribution. To account for the variable energy of β-particles, it must be postulated that another particle is emitted simultaneously, which takes up the remaining energy and momentum:

$$_{0}^{1}n \rightarrow {}_{1}^{1}p + e^- + \overline{\nu}$$

$$_{1}^{1}p \rightarrow {}_{0}^{1}n + e^+ + \nu \tag{12.28}$$

$$_{1}^{1}p + e^- \rightarrow {}_{0}^{1}n + \nu$$

The additional particle, called a *neutrino* ν or an *antineutrino* $\overline{\nu}$, must be neutral, because the charge is already conserved without the neutrino, and it must have zero or almost zero rest mass, because in all β-decays the maximum electron energy observed is practically equal to the total energy available. Since protons, neutrons and electrons each have spin $\frac{1}{2}\hbar$, the conservation of angular momentum requires the neutrino also to have spin $\frac{1}{2}\hbar$. The theory of β-decay, given by Fermi, is based on the analogy of the decay processes (12.28) with photon emission, where a photon is created during a transition caused by perturbation (weak interaction) of the stationary states. Hence, the Fermi golden rule (12.1) can be used to obtain the probability per unit time $N(p)dp$ for the emission of a β-particle, within the momentum range p to $p + dp$, which reads:

$$N(p)dp = \frac{2\pi}{\hbar}|\langle\psi_e\psi_v\Psi_f|\hat{V}|\Psi_i\rangle|^2 \frac{dn}{dQ} \qquad (12.29)$$

where Q is the decay energy, and Ψ_i, Ψ_f are the time-independent wave functions of the nuclear system in the initial and final states. The wave functions ψ_e, ψ_v of the escaping electron and neutrino can be taken in the free-particle form, normalized within a volume V as:

$$\psi_e(\vec{r}) = \frac{1}{\sqrt{V}}e^{i\vec{p}\cdot\vec{r}/\hbar}, \qquad \psi_v(\vec{r}) = \frac{1}{\sqrt{V}}e^{i\vec{q}\cdot\vec{r}/\hbar} \qquad (12.30)$$

Since the interaction only occurs at the position of the nucleus, and $\vec{p}\cdot\vec{r} \ll \hbar, \vec{q}\cdot\vec{r} \ll \hbar$ over the nuclear volume, we may expand the exponentials in a power series, and keep only the first term:

$$e^{i\vec{p}\cdot\vec{r}/\hbar} = 1 + i\vec{p}\cdot\vec{r}/\hbar + \ldots \cong 1$$
$$e^{i\vec{q}\cdot\vec{r}/\hbar} = 1 + i\vec{q}\cdot\vec{r}/\hbar + \ldots \cong 1 \qquad (12.31)$$

As a result of this approximation, the matrix element of the interaction becomes independent of the energies of the electron and neutrino, and is responsible for the *allowed transitions*. It is usually expressed in terms of a constant g of the weak interaction as:

$$\langle\Psi_f|\hat{V}|\Psi_i\rangle = \frac{g}{V}\langle\Psi_f|\hat{M}|\Psi_i\rangle = \frac{g}{V}M_{if} \qquad (12.32)$$

where M_{if} is called the *nuclear matrix element*. The form of the operator $\hat{M}$ is left unspecified, because the form of the weak interaction is as yet unknown. Hence, the probability (12.29) of an allowed transition depends on the electron and neutrino energies only through the density of the final states. The number of available states for either the electron or the neutrino, in the normalization volume V, are determined by the properties of the wave function and the energy of the free particle, which have the form:

$$\psi(\vec{r}) = \frac{1}{\sqrt{V}}e^{i\vec{k}\cdot\vec{r}}, \qquad E = \frac{\hbar^2 k^2}{2m}$$

where $\vec{k} = \vec{p}/\hbar$ for the electron, and $\vec{k} = \vec{q}/\hbar$ for the neutrino. Assuming, for simplicity, that the volume is a cubic cell $V = L^3$, there is a restriction of spatial periodicity on the wave function:

$$\psi(x, y, z) = \psi(x + L, y, z) = \psi(x, y + L, z) = \psi(x, y, z + L)$$

12. Nuclear Radiation

which yields a quantization condition on the wave vector values:

$$k_x = \frac{2\pi}{L}a_x, \quad k_y = \frac{2\pi}{L}a_y, \quad k_z = \frac{2\pi}{L}a_z \quad \text{or} \quad k^2 = \left(\frac{2\pi}{L}\right)^2 (a_x^2 + a_y^2 + a_z^2) = \left(\frac{2\pi}{L}\right)^2 a^2$$

It follows that to each set of integers (a_x, a_y, a_z) there is a corresponding value $k(a)$, and hence, using the wave vector representation, the available $\vec{k}$-states are contained in a spherical volume of radius a, where the number of k values is:

$$n = \frac{4\pi}{3}a^3 = \frac{L^3}{6\pi^2}k^3 = \frac{V}{6\pi^2}k^3$$

We thus obtain for the wave vector interval dk (neglecting spin):

$$dn = \frac{Vk^2}{2\pi^2}dk$$

It follows that an electron, with momentum $\vec{p} = \hbar\vec{k}$, has a number of final states in the momentum interval dp given by:

$$dn_e = V\frac{4\pi p^2}{(2\pi\hbar)^3}dp \qquad (12.33)$$

and for the neutrino, with momentum $\vec{q} = \hbar\vec{k}$, the number of final states in the momentum interval dq is:

$$dn_\nu = V\frac{4\pi q^2}{(2\pi\hbar)^3}dq \qquad (12.34)$$

Hence, the number of final states which have simultaneously an electron and a neutrino with proper momenta can be written as:

$$dn = \frac{16\pi^2 V^2}{(2\pi\hbar)^6}p^2 q^2 dp dq \qquad (12.35)$$

The kinetic energy of the neutrino can be derived from the decay energy Q:

$$Q = T_e + T_\nu \quad \text{or} \quad T_\nu = Q - T_e$$

so that its relativistic momentum reads:

$$q = \frac{T_\nu}{c} = \frac{Q - T_e}{c} \tag{12.36}$$

where it was assumed that the neutrino rest mass is zero. For fixed p, hence at constant T_e, we have $dq = dQ/c$ and Eq.(12.35) becomes:

$$\frac{dn}{dQ} = \frac{16\pi^2 V^2}{(2\pi\hbar)^6 c^3}(Q - T_e)^2 p^2 dp \tag{12.37}$$

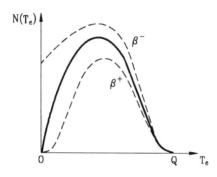

Figure 12.2. Electron energy distribution in β-decay. The dashed lines show the effect of Coulomb interaction on the spectrum shape for negative and positive β-decays

Substituting Eqs.(12.37) and (12.32) into Eq.(12.29) gives:

$$N(p)dp = \frac{g^2}{2\pi^3 \hbar^7 c^3}|M_{if}|^2 (Q - T_e)^2 p^2 dp \tag{12.38}$$

This momentum distribution can be transformed into an energy distribution, using the differential form of the momentum-energy relation of special relativity:

$$c^2 p\, dp = E dE = (T_e + m_e c^2) dT_e$$

so that Eq.(12.38) assumes the form:

$$N(T_e)dT_e = \frac{g^2}{2\pi^3 \hbar^7 c^6}|M_{if}|^2 (Q - T_e)^2 (T_e^2 + 2T_e m_e c^2)^{1/2}(T_e + m_e c^2)dT_e \tag{12.39}$$

A plot of Eq.(12.39) is shown in Figure 12.2. The distribution vanishes at the minimum energy $T_e = 0$, and also at the endpoint $T_e = Q$, which is the maximum energy of the electron. However, at low kinetic energies, the Coulomb effect of the nucleus on the emitted β-particles, which has been neglected so far, becomes significant. The electrons are slowed down by the attractive Coulomb potential, as they leave the nucleus, and the spectrum is shifted toward lower energies. The positrons are accelerated by electrostatic repulsion and there will be a deficiency of low energy positrons in the β^+ spectrum (see Figure 12.2). The Coulomb correction is applied by multiplying Eqs.(12.38) and (12.39) by the Fermi function $F(Z',p)$ and $F(Z',T_e)$ respectively, where Z' is the atomic number of the residual nucleus.

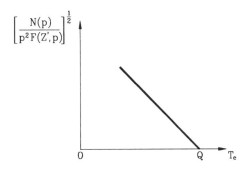

Figure 12.3. Kurie plot for allowed β-transitions

The momentum distribution (12.38) may then be rewritten as:

$$N(p)dp = \text{const} \cdot F(Z',p)(Q-T_e)^2 p^2 dp \qquad (12.40)$$

or:

$$\left[\frac{N(p)}{p^2 F(Z',p)}\right]^{\frac{1}{2}} = \text{const} \cdot (Q-T_e) \qquad (12.41)$$

If the function on the left hand side of Eq.(12.41) is plotted versus T_e, a straight line should be obtained for the allowed transitions, which makes an intercept Q on the energy axis, as in Figure 12.3. Such a plot, called a *Kurie plot*, gives an accurate way of measuring Q and provides a simple experimental test of the theory.

The total decay probability λ is obtained by integrating Eq.(12.40) over all momenta, which yields:

$$\lambda = \int_0^{p_{max}} N(p)dp = \text{const} \cdot \int_0^{p_{max}} F(Z',p)(Q-T_e)^2 p^2 dp$$

$$= \text{const} \cdot m_e^5 c^7 f(Z',\frac{Q}{m_e c^2}) = Cf(Z',\frac{Q}{m_e c^2}) \qquad (12.42)$$

where:

$$C = \frac{m_e^5 c^4}{2\pi^3 \hbar^7} g^2 |M_{if}|^2 \qquad (12.43)$$

Values for both $F(Z',p)$ and $f(Z',Q/m_ec^2)$ are found in tables. Equation (12.42) is usually rewritten in terms of the half-life $t_{1/2}$ of the decay:

$$t_{1/2} = \frac{\ln 2}{\lambda} = \frac{\ln 2}{Cf} \quad \text{or} \quad ft_{1/2} = \frac{\ln 2}{C} \tag{12.44}$$

The quantity $ft_{1/2}$ is called the *ft value* for the transition, can be calculated knowing $t_{1/2}$ and Q (hence f) given by experiment. By combining Eqs.(12.44) and (12.43) we can estimate the magnitude of the β-decay interaction, which is of the order of 10^{-13} weaker than the so-called strong interaction between nucleons.

Problem 12.3.1. Show that, in the case of high-energy β-decay of low-Z emitting nuclei, where $F(Z',p)$ can be approximated by unity, the decay constant is proportional to Q^5.

(Solution): The decay constant depends on Q through the Fermi function $f(Z',Q/m_ec^2)$, Eq.(12.42), which reads:

$$f(Z', \frac{Q}{m_ec^2}) = \frac{1}{m_e^5 c^7} \int_0^{p_{max}} (Q-T_e)^2 p^2 dp$$

$$= \frac{1}{(m_ec^2)^5} \int_0^Q (Q-T_e)^2 (T_e^2 + 2T_e m_e c^2)^{\frac{1}{2}} (T_e + m_e c^2) dT_e$$

where use has been made of Eq.(12.39). For high-energy β-decays, where $T_e \gg m_e c^2$, the last integral simplifies to:

$$\int_0^Q (Q-T_e)^2 T_e^2 dT_e = Q^5 \int_0^Q \left(1-\frac{T_e}{Q}\right)^2 \left(\frac{T_e}{Q}\right)^2 d\left(\frac{T_e}{Q}\right) = Q^5 \int_0^1 (1-x)^2 x^2 dx = \frac{Q^5}{30}$$

such that:

$$f(Z', \frac{Q}{m_ec^2}) \approx \frac{1}{30}\left(\frac{Q}{m_ec^2}\right)^5$$

Substituting into Eqs.(12.42), and using Eq.(12.43), this finally gives the decay constant in the form:

$$\lambda \approx \frac{g^2 |M_{if}|^2}{60\pi^3 \hbar^7 c^6} Q^5$$

Problem 12.3.2. Use the uncertainty relations to show that the electron is not a constituent part of the nucleus.

12.4. Gamma Radiation

An excited state of a nucleus, produced as a result of an earlier α- or β-decay, will decay to a lower energy or to the ground state through emission of electromagnetic radiation. The energy spectrum consists of sharp lines, Lorentzian in shape, which show that the nucleus has discrete energy levels. It is assumed, as in the similar atomic problem, that the energy of the *gamma radiation*, emitted and absorbed in nuclear transitions, is characteristic of the energy difference between nuclear levels:

$$\hbar \omega_{if} = E_i - E_f \tag{12.45}$$

The methods of approximation employed in the atomic case can be used for radiative nuclear transitions, if the wavelength of gamma radiation is large compared to the size of the emitting system:

$$kR = \frac{2\pi}{\lambda} R = \frac{\omega R}{c} = \hbar\omega \frac{R}{\hbar c} = (E_i - E_f) \frac{R}{\hbar c} \ll 1 \tag{12.46}$$

or $E_i - E_f \ll \hbar c / R \cong 20$ MeV. Since, above the binding energy per nucleon (see Figure 10.7 (a)), particle emission is favoured over electromagnetic radiation, the restriction (12.46) is normally satisfied, and the energy range of gamma radiation never exceeds 10 MeV in nuclear decays. Under the assumption (12.46), the perturbation operator assumes the form (11.6), which reads:

$$\hat{V}(r) = \frac{ie\hbar}{m_p} \vec{A} \cdot \nabla = \frac{ie\hbar}{m_p} e^{i\vec{k}\cdot\vec{r}} \vec{A}_0 \cdot \nabla \tag{12.47}$$

and the radiative transition rate becomes:

$$P_{i \to f} = \frac{2\pi e^2}{m_p^2} |\langle \Psi_f | e^{i\vec{k}\cdot\vec{r}} \vec{A}_0 \cdot \nabla | \Psi_i \rangle|^2 \qquad (12.48)$$

The restriction (12.46) implies that the series (11.7), obtained by expanding $e^{i\vec{k}\cdot\vec{r}}$ for the electromagnetic radiation, is rapidly convergent, so that we shall only consider the first two terms of the expansion. This yields:

$$P_{i \to f} = \frac{2\pi e^2}{m_p^2} \left[|\langle \Psi_f | \vec{A}_0 \cdot \nabla | \Psi_i \rangle|^2 + |\langle \Psi_f | (\vec{k}\cdot\vec{r})(\vec{A}_0 \cdot \nabla) | \Psi_i \rangle|^2 \right] \qquad (12.49)$$

The first matrix element can be rewritten, by integrating by parts, as:

$$\langle \Psi_f | \vec{A}_0 \cdot \nabla | \Psi_i \rangle = \int \Psi_f^* (\vec{A}_0 \cdot \nabla) \Psi_i \, dV = -\int \Psi_i (\vec{A}_0 \cdot \nabla) \Psi_f^* \, dV$$

which allows us to write:

$$\langle \Psi_f | \vec{A}_0 \cdot \nabla | \Psi_i \rangle = \frac{1}{2} \int \Psi_f^* (\vec{A}_0 \cdot \nabla) \Psi_i \, dV - \frac{1}{2} \int \Psi_i (\vec{A}_0 \cdot \nabla) \Psi_f^* \, dV$$

Using the identity $\vec{A}_0 = \nabla(\vec{A}_0 \cdot \vec{r})$, one obtains:

$$\langle \Psi_f | \vec{A}_0 \cdot \nabla | \Psi_i \rangle = \frac{1}{2} \int \Psi_f^* \nabla(\vec{A}_0 \cdot \vec{r}) \cdot \nabla \Psi_i \, dV - \frac{1}{2} \int \Psi_i \nabla(\vec{A}_0 \cdot \vec{r}) \cdot \nabla \Psi_f^* \, dV$$

$$= -\frac{1}{2} \int (\vec{A}_0 \cdot \vec{r})(\nabla \Psi_f^* \cdot \nabla \Psi_i + \Psi_f^* \nabla^2 \Psi_i) \, dV$$

$$+ \frac{1}{2} \int (\vec{A}_0 \cdot \vec{r})(\nabla \Psi_i \cdot \nabla \Psi_f^* + \Psi_i \nabla^2 \Psi_f^*) \, dV$$

$$= -\frac{1}{2} \int (\vec{A}_0 \cdot \vec{r})(\Psi_f^* \nabla^2 \Psi_i - \Psi_i \nabla^2 \Psi_f^*) \, dV$$

$$= \frac{m_p}{\hbar^2}(E_i - E_f) \int \Psi_f^* (\vec{A}_0 \cdot \vec{r}) \Psi_i \, dV = \frac{m_p}{\hbar} \omega_{if} \langle \Psi_f | \vec{A}_0 \cdot \vec{r} | \Psi_i \rangle$$

where use has been made of the Schrödinger energy equation for a nucleon. Substituting Eq.(12.45) for $E_i - E_f$ gives:

$$\frac{2\pi e^2}{m_p^2} |\langle \Psi_f | \vec{A}_0 \cdot \nabla | \Psi_i \rangle|^2 = \frac{2\pi}{\hbar^2} \omega_{if}^2 |\vec{A}_0|^2 |\langle \Psi_f | e\vec{r} | \Psi_i \rangle|^2 \qquad (12.50)$$

which is identical to the probability derived in Eq.(11.43) for an *electric dipole transition* (*E*1) in the atomic case, because it is clear from Eq.(11.9) that $|E(\omega_{if})|^2 = \omega_{if}^2|\vec{A}_0|^2$. We have seen that, in calculating the radiative transition rate for the range of wavelengths of the atomic radiation, it is not necessary to go beyond the first term in the expansion (12.49). The second matrix element in Eq.(12.49) can be split into two parts, by using:

$$(\vec{k}\cdot\vec{r})(\vec{A}_0\cdot\nabla) = \frac{1}{2}\left[(\vec{k}\cdot\vec{r})(\vec{A}_0\cdot\nabla) - (\vec{A}_0\cdot\vec{r})(\vec{k}\cdot\nabla)\right] + \frac{1}{2}\left[(\vec{k}\cdot\vec{r})(\vec{A}_0\cdot\nabla) + (\vec{A}_0\cdot\vec{r})(\vec{k}\cdot\nabla)\right]$$

$$= \frac{1}{2}(\vec{k}\times\vec{A}_0)\cdot(\vec{r}\times\nabla) + \frac{1}{2}\left[(\vec{k}\cdot\vec{r})(\vec{A}_0\cdot\nabla) + (\vec{A}_0\cdot\vec{r})(\vec{k}\cdot\nabla)\right]$$

Substituting $\hat{\vec{L}} = -i\hbar(\vec{r}\times\nabla)$ and $\vec{\mu}_l$, as defined by Eq.(7.132), we obtain a first matrix element which gives the rate for the *magnetic dipole transition* (*M*1):

$$\frac{2\pi e^2}{m_p^2}\frac{1}{4}|\langle\Psi_f|(\vec{k}\times\vec{A}_0)\cdot(\vec{r}\times\nabla)|\Psi_i\rangle|^2 = \frac{2\pi}{\hbar^2}|\vec{k}\times\vec{A}_0|^2|\langle\Psi_f|\vec{\mu}|\Psi_i\rangle|^2 \quad (12.51)$$

Then, in a manner similar to that used for the electric dipole interaction, the second matrix element can be reduced to an integral which contains er^2 as a factor, and hence, leads to the rate for the *electric quadrupole transition* (*E*2), which has the form:

$$\frac{2\pi e^2}{m_p^2}\frac{1}{4}|\langle\Psi_f|(\vec{k}\cdot\vec{r})\cdot(\vec{A}_0\times\nabla) + (\vec{A}_0\cdot\vec{r})(\vec{k}\cdot\nabla)|\Psi_i\rangle|^2$$

$$= \frac{2\pi}{\hbar^2}\frac{\omega_{if}^2}{4}|\langle\Psi_f|e(\vec{A}_0\cdot\vec{r})(\vec{k}\cdot\vec{r})|\Psi_i\rangle|^2 \quad (12.52)$$

It is now apparent that the successive terms of the power series expansion can be associated with the interaction of the electromagnetic field with the so-called 2^L *multipole moments*: dipole moments with $L = 1$ (*E*1, *M*1), quadrupole moments with $L = 2$ (*E*2, *M*2), and so on. An expression which is proportional to r^L can be written in terms of the Legendre polynomials of order up to L, which are eigenfunctions of the angular momentum. As a result, gamma radiation emitted by either the electric or magnetic 2^L multipoles carries away an angular momentum equal to $L\hbar$ and can change the angular momentum of the nuclear state by the same amount.

Quantum Physics

EXAMPLE 12.1. Nuclear gamma resonance

It might be expected that nuclear resonant absorption should also occur for γ-radiation, emitted when nuclei in excited states lose their energy by radiation. However, the effect of the recoil momentum, which can be neglected for atomic radiation, becomes dominant for gamma radiation, because of its much higher energy. For emission of radiation from free atoms we simply have the recoil energy $E_R = p_R^2/2m_N = p_\gamma^2/2m_N = E_\gamma^2/2m_N c^2$ where m_N is the nuclear mass, and the recoil momentum is equal to the momentum of the emitted photon $p_\gamma = E_\gamma/c$. This recoil energy prevents the observation of nuclear resonance absorption by free atoms, as illustrated in Figure 12.4 (a). The nuclear gamma resonance, also known as the *Mössbauer effect*, arises when we consider emitting and absorbing atoms which are bound in a solid, such that they are no longer able to recoil individually. In this case, for the system which consists of the atom and the entire solid in which it is embedded, we may consider that the wave function is a product of the nuclear wave function ψ_N and the wave function ψ of the solid, which are not affected by each other, to a first approximation. Before the γ-ray emission, the solid is assumed to be at rest, in a stationary state ψ_i, defined by:

$$\hat{H}\psi_i = E_i\psi_i, \qquad -i\hbar\nabla\psi_i = 0 \qquad (12.53)$$

After emitting a photon of momentum $\vec{p}_0 = \hbar\vec{k}_0$, the solid is in a state ψ_f, which still should be an eigenstate of the momentum operator, with eigenvalue $-\vec{p}_0$:

$$-i\hbar\nabla\psi_f = -\vec{p}_0\psi_f \qquad (12.54)$$

because the recoil is too small to eject the atom from the solid, and hence, the solid as a whole must take up the recoil momentum.

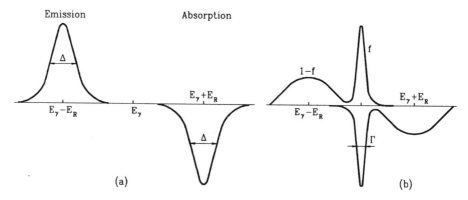

Figure 12.4. (a) Emission and absorption line shift by recoil (b) Resonance absorption in solids

If the state function of the solid is written in the form $\psi_i = \sum_{\vec{k}} c_{\vec{k}i} e^{i\vec{k}\cdot\vec{r}}$, which implies that:

$$\psi_f = \sum_{\vec{k}} c_{\vec{k}i} e^{i(\vec{k}-\vec{k}_0)\cdot\vec{r}} = -e^{-i\vec{k}_0\cdot\vec{r}}\psi_i$$

we may expand the eigenfunction of momentum ψ_f in terms of the complete system of energy eigenfunctions defined by Eq.(12.53), in the form $\psi_f = \sum_n a_n \psi_n$, such that the probability of finding the solid, after the γ-ray emission, in another stationary state E_n is:

$$|a_n|^2 = |\int \psi_n^* \psi_f \, dV|^2 = |\int \psi_n^* e^{-i\vec{k}_0 \cdot \vec{r}} \psi_i \, dV|^2$$

For $n = i$ the solid will remain in the initial state ψ_i, after the γ-ray emission, which means that the photon has carried away the full transition energy, and thus it was emitted without recoil energy loss, with the probability:

$$f = |a_i|^2 = |\int \psi_i^* e^{-i\vec{k}_0 \cdot \vec{r}} \psi_i \, dV| \tag{12.55}$$

called the *recoilless fraction*. A simple expression for f can be obtained by representing the solid as a harmonic oscillator in one-dimension, having the ground state energy eigenfunction and eigenvalue as given in Problem 9.2.1, namely:

$$\psi_0(x) = \left(\frac{M\omega}{\pi\hbar}\right)^{\frac{1}{4}} e^{-M\omega x^2/2\hbar}, \qquad E_0 = \frac{1}{2}\hbar\omega$$

Substituting $\psi_i = \psi_0(x)$ and $k_0 = E_0/\hbar c$ into Eq.(12.55) yields:

$$f = \left|\int_{-\infty}^{\infty} \psi_0^*(x) e^{-iE_0 x/\hbar c} \psi_0(x)\right|^2 = e^{-E_0^2/2Mc^2\hbar\omega} = e^{-E_R/\hbar\omega} \tag{12.56}$$

If the recoil energy E_R is much smaller than $\hbar\omega$, which is the minimum amount of energy, called a phonon, that the solid can accept, a significant fraction f of γ-rays will escape with the full energy of the nuclear transition, and without changing the internal energy of the solid. This implies that the recoilless γ-ray has a Lorentzian lineshape, as defined by Eq.(12.14), which is centred at E_γ, and has the natural linewidth of the transition, equal to the width of the decaying nuclear state. There is no Doppler broadening, as this only comes from thermal excitations of the solid, such that the theoretical resolution of Mössbauer experiments is only determined by the lifetime of the excited state of the nuclear transition. There are a few isotopes where the nuclear gamma resonance may occur, as they possess a suitable Mössbauer transition, with $E_R \ll \hbar\omega$. A crude estimate of the energy $\hbar\omega$ for a solid is obtained by taking $\omega = 2\pi v/\lambda$, where v is the sound velocity and $\lambda = 2a$ (a is the lattice constant). For metallic iron, where $v = 5690$ m/s and $a = 2.9$ Å, we obtain $\hbar\omega = 0.04$ eV and we may compare the decay schemes of the two different isotopes, $^{57}_{26}$Fe and $^{58}_{26}$Fe, given in Figure 12.5. It follows that $^{57}_{26}$Fe is a Mössbauer isotope, actually the best known, whereas nuclear gamma resonance cannot be obtained using $^{58}_{26}$Fe. The fact that gamma resonance is specific to atoms of a particular isotope results in a highly selective form of spectroscopy.

458   Quantum Physics

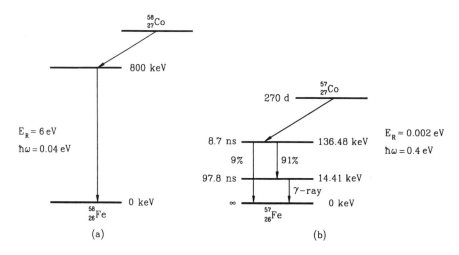

Figure 12.5. Comparative decay schemes for $^{57}_{26}$Fe and $^{58}_{26}$Fe

The Mössbauer spectrum is commonly obtained in the geometry shown in Figure 12.6 (a). The γ-radiation is emitted from a single line source, which consists of a nonmagnetic matrix containing radioactive $^{57}_{27}$Co, and decays to $^{57}_{26}$Fe as shown in Figure 12.5 (b). The transmitted intensity through an absorber containing iron (hence $^{57}_{26}$Fe) atoms in their ground state, is measured as a function of E_γ, which is varied by an amount $E_\gamma v/c$ through Doppler shifting the source relative to the absorber. An intensity minimum (or maximum absorption) occurs at particular velocities, where the source and absorber levels differ by $E_\gamma v/c$. Any increase or decrease in v from the resonance values results in decreasing the absorption (Figure 12.6 (b)).

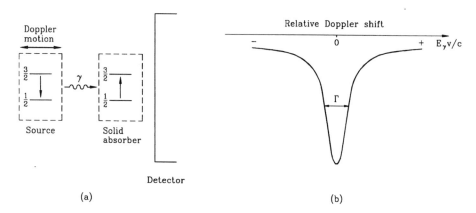

Figure 12.6. (a) Schematic Mössbauer spectrometer in transmission geometry; (b) Mössbauer absorption line produced by Doppler scanning

The natural linewidth of the 14.4 keV level of $^{57}_{26}$Fe is 0.095 mm/s, and the recorded spectrum is obtained as the convolution of the emission and absorption lines, such that the smallest observable Γ is 0.19 mm/s, which correspond to an energy of the order of 10^{-8} eV. This is the order of magnitude of the nuclear hyperfine interactions, which thus can be investigated by this technique.

If the $^{57}_{26}$Fe Mössbauer spectrum is shaped by magnetic interactions only, the nuclear Zeeman levels introduced before in the case of nuclear magnetic resonance, Eq.(11.66), are involved:

$$E_m = -\mu_N g_N B_{hf} m_I \qquad (12.57)$$

where B_{hf} is called the *hyperfine magnetic field*, originating in the atomic electrons. There are six resonance lines, corresponding to the transitions allowed by the $\Delta m_I = 0, \pm 1$ selection rule between the ground $(I = \frac{1}{2})$ and the first excited state $(I = \frac{3}{2})$, which are split into two and respectively four Zeeman levels, as shown in Figure 12.7 (a).

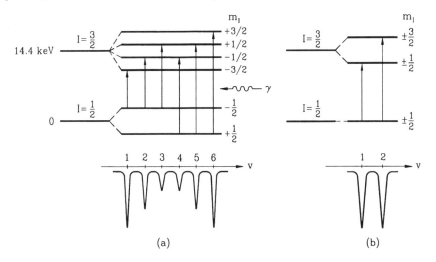

Figure 12.7. The Zeeman (a) and quadrupole (b) splitting, and the corresponding nuclear resonance spectra

Taking into account that the nucleus is a continuous distribution $\rho_N(\vec{r})$ of positive charges within a small volume, rather than a point charge, there is an interaction with the external electric potential of the form:

$$E = \int \rho_N(\vec{r}) V(\vec{r}) dV = V(0) \int \rho_N(\vec{r}) dV + \sum_{i=x,y,z} \left(\frac{\partial V}{\partial x_i}\right)_0 \int \rho_N(\vec{r}) x_i \, dV$$

$$+ \frac{1}{2} \sum_{i,j=x,y,z} \left(\frac{\partial^2 V}{\partial x_i \partial x_j}\right)_0 \int \rho_N(\vec{r}) x_i x_j \, dV + \ldots \qquad (12.58)$$

where the integral is extended over the nuclear volume. The first term defines the *monopole* interaction which remains constant during nuclear transitions and will be disregarded. The second term gives the *dipolar* interaction which vanishes, because the nuclear charge distribution is an even function of coordinates, such that the expectation value of the nuclear electric dipole moment is zero. The third term can be rewritten in terms of the *electric field gradient* components $\partial^2 V/\partial x_i \partial x_j = V_{ij}$, which arise from the atomic electrons and can be assumed to be of axial symmetry about the z-axis of the atomic coordinate system. In general the V_{zz} component makes an angle θ with the nuclear symmetry axis Oz', and it is desirable to express the integrals in Eq.(12.57) as nuclear integrals referred to the nuclear coordinate system. It is common practice to choose, for simplicity, the atomic x-axis along the same direction as the nuclear x'-axis, such that $r^2 = x^2 + y^2 + z^2 = x'^2 + y'^2 + z'^2$ and $z = z'\cos\theta - y'\sin\theta$. We first evaluate the third term in Eq.(12.58) in the atomic system of axes, assumed to define the principal axes of the electric field gradient:

$$E = \frac{1}{2}\sum_{i=x,y,z} V_{ii} \int \rho_N(\vec{r}) x_i^2 dV = \frac{1}{6}\sum_{i=x,y,z} V_{ii} \int \rho_N(\vec{r}) r^2 dV + \frac{1}{6}\sum_{i=x,y,z} V_{ii} \int \rho_N(\vec{r})(3x_i^2 - r^2) dV$$

(12.59)

It is convenient to introduce the mean square nuclear radius:

$$\langle R^2 \rangle = \frac{1}{Ze}\int \rho_N(\vec{r}) r^2 dV = \frac{1}{Ze}\int \rho_N(x',y',z')(x'^2 + y'^2 + z'^2) dV$$

and the nuclear quadrupole moment components $Q_{ii} = \int \rho_N(\vec{r})(3x_i^2 - r^2) dV$, such that Eq.(12.59) simplifies to:

$$E = \frac{Ze}{6}\langle R^2 \rangle \sum_{i=x,y,z} V_{ii} + \frac{1}{6}\sum_{i=x,y,z} V_{ii} Q_{ii}$$

(12.60)

The charge density at the nucleus is only given by the s-electrons, with a spherical distribution, which obeys the Poisson equation $V_{xx} + V_{yy} + V_{zz} = \nabla^2 V = -\rho_e/\varepsilon_0 = e|\Psi_s(0)|^2/\varepsilon_0$, while the electric field gradient originating from other electron shells is subjected to the Laplace equation $\nabla^2 V = 0$. It follows that the first term in Eq.(12.60) is determined by the s-electrons, which do not contribute to the second term, because their spherical charge distribution implies that we may set $V_{xx} = V_{yy} = V_{zz} = q$, and so $\sum_{i=x,y,z} V_{ii} Q_{ii} = q \sum_{i=x,y,z} Q_{ii} = 0$. Hence, we have:

$$E = \frac{Ze^2}{6\varepsilon_0}\langle R^2 \rangle |\Psi_s(a)|^2 + \frac{1}{6}\sum_{i=x,y,z} V_{ii} Q_{ii}$$

(12.61)

where the second term only includes contributions from electrons for which the Laplace equation applies. The first term describes the *nuclear size* effect, called the *isomer shift*, which causes the energy levels of two isotopes with different nuclear radii $\langle R^2 \rangle$, or embedded in different atomic surroundings $|\Psi_s(0)|^2$, to be slightly different, as shown in Figure 12.8 (a) and (d). Because $^{57}_{26}$Fe atoms in the source and absorber have different chemical environments, each Mössbauer absorption line (and hence the centre of the spectrum) is shifted by a constant isomer shift δ.

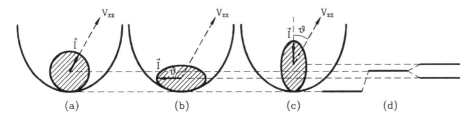

Figure 12.8. Graphical interpretation of the nuclear size (a) and nuclear shape (b and c) effects on the nuclear energy levels (d)

In view of the Laplace equation $V_{xx} = V_{yy} = V_{zz} = 0$, and the axial symmetry $V_{xx} = V_{yy}$ assumed in the atomic coordinate system, which allows us to take $V_{xx} = V_{yy} = -V_{zz}/2 = -q/2$, the second term in Eq.(12.61) reads:

$$E_Q = \frac{1}{6}\sum_{i=x,y,z} V_{ii} Q_{ii} = \frac{q}{6}\int \rho_N(x,y,z)\left[3z^2 - \frac{3}{2}(x^2+y^2)\right]dV = \frac{q}{4}\int \rho_N(\vec{r})(3z^2 - r^2)dV \quad (12.62)$$

The physical significance of this result becomes apparent if we express the integral in terms of nuclear coordinates (x', y', z'), in the form:

$$E_Q = \frac{q}{4}\int \rho_N(x',y',z')\left[3(z'\cos\theta - y'\sin\theta)^2 - r^2\right]dV$$

$$= \frac{q}{4}\int \rho_N(x',y',z')\left[3z'^2\cos^2\theta + \frac{3}{2}(r^2 - z'^2)\sin^2\theta\right]dV$$

$$= \frac{q}{8}\left[\int \rho_N(x',y',z')(3z'^2 - r^2)dV\right](3\cos^2\theta - 1) \quad (12.63)$$

where axial symmetry of the nuclear charge distribution about the z'-axis has been assumed. Equation (12.63) contains the intrinsic *quadrupole moment* Q', as commonly defined in nuclear physics:

$$Q' = \frac{1}{e}\int \rho(x',y',z')(3z'^2 - r^2)dV \quad (12.64)$$

such that $Q' = 0$ for a spherical distribution of nuclear charge, $Q' > 0$ for a distribution elongated at the poles and $Q' < 0$ for a distribution flattened at the poles. Hence, Eq.(12.63) gives the interaction energy of a nucleus, having a quadrupole moment Q', with a field gradient q at the site of the nucleus:

$$E_Q = \frac{eqQ'}{8}(3\cos^2\theta - 1) \quad (12.65)$$

called the *electric quadrupole interaction* energy. It depends on the *nuclear shape*, measured by Q', and on the orientation of the deformed nucleus, as illustrated in Figure 12.8 (b) and (c).

Assuming now that the nuclear spin $\hat{I}$ lies along the nuclear symmetry axis Oz', its component along the atomic symmetry axis Oz defines the angle θ in Eq.(12.65) as $\cos\theta = \hat{I}_z/|\hat{I}|$. Substituting the expectation value $\langle \cos^2\theta \rangle = m_I^2/I(I+1)$ into Eq.(12.65) gives:

$$E_Q = \frac{eqQ'}{8I(I+1)}\left[3m_I^2 - I(I+1)\right] \tag{12.66}$$

The effective quadrupole moment Q, that is experimentally observed in the state m_I, is actually less than the intrinsic value Q', because of the precession of $\hat{I}$ about Oz, caused by the quadrupole interaction. If Q' is assigned to the state $m_I = I$, where E_Q is a maximum, and this maximum energy value is equated to the classical maximum value E_Q, obtained for $\theta = 0$ from Eq.(12.65), one obtains:

$$\frac{eqQ'}{8}\frac{3I^2 - I(I+1)}{I(I+1)} = \frac{eqQ}{8}(3\cos^2 0 - 1) \quad \text{or} \quad Q' = Q\frac{2(I+1)}{2I-1} \tag{12.67}$$

Hence, the quadrupole interaction energy (12.66) finally becomes:

$$E_Q = \frac{eqQ}{4I(2I-1)}\left[3m_I^2 - I(I+1)\right] \tag{12.68}$$

We see that $E_Q = 0$ for $I = \frac{1}{2}$ and the two-level splitting illustrated in Figure 12.8 (d) corresponds to the first excited state of $^{57}_{26}\text{Fe}$, with $I = \frac{3}{2}$ (see Figure 12.7 (b)). The quadrupole splitting is a measure of the asymmetry of the electron charge distribution around the iron nucleus, such that the larger is the splitting of the two-line pattern, the larger is the local distortion from cubic symmetry.

In the presence of both the magnetic and electric quadrupole hyperfine interactions, the energy levels are obtained by combining Eqs.(12.57) and (12.68) as:

$$E_e = -\mu_N g_e B_{hf} m_I + \frac{eqQ}{4I(2I-1)}\left[3m_I^2 - I(I+1)\right], \quad E_f = -\mu_N g_f B_{hf} m_I \tag{12.69}$$

In this case, the separation of the excited state Zeeman levels are no longer equal and the absorption lines are not equally spaced in energy. Normally, the quadrupole interaction is small and the corresponding perturbation is smaller than the perturbation by a magnetic interaction. This results in sextets which are asymmetrically positioned relative to the centre of the absorption spectrum. ♫

Problem 12.4.1. Find the separation and the relative intensities of the two gamma resonance lines (Figure 12.7 (b)) obtained in the presence of the quadrupole splitting of $^{57}_{26}\text{Fe}$ energy levels.

(Solution): The first excited state $(I = \tfrac{3}{2})$ of $^{57}_{26}\text{Fe}$ splits into two levels with energies $E_Q(m_I)$ given by Eq.(12.68) as:

$$E_Q(\pm\tfrac{1}{2}) = \frac{eqQ}{4} \frac{3 \cdot \tfrac{1}{4} - \tfrac{3}{2} \cdot \tfrac{5}{2}}{\tfrac{3}{2} \cdot \tfrac{4}{2}} = -\frac{eqQ}{4}$$

$$E_Q(\pm\tfrac{3}{2}) = \frac{eqQ}{4} \frac{3 \cdot \tfrac{9}{4} - \tfrac{3}{2} \cdot \tfrac{5}{2}}{\tfrac{3}{2} \cdot \tfrac{4}{2}} = +\frac{eqQ}{4}$$

This gives the line separation equal to $eqQ/2$.

The relative intensities in the quadrupole doublet are obtained by adding contributions from the four transitions $|\tfrac{3}{2}, \pm\tfrac{1}{2}\rangle \to |\tfrac{1}{2}, \pm\tfrac{1}{2}\rangle$ and $|\tfrac{3}{2}, \pm\tfrac{1}{2}\rangle \to |\tfrac{1}{2}, \mp\tfrac{1}{2}\rangle$, and respectively the two transitions $|\tfrac{3}{2}, \pm\tfrac{3}{2}\rangle \to |\tfrac{1}{2}, \pm\tfrac{1}{2}\rangle$ between the nuclear Zeeman levels. Since the Zeeman splitting, illustrated in Figure 12.7 (a), is analogous to that found in the case of the $2\,^2P_{\tfrac{3}{2}} \to 1\,^2S_{\tfrac{1}{2}}$ electron transitions, in Problem 11.1.2, similar considerations lead to the same relative intensities for the six allowed transitions, namely:

$$I(\pm\tfrac{3}{2} \to \pm\tfrac{1}{2}) = \tfrac{3}{2}(1 + \cos^2\theta)$$

$$I(\pm\tfrac{1}{2} \to \pm\tfrac{1}{2}) = 2\sin^2\theta$$

$$I(\pm\tfrac{1}{2} \to \mp\tfrac{1}{2}) = \tfrac{1}{2}(1 + \cos^2\theta)$$

where θ specifies the direction of the emitted gamma radiation with respect to the atomic symmetry axis Oz. It follows that:

$$\frac{I_{\tfrac{1}{2}}}{I_{\tfrac{3}{2}}} = \frac{\tfrac{2}{3} + \sin^2\theta}{1 + \cos^2\theta}$$

It is apparent that $I_{\tfrac{1}{2}}/I_{\tfrac{3}{2}} = 1/3$ in the direction $\theta = 0$ and $I_{\tfrac{1}{2}}/I_{\tfrac{3}{2}} = 5/3$ for transversal observation ($\theta = \pi/2$). In most cases, the Mössbauer spectra are recorded from polycrystalline absorbers, where the relative intensities are calculated by averaging over θ:

$$\langle \sin^2\theta \rangle = \frac{1}{4\pi} \int_0^{2\pi}\int_0^{\pi} \sin^2\theta\, d\theta\, d\varphi = \frac{2}{3}, \quad \langle \cos^2\theta \rangle = \frac{1}{3}$$

and hence, both lines have the same intensity, as shown in Figure 12.7 (b).

Problem 12.4.2. Assuming that $v_1, \cdots, v_6$ are the velocities corresponding to the six gamma resonance lines, in the presence of the combined magnetic and electric hyperfine interactions, Eq.(12.69), find expressions for the hyperfine field B_{hf} and the quadrupole shift $eqQ/2$, in terms of the appropriate resonance velocities.

(Solution): The hyperfine field can be directly obtained from separation of the outer lines, corresponding to the transition energies:

$$v_6 = \Delta E\left(\left|\tfrac{3}{2},\tfrac{3}{2}\right\rangle \to \left|\tfrac{1}{2},\tfrac{1}{2}\right\rangle\right) = \frac{1}{2}B_{hf}\mu_N(-3g_e + g_f) + \frac{1}{2}eqQ$$

$$v_1 = \Delta E\left(\left|\tfrac{3}{2},-\tfrac{3}{2}\right\rangle \to \left|\tfrac{1}{2},-\tfrac{1}{2}\right\rangle\right) = \frac{1}{2}B_{hf}\mu_N(3g_e - g_f) + \frac{1}{2}eqQ$$

This gives:

$$B_{hf} = \frac{v_6 - v_1}{g_f - 3g_e} = \frac{v_6 - v_1}{|g_f| + 3|g_e|}$$

where $g_f/g_e = -1.757$. With B measured in Tesla and v_1, v_6 in mm/s, we obtain:

$$B_{hf} = 3.112\,(v_6 - v_1)$$

In a similar way, it is apparent that:

$$\frac{1}{2}eqQ = \frac{(v_6 - v_5) - (v_2 - v_1)}{2}$$

Problem 12.4.3. Show that the recoilless fraction, Eq.(12.56), can be written as:

$$f = e^{-k^2 \langle x^2 \rangle}$$

where k is the wave number of the emitted γ-ray and $\langle x^2 \rangle$ is the mean square displacement, along the photon direction, of the harmonic oscillator representing the solid.

Problem 12.4.4. Assuming that the Mössbauer spectrum is due to the magnetic interaction only, and the atomic symmetry axis Oz is taken along $\vec{B}_{hf}$, show that the relative orientation θ of $\vec{B}_{hf}$ at the nucleus, with respect to the γ-ray direction, can be determined from the relative intensities $I_1, \cdots, I_6$ of the six resonance lines by:

$$\cos^2\theta = \frac{I_1 - I_2 + I_3 + I_4 - I_5 + I_6}{\sum_{j=1}^{6} I_j}$$

Further Reading

General reading in the field of quantum physics:

A.Messiah, *Quantum Mechanics*, North-Holland, Amsterdam, 1962
L.I.Schiff, *Quantum Mechanics*, 3rd ed., McGraw-Hill, New York, 1968
S.Gasiorowicz, *Quantum Physics*, Wiley, New York, 1974
P.A.M. Dirac, *The Principles of Quantum Mechanics*, 4th ed., Clarendon Press, Oxford, 1974
C.Cohen-Tannoudji, B.Din and F.Laloë, *Quantum Mechanics*, Wiley and Herman, Paris, 1977
L.D.Landau and E.M.Lifschiz, *Quantum Mechanics: Non-Relativistic Theory*, 3rd ed., Pergamon Press, Oxford, 1987
S.Brandt and H.D.Dahmen, *Quantum Mechanics on the Personal Computer*, Springer-Verlag, Berlin, 1990

Further reading for each chapter:

Preliminaries
B.d'Espagnat, *Conceptual Foundations of Quantum Mechanics*, 2nd ed., Benjamin, Reading, 1976
R.Eisberg and R.Resnick, *Quantum Physics*, 2nd ed., Wiley, New York, 1985
H.Haken and H.C.Wolf, *Atomic and Quantum Physics*, 2nd ed., Springer-Verlag, Berlin, 1987

Chapter 1
P.C.W.Davies, *Quantum Mechanics*, Routledge & Kegan Paul, London, 1984
F.Mandl, *Quantum Mechanics*, Wiley, Chichester, 1992

Chapter 2
D.Park, *Classical Dynamics and its Quantum Analogue*, Springer-Verlag, Berlin, 1990
F.Battaglia, *Notes in Classical and Quantum Mechanics*, Blackwell Scientific, Oxford, 1990

Chapter 3
W.Pauli, *General Principles of Quantum Mechanics*, Springer-Verlag, Berlin, 1980
J.A.Wheeler and W.H.Zurek, *Quantum Theory and Measurement*, Princeton University Press, Princeton, 1983

Chapter 4
K.Gottfried, *Quantum Mechanics: Fundamentals*, Addison-Wesley, Redwood City, 1989
D.Park, *Introduction to Quantum Theory*, 3rd ed., McGraw-Hill, New York, 1992

Chapter 5
R.Shrankar, *Principles of Quantum Mechanics*, Plenum Press, New York, 1980
A.I.M.Rae, *Quantum Mechanics*, 2nd ed., Adam Hilger, Bristol, 1986

Chapter 6
J.L.Martin, *Basic Quantum Mechanics*, Clarendon Press, Oxford, 1981
D.F.Jackson, *Atoms and Quanta*, Surrey University Press, London, 1989
P.R.Holland, *The Quantum Theory of Motion*, Cambridge University Press, Cambridge, 1995

Chapter 7
L.Biederharn, *Angular Momentum in Quantum Physics*, Addison-Wesley, London, 1981
H.Clark, *A First Course in Quantum Mechanics*, Van Nostrand Reinhold, Wokingham, 1982

Chapter 8
H.F.Hameka, *Quantum Mechanics*, Wiley, New York, 1981
A.Böhm, *Quantum Mechanics*, 2nd ed., Springer-Verlag, Berlin, 1986
T.Y.Wu, *Quantum Mechanics*, World Scientific, Singapore, 1986
D.A.Varshalovich, *Quantum Theory of Angular Momentum*, World Scientific, Singapore, 1988

Chapter 9
J.J.Sakurai, *Modern Quantum Mechanics*, Benjamin, Menlo Park, 1985
H.Kroemer, *Quantum Mechanics*, Prentice Hall, Englewood Cliffs, 1994

Chapter 10
B.H.Bransden, C.J.Joachain, *Physics of Atoms and Molecules*, Longman, London, 1983
L.Pauling and S.Goudsmit, *The Structure of Line Spectra*, McGraw-Hill, New York, 1986
W.Greiner, *Quantum Mechanics*, Springer-Verlag, Berlin, 1989
N.A.Jelley, *Fundamentals of Nuclear Physics*, Cambridge University Press, Cambridge, 1990
R.J.Blin-Stone, *Nuclear and Particle Physics*, Chapman and Hall, London, 1991

Chapter 11
J.C.Townsend, *A Modern Approach to Quantum Mechanics*, McGraw-Hill, New York, 1990
G.Shatz and A.Weilinger, *Nuclear Condensed Matter Physics*, Wiley, Chichester, 1996

Chapter 12
T.E.Cranshaw, B.W.Dale, G.O.Longworth and C.E.Johnson, *Mössbauer Spectroscopy and its Applications*, Cambridge University Press, Cambridge, 1985
W.N.Cottingham, *An Introductory Nuclear Physics*, Cambridge University Press, Cambridge, 1986
K.S.Krane, *Introductory Nuclear Physics*, Wiley, New York, 1988

Index

Absorption of radiation, 410
Action integral, 8
Activity, 439
Adjoint operator, 104
Admissible solution, 254
Alpha decay, 442-445
Amplification, of radiation, 425
Angular momentum, 285-326
    addition of, 308-316
    atomic, 404
    in spherical polar coordinates, 72
    orbital, 12, 296
    spherical harmonic eigenstates of, 292
    spin, 301, 398
    total, 308
    vector model of, 315
Anticommutator, 302
Antilinearity, 54
Antineutrino, 447
Antisymmetric function, 361
Approximate methods, 329-353
Associate Legendre function, 248
Atom, helium, 371-377
    hydrogen, 275, 343, 345
    multielectron, 380-394
    one-electron, 267-274
Atomic radiation, 408-434

Bandwidth theorem, 36
Basis, canonical, 291
    complete, 58
Basis vectors, 56
Beta decay, 447-452
Binding energy, 397
    per nucleon, 397
Binding fraction, 397
Bohr magneton, 277
Bohr radius, 260
Bohr postulate, 6, 7, 196

Boltzmann distribution, 2, 425
Boson, 363
Bound state, 214, 256
Bra vector, 63

Canonical equations, 151
Centre of mass, 267
Central field approximation, 383
Centrifugal potential energy, 254
Characteristic equation, 113
Characteristic polynomial, 114
Classical limit of Schrödinger equation,
    time-dependent, 181
    time-independent, 193
Classical turning point, 204
Classically allowed region, 204
Classically disallowed region, 204
Clebsch-Gordan coefficient, 310
Closure relation, 52, 117
Compatible observables, 160-166
    complete set of, 163
Completeness relation, 127
Compton effect, 19
Compton wavelength, 23
Commutation relations, 144
    for angular momentum, 154
Commutator, 144
Configuration, electron, 385
    of low Z atoms, 386
Confluent hypergeometric function, 258, 400
Constant of motion, 11, 153, 175, 359
Continuity equation, for electric charge, 180
    for mass flow, 180
    for probability, 179
    for spherical waves, 195
Continuum energy states, 353
Coordinates, centre of mass, 267
    spherical polar, 72
Coordinate representation, 66-77, 140

Correspondence principle, 8
Coupling, jj, 392-394
   LS, 390
   spin-orbit, 320
Current density, electric, 180
   probability, 180, 280

Dalgarno perturbation theory, 333-335
Davisson-Germer experiment, 26
De Broglie hypothesis, 25,
De Broglie wave, 25
   equation of, 25
   phase velocity of, 28
Degeneracy, 114
   accidental, 274
   removal of, 162
   spatial, 274
Density, charge, 180
   probability, 40, 41, 179
Density of states, 353
Description, Heisenberg, 171-175
   Schrödinger, 171, 177-179
Deuteron, 407
Deviation, 42
   standard, 42
Diagonalization, 133
Diamagnetic effect, 281
Dipole moment operator, 410
Dirac δ-function, 51
Dirac notation, 62-64
Dispersion, 42, 111
Distribution, Gaussian, 37
   Lorentzian, 440
Distribution function, angular, 272
   radial, 271
Doppler broadening, 442
Doppler shift, 21
Dual space, 63
Dulong-Petit law, 13

Effective Z, 386
Ehrenfest theorem, 174
Eigenfunction, 112
   admissible, 203
Eigenstate, 111
   energy, 202
Eigenvalue, 112
Eigenvalue equation, 112
Eigenvalue problem, 111-121
   energy, 192
   for momentum, 119
   for position, 120
   matrix, 129-133
Eikonal equation, 194
Einstein A and B coefficients, 422-423
Einstein photon hypothesis, 17
Einstein temperature, 13
Electric dipole approximation, 409
Electric dipole transition, 412, 455
   selection rules for, 412-415
Electric field gradient, 460
Electron capture, 447
Electron spin resonance, 429
Emission, induced, 411
   spontaneous, 421-426
Energy band, 225
Energy gap, 225
Energy level diagram, for electron, 385
   for nucleon, 402
Energy spectrum, band structure of, 222
   continuous, 219
   discrete, 213
EPR experiment, 429
Equipartition of energy, 2
Evolution operator, 178, 358
Exchange integral, 375
Excited state, 274
Expansion theorem, 61
Expectation value, 41
   time dependence of, 174
Exponential behaviour, 205

Fermi function, 452
Fermi golden rule, 351-353
Fermion, 363
Fine structure, 322, 391
Finiteness condition, 254
Force, conservative, 173
Fourier coefficients, 34
Fourier integral formula, 35
Fourier series, 33
Fourier transform, 35
Function space, 53
ft value, 452

Gain constant, 425
Galilean transformation, 183
Gamma radiation, 453-463
Gamov factor, 444
Geiger-Nutall law, 443
Generating function, 233, 247

g-factor, electron, 320
　Landé, 323
　neutron, 406
　proton, 406
Ground state, 274
Group velocity, 29, 38
Gram-Schmidt orthogonalization, 116, 344
Gyromagnetic ratio, 392

Half-life, 439
Hamiltonian operator, 71
　in central force field, 75
　in electromagnetic field, 278
Hamilton-Jacobi equation, 182, 193
Harmonic motion, linear, 228-238
Harmonic oscillator, 2, 228
　energy eigenfunctions of, 234
　energy eigenvalues of, 230
　three-dimensional, 239
Hartree equations, 382
Hartree-Fock approximation, 384
Heisenberg description, 171-175
　harmonic oscillator in, 235
Heisenberg inequality, 166
Heisenberg microscope, 46
Helium atom, 371-377
　first excited state of, 372
　ground state of, 372, 374, 378
Hermite polynomial, 232-234
Hermitian operator, 70, 104-106
Hilbert space, 57
Hund rules, 389
Hydrogen, 275
　first excited state of, 345
　ground state of, 334, 343
Hyperfine interaction, magnetic, 403
　quadrupole, 463
Hyperfine coupling constant, 404

Indistinguishability, principle of, 361
Induced emission, 411
Inflection point, 204
Interval rule, in hyperfine splitting, 405
Ionization energy, 384
Isomer shift, 460

Jacobi identity, 145, 152
jj coupling, 392-394

$\bar{k}$-space, 43

Ket vector, 63
Klein-Gordon equation, 31
Kronig-Penney model, 223
Kurie plot, 451

Landé g-factor, 323, 392, 394
Larmor frequency, 13
Laser, 426
Least action, principle, 21
Least time, principle, 20
Legendre polynomials, 246
Legendre equation, 245
　associated, 247
Level width, 440
Lifetime, mean, 424, 439
Linewidth, natural, 425
Lorentzian form, 43, 424, 440
LS coupling, 390

Magic number, atomic, 386
　for neutrons, 398
　for protons, 398
Magnetic moment, atomic, 392
　electron, 276-282
　nuclear, 404
　orbital, 12, 277
　spin, 319-326
Magnetic field, hyperfine, 459
　strong, 325
　weak, 323
Magnetic quantum number, 12, 244
Magnetic resonance, 428-434
　nuclear, 431
Many-particle system, 357-405
Mass defect, 397
Mass number, 397
Mass unit, atomic, 397
Matrix eigenvalue problem, 129-133
　for angular momentum, 285-296
Matrix element, 130
　electric dipole, 414
　nuclear, 448
　of an operator, 129
　reduced, 414
Matrix representation, 137-142
　of orbital angular momentum, 296
Matter wavelength, 25
Momentum observable, 67
Momentum operator, 68, 71
Momentum representation, 83-90
Momentum wave function, 83

472 Index

Monochromatic wave, 34
Mössbauer effect, 456
Motion, electron, 242
    in electromagnetic field, 277
    linear harmonic, 227
    one-dimensional, 202-238
    radial, 253-262
    rotational, 242-250
Multipole moment, 455

Natural linewidth, 425
Neutrino, 447
Neutron, 397
NMR experiment, 431
Norm, 53
Normal frequency, 2
Nuclear binding energy, 397
Nuclear gamma resonance, 456-463
Nuclear quadrupole moment, 462
Nuclear magnetic moment, 404
Nuclear magnetic resonance, 431
Nuclear magneton, 404
Nuclear matrix element, 448
Nuclear radiation, 438-463
Nuclear radius, 445
Nuclear spin, 403
Nuclear spin-orbit interaction, 403
Nucleon, 397
Nuclide, 397

Observable, 66
    canonically conjugate, 166
    compatible, 160
    incompatible, 165
One-dimensional array, 223
One-electron atom, 267-280
    in magnetic field, 280
Operator, adjoint, 104
    angular momentum, 71, 287
    anti-Hermitian, 107
    exchange, 361
    Hamiltonian, 71
    Hermitian, 70, 103-106
    identity, 94
    integral representation of, 96
    ladder, 237, 288
    linear, 93-100
    lowering, 236, 288
    momentum, 71
    non-commuting, 97
    perturbation, 329
    raising, 236, 288
    singular, 99
    spin, 302
    unitary, 108
Orbital, atomic, 383
    equivalent, 385
Orbital magnetic moment, 12, 277
Orthonormal basis, 56
Orthonormality condition, 50, 55, 57
Oscillator, anharmonic, 331
    harmonic, 228
Oscillatory motion,
    energy representation of, 197
    in phase space, 9
    momentum representation of, 88

Parity operator, 108, 121
Paschen-Bach effect, 326
Pauli exclusion principle, 357-366
Pauli matrices, 302
Perturbation theory, stationary state, 329-337
    time-dependent, 348-353
Phase space, 9
Phase velocity, 27
Phonon, 22, 457
Photoelectric effect, 17
Photon, 17
Planck quantum hypothesis, 4
Planck radiation formula, 5
Poisson bracket, 150
Polarizability, electric, 335
    magnetic, 281
Polarization, 415
    circular, 417
    linear, 417
Polynomial method, 228
Population inversion, 425
Position observable, 66
Position operator, 86
Postulate, first, 62
    fifth, 179
    fourth, 154
    second, 70
    third, 125
Potential energy, centrifugal, 254
    effective, 254, 269
    one-particle, 357
    periodic, 224
Potential energy barrier, 204
    of arbitrary shape, 222
    square, 220

Index

Potential energy well, 204
Precession, angular momentum, 311, 316
   magnetic moment, 430
Principal quantum number, 12
Probability, 41
   of direct transition, 351
   reflection, 220
   transmission, 220
Probability current density, 180, 280
Probability density, 40, 41, 179
Projection operator, 126
Proton, 397

Quadrupole moment, nuclear, 462
Quadrupole splitting, 462
Quantum number, azimuthal, 11
   magnetic, 12, 244
   orbital, 245
   principal, 12, 260
   radial, 11, 259
Quasi-classical approximation, 182

Radial equation, 253
   for a free particle, 262
   for one-electron atoms, 256
Radiation, atomic, 408-418
   electric dipole, 409
   gamma, 453-463
   monochromatic, 409
   nuclear, 438-463
Radioactive decay law, 438-440
Radioactive series, 442
Radioactive source, 458
Rayleigh formula, 30
Rayleigh-Jeans law, 3
Recoil energy, 457
Recoilless fraction, 457
Reduced mass, 267
Reflection probability, 220
Regularity condition, 203
Representation, coordinate, 66-77, 140
   energy, 196
   matrix, 137-142
   momentum, 83-90, 139
   observable A, 137
Rigid rotator, 250
Ritz combination principle, 6
Rodrigue formula, 246
Rotation, infinitesimal, 158
   unitary operators for, 159, 360
Rotation properties of vectors, 286

Rydberg constant, 6

Scalar product, 53
Schrödinger description, 171, 177-179
Schrödinger equation,
   time-dependent, 26, 178
   time-independent, 26, 192
Schwartz inequality, 55
Separation of variables, 242
Shell, electron, 385
Shell model, 397
Shielding constant, 386
Single-valuedness, 203, 212, 244
Singlet state, 372
Sinusoidal solution, 204
Slater determinant, 364
Smallness parameter, 329
Sommerfeld quantization rule, 8, 209
   for elliptical motion, 11
   for oscillatory motion, 9
   for translational motion, 10
Spatial quantization, 12, 312
Spectral lines, Doppler broadening of, 442
   natural linewidth of, 425
Spectroscopic notation, 260, 322
Spherical harmonics, 249
Spherical polar coordinates, 72
Spin, electron, 301-306
   nuclear, 403
   nucleon, 398
Spin Hamiltonian, 433
Spin-orbit interaction, 320, 391
   nuclear, 403
Spinor, 303
Spontaneous emission, 421-426
Square-integrable functions, 50
Square well potential, energy levels of, 10
Stark effect, 335
State, degenerate, 274, 335
   excited, 274
   ground, 274
   minimum uncertainty, 167
   spin, 389
   stationary, 191
Subshell, 385
Superposition, principle of, 60
Symmetrization rule, 98

Temporal evolution, 171
   of position and momentum, 172
Thermodynamic equilibrium, 1, 422

Total angular momentum, 308
Trace, 143
Transformation, Galilean, 183
　unitary, 156
Transition, electric dipole, 412, 455
　electric quadrupole, 455
　electron resonance, 432
　magnetic dipole, 432, 455
　radiative, 408-418
Transition rate, 353
Translation, infinitesimal, 157
　unitary operator for, 159, 359
Transmission probability, 220
Triplet state, 376
Tunnel effect, 219
Two-slit experiment, 61

Unbound states, 219, 261
Uncertainty principle, 44-47
Uncertainty relation, position-momentum, 45
　time-energy, 47, 188
Unitary operator, 108
Unitary transformation, 156

Variational method, 341-345
Vector, basis, 56
Vector model of angular momentum, 315

Vector space, 54-58
Velocity, group, 29, 38
　phase, 27
Virial theorem, 77

Wave equation, 2
　radial, 253
　Schrödinger, 196
Wave function, antisymmetric, 361
　for identical particles, 360
　in coordinate space, 66
　in momentum space, 83
　symmetric, 361
　two particle, 365
Wave function space, 49-58
　generalized, 118
Wave vector representation, 449
Wave packet, 33-40
Wave-particle duality, 15, 24
Width of energy level, 440
Wien formula, 3
WKB approximation, 206-209

Zeeman effect, anomalous, 324
　normal, 282
Zero point energy, 230